PEOPLE IN PLACES

PEOPLE IN PLACES

A DOCUMENTARY CASE-STUDY WORKBOOK

Philip E. Steinberg

Kathleen Sherman-Morris

to accompany the eighth edition of

The Cultural Landscape

An Introduction to Human Geography

by James M. Rubenstein

Upper Saddle River, NJ 07458

Executive Editor: Daniel Kaveney
Editor in Chief: John Challice
Vice President/Director of Production and Manufacturing: David W. Riccardi
Executive Managing Editor: Kathleen Schiaparelli
Production Management: Elm Street Publishing Services, Inc.
Manufacturing Buyer: Alan Fischer
Manufacturing Manager: Trudy Pisciotti
Marketing Manager: Robin Farrar
Art Director: Jayne Conte
Art Editor: Sean Hogan
Director of Creative Services: Paul Belfanti
Cover Art: Highway interchange, Skip Nall/Getty Images, Inc.—Photodisc
Compositor: Laserwords

Pearson Prentice Hall
Pearson Education, Inc.
Upper Saddle River, NJ 07458.

Printed in the United States of America
10 9 8 7 6 5 4 3 2 1

ISBN 0-13-142940-X

Pearson Education Ltd., *London*
Pearson Education Singapore Pte. Ltd.
Pearson Education, Canada, Ltd., *Toronto*
Pearson Education—Japan, *Tokyo*
Pearson Education Australia Pty. Limited, *Sydney*
Pearson Education North Asia, Ltd., *Hong Kong*
Pearson Educatión de Mexico, S.A., de C.V.
Pearson Education Malaysia, Pte. Ltd

Contents

Introduction

Summary

This workbook complements the textbook *The Cultural Landscape: An Introduction to Human Geography* by imparting students with:

- an appreciation of the way in which core issues in human geography are experienced by individuals in distinct places; and
- a basic overview of the world's regions.

This is achieved through a series of thirteen modules, each building on a chapter in the *Cultural Landscape* textbook, in which students are presented with:

- selective introductory material on a world region;
- thematic text that expands on the textbook, applying the subject matter of the chapter in question to the relevant region;
- introductory text on a current event/conflict in which the geographic theme covered in the chapter is being played out "on the ground" in the relevant region;
- reprinted documents presenting a range of positions and policy options pertaining to the current event/conflict;
- a series of review questions that can be used by the instructor to assess students' comprehension of the module's text and reprinted readings;
- a series of discussion/essay questions wherein students are asked to integrate material from the textbook and the workbook to suggest a policy for responding to the current event/conflict in the region in question; and
- an annotated list of suggested websites for further research on opinions and options regarding the policy debate.

Why a Workbook in Human Geography?

Traditionally, geography instructors have been forced to choose between two approaches toward teaching the discipline's introductory course. Textbooks adopting a thematic (or topical) approach devote a chapter to each of several key themes (such as industrialization, population, and natural resources) or topics (such as economic geography and political geography), while those adopting a regional approach devote a chapter to each of the world's regions (such as East Asia and sub-Saharan Africa). While neither approach is inherently superior to the other, students exposed to only one approach miss a crucial perspective on geographic concepts and facts. Students who are educated solely through the regional approach probably will amass considerable knowledge about specific regions of the world, but they may end the semester without having gained a conceptual understanding that enables them to explain why—given a set of place characteristics—specific geographic phenomena occur. They also may fail to understand the underlying dynamics of social and physical processes that span (and construct) distinct regions. On the other hand, students who are educated solely through the thematic approach probably will gain a thorough understanding of specific geographic processes and distributions, but they may lack an ability to appreciate the various interrelated factors that give a specific place its unique character and they may fail to appreciate how certain geographic phenomena are encountered and reproduced "in place."

This workbook aims to complement the standard topical-thematic textbook by locating human geographic topics within the context of contemporary social change. This application of human geography is achieved through a problem-based, place-based pedagogy. Our aim is that, after using this workbook, students leave the class with an appreciation of the various problems and choices encountered by "located" individuals around the world.

As an added benefit, students using this workbook receive a thumbnail world-regional geography survey within their human geography course, as the workbook moves systematically from region to region. The level of instruction in regional geography in this workbook is less than what one would receive from a dedicated world-regional textbook (the depiction of each region is necessarily brief and weighted toward the aspects of the region that complement the associated thematic point developed in the corresponding textbook chapter). Nonetheless, by using the workbook, students will be exposed to a survey of the world's regions, a useful benefit considering that at most U.S. universities the majority of students do not proceed beyond an introductory geography class.

Structure of the Workbook

Although this workbook could be used on its own or in conjunction with another textbook, it has been designed specifically for use in tandem with Rubenstein's *The Cultural Landscape: An Introduction to Human Geography*. Each of the thirteen modules in the workbook is paired with a chapter in the Rubenstein text. (The first chapter in the textbook does not have any associated modules, because this chapter is concerned with general conceptual issues that cannot easily be translated into the experiences and issues faced by individuals in place.) Each module develops an issue that lies within the topic of the matched Rubenstein chapter and explicates on that issue with reference to a specific region. The specific theme of each module is foreshadowed in the corresponding textbook chapter's "Global Forces, Local Impacts" box.

The exercise modules in this workbook focus on contemporary problems and issues, within the context of ongoing global processes and transformations. The general intent behind each module is to take a broad theme derived from a chapter in the textbook (such as population), expand upon this theme through an in-depth presentation of a more narrow contemporary problem (such as low fertility in advanced industrial economies), discuss how this problem manifests itself in a specific world region (such as Japan), and present documents that illustrate the range of policies advocated by individuals and institutions seeking to cope with the problem as it is experienced in the region (such as discouraging early retirement and encouraging an increase in the national birthrate). Through these modules, it is hoped that students will gain an appreciation for how the themes of human geography are experienced "on the ground" by people in their specific places, as well as an appreciation of how these individuals in place are forced to confront complex issues as they respond to and shape their local geographies.

Structure of an Individual Module

Each module consists of seven components:

1. *Personal Profile*. Each module begins with a brief profile of an individual person from the region in question who is experiencing the policy question developed in the module. (Sometimes this is an actual person; sometimes it is a composite individual.) For each module, her or his story resurfaces in the Discussion/Essay Questions section, giving students a further opportunity to consider how the policy problem developed in the module is experienced by individuals "in place."
2. *Instructional Text*. Although the precise organization of each module's instructional text varies, in every case the following material is covered:
    a. A topical component develops the specific issue considered in the module, expanding on the discussion in the textbook and directing explanation toward issues relevant to the region and the policy/event presented later in the module.
    b. A regional component presents a brief introduction to the region in question, emphasizing aspects of the region that are relevant to the topic explored in depth in the module.
    c. A policy-debate component introduces the policy issue that is elaborated on in the reprinted readings. This component also explains the institutional context for each of the reprinted readings and provides an overview of how each reading fits into the policy debate.
3. *Readings*. Each module includes five to ten reprinted primary-source readings. Typically, the initial reading(s) are relatively unbiased journalistic accounts, while later readings represent various sides of a debate.
4. *Review Questions*. Following the readings, each module includes four-to-six review questions. These are fairly straightforward questions that may be used by the instructor, or independently by the student, to assess comprehension of the central points of the module.
5. *Discussion/Essay Questions*. Each module includes three in-depth questions in which students are asked to grapple with and make recommendations on a policy dilemma stemming from the issues presented in the module. The first of these questions always builds directly on the personal profile presented at the beginning of the module. (For more on this portion of each module, see the following section on using the discussion/essay questions.)
6. *List of Readings*. As a reference source, instructors and students are presented with a formal bibliography of the reprinted readings, with URLs for locating them on the World Wide Web.
7. *Websites for Additional Research*. Each module concludes with an annotated list of websites that may be consulted for additional research. Typically, these are links to advocacy groups, news clipping services, and other sites that are updated regularly by concerned activists, journalists, and scholars.

Using the Discussion/Essay Questions

Although instructors are encouraged to experiment, for each module the Discussion/Essay Questions are designed to serve as the primary assessment mechanism. The intention behind these modules is not so much to present the student with additional facts as to lead students down a path of critical thinking. Students proceeding through a module assess and evaluate the opportunities and constraints facing individuals and policymakers as they confront geographic phenomena on the ground. Although there are discrete facts covered in the modules and mastery of these can be assessed through short-answer examination questions (hence the review questions section), the principles of abstract thinking and argumentation covered by each module as a whole probably are better assessed through open-ended essays, presentations, and/or classroom discussions.

To this end, the Discussion/Essay Questions are designed to lead students to integrate (1) broad thematic material presented in the textbook, (2) more specific thematic material presented in the workbook, (3) regional characteristics presented in the workbook, and (4) the perspectives and positions presented in the reprinted readings. If formal essays are assigned using these questions, the instructor might choose to grade them based on integration of these four elements, in addition to overall coherence of argument. To minimize grading burdens, an instructor might choose to integrate material from the review questions into short-answer exams given to all students while having only a portion of students contribute essays/presentations on each module (for example, each student could be assigned to write essays on any two modules of his or her choice over the course of a semester). Another exercise using these questions involves assigning specific readings to individual students (or groups of students) and then having students debate each other on a discussion/essay question, representing the views expressed by the author(s) of their assigned reading(s).

Besides serving as questions for evaluative essays and formal debates, the questions also are designed as jumping-off points for informal classroom discussions or in-class presentations. By asking students to make well-reasoned arguments (and by providing them with the materials from which they can construct such arguments), this workbook establishes an environment wherein students can discover, for themselves, how the themes of human geography impact their own lives and the lives of people around the world.

About the Authors

Phil Steinberg is an Associate Professor of Geography at Florida State University. He is the author of *The Social Construction of the Ocean* (Cambridge University Press, 2001) and co-author of *Managing Cyberspace: Governance, Technology, and Cultural Practice in Motion* (Temple University Press, 2005). He has published articles on marine political economy, Internet governance, urban planning politics, and geographic education in journals including *Political Geography, Environment and Planning D: Society & Space, World Bulletin, Urban Geography, Journal of Geography, The Professional Geographer, Review of International Political Economy, Geopolitics, New Media & Society, Info*, and *Geographical Review*.

His experience in geographic education includes serving as lead author of a U.S. Department of Education–funded urban ecology training manual for grade school teachers; designing and conducting U.S. National Park Service–funded workshops training secondary-school instructors in techniques for teaching local industrial-environmental history; and serving as lead academic consultant for Cerebellum Corp., where he assisted in developing concepts and scripts for the Standard Deviants Learn World Geography program that airs nationally on public television. Additionally, he developed the first entirely online offering in Florida State University's College of Social Sciences, a hybrid human–world regional geography course that was the subject of a 2002 article in *The Professional Geographer*.

His current research includes a study of representations of the ocean on fifteenth- to eighteenth-century European maps of the world. During the 2002–2003 academic year, he was a visiting fellow at the Dorothy and Lewis B. Cullman Center for Scholars and Writers at The New York Public Library, and he presently is serving as president of the Association of American Geographers' Political Geography Specialty Group.

Kathy Sherman-Morris is an Instructor in the Department of Geosciences at Mississippi State University. At Mississippi State, she has been involved with the online master's degree program designed specifically for teachers in geosciences. As part of this program, she has led field methods courses throughout the United States and the Caribbean. She is also a member of the Broadcast Meteorology faculty.

Her experience in geographic education also includes an undergraduate degree in geography education and several years' experience working with Florida State's online program in interdisciplinary social science. In addition, she has published on geographic education in *The Professional Geographer*.

Her research interests include hazards, hazard perception, and mass media. Currently, she is focusing her research on local television news and the ways people interact with it.

Acknowledgments

We would like to thank Regan Fawley, John Grimes, Rob McGowan, Janis Paulsen, Darren Purcell, Andy Walter, and Brian Yates, former graduate students who assisted in developing and teaching a series of web-based geography classes at Florida State University that formed the initial concepts for this workbook. We also are grateful to the undergraduate students who took these classes, for their input about what worked and what did not.

We are also fortunate to have had the opportunity to work with Dan Kaveney at Prentice Hall. Dan approached us about this project in response to an article that we authored in *The Professional Geographer*, and he has been a strong supporter ever since. He has proven to be the ideal editor for one's first foray into the world of textbook writing.

We also owe a debt of gratitude to all those teachers, from grade school onward, whose example led us to devote ourselves to education. It was our concern for the practice of education that led us to experiment with new methods for increasing students' capacity for critical analysis of geographic issues. We therefore thank all of those who, through educating us, inspired us to educate others.

Finally, we thank our respective spouses, Donna Jo Hall and John Morris. In addition to providing emotional support and critical feedback, we are grateful for their patience as we tied up household phone lines performing web research. If the lesson of this workbook is that, amid global processes and transformations, embedded and unique relations between people in places still matter, then we are especially fortunate to be engaged in ongoing processes of place-and relationship-construction with Donna and John.

PEOPLE IN PLACES

More than 5 million
1–5 million
Fewer than 1 million
Capital cities are underlined
140°E
N
RUSSIA
CHINA
NORTH KOREA
Sea of Japan
40°N
SOUTH KOREA
Kanazawa
JAPAN
Hitachi
Tokyo
Kyoto
Nagoya
Kobe
Osaka
PACIFIC OCEAN
Nagasaki
30°N
30°N
0
150
300 Miles
0
150
300 Kilometers
130°E
140°E

Cultural and Economic Impacts of the Demographic Transition in Japan

Companion to Chapter 2 Population

As **Kuni Kanbe** fusses over the frail figure of her 84-year-old husband, gaunt and bedridden with cancer, she thinks of her four children, all sympathetic and loving—and a long way away from her home in the village of Omiya.

When she married 56 years ago, she lived with her husband's parents and cared for them as they aged and sickened and died. But now, as she and her husband struggle with age and sickness in this little town in central Japan, her house resounds with the deafening absence of her grandchildren. "I took care of my parents-in-law, but nobody will take care of me," she said resignedly.

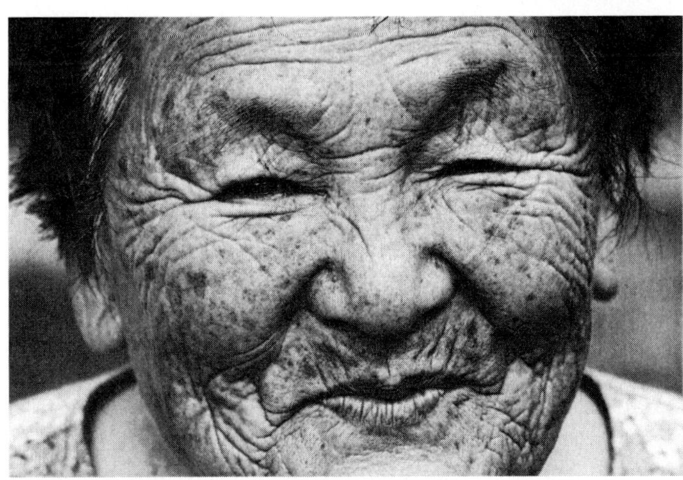

Japan's rural elderly are left to fend for themselves when their children take jobs in the city. *Source:* EPA/Andy Rain/Landov LLC

The sense of unfairness endured by Mrs. Kanbe is common among the elderly in Japan even though, by everyone's standards but their own, the Japanese are models of filial piety. Some 55 percent of Japanese over age 65 live with their children, compared with fewer than 20 percent in the United States and virtually every other industrial country.

Indeed, as the baby-boom generation approaches retirement, threatening the bankruptcy of social security systems across the globe, Japan seems to be the best-positioned of all major countries. It has a flexible and caring system that might be able to cope with retirement of the baby boomers: the family. If retirees can depend on their children for care, the nation is likely to survive the demographic upheaval relatively smoothly.

Yet in the winding alleys of little towns like Omiya, a farming community nearly 200 miles from Tokyo, the mood is one of disquiet. Even if a bit more than half of people over 65 live with their children in Japan, the proportion has plummeted from 80 percent as recently as 1970.

"Young people are scary," Mrs. Kanbe reflected soberly, as she kneeled on the tatami-mat floor a few feet from her sick husband. "The reason young people can kill humans as if they were insects, or fail to understand the feelings of their own parents, is mostly because they haven't had proper moral education."

"I'm scared of young people now."

Such comments are particularly surprising because Mrs. Kanbe is, by Western standards, well cared for by her children. They visit regularly, and her sons have asked her to come and live with them after her husband dies.

Yet she and many elderly women like her are reluctant to move in with their children because they know they would not occupy the traditional throne of the mother-in-law, that of matriarch of the household. Instead they would be guests, staying by the grace of their daughters-in-law.

Adapted from "Once Prized, Japan's Elderly Feel Abandoned and Fearful" by Nicholas D. Kristoff, which appeared in The New York Times, August 4, 1997, http://query.nytimes.com/gst/abstract.html?res=F00A17FD38590C778CDDA10894DF494D81.

Map Source: Map of Japan, p. 425, in *World Regions in Global Context* by Marston et al., ©2002. Reprinted by permission of Pearson Education, Inc., Upper Saddle River, NJ. ■

"Overpopulation" or "Underanalysis"?

Most debates about population begin with consideration of Thomas Malthus's theory about the inevitability of overpopulation crises. As the textbook notes, Malthus has been subjected to numerous critiques. Nonetheless, his theory remains widely accepted in popular thought, probably because of its simplicity and its apparent appeal to common sense.

Malthus argued that, because population increases geometrically (1, 2, 4, 8, 16, 32) while food production at best increases arithmetically (1, 2, 3, 4, 5, 6), at some point the world's population will exceed its carrying capacity, leading to famine, disease, war, and, ultimately, a population crash that will return population to a manageable level. Malthus proposed that we avoid the coming crash by limiting population while we still could. Responding to Malthus's warning, a range of public health initiatives, sponsored by governments and nongovernmental organizations, have emerged to fight the scourge of "overpopulation," which is assumed to be a global problem. One indicator of the pervasiveness of this concept became apparent during the writing of this workbook: the word *overpopulation* receives a check of approval from Microsoft Word's spell-check function, while its opposite, *underpopulation*, is said not to exist in the English language.

A cursory look at the state of the world today suggests that Malthus may have been oversimplifying population/resource/development dynamics (or, perhaps, we have oversimplified his theories in an attempt to define and grapple with the problem of "overpopulation"). High population densities are not always associated with poverty and squalor. If one looks at a list of the countries with the highest population densities (not counting city-states like Singapore), one does find some of the world's poorest countries (such as Bangladesh), but also some of the wealthiest (such as The Netherlands). There certainly are resource scarcity problems in the world, but the fact that some very dense countries are quite well off suggests that the problem is more complex than simply that of people having too many babies.

This finding has led some critics of Malthus to suggest that even if his description of the problem is basically correct his proposed solution should be turned on its head. If overpopulation occurs when there is an excessively high ratio of people to resources, then maybe the problem is not that there are too many people but rather that there are too few resources available. Because, at the global scale, more than enough food is produced to feed all of the world's population, the problem must exist with food distribution. Put another way, perhaps the problem is not that there are too many people in the less developed countries (LDCs) but that people in the more developed countries (MDCs) use resources so inefficiently that not enough resources are left for people in the LDCs. If one follows through on this criticism, one concludes that although global overpopulation is *experienced* as a problem in the poorer countries it can be *solved* only through a change in behavior in the more developed world.

Another critique of Malthus stresses that, although he did not realize it, he was writing from a specific historic context. Malthus first published his *Essay on the Principle of Population* in 1798, just as the Industrial Revolution was getting underway in Europe. Medical advances were beginning to improve life expectancy, but few changes had occurred in agricultural production techniques or in contraception. Since that time, huge advances have been made in both agricultural and contraceptive technology. While these advances could not be expected to eliminate an eventual Malthusian crisis (indeed, advances in agricultural technologies appear limited, as chemical fertilizers and pesticides frequently have long-term detrimental impacts on soil quality), they at least can be expected to postpone the crisis.

Additionally, since Malthus was writing when Britain was at "stage 2" in the demographic transition (see Figure 2-14 in the textbook), the high growth rate that Malthus saw in Britain and assumed to be normal actually may have been an unusual blip as British society advanced from being an agricultural society to one based on industrial production. As proponents of the demographic transition theory note, shortly after Malthus wrote, Europe went through a series of social changes that led people to want smaller families. Among these social changes was a decline in family farms on which parents needed child labor, a realization that due to a decreased death rate more children would survive to help take care of aged parents, and the rise of government-sponsored social security programs, as well as other factors. Around the same time, advances in contraceptive technology helped people follow through on this desire to have fewer children. Accordingly, birth rates have dropped considerably since Malthus's time, and this decline has proceeded to the point where several MDCs now have negative growth rates, something that Malthus never predicted would happen. Some question whether this dynamic of "demographic transition" will transpire in exactly the same manner in today's LDCs, but if the model is even partially correct it will greatly slow down any long-term trend toward global population growth.

Indeed, when one looks at total fertility rates (the number of live births born to a woman in her lifetime) instead of crude birth rates or natural increase rates, it appears that something like the demographic transition is well under way worldwide. Typically, there is a time lag between a drop in fertility rates and a drop in birth (or growth) rates. Even as fertility rates fall, birthrates may continue to rise because there are so many women of childbearing age who had been born in past generations when fertility rates were high. Critics of Malthus who emphasize this point note that, with a few regional exceptions, fertility rates have been declining throughout the world and that we simply must be patient as we wait for the inevitable drop in population growth that will follow. Based on the worldwide decline in fertility rates, many experts predict that the world's population will level off at about 11.6 billion people (close to twice the current population), sometime between 2150 and 2200.

A final critique of Malthus stems from objections to the ethical implications of his theory. Malthus was sure that a terrible time of war and famine would ensue if nothing was done to check population growth. Since, he reasoned, the poor are the least productive members of society (after all, he figured, that's why they're poor!), he suggested that actions be taken to *encourage* death and disease among Britain's poor populations. In the sixth edition of his *Essay on the Principle of Population* (1826), he wrote:

> We should facilitate ... the operations of nature in producing this mortality.... Instead of recommending cleanliness to the poor, we should encourage contrary habits. In our towns we should make the streets narrower, crowd more people into the houses, and court the return of the plague. In the country, we should build our villages near stagnant pools, and particularly encourage settlements in all marshy and unwholesome situations.

Many people (especially, of course, the poor whom Malthus thought should be left to die of disease) question the ethics of statements like these. How could Malthus or the government planners whom he was addressing claim the right to decide who should live and who should not? More generally, Malthusians are criticized because they shift the issue of population control from that of the household to that of the government. Malthusians imply that individuals cannot calculate the number of children that would be ideal in their particular economic situation, and that therefore the government must make this decision for them. However, evidence from many societies suggests that people will adjust the number of children they have to match their socioeconomic situation (the demographic transition is one example of this). Feminists, in particular, point out that in order to empower households to make rational choices about family size one must first empower women within the household, since women usually bear the main responsibilities of childrearing. Thus, a feminist response to the "overpopulation problem" is founded not so much on reducing births as on women obtaining the knowledge, power, and technology to control reproduction.

Localizing Population Issues

None of the preceding discussion is meant to imply that population issues do not matter or that they will just "solve themselves" if left alone. Practically every country is facing some kind of population crisis, whether it's too many people or not enough people, too many young people or too many old people, too high a growth rate or too low a growth rate. Even if one could find a basis for stating that there are "too many mouths to feed in the world," the realities faced by individual countries are much more complex.

A second look at The Netherlands and Bangladesh, and a third country—Niger—shows how different countries are encountering different kinds of population "problems":

- As was noted at the beginning of this module, *The Netherlands*, at first glance, appears greatly "overpopulated." Its population density—1,023 people per square mile—is among the world's highest. But overpopulation is not the country's problem; rather, The Netherlands, like most of Europe, suffers from a problem of low population growth. The country's annual population growth rate (the ratio of annual births to deaths) is only 0.4 percent. After years of very low fertility, The Netherlands is faced with stagnant population numbers and a population that includes a disproportionately large number of people above working age. Because these individuals are also above childbearing age, The Netherlands is likely to face very low population growth (or even population decline) for the foreseeable future. In response, The Netherlands, like much of Western Europe, has been forced to import working-age individuals so that it can continue to support its disproportionate population of pensioners.
- *Bangladesh*, by almost any account, probably *does* have too many people for its land. Not only is its population density more than twice as high as even that of The Netherlands (2,403 people per square mile), but under current economic conditions the country is having great difficulty providing food for its population, either through locally produced crops or through imports. The country's annual growth rate is 2.2 percent—not excessively high (the average for all LDCs excluding China is 1.9 percent), but still a huge burden for a country that begins with such a high population density and such high rates of poverty.
- In the West African nation of *Niger* the population density is only 24 people per square mile. This is not as low as it sounds. Much of the country is desert and is at best inhabitable only by nomadic peoples, which makes the population density number appear artificially low. Still, like most of Africa, Niger is not densely populated by world standards. Niger's problem lies in its 3.5 percent annual growth rate. As in any country where there are so many births, a huge proportion of the population is young; in Niger 50 percent of the population is under age 15 (in the MDCs, by comparison, only 18 percent of

the population is under age 15). Such a situation puts a huge burden on the relatively small working-age portion of the population.

The Graying of Japan

Many MDCs are suffering from population problems similar to those confronting The Netherlands: After years of declining fertility and increased life expectancy, populations are stagnating and an increasing percentage of the population has retired from the workforce and is now making extensive demands on national healthcare and social security/pension systems. Many European governments have coped with this problem by encouraging immigration of working-age individuals and their families. As is discussed in Chapter 3 of the textbook (and in the next module in this workbook), this solution is not a cure-all; immigration brings a host of social and policy problems.

In this module, however, we direct attention away from the migrant-receiving countries of Europe and instead turn to Japan, a country where the phenomenon of declining birthrates has been particularly severe, and where immigration rarely has been considered as a potential solution. An island nation, Japan has been unusually homogenous for thousands of years, with a culture famous for valuing conformity and for looking with suspicion at unique individuals or anomalous sub-cultures. As such, population policy debates in Japan have focused less on how to attract new workers to Japan or how to integrate them into Japanese society and more on how to encourage young Japanese adults to have more children while also enabling them to take care of an ever increasing population of pensioners.

Japan's demographic dilemma is introduced in the first reading, an article distributed via the Japan Information Network (JIN), the on-line service of Japan's Ministry of Foreign Affairs. The reading, published in 1997, announces that Japan's fertility rate has hit yet another record low (1.46 in 1995, the third lowest fertility rate in the world for that year), and the reading presents several other statistics regarding the ongoing decline and "graying" of Japan's population. The article also begins to explore some of the reasons for Japan's declining fertility rate. In the article, the decline is attributed primarily to Japanese couples postponing marriage, which in turn is attributed to the increasing ability of Japanese women to support themselves financially without the help of a husband. This reading is accompanied by a table, from a 2001 publication of the JIN but based on 1997 statistics, that clearly illustrates the projected decline in Japan's population and the projected increase in the percentage of that population aged 65 or over.

The next three readings—all also from the JIN—look at one of the main results of Japan's declining fertility rate: the increase in the percentage of the population that is elderly. As the readings note, the Japanese government has taken a number of steps to alleviate the burdens of Japan's growing "dependency ratio" (the ratio between those who are either too young or too old to work and the number of individuals of working age). Some of the government's efforts are aimed at getting more working-aged individuals into the workforce. Among the efforts noted in the readings are attempts to encourage workers not to retire before age 65 and laws designed to improve working conditions for women workers. If more women and people near retirement age can be brought into the workforce, national productivity will rise and there will be more national income with which to support the growing population of pensioners.

This strategy of increasing female workforce participation could backfire, however. While, in the short term, increased female participation in the workforce could increase the number of workers able to provide for each retiree (and thereby lower the dependency ratio), in the long term it may result in fewer couples having children, which could result in an even higher dependency ratio for future generations.

A second group of policies is aimed at reducing the public burden of caring for the elderly. Efforts noted here include the creation of financial instruments that allow elderly people to remortgage their homes to obtain money for healthcare as well as other measures that provide financial counseling to the elderly. Any changes in Japanese society that will allow the elderly to care for themselves or that will encourage families to care for their elderly relatives will greatly reduce the Japanese government's financial burden.

The Child and Family Care Leave Law, which lets workers temporarily leave their jobs at 25 percent pay with a guarantee that their job will be waiting for them when they are ready to return, serves both ends; it encourages working-age people to remain in the workforce and it facilitates informal care arrangements for those not of working age. The law accomplishes the first goal by making it easier for working-age people to stay in the workforce while caring for a child or elderly relative, and it accomplishes the second goal by freeing the government from having to provide that care. Additionally, this law may encourage two-earner couples to have children, which could limit the decline in the national fertility rate and offset any reduction in fertility resulting from an increase in women entering the workforce.

Despite all of the efforts being undertaken by the Japanese government to cope with the country's declining fertility rate and increasing dependency ratio, the authors of the fifth reading, a 2001 article from the International Monetary Fund's *Finance & Development* magazine, suggest that Japan's demographic situation remains a major factor contributing to the Japanese government's financial woes. The authors recommend a combination of tax increases and reductions in social security benefits as the key to bringing financial stability to the Japanese government.

Recognizing the burden that care for the elderly already places on Japan's large public debt, the Japanese government instituted a series of changes in its national healthcare system (for those aged over 40) in 2000, and these are reported in the sixth reading, an article from JIN's *Trends in Japan* on-line magazine, published in 1997 when the changes were first proposed. Under the new insurance scheme, government funding from the general budget is matched by contributions from individuals and their employers, home-based care is encouraged whenever possible, and healthcare recipients are required to pay a 10 percent co-payment for each service rendered before the insurance scheme covers the rest. As the article notes, the new healthcare system is expected to save the government over 10 billion dollars annually.

The final two readings stress that the aging of the population in Japan and other countries has implications that go beyond the concerns of national budget managers and the elderly themselves. Family members and workers, young and old, are being impacted by the aging population around them, and the social impacts of aging vary greatly depending on cultural context. The first of these readings, an excerpt from a report prepared for the World Health Organization by C.A.K. Yesudian, Professor of Health Service Studies at the Tata Institute of Social Sciences in India, expands the discussion from aging in Japan to problems of aging populations around the world. The final reading, an interview with Makoto Atoh, Deputy Director-General of Japan's National Institute of Populations and Social Security Research, directs attention to cultural aspects of Japanese society that are making the Japanese experience somewhat different from that encountered elsewhere.

Yesudian and Atoh both note that Japan is undergoing a demographic transition that in some ways resembles that seen in the other MDCs (notably, all MDCs have experienced an increase in the age at which couples marry). At the same time, however, the demographic transition in Japan has some unique characteristics due to the Japanese cultural context. Atoh suggests that the drop in fertility in Japan may be so extreme because, unlike in the West, the changing norms that have made it acceptable for women to work outside the home have not been accompanied by a transformation to individualist values. Instead, Atoh states, young men and women in Japan still retain traditional ideals of family. The lack of individualism in Japanese society means that many fewer children are born out of wedlock than is the case in the West. It also means that an urban Japanese woman in her twenties is faced with two choices. One option is for her to live at home, be supported by her parents, and work outside the house while enjoying all the freedoms that come with an urban, single lifestyle. Her other option is to marry, have children, and take on the responsibility of maintaining a traditional Japanese household (where the woman does most of the work: caring for her children, her husband, and her husband's parents) while still maintaining a frenetic urban life with long hours spent at the workplace. Given the two alternatives, it is not surprising that many women (and men as well) choose to postpone marriage and forego childbearing.

The implication of both of these final readings is that the classic demographic transition model, derived from the historic European experience, may not be accurate for predicting what will happen around the world as diverse cultures experience social change. Japan may be entering into a fifth stage of declining population, not anticipated by the classic European-defined model. Indeed, both of these readings suggest that a similar transition to population decline may soon occur in other Asian countries as they move to a post-industrial society without undergoing a parallel transformation to individualist values. This is not to say that the demographic transition model is *wrong*, but rather that geographic principles and concepts are experienced in unique ways by people in their distinct places, a key theme of this workbook.

Readings

from **the Japan Information Network** 1

Declining Fertility Rate

Population to Grow More Top-Heavy

February 14, 1997

The decision by more Japanese families to have fewer children, one of the main factors behind the rapid aging of the population, is turning into a major problem. The total fertility rate, or average number of children women bear in a lifetime, hit another record low of 1.46 in 1995 and is now expected to continue declining until the turn of the century.

Tracking the Fertility Rate

After hovering between 4 and 5 until the start of World War II and around 4.5 from 1947 to 1949, the postwar baby boom years, the fertility rate began to fall steadily, hitting 3.65 in 1950 and 2.04 in 1957. Following this, it stayed between 2.0 and 2.2 until 1974.

The following year, however, it dipped below the 2.0 benchmark to 1.91, and from there it began a rapid downward descent, hitting lows of 1.57 in 1989 and 1.46 in 1993. Despite a slight rebound to 1.50 in 1994, it fell again in 1995 to 1.42. Today, the Japanese rate is higher only than Germany's, which stood at 1.28 in 1993, and Italy's, which totaled 1.33 in 1992, among the industrial countries. To maintain the present Japanese population, a fertility rate of 2.08 is needed, and for more than two decades Japan's figure has failed to make that mark.

Elderly to Outnumber the Young This Year

According to the latest population projections released by the Ministry of Health and Welfare, the fertility rate will continue its slide from today's 1.42 to 1.38 in 2000, after which it will eventually stabilize at 1.61 in 2030.

The growing trend toward having fewer children, coupled with a longer life span, will have a profound impact on the population. According to the ministry's forecast, the number of people aged 65 or older will exceed the number of children under 15 by the end of this year, and the proportion of elderly will reach one in three by 2050. In addition, Japan's total population will drop to less than 100 million in 2051 and less than 70 million in 2095. Finally, the annual number of births in 2050 will be less than 70% of the 1996 figure.

Women's Changing Attitudes Toward Marriage

The explanation usually advanced for the trend toward having fewer children is the growing number of women who are active outside the home and their decision to marry later or remain single. The earnings gap has declined, and today men's salaries are 1.3 times more than women's, far less than the 1972 ratio of 1.9 to 1. A growing number of jobs have opened up to women, and more companies are giving them the same chances as men in hiring. Women today enjoy a degree of independence they have never known before.

Total Population and Age Breakdown

	Total population (million)	Ages 0–14 (%)	Ages 15–64 (%)	65 and over (%)
1995	125.57	16.0	69.5	14.6
2000	126.89	14.7	68.1	17.2
2005	127.68	14.3	66.1	19.6
2015	126.44	14.2	60.6	25.2
2020	124.13	13.7	59.5	26.9
2025	120.91	13.1	59.5	27.4
2030	117.15	12.7	59.3	28.0
2040	108.96	12.9	56.1	31.0
2050	100.50	13.1	54.6	32.3

Note: Statistics as of January 1997. Figures after 1995 are projections.

Source: National Institute of Population and Social Security Research, Ministry of Health and Welfare, *Japan's Estimated Population*

The upshot has been a move to either delay marriage or remain single. The proportion of single women in the 25 to 29 age bracket surged from 30.6% in 1985 to 40.2% in 1990 and 49.0% in 1995, or about half of all women, according to national censuses for those years. At the same time, the average age of first marriage rose from 24.96 in 1960 to 27.17 in 1995, and the proportion of women who remain single throughout their lives, based on the marital status of 50-year-old women, increased from 1.87% to 5.28% during the same period, according to Health and Welfare Ministry findings.

Even after they are married, women are choosing to have fewer children because of growing work opportunities, the increasing amount of money and time needed for raising and educating children, and poor housing conditions.

Government Encourages Families to Have More Children

The trend toward having fewer children, the graying of the population, and population decline will have a profound impact on the country's economic, social welfare, and employment and wage systems if they continue to progress. Without major reforms to the fiscal and social-welfare systems, the tax and social security burden borne by individuals will have to be raised from the current level of 35.8% in fiscal 1996 (April 1996 to March 1997). This larger burden, coupled with a decline in the number of workers, will invariably cause a slowdown in economic growth. The pension system will be hardest hit, with the effect felt most by salaried workers now in their forties. In 1995 each person 65 years of age or older was supported by 5.8 people; in 2050 there will be only 2 workers for every elderly person. Medical insurance premiums will also have to be raised dramatically.

To lessen the impact of these trends and maintain the vitality of economic and social systems through the next century, the government is now promoting sweeping administrative, economic, financial, social-welfare, fiscal, and education reforms. At the same time, it is hammering out countermeasures to stem the trend toward having fewer children and to boost the fertility rate. It is now considering ways to enhance its existing policies of providing child-care allowances, building and improving child-care facilities, and offering child-care leave. It has also begun exploring the possibility of reducing pension payments for families with young children.

Source: "Declining Fertility Rate: Population to Grow More Top Heavy, "in *Trends in Japan*, February 14, 1997, jin.jcic.or/jp/access/social/new.html

from the Japan Information Network 2

Women's Working Conditions

March 2001

The Equal Employment Opportunity Law for Men and Women, which came into effect in April 1986, was revised in June 1997 in order to speed up measures to counteract male-female discrimination. The revised provisions implemented in 1999 prohibit gender-based discrimination in job recruitment, employment, allocation of specific posts, and job advancement; they also make employers responsible for the prevention of sexual harassment. These revisions point mainly in the direction of placing much clearer responsibility for cases of gender-based discrimination on employers.

Along with the revision of the Equal Employment Opportunity Law for Men and Women, the Labor Standards Law was revised in such a way as to remove special restrictions on work done by women during holidays, late at night, and during other nonregular regular working hours. This revision underscored the idea that giving women special treatment was a type of gender-based discrimination. Although labor conditions for men and women have thus been equalized, there are in fact numerous claims that women, for the most part, are still responsible for doing the housework and raising the children. The disappearance of regulations that give women special protection might therefore increase, rather than alleviate, the social burdens women bear.

In 1999, 40.7% of the total number of persons employed by all branches of industry in Japan were women. In 1975, the corresponding percentage had been 32.0%, after which time more and more women found employment, particularly in service and food industries, wholesale and retail outlets, and electrical equipment manufacturing. After the fall of stock prices in 1990 and the end of Japan's "bubble economy" in the period that followed, the environment in which women were forced to seek employment became much more harsh. New female university graduates, in particular, continue to have difficulties in finding suitable jobs. The Ministry of Labor provided employers with guidelines in 1996 urging them to "not use recruitment practices that specifically designate, by gender, separate numbers for men and women sought for employment" and, furthermore, to "not hold company-introduction meetings or distribute written employment information limited to male audiences or readerships." Nevertheless, one sees many enterprises that hold non-restricted company-introduction meetings and carry out other recruitment activities in which female university students take part, although they do not in fact offer suitable employment opportunities for women. The reality reveals that most women sense an invisible wall, yet to be breached. It is hoped that the 1997 revision of the Equal Opportunity Law for Men and Women will improve the current situation.

The percentage of male and female irregular or part-time workers among all working income-earners was 16.2% in 1985, and this figure grew to 24.8% in 1999. In terms of gender, male part-timers represent 11.0% of the total male workforce, while female part-timers, at 45.0%, represent a large and rapidly increasing proportion of the female workforce.

Wage differentials between men and women have not disappeared. While there are not large differences in the initial salaries received by male and female university graduates when they enter a company, beyond the age of 30 or so, women's promotions to higher-paying positions fall greatly behind the promotions given to men. Various reports indicate that in most companies, there is an average difference of anywhere from 10% to 20% in the wages of male and female employees.

Nevertheless, improvements in the working environment are steadily being made. Compared to 10 years ago, there has been an extraordinary change, for example, in the recognition given to the issue of sexual harassment. In October 1993, the Ministry of Labor noted that sexual harassment can result in "definite harm" in the workplace and that "if repeated, will corrupt the working environment considerably." The Equal Employment Opportunity Law for Men and Women revision implemented in 1999 makes it obligatory for employers to make an effort to prevent sexual harassment. Recent years have seen a number of high-profile court rulings in favor of women who have pursued lawsuits over cases of sexual harassment.

3

from the Japan Information Network

An Aging Society

March 2001

Japan's trends toward fewer children and more senior citizens are proceeding at a faster pace than anywhere else in the world. According to the Ministry of Health and Welfare, in 1980 the total fertility rate (the average number of children born to women in Japan) was 1.75. This figure was down to 1.39 in 1997, as compared to the corresponding values of 2.03 in the United States, 1.73 in the United Kingdom, and 1.72 in France. Unless the birthrate rises again, a decrease in the total productive population (ages 15 to 64) is inevitable, generating concern about economic decline and maintenance of the social welfare system.

A major reason for the decline in the birthrate is thought to be the increase in the average age of marriage for both men and women. Working women's reluctance to have children, due to the inadequacy of public systems that would help make holding a job and raising children compatible, is a contributing factor. Added to this, men still participate relatively little when it comes to helping with housework and child care. The government considers the problem of a declining birthrate to require urgent attention, and in 1991 it passed the Child Care Law, which stipulated that employers cannot refuse requests from either men or women to take time off from regular work schedules in order to care for children less than 1 year of age.

As society ages, increasing responsibilities are being borne by working people, especially working women, whose families include senior citizens needing special nursing care. Addressing this issue, family care leave provisions were added to the Child Care Law in 1995, and it was renamed the Child and Family Care Leave Law. This revision, which went into full effect in 1999, enables workers to leave their regular jobs for specified amounts of time in response to a need to give special care to a spouse, a parent, a child, or a spouse's parent.

The law guarantees payment, via employment insurance funds, of 25% of one's regular salary, should an employee be off work for infant or family care. Many larger private enterprises provide additional financial assistance for such purposes. Reinstatement in the employee's regular place of work, following such a work–leave, is guaranteed.

from the Japan Information Network

4

The Rapidly Graying Japan

Coping with an Aging Society Becomes Important National Policy

September 19, 1996

Share of Elderly Doubles in Quarter Century

Japan's population continues to age. According to a Prime Minister's Office study, in 1995 there were 18.60 million people aged 65 or over, accounting for 14.8% of Japan's total population of 125.6 million. The share is 2.8 points above what it was five years earlier and twice the 1970 figure of 7.1%.

The figures were extrapolated from a sample of about 400,000 households covered in the 1995 national census (1% of the total). The study dealt with 11 items, including age distribution of the population and the state of the labor force. Children aged 14 and younger totaled only 19.96 million, or 15.9% of the general population. This is the first time that this age group failed to reach 20 million and provides further evidence of a declining birthrate.

Consequently, the population pyramid has developed two bulges, with one for the generation now in their late forties, born during the postwar baby boom, and one for their children, who are in their early twenties.

860,000 Elderly People Need Care

As society ages, the number of elderly households (consisting only of men over 65 and women over 60 or with the addition of unmarried children under age 18) has increased. A Ministry of Health and Welfare study shows that the number of such households had risen to 5.63 million in 1995, making up 13.8% of the national total of 40.77 million households. Individuals living alone accounted for 46.2% of elderly households, and elderly couples for 49.3%. The ratio of elderly households was 4.9% in 1975, meaning that the figure nearly tripled in two decades.

At the same time, the number of elderly people requiring care at home has reached 861,000. Of these, 284,000 are bedridden, and over half of them have been in the same state for three years or more. By age group, the highest share of those providing care for bedridden elderly persons are in their sixties (28.3%) and seventies (24.2%), indicating that there are many cases of elderly people caring for other elderly people.

Government Measures for an Aging Society

Japanese society seems certain to age even more rapidly in the future, making wide-ranging measures to deal with the situation important national policy goals. In addition to addressing issues like medical treatment, providing care, pensions, and employment, these measures must take into consideration the entire social system and the way it works.

At the beginning of July, the cabinet officially approved the outlines of a policy for dealing with the aging society. It indicates guidelines for integrated measures to prepare for an aging society.

First and foremost, the outline calls for promoting continuing employment until age 65. With regard to a publicly run system of providing care, it indicates that active steps should be taken to create a new system providing care for the elderly using a social insurance method that has suitable public funding built into it.

Other measures include: the introduction of a system of reverse mortgaging that provides financing to elderly people, using their homes or other property as collateral; consideration of asset management support for those who are senile; the creation of housing and communities that eliminate steps and other physical obstructions in the living area; and the creation of a working environment that makes it easier to continue working while caring for elderly people or small children.

The government will keep watch over the implementation of policies in keeping with the outline, which it plans to review in about five years.

Source: "The Rapidly Graying Japan: Coping with an Aging Society Becomes Important National Policy," *Trends in Japan*, 9/19/96

5 from **the International Monetary Fund**

Japan: Population Aging and the Fiscal Challenge

by Martin Mühleisen and Hamid Faruqee
March 2001

With Japan facing a demographic crisis, government finances—stretched to the limit to keep the economy afloat—have to cope with the rising strain on public pension and health systems. This article looks at the economic and fiscal costs of aging in Japan.

The populations of all industrial countries are aging. As prosperity has increased, birth rates have declined, and, with the baby-boom generation about to enter retirement, public pension schemes have come under pressure to raise contribution levels or cut the size of benefits. Japan, whose population enjoys the greatest longevity worldwide, will be particularly affected. The share of elderly people as a percentage of the working population in Japan is already one of the highest in the world, whereas the fertility rate is among the lowest, implying that the age distribution of the population will shift rapidly in the coming decades. By 2025, there will be roughly one elderly person for every two persons of working age, which will leave Japan with a higher old-age dependency ratio than any other major industrial country (Table 1).

Table 1
Old-age dependency ratios: comparative projections[1]

	Canada	France	Germany	Italy	Kingdom United	States United	Japan
2000	20	28	25	28	27	21	27
2025	36	41	36	43	36	33	47

Table 2
General government finances, 1999 (percent of GDP)

	Canada	France	Germany	Italy	United Kingdom	United States	Japan
Actual balance	2.8	−1.8	−0.7[2]	−1.9	0.3	0.01[2]	−9.21[2]
Structural balance	3.3	−0.8	−0.7[2]	−0.5	0.1	0.21[2]	−8.11[2]
Gross debt	88.1	58.6	61.1	114.9	44.8	62.4	125.4
Net debt	56.7	49.0	52.4	108.8	39.0	50.6	38.1
Net debt, excluding social security[3]	...	...	53.1	...	...	59.1	87.9

Source for Tables 1 and 2: United Nations, 1996, World Population Prospects 1950-2050 (New York).
[1] *Number of elderly (65 years and older) as a percentage of the working-age population (20–64 years).*
[2] *Excluding social security.*
[3] *... denotes data not available.*

As this demographic shock unfolds, Japan will face substantially greater fiscal challenges than other countries. While most industrial nations have successfully reduced government deficits during the past decade, Japan's fiscal situation deteriorated dramatically following the government's efforts to resuscitate the economy. The immediate task will be to return the government deficit to a sustainable level, which will be complicated by rising social security benefit payments and the need to avoid an abrupt shift in the fiscal stance that could jeopardize growth prospects. Even after the overall fiscal situation has stabilized, however, the government still faces the long-term task of preserving the solvency of the social security system.

The Expected Population Decline . . .

A central feature of Japan's demographics is a long-run decline in fertility. In the postwar era, the total fertility rate—defined as the average number of births per woman—experienced a sharp decline in a single 10-year span, falling from over 3½ births in 1950 to just 2 births in 1960 (see chart). The decline has since continued at a somewhat slower pace, but the fertility rate has now dropped well below the replacement rate. This decline is often attributed to three economic factors: a decrease in the salary gap between men and women, difficulties reconciling work with child rearing, and the generous social security system that has increased the financial independence of the elderly and made them less reliant on the support of their children. Most projections have assumed that the fertility rate would eventually stabilize and increase toward the replacement rate, so that the population would be stationary by the latter half of this century. However, earlier projections have proved somewhat overoptimistic, and the time frame in which the fertility rate is expected to recover has been pushed back repeatedly in recent years.

The implication of the dramatic fall in fertility is that, for some time to come, fewer young Japanese adults are expected to enter the economy. Redefining "births" as the inflow of young adults into the economy, the middle panel of the chart shows the historical

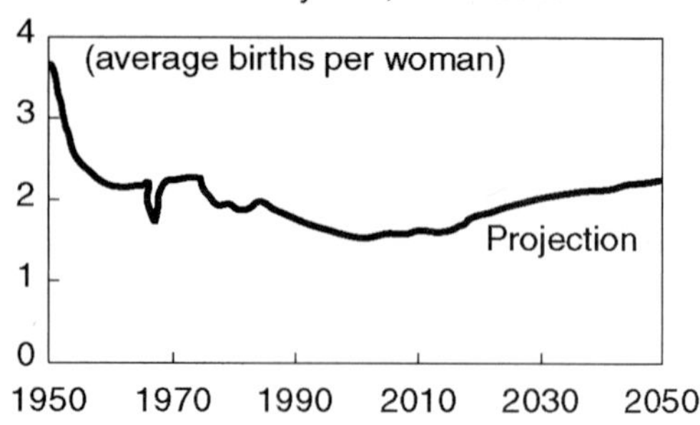

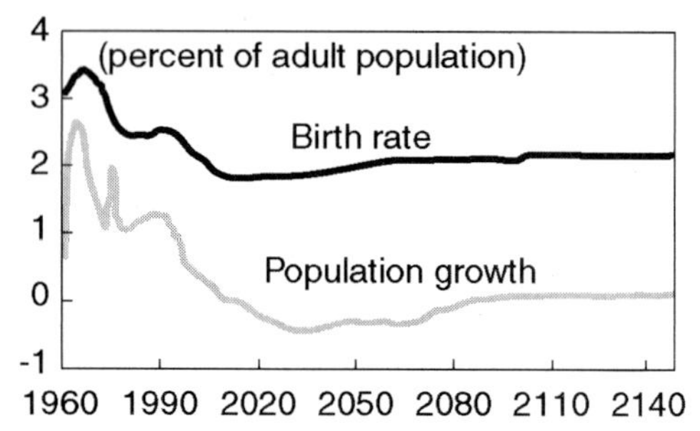

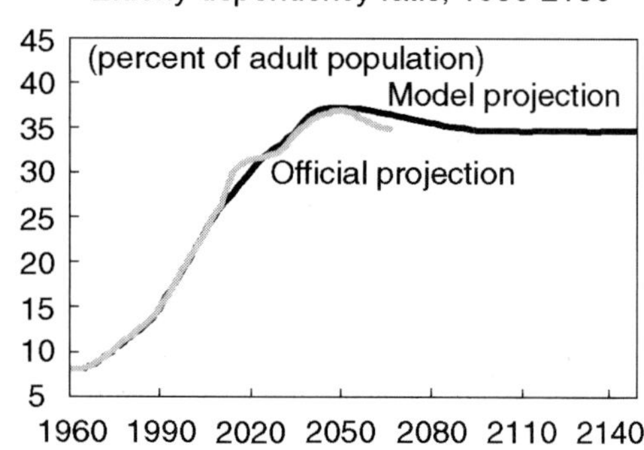

and projected evolution of Japan's "birth rate" since 1960. The inflow of young adults as a share of the adult population (age 15 and older) has declined significantly over the past 40 years; even assuming a modest recovery in fertility over the next 50 years, the "birth rate" is projected to remain far below historical levels well into the next century. This does not assume any population increase from immigration, which has alleviated demographic pressures elsewhere.

Japan's demographics imply not only an aging population but also a declining one. With fewer young adults expected in the future, the proportion of working-age people will diminish over time and the population's average age will increase. In addition, fertility rates have already fallen to such an extent that the adult population and workforce can be expected to decline for much of the twenty-first century despite an expected increase in life expectancy (from 81 years at present). The bottom panel of the chart shows the ratio of Japan's elderly dependents (65 years and older) to the adult population, which is projected to rise dramatically and cause Japan's population to age more rapidly than the populations of the other major industrial countries.

. . . Will Lead to Shrinking Economic Output

There is a broad consensus among economists that demographic changes will reduce output growth and limit increases in economic welfare. A shrinking population is associated with lower employment and output, although rising capital intensity, productivity increases, and higher labor participation rates could mitigate the impact on per capita incomes. On the basis of long-term simulations produced by the IMF's world macroeconomic model (MULTIMOD), Japan's demographics imply that the level of real GDP will fall by a cumulative 20 percent over the next century compared with a baseline simulation with a stationary population. The output costs of aging will reach almost $\frac{1}{2}$ of 1 percentage point in lower annual GDP growth between 2025 and 2075, when the demographic changes are expected to be most pronounced. In per capita terms, GDP is expected to drop by about 5 percent relative to the baseline scenario, primarily because the percentage decline in effective labor is larger than the fall in the number of workers, given the aging of the workforce and the differences in labor productivity and labor supply across age groups implicit in Japanese age-earnings profiles.

As in previous studies, the simulations suggest that investment and saving levels decline with GDP in the long run. The fall in investment reflects a desire to shed capital in the face of declining labor and output in the economy; the rate of investment as a share of GDP, though, is more or less unchanged. Contrary to earlier findings, however, saving rates do not necessarily decline as the population ages. Despite a higher proportion of elderly, who tend to save less, the decline in the inflow of young adults (who tend to consume at higher rates) and the increase in longevity (which tends to lengthen planning horizons) act to raise private saving rates. Consequently, as long as the government is able to keep the fiscal deficit under control, Japan's current account surplus need not decline appreciably as its demographic changes continue to unfold.

The Fiscal Problem

Japan's public finances worsened dramatically during the 1990s, however, and the government is not in a position to cope well with an aging population. Tax revenues plunged after the collapse of the asset-price bubble in the early 1990s, and expenditures have been driven upward by economic stimulus measures—above all, public works spending—and by the costs of assuming the liabilities of a number of failed financial institutions. Although the pension system still runs a surplus and owns substantial assets (around 50 percent of GDP), these are more than offset by future pension claims. Excluding the social security system, the general government deficit amounted to 9 percent of GDP in 1999, and net debt has reached close to 100 percent of GDP, which is among the highest for industrial countries (Table 2). The ratio of debt to GDP has roughly doubled over the past decade, and markets have been concerned that its true level could be even higher, given that the government might have to cover contingent liabilities from loan guarantees and losses by public sector enterprises.

The government has taken some steps to prepare for the demographic change. The recently legislated pension reform (approved by the Diet in March 2000) contained provisions to cut lifetime pension benefits by around 20 percent for future retirees, particularly through a 5 percent reduction in benefit levels; a gradual increase, beginning in 2013, of the age of eligibility from 60 to 65 for earnings-related pension payments; and full indexation of pension increases to the consumer price index instead of disposable income. Although government transfers to the basic pension scheme are to be raised to one-half of basic pension benefits from one-third starting in 2004, this has reduced the government's unfunded pension liabilities (previously estimated at around 100 percent of GDP) by a third. However, further contribution rate increases will be necessary to prevent pension assets from being depleted, and substantial reforms will also be needed to cope with fast-growing public health expenditures.

On the basis of simulations using the current benefit and transfer structure, contributions to the main wage-based employee pension system are expected to increase from $17\frac{1}{2}$ percent currently to almost 30 percent over the next 50 years. Medical contribution rates would have to rise by a similar proportion, mainly as a result of the steep rise in old-age medical costs. Despite these increases, financial balance in the social security system would depend on a doubling of government transfers to more than 5 percent of GDP a year, which illustrates that the fiscal outlook will become considerably more difficult as the demographic transition sets in.

Once the economy has recovered, the government will need to accord high priority to fiscal consolidation and the stabilization of public debt before population aging intensifies sometime after 2010. Even assuming that the government manages to reduce its fiscal deficit to about 3 percent of GDP within a decade, however, net debt is likely to stabilize only after reaching around 120 percent of GDP—a level that would leave public finances vulnerable to interest rate shocks and policy slippages. While a faster pace of consolidation would be desirable, the problem for the authorities lies in addressing the fiscal imbalance in a way that will not jeopardize the nascent economic recovery. In 1997, the last attempt at fiscal consolidation had to be abandoned because raising the consumption tax rate and cutting public works spending, among other factors, weakened Japan's economic recovery.

Policy Options

Although the results of long-term simulations are tentative at best, it is clear that a policy of slow debt consolidation and gradual social security reform would be a long-term drag on living standards, because of economic disincentives posed by high social security contribution rates and indirect tax increases necessary to finance rising government transfers and interest payments. What could be gained by pursuing different policies and at what cost?

Social Security

Macroeconomic simulations suggest that reducing pension and health benefits, coupled with lowering social security contributions, would be preferable to maintaining the present generous benefit levels, in terms of both growth and economic welfare measures. While benefit cuts would imply short-term output costs (because consumption would drop as forward-looking consumers increased their saving for retirement), the positive long-term effects on output would be substantially larger, owing to a fall in interest rates that would boost investment. Moreover, a solution that reduced payroll deductions would clearly be more equitable from the point of view of younger workers.

In contrast to benefit cuts, a reduction in social security contributions financed through a consumption tax increase would have mixed effects on private consumption and wealth. There would be less of a buildup in private savings, because pension financing would be partly shifted from workers to pensioners, who generally tend to save less. Growth would be somewhat higher than in a system with higher social security contributions, mainly because the labor supply would increase as a consequence of lower payroll taxes. (Historical data indicate that female employment, in particular, reacts to changes in disposable income.) However, a rise in interest rates would lead elderly consumers—who would enjoy higher returns on their assets—to increase consumption. The resulting decline in the current account balance would depress financial wealth through a drawdown of foreign assets, which would eventually imply a drop in consumption relative to the baseline, although only toward the end of the century.

Fiscal Consolidation

In view of Japan's long-term fiscal imbalance, any move to address the fiscal deficit will not only have a direct impact on aggregate demand, but will also affect individuals' future saving behavior. Given the slow revival of confidence after a long period of stagnation, any recovery in Japan is likely to be fragile and prone to reversals. The choice of policies for stabilizing the fiscal situation is therefore important, especially if the government decides to pursue a somewhat faster pace of consolidation to reduce its exposure to fiscal risks.

Since Japan's government sector is already relatively small (the large Fiscal Investment and Loan Program is not part of the general government sector and is thus excluded from the simulation), there are essentially three options for reducing the fiscal deficit: (1) cuts in public investment (which is significantly higher as a share of GDP in Japan than in other major industrial countries); (2) a broadening of the direct tax base (with taxpayers benefiting from large income tax deductions and exemptions); and (3) increases in the consumption tax rate (which, at 5 percent, is still relatively low by international standards). Again, our macroeconomic simulation framework can be used to look at the costs and benefits of these strategies.

The results suggest that ambitious debt stabilization could lead to substantial short-term output costs. This is particularly true for public investment cuts, although these would also generate the strongest long-term welfare gains. From a political point of view, investment cuts would appear to offer an easier way of reducing the deficit, given increasing public dissatisfaction with public works projects. However, economically, it would be less risky to pursue consolidation through a careful policy mix that would need to include tax increases to limit the impact on aggregate demand. In terms of tax measures, the simulations indicate that increasing the consumption tax would be somewhat less beneficial than broadening the direct tax base. This highlights the fact that the value-added tax—although generally considered to be close to an optimal tax—is distortionary in that it introduces a wedge between producer and consumer prices. The estimated efficiency losses that would result from broadening the direct tax base are smaller, although only slightly.

Conclusion

Demographic changes will be a defining feature in Japan for the foreseeable future. A sustained decline in fertility rates underlies a rapid aging and dwindling of Japan's population that can be expected to continue well into this century. This dramatic demographic shift is likely to have profound social and economic implications, including slower growth in output for some time. High public debt and adverse population dynamics increasingly constrain the government's room for maneuver, suggesting that strong policy adjustments will eventually be required to put public finances back on a sustainable footing. Reforms currently being implemented in the pension and health systems are a step in the right direction, but further measures will be needed to avoid a large increase in payroll taxes and government transfers that would distort incentives and hamper growth.

Ambitious debt stabilization could potentially result in substantial short-term output costs, posing a risk to the recovery. Therefore, as long as private demand remains fragile, fiscal adjustment policies should be implemented cautiously. However, once the recovery is on a sound footing, Japan will need to implement a long-term fiscal strategy to make public finances sustainable. Public investment cuts, measures to broaden the income tax base, some increase in the consumption tax, and reductions in social security benefits are likely to be the key building blocks of a longer-term solution.

Source: Martin Muhleise & Hamid Faruqee, "Japan: Population Aging and the Fiscal Challenge," *Finance & Development*, March 2001, imf.org/external/pub/ft/fandd/2001/03/Muhleise.htm.

from the Japan Information Network

6

Caring for a Graying Japan

Elderly Nursing-Care System Is In the Works

June 20, 1997

As part of the government's response to the rapid aging of Japan's population, a range of bills aimed at establishing a system of public nursing-care services are moving toward approval in the Japanese Diet, which ended its regular session on June 18. Discussion on the nursing-care insurance bills will carry over to the extraordinary Diet session scheduled to convene in the fall. Pending passage, they are slated to take effect from April 2000.

Participation from Age 40

The public nursing-care insurance scheme being discussed is a system, supported by monthly contributions, to provide low-cost nursing-care service to the bedridden and others in need. All people over the age of 40 will contribute to the system, which will be operated at the municipal level. To receive nursing care, an applicant must meet a level of need prescribed by the administering body, and as a rule must be at least 65 years old (or 40 years old for those with conditions relating to aging that necessitate care).

The monthly payments will vary slightly according to the contributor's municipality of residence and income, but the Ministry of Health and Welfare estimates that the average monthly contribution will be about 2,500 yen (22 dollars at 115 yen to the dollar) over the first three years of the program. Salaried workers, who are enrolled in Health Insurance Societies, will split the cost of the payments with their employers. The recipients of nursing-care services will be required to pay 10% of the costs incurred, with the rest to be covered by insurance premiums and public funds.

The services included in the new nursing-care system will be broadly divided into two categories: those provided at patients' homes and those provided in nursing homes and other care facilities. Based on the nursing needs of the elderly care recipients, home-care services are to range from assistance with household chores to the dispatch of nurses to patients' homes, as well as day-care services—such as bathing and rehabilitation—at care facilities. According to the Ministry of Health and Welfare's program model, those needing the least care will receive visits from nurses, home help, and day-care service once or twice a week; those most in need of nursing care will be looked after every day. It will now be up to the national and local governments to boost staff levels and the number of facilities in order to provide this standard of service.

Preparing for a Gray Age

The need for the introduction of this sort of nursing-care insurance stems from the unprecedented speed with which Japan's population pyramid is growing top-heavy. The number of Japanese aged 65 or older has increased by more than 600,000 annually for the last several years, and last year the portion of the population in this advanced age bracket reached 15%. The growth of this percentage from 7% to 14%, which took 45 years in Germany and all of 130 in France, took place over a mere 24-year span in Japan.

The ratio of Japan's population aged 65 or older is expected to reach 17% in 2000 and a world-high 21% in 2010; it is estimated that it will go as high as 28% in 2040. This rapid graying of society is leading to an annual increase of over 100,000 elderly people that need nursing care or assistance with their day-to-day activities, a population that has already topped 2 million.

Insufficient nursing-care services are also straining the financial health of the health insurance system by causing many elderly patients to resort to long-term, medically unnecessary "social hospitalization" instead. Unless nursing services are expanded, the estimated increase in the population of elderly needing nursing care, to 2.8 million by 2000 and to 4 million by 2010, will only exacerbate this situation.

These worries provided the impetus for the decision to separate nursing services from medical care and introduce a separate social insurance scheme for the former. The Health and Welfare Ministry estimates that the plan will cost 4.2 trillion yen (36.5 billion dollars) in fiscal 2000, its first year of operation. But the burden on the health insurance system should be lessened by about 1.2 trillion yen (10.4 billion dollars) as socially hospitalized patients' coverage is switched to nursing-care insurance.

Northern European Model

Similar systems of nursing-care services are in place in Germany and the nations of Scandinavia. Germany's scheme began offering home-care nursing in 1995 and introduced facility-based care the following year. All citizens aged 18 or older contribute to the fund supporting the plan, which means it relies on no public monies. Care recipients and their families can opt to receive services directly or to get cash vouchers for their nursing care needs.

The municipally operated Scandinavian schemes, on the other hand, draw on public funds to provide nursing care for the needy elderly. Medical and welfare services are unified and many care facilities offer 24-hour service.

The Japanese nursing insurance scheme is a melange of these systems, featuring some of the strong points of each of them. But its introduction is not expected to be completely smooth. At present, Germany is the only nation with a nursing insurance scheme in place, and that program is still young. A period of trial and error is therefore expected to continue for some time. The nursing-care insurance system is scheduled for review five years after its implementation in fiscal 2000.

Source: "Caring for a Graying Japan: Elderly Nursing-Care System is in the Works" *Trends In Japan*, June 20, 1997, jin.jcic.or.jp/trends96/honbun/tj970618.html

from **the World Health Organization** 7

Socio-economic Implications of Ageing

by *C.A.K. Yesudian*

Introduction

For the first time in human history, we are going to enter into a new millennium that will be very different from the earlier ones. The demographic composition of the world is undergoing transformation, which will change the face of the world in the next millennium. The world is graying and this process is accelerated and intensified by the decreasing birth rate and mortality rate, and increasing longevity of life. This change is going to affect the social and economic lives of human beings to a great extent. However, the world is not ageing uniformly. The demographic transition varies from one region to another and from one country to another in the same region. Further, the impact of ageing process on the social and economic lives of people too is going to be different for different countries depending on their social fabric, cultural values and the economic structure of the society.

This paper attempts to probe the implications of the ageing process on the social and economic lives of human beings in the 21st century. The paper initially describes the demographic change that affects the world, different regions and countries. Then it analyzes the impact of such a transition on the social and economic lives in the society.

Population Growth and Change

Population growth in the first quarter of the next century will be tremendous. According to the WHO estimate, the global population will increase from 5.68 billion in 1995 to 8.03 billion in 2025. However, in the second quarter of the century, the population will reach around 10 billion. Most of the population growth is going to take place in the less developed and least developed region of the globe. On the other hand, population of the more developed countries will increase marginally from 1.17 billion in 1995 to 1.22 billion in 2025. Apart from the regional variations, there are differences in growth of population in different age groups. While the population of children below the age of 15 will be decreasing in the first quarter of the next century, the aged population of 65 years and above will increase. This will happen in all the regions of the world. This means an increase in the elderly share of the population. It is expected that the proportion of aged 65 years and above will increase in the developed world in the first quarter of the next century due to low birth and death rates and increase in longevity of life. Further, the proportion of the very old, i.e., 80+ years will also increase in developed countries. On the other hand, in the developing world, there will be substantial increase of older persons in absolute numbers. Especially in countries like China and India, the older population is going to increase to the extent of 194 million in China and 119 million in India in the year 2025. Therefore, the early part of the next century will see the intensification of ageing in the developed world and rapidity of ageing in the developing world.

Ageing and the Family

Family is the key institution that provides social and economic support to the individual at different stages of life. The structure of family has undergone changes in different societies differently at different stages of human history. Intergenerational relationship and the role of women in the family are changing that affect the care of the aged in the family. This is a major concern for the policy makers who are planning the welfare of the aged. These two aspects are discussed in detail in this section of the paper.

(a) Intergenerational Relationship

Industrialization and urbanization have brought changes to family structure in this century to a great extent. The extended family that existed in the feudal society has changed to nuclear family. This has affected the position of the elderly in the family as well as the family's capacity to take care of the aged. Hess and Markson summarize this change in the following manner. "The disintegration of the feudal state, the decline of clan or family as a basis for political power, the rise of cities and the increased mobility of populations moving from the country to cities to seek work all had profound implications for the structure of the family. With the emergence of capitalism, division of labour, and industrialization, added emphasis was placed on economic or market relationship rather than on traditional ruler-ruled and familial ties. With greater potential for individual self-expression and freedom came a loosening of the family as a basis for social stability and regulation. One's position in the economic structure rather than place in the family became a crucial indicator of status and life chance." The implication of such a change was felt more in the Western world than in other countries.

Today, the aged in the Western world feels the need for independence and prefers to live alone or with the spouse in their house. For children, it is no more an obligation to take care of the aged but an emotional bond to respond to their needs. The preferred pattern of intergenerational relationship is intimacy at a distance and by choice. Thus the family ties in the Western world are far from dead but has changed in quality and increased in quantity. Dooghe observes that intergenerational linkage is the salient feature of kinship networks. He found that children kept close contact with the aged parents to a great extent in the Western world. He concludes that intergenerational alienation is more a myth than reality in the Western world.

Since individualism, independence, and achieved position in the family are emphasized in the Western culture, the change in the intergenerational relationship is smoother. The aged would like to live independently as long as possible and the children do not feel guilt of being away from the parents. "Older people may view dependence on informal helpers as a deviation from cultural norms of independence, leading to devaluing self-perceived concepts and lowered morale." On the other hand, in the Eastern cultures, the change is still not fully accepted within the cultural framework. This leads to stress and strain in the intergenerational relationship. Here the aged would like to live with the children and the children have the obligation to take care of the aged in the family. For example, 66.0 percent of the aged in China in 1991and 65.0 percent of the aged in Republic of Korea in 1990 were living with their families. On the other hand, in the United Kingdom, only 15.2 percent of the aged were living with their families in 1990. But industrialization and urbanization are changing the extended family structure and making it difficult for the children to fulfill their obligation towards their parents. This creates stress, strain, and guilt among children in the Eastern cultures.

Like urbanization and industrialization, the ageing process itself is going to affect the family structure. With the declining fertility rate and increasing longevity of life in the new millennium, we will have a narrow vertical family structure, of three, four or even five generations. This means that there will be more generations surviving but will have fewer children in each generation. This will happen to the developed world first, where there is already a large proportion of aged population. In such a structure, there will be fewer younger persons to take care of too many aged persons. This is going to bring more pressure on the younger generation. Even if the aged lives alone independently, the need for care during emergencies will be difficult for the smaller number of younger persons in the family. This may call for more professional and formal support for the aged at home in the developed world.

On the other hand, in the developing countries the fertility rate will not decline dramatically in the early part of next century. Therefore, the family structure will be still horizontal with more children and fewer generations above. This will be an advantage for the family in the developing world to take care of the aged in the family. The informal family support system can be better sustained in the horizontal family structure. However, modernization is likely to replace family commitments with individualism, and the segmentation of urban living and the less commitment of the elderly towards their children are likely to affect the informal family support system in the developing world.

Challenges Ahead: The above analyses of intergenerational relationships have implications for the care of aged in the future. First of all, the tendency of the aged to live alone or with his or her spouse in their own homes has implications for the housing policy for the aged. Most of these houses are large and do not have amenities for the aged to live safely and comfortably. At the same time the aged does not have the resources to maintain and create facilities for him or her in the house. Therefore, appropriate housing for the aged will be an important issue to be addressed in the welfare policy of the aged in the Western world. The policy should emphasize the role of the government and the private sector to provide safe and comfortable housing for the aged.

Due to the vertical growth of family structure, there are fewer younger children to give care for the aged and many of them are likely to live away from the aged. The aged would also like to live independently in his or her own home. Therefore, there is going to be diminishing informal care giving and increasing demand for formal home care. Therefore, the challenges would be developing a variety of home care services to maintain the supported autonomous life style of the aged. Here the technologies of the developed world (robotic, portable, easy to operate and communication technologies) are going to play an important role to bring the health and social services at the doorsteps of the aged. On the part of the children, the responsibility would be to maintain intimacy at a distant by way of communicating and supporting them in times of need and be at their side in times of emergencies. Here again the modern communication and transport technologies will play an important role to fulfill these responsibilities.

Since there will be rapid urbanization in developing countries in the early part of the next century, there will be more stress and strain to maintain the present living arrangement of the aged. In this transition, how far the informal support system can be sustained will be a challenge for the changing family structure of the developing countries. If it cannot sustain, what will be the alternative support system for the aged in the developing world?

(b) Women and Ageing

Traditionally, women are the caregivers for the aged in the family. In the Western countries, normally the daughter is the caregiver, where as in Asian countries, the son is responsible for the aged parents, which means the daughter-in-law is the caregiver. Today, women are more educated and would like to work outside the home. The trend is increasing fast in the developed countries. This working role sometimes brings in conflict with the care-giving role, especially if there is a very old parent to take care of. Many women have to forego their jobs to take care of the aged. This dilemma is going to increase in the future, as the family structure becomes more and more vertical in shape. In a family, there may be more than four old persons living at a time, of whom more than one may need constant care.

In the developing countries of Asia and Africa, still a majority of the population is living in villages involved in agriculture. Most of the women work in the agriculture field, which are closer to their homes. Therefore, women can provide care to the aged, even if they are working in the agricultural field. However, due to urbanization, industrialization and rural poverty, many migrate to cities in search of job. Normally, the weak and frail are left in the villages. This leads to destitution of the aged in developing countries. In cities of the developing countries, the women often have to work to supplement the family income. In such situation, the care of the aged in the city becomes a problem. It puts additional burden and strain on the women as caregivers. As industrialization and urbanization advances in the next centuries in the developing countries, this problem is going to increase in cities. In the absence of any home care facilities and the cultural pressure to take care of the aged, the women are going to face more physical and emotional strain in the future.

Life expectancy of women is longer than men. According to the WHO estimates, at the global level, the life expectancy at birth for women is 67.7 and for men 63.4. This gap was found to be wider in developed countries than in developing countries. A large proportion of women also cross the age of 85 years, especially in developed countries. For example, in Japan there are 8.51 million males and 11.98 females aged 65 years and above, i.e., 41.53 percent males and 58.47 percent females. The picture changes dramatically at the age of 85 years and above. There are 580,000 males and 1.39 million females at the age of 85 years or above, i.e., 29.44 percent males and 70.56 percent females indicating the high survival rate of women over men.

The implications of longer life of women are manifold. Larger proportion of older women is likely to become widows. In the case of men, many remarry and hence the proportion of widowers in Europe and North America was found to be low. For women, widowhood signifies loss of role in relation to the spouse. With the emergence of the vertical family structure in the developed countries, it was found that a larger proportion of widows was living alone. Further, a study in 1980 showed that among the very old, the poor and the very needy were women. Often loss of spouse also means loss of income and support. Hess and Markson too found that more women were living alone and their income was found to be very low for good housing. Since women live longer, they are likely to suffer from more chronic illnesses and disabilities. Institutionalization is more likely for women than men due to their longevity of life and widowhood.

Widowhood in developing countries can be still more difficult for women. In India, widowhood can push the aged woman into oblivion. She is excluded from all the social functions in the family and financially also her dependency increases on her sons. They constitute the poorest among the elderly. A study conducted in a southern state of India shows that widows have the lowest financial resources for their living. Widowhood involves loss of roles for the women to a great extent. However, with the modernization, widows are gaining their place in the society, especially in the cities.

Challenges Ahead: As more and more women take up jobs in the organized sector, it has implications on the care-giving function of the women. Especially in developing countries, where the aged mostly lives with the family, there is tremendous pressure on the working women to play the dual role of work and care giving. Are there any alternatives in the informal support system to take care of the aged, while the woman is working? There may be community based voluntary support available for the aged. If there is no alternative in the informal support system, the urban families in the developing world may have to look for viable formal support system to take care of the aged.

Though larger number and proportion of women live longer than men, old age social security and benefits are fewer for women. That is why there are more poor and needy among the female aged widows than among the male aged. They also suffer from more chronic diseases more intensely and also from disabilities. Ultimately, they may end up in institutions. All these have heavy financial implications for the health and social service sectors. Therefore, it will be a challenge for the governments to find a viable social security system for women that will meet their health and other needs.

Source: Excerpts from C.A.K. Yesudian, "Socio-economic Implications of Ageing," Introductory report prepared for the World Health Organization's Symposium on Ageing and Health, held November 10–13, 1998.

from the National Institute for Research Advancement

8

Roundtable on Japan's "Depopulation"

A discussion with Makoto Atoh and Shinyasu Hoshino

Autumn 1998

Makoto Atoh is Deputy Director-General of the National Institute of Populations and Social Security Research, and a visiting professor in the Graduate School of Education at the University of Tokyo. Shinyasu Hoshino is President of NIRA.

Hoshino: The National Institute of Population and Social Security Research, where you serve as Deputy Director-General, published an intriguing report in early 1997. This report, Population Projections for Japan, projects trends going all the way to the year 2050. It was full of surprises.

First of all, the trend toward fewer children continues. According to the projections, in 1995 Japan's total fertility rate (TFR) reached 1.42, in a total population of 125.57 million. The population is predicted to decrease after peaking in 2007, which seems much earlier than previous projections if my memory serves me. That finding also means that the "aging society" we keep hearing about will come much earlier.

Moreover, Japan will go through a "depopulation" period after a constant increase for many generations. Nothing and no one, it seems, can stop the trend toward fewer children. I would like to ask you to explain the reasons for the trend toward fewer children in Japan.

Atoh: The trend toward fewer children in Japan is basically the result of the decline in fertility that started in the mid 1970s. From the postwar period to the end of the 1950s, there was a quick transition from big families to smaller ones, that is, from four or more children to two. From around 1960 to 1975, the average number of children per family plateaued at almost exactly two. Japan's population growth stabilized as a result, and it seemed as if problems associated with the aging of the population would arise, if they were to arise at all, only far in the distant future. Two children per couple seemed about right to most people, and nobody paid much attention to the birth rate.

After the mid 1970s, however, the fertility level—the "replacement level" in population jargon—of Japan began to drop, and rapidly. Their replacement level is simply the fertility rate required for a stationary population, given a certain mortality rate. The recent generally accepted figure in Japan is 2.08 (that figure changes with changes in the morality rate). Japan's fertility rate started dropping sharply in the mid-1980s, to the shocking figure in 1989 of 1.57, a record low. And it continued to drop, hitting 1.39 in 1997.

Why Are Japanese Women Having Fewer Children?

Atoh: What demographic factors caused the decline? Every developed nation has its own reasons, but Japan's are very obvious. Childbirth patterns among married women have not yet changed visibly; it's that people, especially women, are postponing marriage itself. The decline in the number of married people has predictably led to the trend toward fewer children.

Many Western countries have also experienced this trend over the past two or three decades. The similarities to Japan in that regard are striking: an increase in "later" marriages, postponed childbirth, and the trend toward fewer children. One noted difference between those countries and Japan is that, in the West, it is not uncommon to see unmarried couples living together, and many of those even have children outside the sanctity of marriage. Thus, a decline in the married population doesn't necessarily mean an immediate decrease in the number of children. In Japan, the ratios of couples living together and children born outside marriage are negligible. Once the number of marriages declines, it automatically means a decrease in the number of children. This simple and obvious reason might be pushing down the fertility rate of Japan.

But why do people stay single so long? Two reasons are Japan's rapid economic growth and its resultant transition to affluence. Another is the higher education levels of Japanese women and their relative advancement in the work place. No doubt, these two factors have influenced young people; their behavior attests to it.

It is surely a good thing that Japanese women are now better educated and have better careers. The wage gap between men and women is indeed narrowing, but one unfortunate result is that it is now much harder for women to reconcile the two worlds of "work" and "marriage, childbirth, and child rearing." Economists might analyze it this way: When women have abundant and attractive employment opportunities with good pay, the cost (or value) of time spent not working will skyrocket; that necessarily means that the opportunity costs for "marriage, childbirth, and child rearing" also rises, since these things must be done outside company time. In short, the advantages of getting married have declined, and women now tend to work as long as they can.

But there's another side to the coin. I make no value judgment here, but we must come to terms with the realization that the tastes and habits of young people in Japan's mass consumer society have become outrageously extravagant. Many don't want to leave the family home; why pay rent and all the other bills that come with keeping an apartment when you can lead a luxurious, worry-free life as a carefree single? Some wags have dubbed these people as "parasite singles," (laughs) and point out the almost raw exploitation of their parents.

A recent survey revealed that about 85 to 90 percent of employed women in their 20s throughout Japan live with their parents. During the rapid economic growth period, for young people who wanted to undertake further study or find a job, there was hardly any alternative to living separately from parents, many of whom lived in rural areas far from where the jobs were. Nowadays, many young people are born in urban areas and grow up there; most can go to school and get a job in the same general vicinity. So they don't leave home. These people have much higher "net" income than the former generations, and they have a vast smorgasbord of consumption opportunities and choices. Thirty years ago, there were clear differences between the consumer behaviors of men and women, but not any more. Today we live in an age of freedom: men and women can both go out on the town, enjoy the nightlife of the big cities, and take solo trips abroad at will.

It's in ways like those that a big gap has been created between the single life and the married life. If you weigh the advantages and disadvantages, it's understandable why more and more people postpone marriage. (laughs)

Hoshino: What a fascinating explanation. Japan's unique prewar or "family" system, in which the elder male child inherited authority and power from the family, was abolished after the war, when it morphed into the so-called shinzoku, or "relative" system. As home electronics and other labor-saving devices became widely available and affordable, women were liberated from many of the heavier and time-consuming housekeeping duties. And as women become better educated and more advanced in the work place, the gap between men and women continues to gradually shrink. The result: a trend toward fewer children.

Family Forms and Population Change

Hoshino: You wrote in one of your books that people in Europe have actively cultivated the concept of individualism. But in Japan, unfortunately, individualism has not taken firm root. The concepts of "living together" and children born outside marriage remain foreign. If individualism does eventually take root here, it will certainly take some time, maybe as much as 30 years or so. Even if that does happen, Japan will still suffer from fewer children, won't it?

Looking at it from a different angle, I wonder if we could expect couples to raise children based on a more developed sense of individualism, by following one's own sense of values and moral codes. Some may even start thinking that it's nice to have three children! But women in Japan have just broken their chains, and are running for freedom, so it's not too likely that many would agree with that speculation (laughs).

Atoh: Individualism is indeed a key word. Its opposite is what I call "familism." In Japan it is said that the prewar system changed after the war and that the system transformed into a nuclear-based family model.

But how exactly did that transformation take place? My guess is that it just appeared that way; as the population ratio of urban areas rapidly rose, young people from rural areas made nuclear families in the urban areas. The extended family had transformed to a nuclear family. In much of the West, the trend has been away from nuclear families and toward individualism. But that doesn't seem to be happening in Japan. Several recent surveys and opinion polls have revealed that young urban people still have a surprisingly traditional sense of values. Just a third of Japanese women really seeks to reconcile work and family lives and to pursue careers. Another third wishes to reconcile work and family lives too, but the women in that group claim they will quit their jobs, have and raise children, and, after the children reach young adulthood, the women will resume working, part-time. The women in the final third of those surveyed plan to quit their jobs when they get married, and will be content as full-time housewives. In other words, two-thirds of Japanese women still want to marry and become housewives. This is a big difference from Western countries.

Hoshino: Individualism seems to have a long way to go before it takes root here, doesn't it.

Atoh: Well, typical individualism is seen in English-speaking countries such as the United States, England, and Australia, and in northern European societies, where women's social advancement has developed. On the other hand, especially in southern Europe, a deep-rooted familism exists, but it is different from Japan's. Many people there have an image of a mother who puts on an apron, holds her children, and makes delicious dishes in the kitchen, where men are not allowed to enter. Germans use the term "3K," for kuchen, kirche, and kinder (kitchen, church, and children). In all these three areas, the women are in charge. I think such a sense of values, a view of family, and a tradition of familism exist simultaneously, even if they diversify in different cultures.

Given that thinking, once you have married, you have to adjust yourself to those values. Japanese women tend to keep those traditions at arm's length by staying single.

Hoshino: I once heard that a book for housewives published in the Victorian period said the duty of a good housewife is to "keep house neatly."

Atoh: Yes, and that referred to "modern" families! The Victorian housewife clearly had a full-time job on her hands, as does the contemporary housewife in Japan. The division of labor in which husbands work outside the home and wives run the house took root in the middle class, when the Industrial Revolution started to divide working places and homes. It began to spread in the early 19th century.

In Japan, these concepts filtered into the middle class in big cities during the Taisho era (1912–1926). Consequently, they exploded in the rapid economic growth period after the war and seem to have peaked in the mid-1970s. The definition of "full-time housewife" varies, but according to the national census, the ratio of full-time housewives was highest in 1975; it has declined ever since. So Japan experienced a peak in "familial modernity" in the mid-1970s. This is a very different pattern from what has gone on in the West.

Hoshino: Doesn't Japan's overcrowding also have an effect on the trend toward fewer children?

Atoh: That is very hard to answer. The trend toward fewer children took place evenly all over the country in the mid-1970s. Differences existed among the figures, but the rates of decrease were remarkably similar. A few differences might be found between Hokkaido and Tokyo, for example. Hokkaido obviously has more space for living environments, so that explanation alone isn't satisfactory. It is certain that the size of a house correlates extremely well with the number of children a couple has. But the relation of cause and effect could possibly be the opposite. People might choose to move into a more spacious house because they will have three children. It doesn't necessarily mean that a smaller house causes a couple to have fewer children, any more than a bigger house causes a couple to have more children.

Japan's Population Decrease: Three Perspectives

Hoshino: I would like to get your perspective on three points.

First, Japan's overcrowding. We don't have a lot of land, so perhaps there are some good aspects to the country's population decline?

Second is a preconception that less population somehow means less national power. I think that notion is probably deeply rooted in the Japanese mentality; it's an idea that was created by the Meiji leaders for enforcing the wealth and military strength of the country. But isn't that way of thinking out of date?

And third, although Japan is clearly becoming an "aging society," and today's working-age population must look after the nonworking elderly, can we say that the dependent population (the population supported by the working-age population) has not changed? Those who require support have changed, from children to the elderly, but those over 65 have more assets than the working generations do. An important factor seems to me to be the question of how to transfer the assets of the elderly generation to the next generations.

Atoh: Japan is a small island-nation with few resources. After the war, the country gave high priority to education, to improving its human resources, and to becoming an exporting nation. Many most correctly believed these priorities to be the road to wealth. If the country is overcrowded from a demographic point of view, we're a little late to do much about it (laughs).

I conducted a survey some time ago on the question of overcrowding. It turned out that about 50 percent of Japanese nationals felt that Japan is overcrowded. But over the longer term, many felt much more anxiety toward a population decrease and toward population aging.

It is very hard to evaluate the full meaning of the population decline or the decrease in population density in Japan during the next 100 years. There will always be negative and positive things to say. On the positive side, a low population density means more space: more available land, greenery, and housing. These benefits might even create a spiritually affluent society. Furthermore, it's probably better for the environment. I once attended a lecture by someone who said that a single person in a developed country uses roughly 20 times more energy than one in a developing country. He wryly cracked that the death of a person in a developed country would save enough energy for 20 people in a developing country!

Still, it would work negatively on economics. Remember that each person is a consumer as well as a producer. In our projections, Japan will lose some 800,000 people annually around the year 2040. That means the national market will radically shrink. I can't help but wonder whether Japan's industries and the economy can take it. In a market shrinking at that speed, economic management would be hard, economic growth would plummet, and living standards would worsen.

I think that the populations in developed countries would be better off by declining slowly. Not radical declines, but gradual ones. A one percent annual decline would have too big an influence. It should be less than that.

Hoshino: A slow decline is clearly better than a rapid one.

Atoh: The second point—that a population decline translates into national weakness—is obviously an outdated notion. But when we examine a limited area, the size of Japan's population would have an influence on its relations with the outside world. For example, before World War II, Japan's population was three times as large as that of the Korean Peninsula. It still is twice as large today, but the gap is getting smaller and smaller. Beyond the year 2050, they will probably even out.

Your third point, on the problems of population structure, is interesting. The ratio of the working-age population versus the dependent population before the war was around 10:7, which is similar to the projected ratio of the coming aging society. The population structure of the 40 years from 1960 to 2000 is actually the most economically unburdened one: Its ratio is approximately 10:4. It was a happy age, one that embraced historically the highest proportion of working age population.

Under these circumstances, if, as you mentioned, the number of children equaled the number of elderly, we'd be shelling out some very big cash (laughs). Besides, before World War II, the burden for child support was far less than in the postwar period. Children started to work immediately after finishing compulsory education.

If the elderly stay healthy, work as long as possible, contribute socially (by being employed, as well as by ways other than earning a wage), and continue to hold assets, we can expect some positive developments. But will these developments last? As preventive medicine and medical technologies progress, the breakouts of cancer, stroke, and cardiac disease will probably decline in the developed countries. Under those circumstances, life span continues to extend, and it is a healthier life span.

Long-term care is a high-priority issue in an aging society; it is directly connected to the transformation of the family structure. As more and more women have begun to work outside the home, the number of households where three generations live together has dropped. Who will take care of the elderly who need support? It has become a big problem in most developed countries. Population aging is related to problems concerning pensions and medical care, but the biggest issue on aging is this problem of long-term care. Women used to take the responsibility as full-time housekeepers, but few can do it today.

Hoshino: It's a new problem.

Atoh: Indeed; it requires some very serious thought and planning. But there's one more issue: the assets of the elderly people. As we've discussed, many workers today have decent jobs, and aren't shy about spending their hard-earned money. But the structure will change and the supporting burden will fall heavier and heavier on the shoulders of younger generations. Sooner or later that burden will become unbearably heavy, and a social security system will have to be restructured. Under these circumstances, will elderly people lead affluent lives? Will their assets be enough to keep them afloat? I'm slightly pessimistic about it; continuing the current cycle may be difficult.

World Population Policies

Hoshino: You said a very interesting thing that would never have occurred to me. We are all going to need to become much more keenly aware of the relation between population decline and economics over the next two or three decades.

It's not hard to imagine, say in the year 2025, most of our Asian neighbor countries gaining a purchasing power standard similar to Japan's. Simultaneously, growth rates would slow, urbanization would continue, nuclear families would emerge, and birth rates would decrease considerably. In that scenario, populations are not likely to explode, and economic activities would be far from brisk. In short, the economy would go from being a production-oriented one to one trending toward more consumption. Natural resources might even be more highly valued than they are today, and recycling could take off big time. The transition won't be easy, especially as the relationship between economic growth and population decline grows more complex.

Atoh: Actually, the birth rates in Asian Newly Industrialized Economies have been similar to Japan's. The total fertility rates of Taiwan and Singapore, to name just two, have been less than 2.0. In the 1980s, Singapore suffered a decline to 1.4; the total fertility rate in Thailand is now 2.0. The fertility transition or the demographic transitions have been accomplished. In China, the Philippines, Indonesia, India, and Bangladesh, the birth rate is declining.

Hoshino: Population policies have succeeded, then?

Atoh: Yes. Africa's birthrate is still very high, but we do have reason for optimism. The population projections by the United Nations are revised every other year, and they have been moving constantly downward. The UN used to say that world population would reach 10 billion in 2050; it then revised that figure to 9.4 billion in the current projections. That was because the birthrates in developing countries, especially in poorer countries such as Bangladesh, were proved to be affected by population policies. Many people used to hold the view that economics and population are in a vicious circle, that until modernization progresses the birthrate will not decrease. But now these countries have shown that breakthroughs are possible.

All in all, the population policies of most governments around the world, with some help from the United Nations and the developed countries, should be considered successes. That to me seems reason to be hopeful.

Hoshino: In a survey of the women at your institute, most answered that they would like to have two or three children. That's encouraging (laughs). In short, given good conditions, it seems that most people would answer similarly, or at least as high as the population's replacement level of 2.08. A family of a husband, wife, and two children seems to be the archetype common throughout the world.

Atoh: But what prevents the realization of the ideal? It brings us back to what English-speaking countries, like the United States and England, and northern European societies have accomplished . How different are these societies from those of other developed nations? Gender equality is a big theme in those places; most men and women aspire to strong individualism. Most choose either to live together or to build a traditional family, and have as many or as few children as they like, whenever they like. At the International Conference on Population and Development held in Cairo in 1994, the subject of reproductive rights was a main theme. This is a basic right for every couple and individual. The concept applies to all countries, developed and developing; I think it represents a kind of hope for developed nations. At least in northern Europe and in the United States and England—despite the higher rates of women's advancement—birthrates are high.

It used to be said that if a national government operated a population policy in a "top-down" order, the birthrate would diminish. The Cairo Conference, however, indicated that excessive reliance on such a policy sometimes backfires. In the end, most agree that providing family planning while paying utmost respect to women's choices regarding birth is the best way to go. And that applies to developed and developing countries alike.

Hoshino: It is indeed a good and encouraging concept. Thank you very much for taking the time to share your thoughts with us today.

Source: "Roundtable on Japan's 'Depopulation'" A discussion with Makoto Atoh and Shinyasu Hoshino in *Review,* Autumn 1998, nira.go.jp/publ/review/98autumn/round.html.

Review Questions

1. Why might Malthus's prediction of inevitable and catastrophic overpopulation be incorrect?
2. Why are so many young adults in Japan postponing or foregoing marriage?
3. What does Makoto Atoh mean by the term "familism"? Why does he say that it, instead of individualism, is the guiding principle of Japanese society?
4. Why does a country's total fertility rate typically drop more rapidly than its crude birthrate or natural increase rate as it goes through the demographic transition?
5. Why might Japan's efforts to make the workplace more appealing to women have the short-term impact of easing Japan's burden of providing care for the elderly while having the long-term impact of increasing its dependency ratio?

Discussion/Essay Questions

1. Consider the case of Kumi Kanbe, the elderly woman profiled at the beginning of this module. In traditional Japanese culture, sons (with the assistance of their wives) are supposed to take care of their parents in the parents' homes. And yet in modern Japan, sons leave ancestral homes for the city, many never marry, and even when they do marry their wives typically are working women who do not have the time or desire to relocate to their in-laws' homes. This situation will get even worse in future generations, as today's young adults remain childless, leaving them with no one to care for them when *they* get old. In the final reading, Makoto Atoh implies that the only way out of this situation is a wholesale change in Japanese cultural norms. Is this practical or even desirable given that, according to the author of the profile of Mrs. Kanbe, "By everyone's standards but their own, the Japanese are models of filial piety?" Given the expectations of Japan's elderly and the desires of its young, what policies might be promoted by a government agency or nongovernmental organization to improve care for the elderly?

2. With 873 people per square mile, Japan has one of the highest population densities in the world. Not counting countries with fewer than 500,000 people, only nine countries have higher population densities: Bahrain, Bangladesh, Belgium, Lebanon, Mauritius, The Netherlands, Palestine (or Israel, if one includes the Palestinian territories as part of Israel), Singapore, and Taiwan. Given Japan's high population density, its projected drop in population from 127.68 million in 2005 to 100.50 million in 2050 may be seen as a positive development. How might Japan design a population policy that retains this population drop but avoids some of the drawbacks discussed in the readings?
3. When you served as your state's health commissioner, you instituted a program of classroom-based sex education, adult health education, easy access to contraceptives, and mandatory premarital counseling. Now, for the first time ever, your state's natural rate of population growth (the difference between the birthrate and the death rate) has fallen to 0.0, meaning that the population is staying exactly stable. The National Committee to End Population Growth is thrilled with your achievement and is throwing a banquet in your honor, at which it is presenting you with the Thomas Malthus Achievement Award. Write a speech, to be delivered at the banquet, at which you politely accept the award and explain how your public health program is and/or is not in the spirit of Malthus.

List of Readings

1. "Declining Fertility Rate: Population to Grow More Top-Heavy," in *Trends in Japan*, February 14, 1997, posted on the website of the Japan Information Network, *http://web-japan.org/trends96/honbun/tj970202.html*. Table is reprinted from "A New Look at the Social Security System," in *Japan Access*, posted March 2001 on the website of the Japan Information Network, *http://web-japan.org/factsheet/social/new.html*.
2. "Women's Working Conditions," in *Japan Access*, posted March 2001 on the website of the Japan Information Network, *http://web-japan.org/factsheet/woman/work.html*.
3. "An Aging Society," in *Japan Access*, posted March 2001 on the website of the Japan Information Network, *http://web-japan.org/factsheet/woman/aging.html*.
4. "The Rapidly Graying Japan: Coping with an Aging Society Becomes Important National Policy," in *Trends in Japan*, September 19, 1996, posted on the website of the Japan Information Network, *http://web-japan.org/trends96/honbun/tj960903. html*.
5. Martin Mühleisen and Hamid Faruqee, "Japan: Population Aging and the Fiscal Challenge," in *Finance & Development*, March 2001, posted on the website of the International Monetary Fund, *http://www.imf.org/external/pubs/ft/fandd/2001/03/muhleise.htm*.
6. "Caring for a Graying Japan: Elderly Nursing-Care System Is in the Works," in *Trends in Japan*, June 20, 1997, posted on the website of the Japan Information Network, *http://web-japan.org/trends96/honbun/tj970618.html*.
7. Excerpts from C.A.K. Yesudian, "Socio-economic Implications of Ageing," introductory report prepared for the World Health Organization's Symposium on Ageing and Health: A Global Challenge for the 21st Century, held November 10–13, 1998 in Kobe, Japan, posted on the website of the World Health Organization's Kobe Centre for Health Development, *http://www.who.or.jp/ageing/introduction/background/6papers2.html*.
8. "Roundtable on Japan's 'Depopulation': A Discussion with Makoto Atoh and Shinyasu Hoshino," in *Review*, Autumn 1998, posted on the website of the National Institute for Research Advancement, *http://www.nira.go.jp/publ/review/98autumn/round.html*.

Websites for Additional Research

1. The Population Research Bureau is a nongovernmental organization, based in Washington, D.C., that issues frequent reports on global population issues. Most of these reports, as well as links to numerous population-related statistics, can be found at the PRB's website, http://www.prb.org. An excellent starting point for basic population statistics is its annual World Population Data Sheet, accessible on the website.
2. There are numerous nongovernmental organizations devoted to issues concerned with aging. Foci of individual organizations include efforts to reform rules for eldercare facilities, health insurance and social security reform, accessibility issues, efforts to increase respect for the elderly and improve their integration into mainstream society, medical research on elderly health issues, and demographic analysis. A good place to start is the website of the United Nations Programme on Ageing, http://www.un.org/esa/socdev/ageing/.
3. The Japanese government's Ministry of Foreign Affairs maintains a host of English-language information on Japan at http://web-japan.org.

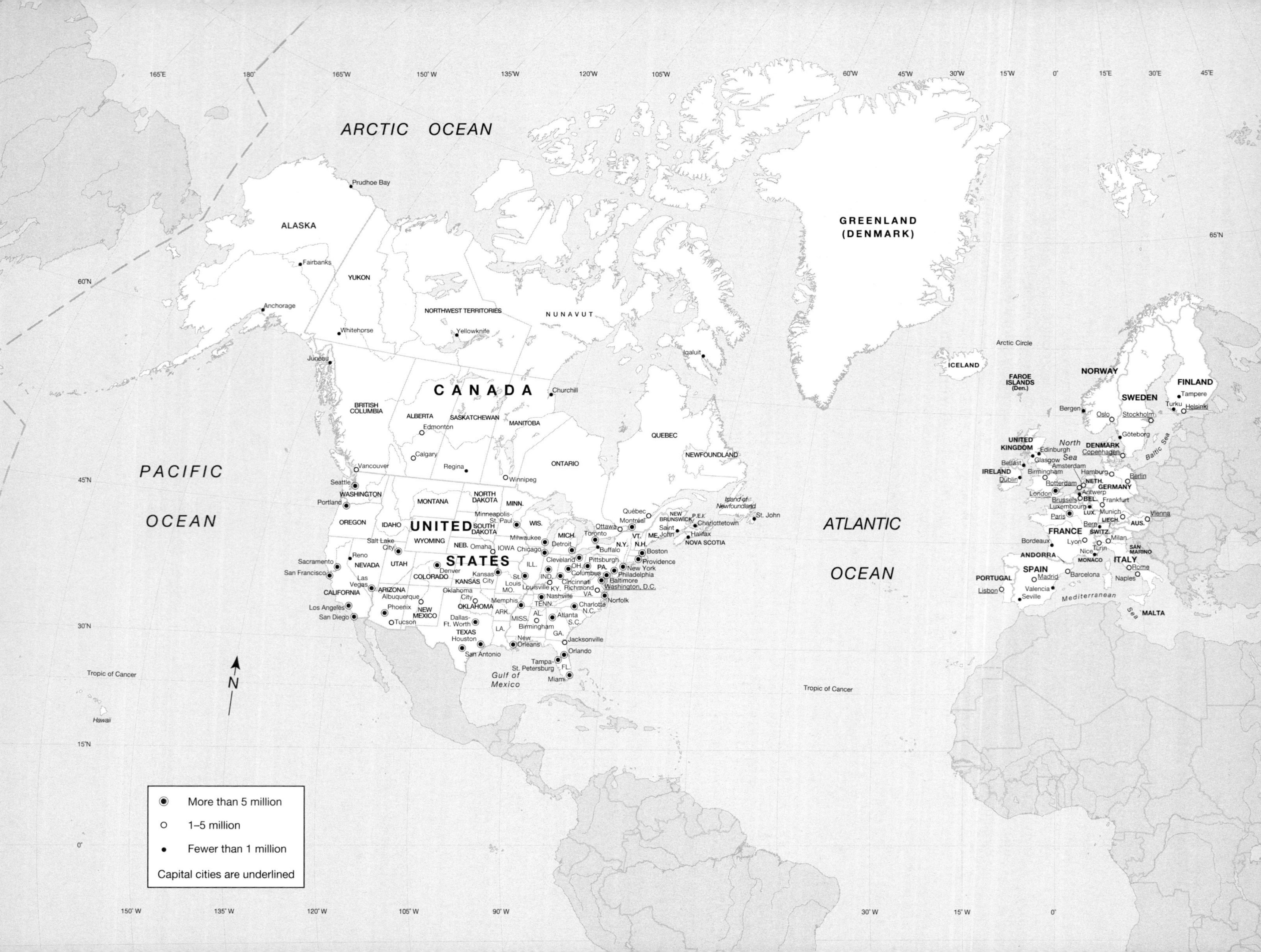

ARCTIC OCEAN
PACIFIC OCEAN
ATLANTIC OCEAN
GREENLAND (DENMARK)
ALASKA
CANADA
UNITED STATES
YUKON
NORTHWEST TERRITORIES
NUNAVUT
BRITISH COLUMBIA
ALBERTA
SASKATCHEWAN
MANITOBA
ONTARIO
QUEBEC
NEWFOUNDLAND
NEW BRUNSWICK
P.E.I.
NOVA SCOTIA
Island of Newfoundland
Prudhoe Bay
Fairbanks
Anchorage
Juneau
Whitehorse
Yellowknife
Iqaluit
Churchill
Edmonton
Calgary
Regina
Winnipeg
Vancouver
Toronto
Ottawa
Montréal
Québec
Saint John
Charlottetown
Halifax
St. John
WASHINGTON
OREGON
IDAHO
MONTANA
NORTH DAKOTA
SOUTH DAKOTA
MINN.
WIS.
MICH.
WYOMING
NEB.
IOWA
ILL.
IND.
OH.
PA.
N.Y.
VT.
N.H.
ME.
NEVADA
UTAH
COLORADO
KANSAS
MO.
KY.
VA.
CALIFORNIA
ARIZONA
NEW MEXICO
OKLAHOMA
ARK.
TENN.
N.C.
S.C.
TEXAS
LA.
MISS.
AL.
GA.
FL.
Seattle
Portland
Sacramento
San Francisco
Reno
Las Vegas
Los Angeles
San Diego
Salt Lake City
Phoenix
Tucson
Albuquerque
Denver
Minneapolis-St. Paul
Omaha
Kansas City
Oklahoma City
Dallas-Ft. Worth
Houston
San Antonio
Milwaukee
Chicago
St. Louis
Memphis
New Orleans
Detroit
Cleveland
Columbus
Cincinnati
Louisville
Nashville
Birmingham
Atlanta
Buffalo
Pittsburgh
Boston
Providence
New York
Philadelphia
Baltimore
Washington, D.C.
Richmond
Norfolk
Charlotte
Jacksonville
Orlando
Tampa-St. Petersburg
Miami
Gulf of Mexico
Hawaii
Tropic of Cancer
Arctic Circle
ICELAND
FAROE ISLANDS (Den.)
NORWAY
SWEDEN
FINLAND
DENMARK
UNITED KINGDOM
IRELAND
NETH.
BEL.
LUX.
GERMANY
FRANCE
SWITZ.
LIECH.
AUS.
ITALY
SAN MARINO
MONACO
ANDORRA
SPAIN
PORTUGAL
MALTA
North Sea
Baltic Sea
Mediterranean Sea
Bergen
Oslo
Stockholm
Göteborg
Turku
Tampere
Helsinki
Copenhagen
Edinburgh
Glasgow
Belfast
Dublin
Birmingham
London
Amsterdam
Rotterdam
Hamburg
Berlin
Antwerp
Brussels
Luxembourg
Frankfurt
Paris
Munich
Vienna
Bern
Lyon
Milan
Turin
Nice
Bordeaux
Barcelona
Madrid
Lisbon
Valencia
Seville
Rome
Naples
165°E
180°
165°W
150°W
135°W
120°W
105°W
90°W
60°W
45°W
30°W
15°W
0°
15°E
30°E
45°E
65°N
60°N
45°N
30°N
15°N
N
More than 5 million
1–5 million
Fewer than 1 million
Capital cities are underlined

Immigration in North America and Western Europe

Companion to Chapter 3 Migration

Pedro was born in southern Mexico. He dropped out of school at age 14 to make money for his family, working for a timber harvesting company that employed many local men. Soon, however, he found that his true knack was as a mechanic. By age 19 he had opened a small shop in his village, where people brought their cars, their radios, their refrigerators, anything that possibly could be fixed.

Thousands of undocumented Mexicans cross into the United States every month. *Source:* AP/Wide World Photos

By the time he was 22, Pedro was relatively successful. He had a wife and two children, and the shop brought him a steady stream of business. But then the timber company left town. In response, villagers took up and left, some for Mexico City but many more for the United States. As people left, and those who stayed became poorer, Pedro's business slowed to a trickle. Soon he was left with no decision but to leave himself.

A cousin offered him a place to stay in New York City, a city with a growing community of Mexican immigrants. He found a job (where he is paid in cash, off the books) at a small, independent, Anglo-owned hardware store in a predominantly African-American neighborhood. His counter at the hardware store is littered with business cards (in English and Spanish) promoting his skills as a handyman, and Pedro's side business is flourishing. Almost every evening, after the hardware store closes at 6:00, he walks a few blocks to one home or another to fix someone's plumbing, retile someone's bathtub, or rekey someone's locks. After finishing the job, he makes a one-hour commute on the subway to his home in a predominantly Mexican neighborhood.

Pedro's dream is to quit the hardware store job and work full-time as a handyman. Because of his undocumented status, though, receiving a bank loan for his handyman business is out of the question. He might have saved enough by now, except that several times a year he sends hundreds of dollars back to his family in Mexico. If he could obtain legal residency status, he *might* bring his family to the United States, but his wife does not want to move. Even for Pedro, Mexico still feels like home, and it is there that he would like to raise his family. Pedro goes back once a year for Christmas, and, although he has never been caught, slipping through the border is always a nerve-racking experience. Not quite at home in New York, but not able to make a home in Mexico, Pedro sometimes feels like his home is the gap in the fence that marks the border. ■

Immigration as a Response to Population Problems

The movement of people from areas with few employment options to those with (or perceived to have) many opportunities illustrates the most common type of migration: economic migration. Like Ilhami Elhoussin in the story at the beginning of Chapter 3 in the textbook, Pedro (from the beginning of this module) came from an LDC (less developed country) to find work in an MDC (more developed country). Ilhami and Pedro share other similarities as well. Both left their families behind when they migrated and both face obstacles related to their living in Spain or in the United States without proper documents. Their stories are representative of countless workers, some with legal documentation and some without, who have moved to Western Europe or North America in hope of finding opportunities not available to them in their native countries.

Pedro and Ilhami are doing jobs that others in their host countries do not want to hold for the wage that employers are willing to pay. The money that Ilhami earns picking vegetables and that Pedro earns working as a clerk in the hardware store is less than what other jobs in the area pay. For both Ilhami and Pedro, the chance to work at what they are doing presents an opportunity not available to them in their native countries. Because both are making more money than they ever could have in Morocco and Mexico, Ilhami and Pedro send money home to relatives there to help give them a better life. This is a common occurrence when a large number of people from LDCs migrate to MDCs, and the money sent home can sometimes contribute a significant amount to their home nation's economy.

As Chapter 3 of the textbook stresses, patterns of migration between countries are strongly linked with population growth and decline. This is especially true of economic migration. A key concept covered in Chapter 2 of the textbook (and in the accompanying module in this workbook) was the demographic transition. The precise pattern differs from country to country, but in general demographic transition theory holds that as countries move away from subsistence agriculture to commercial industrial and service activities, they tend to go through a period of rapid population growth (stage 2), followed by one of slower growth (stage 3), followed by one of stability or even population decline (stage 4). International migration, in part, results from the fact that at any given time different places are reaching different stages in their demographic transition. Or, to put it bluntly, some places (LDCs at stage 2 in the demographic transition) have "too many" people while others (MDCs at stage 4) have "too few."

For countries in the fourth stage, slow population growth hits especially hard because it usually is accompanied by a decline in the number of working-age people relative to the number of people beyond the working age. Because a country's working-age population is also its childbearing population, countries with a shortage of working-age people face a never-ending problem: Not only are there relatively few people in the workforce to support a large population of retirees, but relatively few children are being born to become the next generation of workers.

Many of these "stage 4" countries are searching for ways to cope with low population growth. However, the problem (and the choice of prospective remedies) is considerably different in places like Europe and Japan than in other more developed areas like the United States, Canada, and Australia. Europe and Japan have long had very high population densities and, after a period of rapid growth in the nineteenth and early twentieth centuries, they have recently seen their populations stop growing or decline. Historically, these countries more often have had to deal with overpopulation than underpopulation, and until recently they were exporting substantial numbers of people to settler colonies in the Americas, Australia, New Zealand, and parts of Africa. Now, suddenly faced with a relative shortage of working-age people due to slow population growth, the countries of Western Europe are engaged in spirited national debates over immigration: To what extent should immigration be encouraged, and how should immigrants be assimilated into what had been fairly homogeneous cultures?

In the "settler colonies" of North America the situation is quite different. When Europeans first came to this region, they perceived it as being almost empty of people. Indigenous population densities in much of the continent were very low, and they plummeted further shortly after the Europeans' arrival, due to European-borne disease and violence. Colonizing powers (and, after independence, the governments of the United States and Canada) responded by encouraging immigration from the "parent" countries: Britain, in the case of the United States, and Britain and France, in the case of Canada. This migration was soon joined by migrations from nearby European areas where people had relatively complementary, Protestant cultures (such as Germany, Scandinavia, and The Netherlands). At the same time, settlers increased the region's population density through the forced immigration of slaves from Africa. After the first decades of the nineteenth century, as the Industrial Revolution hit North America and its interior became settled, the continent faced another population shortage. This time, the immigrants came from Ireland and East Asia, followed, by the end of the century, by large numbers from Southern and Eastern Europe. Immigration to North America slowed down considerably from around 1920 through 1960 (due, in large part, to the anti-immigrant legislation discussed in Chapter 3 of the textbook), but since the 1970s North America has seen a new wave of immigrants, primarily from Central America, the Caribbean, and Asia.

Despite their common history as recipients of immigrants, the two countries of North America—the United States and Canada—have begun to diverge somewhat in their contemporary immigration policies. Even though the United States and Canada have similar visa-granting policies, in practice it is often more difficult to actually migrate to the United States than it is to move to Canada. Canada's rules about who can enter and work there are less specific than the rules for entry into the United States, which recently has led some in the United States government to criticize Canada's immigration policy. Critics argue that because Canada places few restrictions on who can come into the country, the country provides a "safe haven" for terrorists, who then would be able to easily cross into the United States. Although the United States is powerless to change Canada's immigration policy, it can control border crossings from Canada to the United States, and shortly after the terrorist attacks of September 11, 2001, the United States and Canada agreed to a list of 30 items that would tighten security at the U.S.–Canada border.

Despite their current policy differences, the two countries of North America have similar immigration histories, which provides them with a certain common perspective on current migration issues. This perspective differs markedly from the countries of Western Europe, which has had a very different migration history. As has been noted, historically Western Europe has been an exporter of people while North America has been an importer. Although the two regions presently are faced with similar immigration needs, because of their divergent histories they confront immigration policy from fundamentally different perspectives.

As a region with historically little in-migration, European countries tend to view themselves as homogenous with distinct national cultures, and Europeans fear that waves of immigrants (whether from Eastern Europe, Asia, Africa, or even other Western European countries) may break the bonds of national cohesion. As one policy option, several European countries (most notably Switzerland and Germany) in the 1970s implemented "guest worker" programs. These programs typically permitted only temporary migration by single (usually male) workers. The intention was that a worker would come to, say, Switzerland, to work on a specific job, and then go back to, say, Turkey, to raise a family. Turkish influence on Swiss culture would be kept to a minimum. To further enforce this segregation of guest workers from European society, a guest worker's child born in the host country generally would not be eligible for citizenship there. Family unification in the host country also was discouraged so as to prevent the formation of a permanent immigrant community.

This strategy was perhaps destined to fail. Depriving immigrants of the institutions of their traditional society *and* depriving them of access to the society in which they are resident is almost a prescription for creating a subordinate, minority community plagued by crime and unemployment. The irony is that by creating immigrant "underworld" communities, within the host country's borders but beneath the national net of social services and institutions, these European governments may have done more to strain the fabric of their European cultures than would have happened had the immigrants simply been allowed to establish more "normal" immigrant societies within the host country's borders.

A second strategy, adopted by Sweden among other countries, was based on an entirely different principle. Instead of attempting to isolate immigrants from the European society, these countries attempted to thrust assimilation on the immigrants. Immigrant families were provided with generous packages of social benefits, language lessons, job placement services, and so on. In short, the goal was to transform, say, Ethiopians into Swedes as quickly as possible, thereby minimizing the destabilizing influence that the immigrants might have on Swedish society.

This strategy of permanent migration and assimilation, however, has proved as problematic as that of temporary residency and isolation. While Ethiopians might want to work in Sweden, and they might even recognize many of the good things about Swedish culture, they might not be willing (or, perhaps, even able) to completely abandon Ethiopian culture or their relatives who remain in Ethiopia. Additionally, by showering immigrants with extensive benefits, countries like Sweden might only be encouraging further immigration by those most in need (including family members who would not be entering the workforce but instead would place further demands on social services). Thus, a frequent result of this migration-assimilation strategy has been resentment, both among the native population (who resent the waves of new immigrants demanding access to social services) and among the immigrants themselves (who resent the host country's refusal to recognize their distinct cultures and needs).

In North America, questions concerning immigration policy generally are not quite so loaded politically. For instance, while the identities of several major parties in Western Europe revolve around their stance on immigration, few parties or politicians in North America are so defined. In the United States, for instance, the low level of attention paid to the issue has allowed the Republican Party to simultaneously accommodate pro-immigration business interests and anti-immigration cultural nationalists. North America has a long history of encouraging immigration, and, while there certainly is a sense of an "all-American" culture associated with white Anglo-Saxon Protestantism (that is, English-German heritage), most residents of the United States and Canada likely would acknowledge that their national culture is a result of immigration from many countries and that American citizens, regardless of their descent (such as Italian-Americans, Japanese-Americans, and African-Americans), are all "real" Americans.

While some North Americans of European descent fear that increased numbers of immigrants from non-European countries might tarnish the American (or Canadian) character, North American anxiety concerning immigration usually becomes serious only during times of economic downturn, when there is resentment against immigrants for taking "American" or "Canadian" jobs. This anti-immigrant sentiment also has been strong in North America (especially in the United States) during times of international tension (for example, official and unofficial acts of discrimination and violence against Japanese-Americans during World War II and against Arab- and Muslim-Americans in the wake of the September 11, 2001, attacks).

Cultural and Economic Arguments for and against Immigration

Complex decisions about immigration policies cut across caricatured divisions of "Left" and "Right" or "liberal" and "conservative," as is demonstrated by this module's readings. In the United States, a number of anti-immigrant movements have risen among "cultural conservatives" on the Right, from the anti-Catholic "Know-Nothings" of the 1850s to the Ku Klux Klan (which was anti-Jewish and anti-Catholic as well anti-Black). Some anti-immigrant groups continue to base their views on claims of racial superiority, as exemplified in the first reading, a statement by Great Britain's National Front.

In the second reading, the Austrian Freedom Party (FPÖ), which has been a member of Austria's ruling coalition since 2000, steers clear of overtly racist language and even recognizes a place for distinct, *historic* minority communities in Austria. Nonetheless, the Party Program (excerpts from which are reprinted here) asserts both that "The legal system in Austria presupposes that the overwhelming majority of Austrians is of German origin" (Chpt. IV, Art. 1) and that "Austria is not a country for immigration" (Chpt. IV, Art. 4), leading some to charge that the FPÖ's immigration policy is a descendant of Nazi racialism.

Before one assumes, however, that resistance to immigration for cultural reasons always comes from the Right, one should consider the case of the Dutch anti-immigrant politician Pim Fortuyn, who is profiled in the third reading. In 2002, the party led by Fortuyn (who was assassinated just prior to the election) became the second most powerful party in The Netherlands. Fortuyn, who was openly gay, opposed immigration, in part, because he feared that immigrants from conservative Islamic countries would endanger The Netherlands' famously liberal social laws and norms.

In the United States and Canada, as has been noted, it is more difficult than in Europe to make the argument that immigrants may taint an American or Canadian cultural purity, since these nations themselves are multicultural creations, forged through waves of immigration. Nonetheless, in the fourth reading, Pat Buchanan, a former speechwriter for Richard Nixon and frequent Republican Party candidate for president, makes the argument that unless immigration is restricted, immigrants will continue to reside in unassimilated marginal communities that, in the long run, will threaten America's unity.

This cultural argument against immigration to core countries frequently is paired with an economic one. Opponents of immigration point out that although immigrants hurt native workers by depressing wages and taking their jobs, that is only the short-term wound. In the long term, this argument goes, immigrant workers hurt the national economy by reducing the pressure on companies to achieve maximum productivity, which is required for global competitiveness. This argument is made explicitly in the statement of the Coalition for the Future American Worker, an umbrella organization of U.S. groups that lobby for restrictive immigration policies.

Contrasting this economic argument is the position taken by many business groups that, given the demographic issues discussed earlier in this chapter, immigration is necessary for the economies of MDCs. In the sixth reading, the case for immigration of unskilled and semi-skilled workers is made by the Essential Worker Immigration Coalition (EWIC), a group funded by the U.S. Chamber of Commerce (one of the country's largest pro-business lobbies) and the trade organizations of many service-sector industries that depend on low-wage immigrant labor. In the seventh reading, a similar position is taken by another pro-business group, the Employment Policy Foundation (EPF), although with reference to skilled as well as unskilled immigrant workers.

Business interests that argue for increased immigration have been joined by some rather unlikely allies. Civil liberties and immigrants' rights groups typically support immigration as part of their antidiscrimination agendas. Governments of LDCs also generally encourage MDCs to relax their restrictions on immigration. Although these governments may fear a "brain drain" as their college-educated professionals move to the core, these fears generally are outweighed by the benefits that the emigrant-sending country will receive from reduced unemployment and from money sent home by overseas workers.

Labor unions have had a particularly complex relationship with immigration policy. Historically, notwithstanding proclamations about international solidarity, unions often have taken an anti-immigration stance, fearing that immigrant workers will undercut the wages and jobs of union members in the more developed countries.

Recently, however, unions have acknowledged that vast numbers of non-unionized immigrants already have come to MDCs from LDCs and that they apparently are here to stay. In response, unions have begun to couple a fairly ambivalent stance on *new* immigration with a strong push for granting permanent resident status and full civil liberties (including, of course, the right to organize) to immigrants who have already arrived. This stance is exemplified in the eighth reading, from the AFL-CIO (the U.S. confederation of trade unions).

Practically all of these arguments for and against immigration to the United States, both on cultural and economic grounds, are made in the course of an episode of CNN's *Crossfire* from 2001, which featured Pat Robertson and Raul Yzaguirre, president of the National Council of La Raza, a group dedicated to promoting the interests of Mexican-Americans. A transcript of this program is reprinted as the ninth reading.

The last two readings offer a snapshot of the issues surrounding the U.S.–Canadian border. In the tenth reading, Herbert London of the Hudson Institute, a U.S. research institute, argues that immigration restrictions in Canada should be stepped up to U.S. standards. This view is contrasted in the final reading, from the Canadian Council for Refugees, which argues that the "threat" of terrorists seeping into the United States as a result of Canada's immigration policy on refugees is exaggerated and xenophobic.

Readings

from the National Front 1

Patterns of Prejudice

One of the favourite accusations thrown at the National Front by its multiracialist critics is that we are simply a bunch of bigots, that our stance on race, at the core of our political philosophy, is just ignorant prejudice against people whose skin colour is no more than a superficial manifestation. Is this so? Are our racial policies merely the product of prejudice, or are they instead based on sober realism and the courage to face facts?

Let us start by looking at what 'prejudice' means. The word comes from Latin roots meaning to judge before, or less literally but more usefully, to judge a case ahead of the facts. In modern usage it is generally taken to mean forming an opinion, especially about an issue, person or group of people without knowing, or without taking into account, all the relevant facts. Someone is 'prejudiced' if they judge a case without hearing the evidence, or they ignore the evidence before them because it does not accord with their preconceived opinion.

Certainly, the NF has judged the case as far as the coloured population in Britain is concerned. We know that they were brought here against the best interests, and indeed the wishes, of the vast majority of the native British people.

But has this verdict been arrived at as a result of prejudice, or is it a fair judgement on the basis of the evidence? Are the arguments we use to justify our conclusion merely expressions of bigoted dislike, or are they solidly founded in reality?

Never Fit

We argue, for example, that West Indian and other Negroes will never fit in, as multiracialists claim, to become equal and integrated members of a predominantly White society. This is because they are inherently unfitted to do so intellectually, and are thus condemned to exist in White society as a permanent underclass, confined to the lower social strata and, not unnaturally, bitterly resentful of the alien society in which they are thus trapped. This resentment will inevitably explode into violence, rioting, and crime.

To seek to remedy this by artificially promoting Blacks to high office, ignoring their handicaps, is a recipe for chaos. Blacks cannot fulfill these roles competently and Whites quite naturally resent being discriminated against in their own country in favour of aliens.

What are the facts? Over almost seventy years, in study after study, conducted by scientists and educationalists in numerous countries, studies conducted by such bastions of racial rationalism as the Inner London Education Authority, the U.S. Army, and Harvard and Oxford Universities, on every measure of intellectual ability and educational attainment Blacks perform significantly worse, on average, than Whites. In the case of average IQ, for example, the average Negro figure is only 85% of the White average. In fact, the higher the proportion of White genes the higher the intelligence: a pure-bred Negro fresh out of Africa scores nearer 70%. If a married couple, perhaps of the sort that would criticise us, parrot-like, for our views, had a child with this order of intelligence they would be seriously concerned, and seek remedial treatment for it. The aboriginal Australian fares even worse than the Negro.

This is not the place for an extensive review of studies of intelligence. Readers can consult *Race* by Dr. John R. Baker, former Reader in Cytology at Oxford University, published by the Oxford University Press, or *The Testing of Negro Intelligence*, an exhaustive review of hundreds of studies demonstrating racial differences in intellectual ability by Dr. Audrey M. Shuey, and of course there is *The Bell Curve* by Herrnstein and Murray.

Worse still for the "We're all equal" crowd, Negroes perform just as badly in comparison with Whites on evoked potential tests, where a light is flashed in a subject's face and the speed and density of the evoked brainwave response is measured on an electro-encephalograph—a test so culture-free it can just as well be given to a dog or cockroach as to a man.

Faced with such facts, multiracialists make various attempts to wriggle off the hook, all of them futile. They argue that all these tests are unfair, that they are written by White people in a White society and thus are biased against non-Whites. Alas for them, Chinese and Japanese, who are not noticeably more Caucasian and are often very much more culturally distinct from White society than Negroes, actually do as well or slightly better on average than Whites on these "White mens' tests."

At a deeper level of intelligence testing, however, there are different forms of intelligence. For example, it is possible to score high according to conventional intelligence scales but have no capacity for creative invention, as a proportion of Whites observably do. In fact it is this creative intelligence which the British possess in abundance: around 80% of everything ever invented was invented by British men.

So the multiracialists fall back on conceding the reality of lower average Negro intelligence, but blame it on environment rather than on innate heredity. "Enough positive discrimination in favour of Blacks will make them our equals" runs their argument, though they are rarely honest enough to state it bluntly.

If "social deprivation" and "racial discrimination" are responsible for the poor performance of Negroes, groups such as the American Indians, who score considerably worse than American blacks on every measure of social deprivation, would be expected to do worse, or at least as badly, as blacks. In fact they do a lot better in the same tests.

Advocates of racial equality who argue that differences in intelligence are not innate ignore numerous studies demonstrating that at least 80% of differences in intelligence are inborn, the product of genes, not environment. Read *The Inequality of Man* by H. J. Eysenck, former Professor of Psychology at the University of London, for the facts here.

So-called liberals, when confronted with arguments and facts they cannot defend, resort to underhand and quite despicable tactics, revealing themselves as the most illiberal and intolerant clique of all. An associate of Eysenck in the field of personality and twin studies (studies of identical twins who were separated at birth, etc.) was Sir Cyril Burt. When one statistic out of hundreds was discovered to be suspiciously similar to another, such a hue and cry was made, with accusations and connotations of fraud, that even now the stigma lingers, even though Burt's work has been cleared of improper practice and verified in scores if not hundreds of other studies. Eysenck himself, who was half-Jewish, was accused of being a "Nazi." More recently Chris Brand lost his lectureship at Edinburgh University and a book contract for reaching the "wrong" conclusions of his research.

Note that we are not talking about denying the "Holocaust" or some other strongly-held religious or moral doctrine—but about the results of tests that can be made and verified in any psychology laboratory, if not any living room.

Real World

Stepping out from the arcane field of intelligence measurement, what do we see in the real world? We see exactly what honest psychologists' conclusions would lead us to expect. Negroes, innately less intelligent, are at the bottom of every White social heap, but this is blamed by advocates of the dogma and their tan-skinned imitators on "White racism."

So what about Black performance when there are no Whites to be "racist"? What did the Negro accomplish in Africa before the White Man came? As Baker illustrates in *Race*, virtually nothing.

When our ancestors, hundreds of years ago, spread across the world as explorers, traders and conquerors, what did they find? In Asia, and in the Americas, there were great cities, vast and sometimes ancient civilisations and cultures which featured emperors, poets, law-givers and philosophers, priests, generals, and architects. Sometimes there were writings of minds as profound as any in Europe. Mighty buildings, fields sown with native crops, rice in Asia and maize in America, tilled or grazed by domestic native animals. What did they find on the Negro's home turf? Primitive tribes, living in mud huts in the African jungle, frequently eating one another. Negroes without the wheel, without the written word, without a history.

Professor Arnold Toynbee, one of this century's leading historians, summed it up by concluding that of twenty-one civilisations in world history to date not one had been founded by Black men.

Evidence

On the Negro question the evidence is overwhelming. Blacks are, on average, less intelligent than Whites, and they were born that way. Since measured intelligence correlates well with social and educational attainment, it is no wonder that Blacks consistently fail to rise. They always will be at the bottom, unless they are artificially propped up in jobs they cannot do in a pathetic and indeed patronising bid by Whites in a perverse attempt to show how "goody goody" they are.

Unless we actually want ghettos of miserable, frustrated and angry Blacks seething in our inner cities so that middle-class suburban types can drive past in rich synthetic smugness, congratulating themselves on how tolerant they are and how wonderfully multiracial Britain is, we will do the only rational thing, the kindest and fairest thing for the Blacks themselves, and send them home to their own kind. Western-educated and trained Blacks would doubtless rise high in an Africa that seems to be sinking inexorably back to the jungle, perhaps making a vital contribution to Negro welfare.

On this issue our verdict is based on the facts. We have judged the case on the evidence, fairly, and come to the only logical conclusion. It is the multiracialists, who have a hidden agenda, who stubbornly maintain the fiction of racial equality despite overwhelming evidence to the contrary.

They cannot cite a mass of objective evidence to support their beliefs. At best they can only snipe and smear that which proves our case. Or simply seek, through the Race Acts, to suppress it by imprisoning honest men. It is they, not we, who hold to their opinions in defiance of reality. Dare we say it—it is they, not we, who are prejudiced.

Source: Steve Brady, "Patterns of Prejudice" posted at www.natfront.com/prejudic/html. This is a revised and updated version of an article that originally appeared in *Vanguard*, 1987.

2

from Die Freiheitliche Partei Österreichs (the Austrian Freedom Party)

Party Programme

Adopted October 30, 1997

Accompanying information concerning the English translation of the FPÖ Party Programme:

The following translation of the FPÖ Party Programme tries to reflect, as far as possible, the content and meaning of the original German version. It should be borne in mind, however, that some phrases and words do not have an exact equivalent in English. In these cases the editors have tried to convey the meaning of the ideas rather than present a word-for-word translation. Where this has been done the original German has been included in brackets, to enable direct reference to the German language edition of the programme.

Chapter III: Austria First! "For Us, Austria and Its People Take First Place."

Article 1

Austria is more than a simple administrative unit. Its people are linked by a will for independence and belong together in regional diversity. This will is expressed by the democratic federal and constitutional Republic of Austria.

1. Austria's self-image [*Österreichpatriotismus*] is expressed in the will for the independence and unity of Austrians, in the will to preserve democracy, human rights, the rule of law and federalism, and in the will to cultivate Austria's cultural heritage and protect its environment, countryside, and nature.
2. Austria's identity is formed by a variety and multitude of regional identities. After a painful past the Austrian people have demonstrated a will to be together within the bounds of regional peculiarities.

Article 2

This dedication to Austria underlines a permanent task to preserve and develop democracy as a basis for Austrian patriotism. Beyond that it means an obligation to stand up for Austria's independence and to preserve its constitutional principles.

According to the Freedomite understanding of an Austrian self-image [*österreichischer Patriotismus*] related to a democratic society there is an ongoing commitment to develop and preserve democracy for the people. This commitment includes the preservation of federal, social, and liberal constitutional principles.

Article 3

As all Austrian people belong together we do not only have civil rights but also civil duties, especially that of solidarity for the preservation of a functioning state unit and in making a contribution to internal and external security.

Included in the duties mentioned above is, above all, the duty of being solidly behind one's compatriots—as for instance, in supporting the elderly and infirm, in avoiding social cases of hardship and so on, by making a contribution to upholding the functioning of the state—for instance, by taxation to a statutory upper limit or through a personal contribution to the defense of the country or the preservation of internal security in the field of civil defense and disaster control.

Article 4

Austria's social and cultural heritage justifies pride in the results, traditions, and achievements. From this arose a positive self image [*Patriotismus*], which calls for self-assured Austrian politics and a resistance to a decline in the cultural level, and increasing efforts to revile traditions and to willfully disparage Austria.

1. As Austria played a decisive role in the all-German and all-European history and in view of the resulting cultural heritage, it is legitimate to appear on the international stage with self-confidence and pride.
2. We reject the politics that ends in–especially since Austria's entry into the EU—massive efforts to standardize and level down, to the detriment of Austria's intellectual and cultural substance.
3. The modern trend to get some public response by abusing Austria and disparaging Austrian qualities needs resolute intellectual resistance on the part of all people who believe in Austria [*erfordert einen entschlossenen geistigen Widerstand aller patriotischen Kräfte*].
4. Especially in the media, a decline in the cultural level has been obvious for years; so we need a new intellectual and cultural move to keep Austrian traditions and regional peculiarities alive.

Chapter IV: The Right to a Cultural Identity [*Heimat*] "Dedication to Our Cultural Identity Is the Basis of Our Political Activity."

Article 1

Our country [*Heimat*] is the democratic Republic of Austria and its federal states, including the historically settled indigenous groups (Germans, Croats, Roma, Slovaks, Slovenians, Czechs, and Hungarians) and the culture moulded by them. The legal system in Austria presupposes that the overwhelming majority of Austrians is of German origin. [*Wobei von der Rechtsordnung denklogisch vorausgesetzt wird, daβ die überwiegende Mehrheit der Österreicher der deutschen Volksgruppe angehört.*]

1. The term "cultural identity" [*Heimat*] is defined in spatial, historical groupings [*ethnische Volksgruppen*] and cultural ways.
2. So our country [*Heimatland*], with its centuries-old historical groups and their cultural traditions and achievements enshrining civilization, are subjects to be protected.
3. In Austria the law on historical groups lists the individual historically settled (autochthone) cultural groups as subjects to be protected.

Article 2

The cultural identity [*Heimat*] is to be preserved, protected and developed in these spatial, historical groupings [*ethnische Volksgruppen*] and cultural aspects.

1. This means a special commitment to preserve a viable environment and to protect and develop the cultural traditions of civilization, within the framework of a freedomite-democratic state bound by the rule of law. It means the protection of the population and the cultural identity of the indigenous (autochthonous) cultural groups as provided for in Art. 19 of the Basic Law (which has the rank of a constitution) concerning citizens' general rights dated Dec. 21, 1867.
2. The coexistence and working together of the different cultural groups have formed Austria's characteristic identity, which can only be preserved through a guaranteed continuing existence of these historically settled cultural groups [*historisch ansässige Volksgruppen*]. This is something which is particularly necessary in times of developing supraregional entities.

Article 3

Every Austrian has the basic right to decide independently and freely his identity and the cultural group [*Volkstumszugehörigkeit*] to which he belongs. No disadvantage must result from this attachment to a cultural tradition.

1. The basic right to decide one's own identity and culture has to be guaranteed to all Austrians.
2. This right is not limited to the historic indigenous groups [*verschiedene Volksgruppen*], as mentioned in article 1, as subjects to be protected under the term "cultural identity" [*Heimatbegriff*]. It is up to every citizen to decide if he regards himself as belonging to a cultural [*ethnisch*] group. Consequently, every citizen has the right to decide on his own to which cultural [*Volksgruppe*] group he wants to be assigned according to his identity. But he only can derive individual rights from his attachment to a cultural tradition concerning the historic indigenous groups.
3. On the other hand, the state does not have the right to regulate and determine how the citizen is to regard himself. There must be no public disadvantage or private discrimination of any Austrian because of his freely self-determined cultural tradition [*Volkstumszugehörigkeit*].
4. The free acknowledgement of one's own cultural tradition [*Volkstum*] is a basic principle for the preservation and further development of the cultural values and the historical-cultural self-awareness of every community [*ethnische Gemeinschaft*]. The awareness of the special qualities of one's own people is inseparably linked to the willingness to respect what is special about other peoples.

Article 4

Because of its topography, its density of population, and its limited resources Austria is not a country for immigration.

1. The basic right to a homecountry [*Heimat*] does not allow for an unlimited and uncontrolled immigration to Austria. The protective requirement of this fundamental right to a homecountry makes clear that Austria because of its small size, its density of population and its limited resources cannot be a country of immigration.
2. Unlimited immigration would demand too much of the resident population as far as an active capacity for integration is concerned. It would endanger the right to preservation and protection of cultural identity [*Heimat*]. We reject multicultural experiments that bear social conflicts with them.
3. To protect the interests of the Austrian population requires full sovereignty in matters concerned with the rights of immigrants.
4. Austria must give asylum to people who are persecuted for racist, religious, or political reasons, if they do not come to Austria through a secure third country. Every persecuted person has the right to profess his own cultural tradition [*Volkstum*] and to return to his homecountry [*Heimat*], especially the numerously displaced persons, driven out of their countries who have been denied their basic right to a homecountry [*Heimat*] by violence and expulsion in the course of the tragic events of the last decades. They do not lose this fundamental right and keep the right of return to their country of origin [*Heimat*].

Chapter V: Christianity—The Foundation of Europe "From Our Traditions We Gain Strength for Something New. It Was Christianity That Decisively Formed Our Traditions."

Article 1

The world order formed by Christianity and the ancient world is the most important intellectual foundation of Europe. The prime intellectual movements from humanism to the enlightenment are based on them. The cultural character of Christian values and traditions even embraces members of non-Christian religions and people without any confession.

1. European civilization has its oldest roots in antiquity. The face of Europe is decisively formed by Christianity in all its denominational diversity. Beyond that Europe was also influenced by Judaism and other non-Christian religious communities.
2. Europe's established principles of law are based on the Christian consensus of common values.

Article 2

The preservation of the intellectual foundations of the West necessitates a Christianity that defends its values. In striving for the preservation of these European foundations the Freedom Party sees itself as an ideal partner of the Christian churches, even if they sometimes take other positions on political issues.

1. The intellectual foundations of the West include the idea of human dignity and basic liberties as well as ideas deriving from this, such as the idea of democracy, codetermination and the rule of law, solidarity and respect for life and creation.
2. But these foundations are endangered by different streams of thought. The increasing fundamentalism of radical Islam which is penetrating Europe, as well as hedonistic consumption, aggressive capitalism, increasing occultism, pseudo-religious sects and an omnipresent nihilism threaten the consensus of values which is in danger of getting lost.
3. The big Christian churches play a decisive part in preserving the European consensus of values. This is also a concern of the Freedomite movement that sees itself as a natural partner of Christian churches. Therefore the FPÖ supports religious instruction in state schools and decisively rejects efforts to adopt in its place "ethics," an unfounded and questionable subject because of its philosophic and ideological position.
4. However, many people in Europe expect the churches to make more effort in decisively fighting this intellectual menace and not to be content by just playing the role of social welfare societies.
5. Liberalism throughout its history has always opposed ideological and religious intolerance often exercised by religious institutions. During this historical phase anticlericism grew which is now, in view of the changed role of ecclesiastical and religious institutions, outdated.

Article 3

The protection of autonomy for the churches and recognized religious communities requires an institutional separation of the church and the state. This separation also provides an essential guarantee to safeguard the freedom of every individual.

1. The protection of church autonomy and the recognized religious communities requires an institutional but not an intellectual separation of the church and the state. Above all this is necessary to avoid party political abuse of the churches and to avoid appealing to someone's conscience with the ideology of a party.
2. The institutional separation of the church and state has contributed to creating a situation which was decisive for the freedomite tradition in Europe.
3. The religious commitment and the values of the churches and recognized religious communities needs autonomy to guarantee the best conditions in which they can develop. As historical experience shows, this is another contribution to the freedom of the individual which cannot be renounced.

Source: Excerpts from "Program of the Austrian Freedom Party," adopted October 30, 1997 posted at www.fpoe.at/programm/ parteiprogramm_eng.pdf.

3

from the Cable News Network

Pim Fortuyn: Man of Paradox

May 9, 2002

AMSTERDAM, Netherlands—Dutch maverick politician Pim Fortuyn was a colourful figure in what many voters complained was a bland political landscape in the Netherlands.

The 54-year-old courted controversy with his robust style, being blunt, outspoken and flamboyant, an approach formerly unseen in Dutch politics.

The former Marxist, sociology lecturer and newspaper columnist stood out with his shaven-head and bright, colourful ties and was also conspicuous travelling around in a car with blacked-out windows.

Proudly homosexual, he spoke out against immigration and high taxation and accused the Dutch government of poor performance.

He also described Islam as a "backward culture" in his book, *Against the Islamicisation of Our Culture*.

He attracted a wide following, but was shot dead just nine days before a scheduled May 15 election in which opinion polls had forecast his newly-formed Lijst Pim Fortuyn party, which won a stunning 35 percent in local elections in Rotterdam in March, would get between 25 and 28 seats in the 150-member parliament, at the expense of the socialists.

In November last year he became leader of the Leefbaar (livable) Nederland party. He guided the party to the right, but in February he was expelled after criticising Muslims in the newspaper *De Volkskrant*, and suggesting an article in the Dutch constitution banning discrimination should be changed.

Although Fortuyn was part of a new wave of Dutch politicians and often perceived as an extremist, he insisted he was not like Jean-Marie Le Pen and wanted nothing to do with the French far-right leader.

He was at pains to point out that he was not against immigrants, but he questioned their ability to assimilate into a liberal and racially tolerant culture and argued immigration had to be curbed in order for the Netherlands' liberal social values to survive.

"My policies are multi-ethnic and certainly not racist," he said. "I want to stop the influx of new immigrants. This way, we can give those who are already here the opportunity to fully integrate into our society."

In a recent interview, he argued: "In Holland, homosexuality is treated the same way as heterosexuality. In what Islamic country does this happen?"

Fortuyn's platform seemed out of place in the ultra-liberal Netherlands, which he argued was full up with 16 million people.

While not advocating deportation, he criticised the country's estimated 800,000 Muslims for not embracing Dutch life and said government benefits should be restricted to Dutch speakers.

Though tolerant of such subcultures, Fortuyn targeted a deep vein of suspicion of immigrants and also blamed them for a rising crime wave.

He said: "I'm not anti-Muslim. I'm not anti-immigrant. I'm saying we've got big problems in our cities.

"It's not very smart to make the problems bigger by letting in millions more immigrants from rural Muslim cultures that don't assimilate. This country is bursting. I think 16 million people is quite enough."

Fortuyn also cautioned about the expansion of the European Union to include Eastern European countries, and lamented the loss of Dutch national identity within the EU.

Source: Pym Fortuyn: Man of Paradox, posted on CNN.com, 5/9/02.

from **Patrick J. Buchanan**

4

To Reunite a Nation

January 18, 2000
Richard Nixon Library
Yorba Linda, CA

Let me begin with a story: In 1979, Deng Xiaoping arrived here on an official visit. China was emerging from the Cultural Revolution, and poised to embark on the capitalist road. When President Carter sat down with Mr. Deng, he told him he was concerned over the right of the Chinese people to emigrate. The Jackson-Vanik amendment, Mr. Carter said, prohibited granting most favored nation trade status to regimes that did not allow their people to emigrate.

"Well, Mr. President," Deng cheerfully replied, "Just how many Chinese do you want? Ten million. Twenty million. Thirty million?" Deng's answer stopped Carter cold. In a few words, the Chinese leader had driven home a point Mr. Carter seemed not to have grasped: Hundreds of millions of people would emigrate to America in a eyelash, far more than we could take in, far more than our existing population of 270 million, if we threw open our borders. And though the United States takes in more people than any other nation, it still restricts immigration to about one million a year, with three or four hundred thousand managing to enter every year illegally.

There is more to be gleaned from this encounter. Mr. Carter's response was a patriotic, or, if you will, a nationalistic response. Many might even label it xenophobic. The President did not ask whether bringing in 10 million Chinese would be good for them. He had suddenly grasped the real issue: How many would be good for America? Mr. Carter could have asked another question: Which Chinese immigrants would be best for America? It would make a world of difference whether China sent over 10 million college graduates or 10 million illiterate peasants, would it not?

Since the Carter-Deng meeting, America has taken in 20 million immigrants, many from China and Asia, many more from Mexico, Central America and the Caribbean, and a few from Europe. Social scientists now know a great deal about the impact of this immigration.

Like all of you, I am awed by the achievements of many recent immigrants. Their contributions to Silicon Valley are extraordinary. The over-representation of Asian-born kids in advanced high school math and science classes is awesome, and, to the extent that it is achieved by a superior work ethic, these kids are setting an example for all of us. The contributions that immigrants make in small businesses and hard work in tough jobs that don't pay well merits our admiration and deepest respect. And, many new immigrants show a visible love of this country and an appreciation of freedom that makes you proud to be an American.

Northern Virginia, where I live, has experienced a huge and sudden surge in immigration. It has become a better place, in some ways, but nearly unrecognizable in others, and no doubt worse in some realms, a complicated picture over all. But it is clear to anyone living in a state like California or Virginia that the great immigration wave, set in motion by the Immigration Act of 1965, has put an indelible mark upon America.

We are no longer a biracial society; we are now a multi-racial society. We no longer struggle simply to end the divisions and close the gaps between black and white Americans; we now grapple, often awkwardly, with an unprecedented ethnic diversity. We also see the troubling signs of a national turning away from the idea that we are one people, and the emergence of a radically different idea, that we are separate ethnic nations within a nation.

Al Gore caught the change in a revealing malapropism. Mr. Gore translated the national slogan, "E Pluribus Unum," which means "Out of many, one," into "Out of one, many." Behind it, an inadvertent truth: America is Balkanizing as never before.

Five years ago, a bipartisan presidential commission, chaired by Barbara Jordan, presented its plans for immigration reform. The commission called for tighter border controls, tougher penalties on businesses that hire illegal aliens, a new system for selecting legal immigrants, and a lowering of the annual number to half a million. President Clinton endorsed the recommendations. But after ethnic groups and corporate lobbies for foreign labor turned up the heat, he backed away.

The data that support the Jordan recommendations are more refined today. We have a National Academy of Sciences report on the economic consequences of immigration, a Rand study, and work by Harvard's George Borjas and other scholars. All agree that new immigration to the United States is heavily skewed to admitting the less skilled. Unlike other industrialized democracies, the United States allots the vast majority of its visas on the basis of whether new immigrants are related to recent immigrants, rather than whether they have the skills or education America needs. This is why it is so difficult for Western and Eastern Europeans to come here, while almost entire villages from El Salvador have come in.

Major consequences flow from having an immigration stream that ignores education or skills. Immigrants are now more likely than native-born Americans to lack a high school education. More than a quarter of our immigrant population receives some kind of welfare, compared to 15 percent of native-born. Before the 1965 bill, immigrants were less likely to receive welfare. In states with many immigrants, the fiscal impact is dramatic. The National Academy of Sciences contends that immigration has raised the annual taxes of each native household in California by $1,200 a year. But the real burden is felt by native-born workers, for whom mass immigration means stagnant or falling wages, especially for America's least skilled.

There are countervailing advantages. Businesses can hire new immigrants at lower pay; and consumers gain because reduced labor costs produce cheaper goods and services. But, generally speaking, the gains from high immigration go to those who use the services provided by new immigrants.

If you are likely to employ a gardener or housekeeper, you may be financially better off. If you work as a gardener or housekeeper, or at a factory job in which unskilled immigrants are rapidly joining the labor force, you lose. The last twenty years of immigration have thus brought about a redistribution of wealth in America, from less-skilled workers and toward employers. Mr. Borjas estimates that one half of the relative fall in the wages of high school graduates since the 1980s can be traced directly to mass immigration.

At some point, this kind of wealth redistribution, from the less well off to the affluent, becomes malignant. In the 1950s and '60s, Americans with low reading and math scores could aspire to and achieve the American Dream of a middle-class lifestyle. That is less realistic today. Americans today who do poorly in high school are increasingly condemned to a low-wage existence; and mass immigration is a major reason why.

There is another drawback to mass immigration: a delay in the assimilation of immigrants that can deepen our racial and ethnic divisions. As in Al Gore's "Out of One, Many."

Concerns of this sort are even older than the Republic itself. In 1751, Ben Franklin asked: "Why should Pennsylvania, founded by the English, become a Colony of Aliens, who will shortly be so numerous as to Germanize us instead of our Anglifying them?" Franklin would never find out if his fears were justified. German immigration was halted by the Seven Years War; then slowed by the Great Lull in immigration that followed the American Revolution. A century and half later, during what is called the Great Wave, the same worries were in the air.

In 1915 Theodore Roosevelt told the Knights of Columbus: "There is no room in this country for hyphenated Americanism. . . . The one absolutely certain way of bringing this nation to ruin, of preventing all possibility of its continuing to be a nation at all, would be to permit it to become a tangle of squabbling nationalities." Congress soon responded by enacting an immigration law that brought about a virtual forty-year pause to digest, assimilate, and Americanize the diverse immigrant wave that had rolled in between 1890 and 1920.

Today, once again, it is impossible not to notice the conflicts generated by a new "hyphenated Americanism." In Los Angeles, two years ago, there was an anguishing afternoon in the Coliseum where the U.S. soccer team was playing Mexico. The Mexican-American crowd showered the U.S. team with water bombs, beer bottles, and trash. The Star Spangled Banner was hooted and jeered. A small contingent of fans of the American team had garbage hurled at them. The American players later said that they were better received in Mexico City than in their own country.

Last summer, El Cenizo, a small town in south Texas, adopted Spanish as its official language. All town documents are now to be written, and all town business conducted, in Spanish. Any official who cooperates with U.S. immigration authorities was warned he or she would be fired. To this day, Governor Bush is reluctant to speak out on this de facto secession of a tiny Texas town to Mexico.

Voting in referendums that play a growing part in the politics of California is now breaking down sharply on ethnic lines. Hispanic voters opposed Proposition 187 to cut off welfare to illegal aliens, and they rallied against it under Mexican flags. They voted heavily in favor of quotas and ethnic preferences in the 1996 California Civil Rights Initiative, and, again, to keep bilingual education in 1998. These votes suggest that in the California of the future, when Mexican-American voting power catches up with Mexican-American population, any bid to end racial quotas by referendum will fail. A majority of the state's most populous immigrant group now appears to favor set-asides and separate language programs, rather than to be assimilated into the American mainstream.

The list of troubling signs can be extended. One may see them in the Wen Ho Lee nuclear secrets case, as many Chinese-Americans immediately concluded the United States was prosecuting Mr. Lee for racist reasons.

Regrettably, a cultural Marxism called political correctness is taking root that makes it impossible to discuss immigration in any but the most glowing terms. In New York City billboards that made the simple point that immigration increases crowding and that polls show most Americans want immigration rates reduced were forced down under circumstances that came very close to government-sponsored censorship. The land of the free is becoming intolerant of some kinds of political dissent.

Sociologist William Frey has documented an out-migration of black and white Americans from California, some of them seeking better labor market conditions, others in search of a society like the one they grew up in. In California and other high immigration states, one also sees the rise of gated communities where the rich close themselves off from the society their own policies produce.

I don't want to overstate the negatives. But in too many cases the American Melting Pot has been reduced to a simmer. At present rates, mass immigration reinforces ethnic subcultures, reduces the incentives of newcomers to learn English; and extends the life of linguistic ghettos that might otherwise be melded into the great American mainstream. If we want to assimilate new immigrants—and we have no choice if we are to remain one nation—we must slow down the pace of immigration.

Whatever its shortcomings, the United States has done far better at alleviating poverty than most countries. But an America that begins to think of itself as made up of disparate peoples will find social progress far more difficult. It is far easier to look the other way when the person who needs help does not speak the same language, or share a common culture or common history.

Americans who feel it natural and right that their taxes support the generation that fought World War II—will they feel the same way about those from Fukien Province or Zanzibar? If America continues on its present course, it could rapidly become a country with no common language, no common culture, no common memory, and no common identity. And that country will find itself very short of the social cohesion that makes compassion possible.

None of us are true universalists: we feel responsibility for others because we share with them common bonds—common history and a common fate. When these are gone, this country will be a far harsher place.

That is why I am proposing immigration reform to make it possible to fully assimilate the 30 million immigrants who have arrived in the last thirty years. As President, I will ask Congress to reduce new entry visas to 300,000 a year, which is enough to admit immediate family members of new citizens, with plenty of room for many thousands with the special talents or skills our society needs. If after several years, it becomes plain that the United States needs more immigrants because of labor shortages, it should implement a point system similar to that of Canada and Australia, and allocate visas on a scale which takes into account education, knowledge of English, job skills, age, and relatives in the United States.

I will also make the control of illegal immigration a national priority. Recent reports of thousands of illegals streaming across the border into Arizona, and the sinister and cruel methods used to smuggle people by ship into the United States, demand that we regain control of our borders. For a country that cannot control its borders isn't fully sovereign; indeed, it is not even a country anymore.

Without these reforms, America will begin a rapid drift into uncharted waters. We shall become a country with a dying culture and deepening divisions along the lines of race, class, income, and language. We shall lose for our children and for the children of the 30 million who have come here since 1970 the last best hope of earth. We will betray them all—by denying them the great and good country we were privileged to grow in. We just can't do that.

With immigration at the reduced rate I recommend, America will still be a nation of immigrants. We will still have the benefit of a large, steady stream of people from all over the world whose life dream is to be like us—Americans. But, with this reform, America will become again a country engaged in the mighty work of assimilation, of shaping new Americans, a proud land where newcomers give up their hyphens, the great American melting pot does its work again, and scores of thousands of immigrant families annually ascend from poverty into the bosom of Middle America to live the American dream.

Source: Patrick Buchanan, "To Reunite a Nation," speech delivered at the Richard M. Nixon Library, January 18, 2000.

5

from the Coalition for the Future American Worker

What Is the CFAW?

The Coalition for the Future American Worker (CFAW) is an umbrella organization of professional trade groups, population/environment organizations, and immigration reform groups. CFAW was formed to represent the interests of American workers and students in the formulation of immigration policy.

CFAW believes that American workers are the best, most creative, most productive and highly motivated in the world. Americans, in large numbers, have responded to the challenges set forth by the new global and technological economy. In the space of one generation, American workers have made the transition from the industrial economy of yesterday to create the most advanced and productive economy in the world.

If the American workforce has been able to reshape the American economy so successfully, American workers can also meet any future challenges. CFAW believes that in the fast-paced economy of the future, American companies must commit to providing American workers and students with the training and opportunities they will need to continue to contribute to America's prosperity.

Increasingly, corporations operating on U.S. soil are seeking permission to import foreign workers on a temporary or permanent basis. In the case of lower-skilled occupations, this importation removes the incentive for businesses to improve the productivity of the jobs and raise wages. Importation of higher-skilled workers tends to lead to discrimination against older American workers and slams the door of opportunity on many American students who otherwise would have a chance at those jobs if the corporations were willing to provide minor re-training. The importation of foreign workers is especially harmful to those Americans who are under-represented in the high-tech field: blacks, Hispanics, and women. But the labor-importing corporations are very well organized and well represented before Congress and the Administration in their efforts to secure access to foreign labor. CFAW was established to be a counterbalance to these well-financed industry lobbyists.

CFAW has been formed to help government officials understand the needs of American industry and commerce, and to tailor government policies to ensure the best outcome for both workers and industry. The growing practice of seeking large numbers of workers abroad, rather than committing to the training and advancement of American workers, is ultimately harmful to both labor and industry. It is CFAW's goal to work with government and business to find solutions that will ensure that employers can gain access to workers they truly need, while protecting job opportunities for American workers and students and allowing them to advance their skills and maintain high wages.

Source: "What Is the Coalition for the American Worker" Copyright ©FAIR. Used with permission.

from the Essential Worker Immigration Coalition

6

Documenting the Labor Shortage

March 2002

The Essential Worker Immigration Coalition (EWIC) is a coalition of businesses, trade associations, and other organizations from across the industry spectrum concerned with the shortage of both skilled and lesser skilled ("essential worker") labor. EWIC stands ready to work with the Administration and Congress to push forward on important immigration policy issues.

EWIC supports policies that facilitate the employment of essential workers by U.S. companies and organizations. Current immigration law largely prevents the hiring of foreign essential workers. EWIC supports reform of U.S. immigration policy to facilitate a sustainable workforce for the American economy while ensuring our national security and prosperity.

Bureau of Labor Statistics. (BLS) projections indicate that the United States will create 17 million new jobs by 2010, 58 percent of which will not require a four-year college degree. The service-producing sector will add 20.5 million jobs with a total projected increase in labor force of 17 million.

Meanwhile, the United States is not producing enough new workers to sustain such growth and our current workforce is aging. By 2010, the labor force ages 46–64 will have the fastest growth rate. More than 60 million current employees will likely retire over the next 30 years. Testifying before a House Subcommittee in 2000, Dr. Richard Judy of the Hudson Institute said, "After 2011, the year in which the first of the baby boomers turns 65, their flight to retirement will reach proportions so huge as, barring unforeseen increases in immigration and/or participation rates among the elderly, to reduce the total size of the nation's workforce."

The following are examples of sector-specific information on present and future labor needs:

From the American Health Care Association

On February 18, 2002, *The New York Times* reported that more than 90% of all nursing facilities do not have the number of staff necessary to provide good care. The U.S. Dept. of Health and Human Services recently reported that nursing homes currently need 181,000 to 310,000 nurse aides (an entry-level position) to reach full staff levels. This number is expected to grow to over 800,000 by the year 2008 as more baby boomers need long term care. Nursing homes have hired over 100,000 people from the welfare roles, wages are higher than in the service sector generally, and the facilities generally provide a training and certification course (paying wages while attendees take classes) to become a Certified Nurse Aide (CNA).

From the American Hotel & Lodging Association

A recent report by the American Economics Group estimated current lodging industry employment at 1.9 million with projected growth to over 2.6 million in 2010, meaning that the industry will require more than 700,000 additional workers this decade.

From the American Meat Institute

According to the BLS, the meat and poultry packing and processing industry employed more than 500,000 workers in 2000 (compared to 235,000 in 1975), and is projected by BLS to employ more than 540,000 workers, a 7.6 percent increase, by 2010. Most large meat and poultry packing plants are located in low-population, rural areas, which pose unique challenges to this labor-intensive industry. Not only does a labor shortage reduce productivity and efficiency in meat and poultry plants, it reduces capacity to buy and process poultry and livestock, hurting farmers as well.

From the Associated General Contractors

December 2000 "Insights in Construction" Survey by AGC and Deloitte & Touche listed a shortage of skilled labor as the biggest challenge facing the construction industry over the next five years. The BLS recently estimated that more than 2 million workers will be needed in construction trades and related fields between 2000 and 2010 due to job growth and net replacements for retiring workers.

From the Building Service Contractors Association International

Current employment in building services is over 1 million, and has grown steadily in the last year. According to a survey by the Association, all responding members reported they expect to increase employment in the next year and all reported difficulty filling vacant positions. These vacancies have resulted in curtailment of seeking additional service contracts and expansion plans. Notably, these shortages are affecting an industry that employs anywhere from 40 to 99% women and minorities.

From the National Association of Home Builders

Finding skilled workers has become increasingly difficult for home builders. Recent surveys of local home builder associations have consistently ranked labor availability as one of the most critical issues facing the industry. Other NAHB surveys have reported that the labor shortage has added 20 days to the time needed to build a single-family home, significantly adding to its cost. According to BLS, over 200,000 new workers are required by the industry each year to meet consumer demand for housing.

From the National Restaurant Association

Restaurants are the largest private-sector employer with over 11.3 million employees. By 2010 the industry expects to employ an additional 2 million workers. Labor shortages consistently poll among the top issues for restaurants/small business. According to the National Council of Chain Restaurants, workforce shortages, particularly in metropolitan areas, are among the most significant short and long term challenges to the industry.

From the National Roofing Contractors Association

The lack of qualified workers is the single biggest problem facing roofing contractors today. In a recent on-line survey of members, over 50% responded they could hire up to five additional employees right now if qualified workers were available. The BLS projects an additional 50,000 roofers will be needed over the next decade to keep pace with demand.

From the U.S. Chamber of Commerce

The U.S. Chamber of Commerce's Center for Workforce Preparation did a survey of local chambers of commerce in the summer of 2001. Ninety-nine percent of leading chamber of commerce CEOs reported workforce development as a priority issue among employers. Eighty percent of survey respondents listed a "shortage of workers/low unemployment" as a key workforce development issue in their community. For example, a representative from Chamber member Ingersoll-Rand, a multinational manufacturing company, testified in Congress in 2001 that one of their key workforce problems is finding workers for skilled trades, including welders, tool-and-die makers and skilled machinists, and even though the company operates its own training facilities for these jobs, it cannot find enough applicants for the training.

Source: Copyright ©2003 Essential Worker Immigration Coalition. Used with permission.

from the Employment Policy Foundation

7

Immigrant Workers Are Critical to America's Future Growth

Foreign-born workers with the right skills will be crucial to fill jobs left vacant by retiring Americans

Washington, D.C., June 11, 2001—As more than 60 million Americans in their prime working years grow older and retire in the next 30 years, foreign-born workers will be needed to fill job openings and provide critical skills to the U.S. labor market, according to an analysis by the Employment Policy Foundation.

"Immigration has always been important to the U.S. workplace," said EPF President Ed Potter. "Today, foreign-born workers provide 12 percent of the total hours worked in a week. Without their contributions, the output of goods and services in our nation would be at least $1 trillion less. Future economic growth will depend on having enough immigrant workers with the needed skills."

Between 1994 and 2000, the total U.S. labor force grew by 10 million, with nearly 4.7 million supplied by foreign-born residents, according to EPF review of the Census Bureau's Current Population Survey. During that time, unemployment for both groups fell dramatically. The native-born unemployment rate went from 6.3 percent to 4 percent, and foreign-born unemployment improved from 8.3 percent to 4.4 percent.

American and immigrant workers have benefited from bigger paychecks as well. For both groups, earnings grew more than 32 percent in the past six years. In March 2000, the typical full-time employee earned $34,990 per year, while the typical foreign-born worker earned $31,971.

What was striking, Potter said, was that foreign- and native-born workers tended to complement each other in the job market. As native-born Americans seek better education and careers, they tend to avoid jobs in the manual labor service industries. Census and Bureau of Labor Statistics data show that immigrants fill a growing share of low-skill jobs. About 1.1 million new lower-skilled immigrants in the labor force have filled the gap since 1994 as the native-born population attracted to such jobs has declined from 9 million to 7.6 million.

Immigrants also provide an important resource to meet the growing demand for highly skilled college-educated managers, technical specialists, and professionals. Since 1994, the number of foreign-born college graduates in the labor force has increased 43.8 percent to 4.6 million.

"The long-term trend shows that there will be continuing job opportunities for both native- and foreign-born workers over the next 30 years," Potter said. "Employment policy, however, has failed to keep up with this new economic reality."

U.S. immigration policy continues to limit the number of permanent and temporary workers into the country. That policy was created based on a presumption that foreign-born labor is a threat to the job security and incomes of native-born workers. The application process to enter the United States often discourages potential employers from recruiting foreign-born employees despite the need for more workers.

To read EPF's analysis: http://www.epf.org/research/newsletters/2001/pb20010608.pdf

Source: "Immigrant Workers are Critical to America's Future Growth" June 11, 2001 news release.

8

from the American Federation of Labor-Congress of Industrial Organizations

Building Understanding, Creating Change

Defending the Rights of Immigrant Workers

What Union Members Should Know about the AFL-CIO Policy on Immigration

The AFL-CIO proudly stands on the side of immigrant workers. Immigrant workers are an extremely important part of our nation's economy, our nation's union movement and our nation's communities. In many ways, the new AFL-CIO immigration policy signals a return of the union movement to its historical roots. It is increasingly clear that if the United States is to have an immigration system that really works, it must be simultaneously orderly, responsible and fair. The policies of both the AFL-CIO and our country must reflect those goals.

The United States is a nation of laws. This means the federal government has the sovereign authority and constitutional responsibility to set and enforce limits on immigration. It also means our government has the obligation to enact and enforce laws in ways that respect due process and civil liberties, safeguard public health and safety and protect the rights and opportunities of workers.

The Current System Is Broken

Unfortunately, the current system of immigration enforcement, while failing to stop the flow of undocumented people into the United States, is causing workplace discrimination against immigrants and minorities, particularly undocumented workers. The current system leaves unpunished unscrupulous employers who exploit undocumented workers and retaliate against them when they join with other workers to assert their rights, thus denying labor rights for all workers.

This system of workplace immigration enforcement in the United States, with its emphasis on the I-9 system, is broken, targets workers instead of the egregious employers who exploit them and needs to be fixed.

Labor's Principles

We believe the following principles should form our national immigration policy. Specifically:

- Undocumented workers and their families make enormous contributions to their communities and workplaces and should be provided permanent legal status through a new legalization program;
- Employer sanctions and the I-9 system should be replaced with a system that targets and criminalizes employers who recruit undocumented workers from abroad for economic gain;
- Immigrant workers should have full workplace rights, including the right to organize and protections for whistle-blowers;
- Government safety net benefits are important for all workers, and those unfairly taken away by Congress in 1996 should be restored;
- Labor and business together should design mechanisms to meet legitimate needs for new workers without compromising the rights and opportunities of workers already here; and
- Guest worker programs should be reformed but not expanded.

The AFL-CIO supports a broad legalization program that makes no distinction based on country of origin and that allows undocumented workers and their families who have been working hard, paying taxes and contributing to their communities the opportunity to adjust to permanent legal resident status. We should recognize that one of the reasons for undocumented immigration is that our current legal immigration system for family members and for workers is in shamefully bad shape. A broad legalization program providing permanent residence status, rather than a large new guest worker program, should be the focus of our efforts.

The AFL-CIO and its affiliated unions will work vigilantly with our coalition partners representing the immigrant, ethnic, faith, and civil rights communities to ensure that comprehensive legislation providing for legalization and the enforcement of workplace rights for all workers is introduced in Congress and ultimately signed into law.

History has proven that mistreatment of one group in a workplace ultimately will lead to the mistreatment of all workers. We must be mindful of and learn from the history of oppression that many U.S. workers have faced, in particular the long struggle of African American workers. All workers must understand the difference that unions make for workers, whether it is a living wage, better benefits or a safer work environment.

Source: "Building Understanding, Creating Change: What Union Members Should Know about the AFL-CIO Policy on Immigration."

9

from the Cable News Network

Should the President Grant Amnesty to Mexicans Working Illegally in the U.S.?

Aired September 7, 2001—19:30 ET

Tucker Carlson, Co-Host: Tonight, seeing red over green cards. Why are some so angry at efforts by President Bush and President Fox to change immigration laws? Should millions of undocumented Mexicans become U.S. citizens?

Announcer: Live from Washington: CROSSFIRE. On the left, Bill Press; on the right, Tucker Carlson. In the CROSSFIRE, former presidential candidate Pat Buchanan, and Raul Yzaguirre, president of the National Council of La Raza.

Carlson: Good evening and welcome to "CROSSFIRE." President Fox has returned to Mexico, but millions of his countrymen remain in America, many of them illegally. Should they be granted amnesty and allowed to become citizens? That's the position of the Mexican president. A surprisingly broad spectrum of Americans agree, for different reasons.

Big business sees cheap labor. Unions see new members. And both parties see potential voters. Just yesterday, the Senate extended by a year the deadline for illegal immigrants to apply for visas. Liberalized immigration laws may be on the way. But in the background there are rumblings and questions. How would millions of new citizens affect the economy, government social services, and schools? And will the tide stop here? If Mexico, why not the rest of the world? Bill?

Bill Press, Co-Host: Pat Buchanan, welcome back to "CROSSFIRE".

Pat Buchanan, Former Presidential Candidate: Thank you, Bill.

Press: We have seen the dos amigos tour, Jorge and Vicente traveling around the country. What they are saying is there are 3.5 to four million Mexican immigrants here. These are people who own the homes, a lot of them own businesses, their kids are in schools, they're paying taxes, they've got—they're doing good jobs. They are model citizens, in fact, in every way but one, Pat. So why not recognize reality and give these people an opportunity to be here legally?

Buchanan: First of all, there are 21 million Mexican Americans who are here legally and are contributing to this country. You are talking about three to six million illegal aliens. Bill, Mr. Bush—the White House was rolled. For last 10 years there has been an invasion of this country, conducted, aided, and abetted by the government of Mexico, wherein 15 million individuals have been apprehended on our border. Six million have gotten through.

The president of the United States should have called the president of Mexico and told him this is going to stop. The invasion of our country is going to stop, and you are going to help us or you're going to pay a price. Bill, we cannot have an illegal invasion of this country overrunning your borders and basically taking jobs that belong to Americans, 500,000 of which were lost last month.

Press: But Pat, whatever fence is there is obviously not working. Whatever border guards are there are not working. There are four million here. So what are you proposing as an alternative? Rounding them up, turning this into a police state—rounding these people up sending them back in cattle cars?

Buchanan: Bill, when I ran in '92 I recommended a Buchanan Fence at San Diego, remember?

Press: I do.

Buchanan: I was called a nativist xenophobe.

Press: It's not working.

Buchanan: *The New York Times* says that the fence is working. Imperial Beach—they say property values are up. Illegal aliens are not coming across in thousands anymore. They are not coming across at all. They have been moved into the desert. Move the security fence, the border patrol, to the seven major crossing points. All we are asking, Bill, is—we've got the most generous legal immigration policy in the world—defend the borders of the United States. And the chief law enforcement officer, the president, is obligated to do that. And instead of playing footsie with someone who is violating those borders, and that's the president of Mexico and his foreign minister Mr. Castellaneda.

Carlson: Amen. Hello, Mr. Yzaguirre. Thank you for joining us.

Raul Yzaguirre, President, National Council of La Raza: My pleasure.

Carlson: Now, picture this. You are waiting in line in the rain for a play. You have a ticket. As you are standing there, the ticket agent comes out and says, "I'm sorry, you can't come in." Because, it turns out, dozens of other people have snuck in the fire door and to award them we have given them the best seats in the house. So though you have a ticket, go home. That is essentially the position that millions of people in other third world countries who want to come to United States find themselves. They have waited, trying to get here legally. And all of a sudden the president of Mexico trying to convince our president, "end run around them."

Yzaguirre: I think the analogy is wrong. What we are talking about are people not getting in the front of the line, we are talking about people—not an illegal invasion as Pat has consistently said, in trying to awaken fears in the American public—what we are talking about are hardworking people who are paying taxes, who are helping this economy. They are benefiting you and I. They're making it possible for you to send your grandmother to a nursing home to get taken care of.

These are people who are—and everybody agrees. Alan Greenspan agrees, *The Economist* agrees that we have an economy that grew precisely because we had these folks coming into work. I testified this morning before the Senate of the United States with the U.S. Chamber of Commerce. They indicated that we will have a shortfall of five million workers in the near future. And if you stretch it out, 35 million workers, new—more workers, including the ones we already have here.

Carlson: Then wait a second. Why not fill those jobs with people who deserve them? Say—let's say people in Sierra Leone, who are waiting outside the American embassy, who have taken lotteries trying to get here legally. How insulting—how outrageously insulting it is to them to say that people who snuck over the border illegally get your spaces and you don't. You still have to wait in line.

Yzaguirre: Well, what we are talking about are people who are attracted to this country because there are willing employers willing to give them jobs. As long as there is a demand in this country, whether they come from Mexico, Algeria—Canada, which is also a contributor to the flow of workers into this country—as long as there is a demand here, as long as there is willing employers, there will be willing workers.

Press: Pat. Go ahead. Did you want to . . .

Buchanan: Let me respond to that. Look, they are attracted to here, but also they are driven here. The Mexican society and the Mexican government is an absolute failure. You know as well I do three times they devalued their savings. Those people's minimum wage is 33 to 50 cents [an hour]. Mexico is failed, its government is failed, and it is solving its problem by using my country as a spillway. . . .

Yzaguirre: It is my country also.

Buchanan: All right, our country—as a spillway and a flophouse, pushing its unemployed and pushing its jobless and pushing its poor into our country as a safety valve because they can't deal with their own problem. My friend, they can't use our country that way. We've got the most generous immigration policy in the world. 250,000 Mexicans come in legally every year. Now we've got 21 million of them. Good citizens. All we are saying is, if we got immigration laws, for Heaven's sakes enforce them.

Yzaguirre: You know what's happened? We've supported the continual increase of the border patrol. It hasn't stopped the flow immigration. You supported employer sanctions. We were opposed to it. It created massive levels of discrimination against Hispanics. You said that that would stop the flow of immigration into this country. It didn't do so.

Buchanan: Look, Raul, you know as well as I do they are not enforced. I went out there to IBP, the beef packers. Wages have fallen 40 percent, Iowans won't take the jobs because these guys, Bill—and you're a liberal, you ought to be here in on this. These guys have cut the wages 40 percent. So Iowans say, "We don't want them." So these poor fellows come from Mexico and they work on them and they can't do anything because they are illegal. Let wages go up.

Press: Let me come back to this. You know, as a "CROSSFIRE" guest you have just done one of the things—just a couple minutes ago—that most "CROSSFIRE" guests do. They don't answer the question. As a "CROSSFIRE" co-host, Pat, I recognize it and I'm not going to let you get away with it. My question to you was, is all this talk about borders beyond the point. There are $3\frac{1}{2}$ to 4 million here, now. Illegally. What are you going to do about them, Pat? Are you going to turn this into a police state, round them up, send them home in cattle cars? I ask you again.

Buchanan: First, you have understated the number. There are nine to eleven million illegal aliens in this country. Probably six million of them are Mexicans. What do we do here? Here is what you do. First thing you impose employer sanctions. Take one of these big businesses and put the blocks to them that have hired illegals. Fine them big money.

Secondly, anyone who is arrested, put them in prison, send them back immediately. Third, anyone arrested—I don't care if they're spitting on a sidewalk—illegal—send them back. We don't need to do anything right now but pursue the present policy and gradually move them out. Bill, you've got—you cannot let these people who are here illegally compete with five, six million unemployed Americans for jobs.

Press: Here is—that is the fallacy, Pat, to your argument. That you know, most of these people are in low wage and they're in low-skilled jobs that Americans don't want. They're washing cars, they are working in restaurants, they're cutting lawns, they are making beds in houses, those are the jobs ...

Buchanan: Who did those jobs when you and I grew up.

Press: I want you to listen to—I don't know who did it when we grew up. I know who's doing it now.

Buchanan: They were young, weren't they? We did them.

Press: I want you to listen to what President Bush said. Your buddy George Bush said today as President Fox was leaving the White House. He said, "The truth of the matter is that if somebody is willing to do jobs others in America are not willing to do, we ought to welcome that person to the country, and we ought to make that a legal part of our economy."

Buchanan: That is insane. What he is saying is we got open borders. Let me tell you something. Every poor—there are 20 million Mexicans who live on less than $2 a day. He is talking open borders. You bet he can be hired in the United States for three bucks an hour. That is the end of your country, Bill. Your country is more than an economy. It is our family, it is our national home. In this country—your home too—is being overrun by illegals.

Yzaguirre: Why are we calling them illegals?

Buchanan: Because they broke—they broke in line, they broke the law, they broke into the country, they are freeloading on a lot of social welfare benefits. That is why I call them illegals.

Yzaguirre: If you're calling them illegals you've got to call every employer an illegal, every consumer an illegal, everybody who hires them an illegal. Let's—if you are going to paint somebody with that brush, let's paint it across the—let's be equal about it. You are talking about a police state.

Buchanan: I'm not talking about a police state.

Yzaguirre: Of course you're talking about a police state. Draconian steps that would create a police state based on fear-mongering, based on ...

Buchanan: Cut it out, Raul. What I'm talking about is either enforce the laws that are on the books against employers who hire illegals and against law breakers. If you don't enforce, then repeal them. What we cannot do is have the president of the United States, who claims to be the chief law enforcement officer of America, say, "I'm going to get blanket amnesty to three million people who happen to evade the border patrol."

Yzaguirre: You are missing the issue. The issue is what's in the American interest. If the AFL-CIO, the U.S. Chamber of Commerce, *The Wall Street Journal*, the religious community, civil rights organizations, are all united on a comprehensive immigration plan, doesn't that say something about ...

Carlson: If that were true, it might. Let me ask you a question quickly about President Fox, here. President Fox clearly wants United States immigration policy to change, become more liberal. He said the other day he wants this change to take place immediately. Of course, it will help him domestically. What would happen if say three million Salvadorans and Guatemalans and Nicaraguans massed at [the] southern border of Mexico.

Yzaguirre: Why are [you] creating these situations, hypothetical?

Carlson: Because it is an intensely revealing question. I would really like your answer. Because you know as well he would send the army out, wouldn't he? Of course he would. Wouldn't he?

Buchanan: If the army was there.

Yzaguirre: Why, you know, if—if three million Americans ...

Carlson: Open borders with the United States but not with Guatemala. Why? ... I will—I'll tell you exactly why.

Yzaguirre: Ridiculous, an incredibly ridiculous situation. It has no fact in reality. It has no basis in reality. This is—this is part of the scare tactics that have contaminated ...

Carlson: It's a valid question and you are not answering it. The question really is—let me boil it down real quick. The question is, why doesn't Mexico pursue the same immigration policies that it is asking the United States to pursue? Virtually open borders with its neighbors. It is not. Why not?

(Crosstalk)

Yzaguirre: No, the president is very clever. He wants regulated immigration into this country, he wants a comprehensive program that includes a guest worker program that includes legalizing and regularizing folks who are already here, it is ...

Carlson: Who are here illegally.

Yzaguirre: You—you know what?

Carlson: Why do you have to regularize them if they're not here illegally?

Yzaguirre: You defined them as illegally. They did [the] same thing that your ancestors did.

Buchanan: My ancestors came here legally.

Yzaguirre: What's the difference? You can change the laws?

Buchanan: We've already changed the laws. But enforce the laws on the books. Don't you agree with that?

Press: All right, gentlemen. Hold your breath just a second here because we are going to have to take a break. When we come back: as always, there is a political dimension to this issue. Someone asked, [if] George Bush is successful in obtaining amnesty for all these people, will that be a huge political windfall for the Republican party?

(Commercial Break)

Press: Welcome back to "CROSSFIRE." President Fox may have gone back to Mexico but while he was here, he sure has stirred up a cloud of dust over illegal immigration, challenging us to fix the problem by the end of the year by making legal all those now here illegally. His amigo Jorge Bush seems to like the idea. But will Congress go along? And is it best for the United States? Former presidential candidate and "CROSSFIRE" co-host Pat Buchanan says, "No way, Jose." National Council of La Raza President Raul Yzaguirre says, "Bienvenidos." Welcome, Tucker.

Carlson: Welcome, Bill. Raul Yzaguirre, you pointed out earlier that there is something of a consensus forming on immigration reform and that labor is behind it, obviously. Pathetic and marginalized, looking for members. That both parties looking for votes are for it. Big business, which is always looking for cheaper labor, is for it. But it turns out the average person isn't for it. There is [a] new Harris Poll, fascinating results. Let me just read two of them to you. 65 percent of people asked believe that immigration was a net drain on the country and its social services. Only 30 percent of Americans—ordinary Americans asked—were for a kind of amnesty for Mexican immigrants. What do ordinary people know that union bosses don't?

Yzaguirre: Well, ordinary people know what's right. And we do you a different kind of a poll when you explain to them the facts as Celinda Lake and others have done, where they said, "this is what immigrants contributed to this country." They are creating a—contrary to what Pat Buchanan says—they are creating jobs, not taking away jobs. They do so because, they save a common industry that allows it to be stay—to stay in this country and occupy accountants and managers and so forth, so they do create jobs.

But contrary to that poll, we find that when you explain to the American people what a limited legalization program is about, what immigrants do and that when the facts—what the facts are surrounding this issue, over two thirds support immigration and support a legalization.

Carlson: But wait a second. Look, people sense that a recession is coming. A downturn is already here. Why is it unreasonable for them to think, gee, the jobs immigrants take could be my job? This year over 195,000 high-tech H1V visas were issued, as you know. In the same year, the tech sector goes poof. Is this a good thing for American citizens? Of course not.

Yzaguirre: Tucker, the research shows very, very clear that we are talking about paltry jobs, labor jobs, field jobs that Americans do not want and will not take. So [they] are helping our economy. And by the way, the 4.9 unemployment rate used to be considered full employment.

So even in bad times, even in high-unemployment times, there is going to be a great deal of need for these workers to continue to sustain our economy. And we also need to look at the long run. We are going to have a worker shortage. And Mexico, contrary to what Pat is saying, in fact is going to be approaching a time when its birthrate is decreasing and its job creation rate is increasing, so they will be having some significant shortage of labor themselves.

Buchanan: Let me say here that the population of the world is going to increase by three to four billion in the next 50 years. That is 30 to 40 Mexicos in the next 50 years. If we don't get control of our borders, Raul, we are going to lose our country.

Press: I want to ask you about the politics of this a second. I remember in California, in particular, when I got started in politics the Latino vote was a Republican vote. They're small businessmen, they're homeowners, they're conservative, they're Catholic. And then Pete Wilson came along and declared war on them with Proposition 197. And now, for Democrats, the Latino vote is the gift that keeps giving.

If the Republicans in Congress and conservatives in Congress wage war on the Mexican community again and refuse to grant amnesty, aren't they just saying, "Here Democrats, [you can have] this gift for the rest of your political life."

Buchanan: I hope Republicans will put their country ahead of their party. However, Pete Wilson was the last Republican to win statewide—if I recall correctly—in California. He came back—what [was] he, down to 20 percent when you guys had him? And he came back and won that race big time.

Press: And a Republican hasn't won since. That's the point.

Buchanan: Here's the reason. Dole and Kemp, running against Mr. Clinton. First time Hispanic voters, those who were poor and just came into the country recently—Clinton won them 91 to six, first-time Hispanics because they are very, very poor and they worry about social services.

Six of the seven largest states in the country [have] the most immigrants: Massachusetts, New York, Illinois, New Jersey, California, Texas, Florida. Clinton won six out of seven. Out of 15 states with the smallest immigrants, Bush won 14 out of 15, losing only Maine. The Hispanic vote—if they keep this up—is going to kill the Republican party.

Press: You said something that I want to come back to, which is what's good for this country. I think it took President Fox today maybe to remind us what's at the very heart of this issue, and that is who we are as Americans. I would like you to listen to what President Fox told a joint session today—or yesterday. Here it is.

(Begin Video Clip)

Vicente Fox, President of Mexico: Many among you have a parent or a grandparent who came into this country as an immigrant from another land. Therefore, allow me to take this opportunity to pay homage to those brave men and women who in the past took on the challenge of building a new life for themselves and for their families in this country.

(End Video Clip)

Press: Isn't that the point, that we are all immigrants? Why don't we recognize that and welcome these people?

Buchanan: We're not all. We're the children of immigrants. We have the most generous immigration policy in the world, we take in more legal immigrants than all other countries in the world combined. Mr. Fox—I'll tell you what he ought to do. He ought to put his army on that border [and] stop those poor Mexican folks who are dying in the desert of Mexico and the United States trying to evade our laws and our border patrol. They are dying!

What has he and Castaneda done for them? Nothing. You know what they're doing? They are giving them survival kits with condoms and meat and water so they can sneak into the United States. They are dying. If he cares about them, why doesn't he take care of them?

Press: Raul?

Yzaguirre: He is taking care of them. He has identified 250 areas that send immigrants into the United States. He has an economic development plan, he talked it over with President Bush about that. He is using remittances back to the country to do development, he is on the path of job creation. And he's making a very fundamental point. We need to control these borders jointly, as a partnership. We need to have a policy that is a united policy.

Buchanan: Tell him to give us a hand.

Carlson: OK. We are going to have to leave it there. Raul Yzaguirre, Pat Buchanan, thank you both very much. Bill Press and I (are) already safely in this country. We'll debate—we'll be back in a moment with closing comments.

(Commercial Break)

Carlson: You know who I feel sorry for, Bill? I feel sorry for the millions of people in Africa and Asia, clustered outside American immigration centers, believing that if they play by rules and do everything right, they can come to this land of promise—who are about to find out no dice. You don't have a political lobby powerful as Mexico does. You are just going to be bumped out of line by better lobbyists.

Press: There's a little problem you're overlooking, which is an ocean they can't walk across. I have one thing to say, Tucker. Give me your tired, your poor, your huddled masses yearning to be free. That is what this country is all about.

Carlson: That's right. And don't give me your lawbreakers. Send them back! Give me those people who want to be here, who want to go by the rules, who want to obey the law. Those are the immigrants I want.

Press: Isn't it obvious to you the fact that four million people are here illegally means the law ain't working. Law enforcement is not the answer. You cannot solve it by "I am sending them all back."

Carlson: So people who shoplift, so they ought to be able to keep the goods because shoplifting has always been very effective.

(Crosstalk)

Press: Here we go, no? Welcome. From the left, I'm Bill Press. Have a good weekend. Good night from "CROSSFIRE."

Carlson: From the right I'm Tucker Carlson. See you next week on "CROSSFIRE."

Source: Transcript of "Should the President Grant Amnesty to Mexicans Working Illegally in the U.S.?" Crossfire, September 7, 2001.

10

from the Hudson Institute

Good Fences

By Herbert I. London
July 1, 2003

Recently I attended a Washington, D.C. meeting devoted to American-Canadian relations. A former Canadian ambassador began the proceedings by describing how alike the two nations are and in "big think" terms recommended the harmonization of laws in order to promote even closer ties than now exist.

His speech was elegantly delivered and well received. He noted the recent tension over foreign policy and the longstanding ties between friendly neighbors. He spoke of the successful implementation of NAFTA and the manifold ways in which Canadians and Americans interact and trade.

Yet when all of these points were considered, I found myself unpersuaded and confounded. The United States and Canada are indeed alike in many ways, yet different somehow. Their respective histories came from the same root, but the branches went in different directions.

The Friendship Arch on the border at Blaine, Washington, reads, "Children of a Common Mother." However, children from the same family are not the same. Canada sought to legitimate its link to British monarchy, while the United States has generally been suspicious of state power.

Yet it is remarkable that the 4,000-mile border has been peaceful throughout the history of the two nations. It is the longest undefended border in the world. Eighty percent of the Canadian population lives within 100 miles of the United States. That's the good news.

Since September 11, 2001, a lot has changed. The border of peace also represents a portal for terrorist activity. Lax immigration laws make Canada a congenial center from which to launch attacks against the United States. Moreover, despite efforts to increase security at border points, there is still disagreement over how much security is necessary.

At the moment, Canada accepts twice as many immigrants and four times as many asylum seekers—as a proportion of its population—as the United States. Canada allows consulate members abroad to issue visas—a policy the United States reproves. New arrivals often claim asylum status without any documentation and, in a matter of weeks, can obtain a driver's license and insurance.

In addition, British Columbia has introduced legislation decriminalizing marijuana use. How would Canada and the United States harmonize drug abuse policies in the northwest? The Canadian courts are imposing gay marriage on the country. Does that mean the United States would be obliged to follow suit if laws were harmonized?

If the Americans look north, what they see is a nation that criticizes U.S. foreign policy, yet relies on American defense expenditures for its own protection. Canada routinely excoriates U.S. health-care policies, but when a wealthy Canadian needs surgery, he invariably heads south in order to avoid his own socialistic system.

What Americans also see is a nation that often disregards document fraud, lax visa practices, and an asylum policy devoid of restrictions. Terrorist groups have the freedom to raise money and melt into the many immigrant enclaves. Canadian legal practice usually involves endless litigation favoring criminals.

Most significantly, none of these conditions are likely to change. Canadians jealously guard their sovereignty. In discussion of Canadian policies, Americans are told "we will cooperate to the degree a policy shift makes sense for Canadians. Washington doesn't make policy north of the border." Indeed it does not and that is the problem with legal and policy harmonization.

There is only so far you can go before you run headlong into a dispute. Canada may look and often act like the United States, but it is different. Americans should be thankful that the border is peaceful, but we are no longer thankful it is porous.

When it comes to diplomatic relations, conditions are vastly better than with Western Europe, but they are not what they once were and not where Americans would like them to be. It is one thing to raise the issue of legal harmonization; it is quite another matter to put it into effect. As this American listened intently to the harmonization proposal, all I could think was, "not for us."

Source: Herbert London, "Good Fences," *Ameican Outlook Today*, July 1, 2003 from The Hudson Institue, hudson.org/index.cfm?fuseaction=publication_details&id=2911

from Canadian Council for Refugees 11

Questions about Canada's Security Agenda ... and its Impact on Refugees & Immigrants

Security measures adopted in the wake of September 11, 2001, have made a difference in the lives of refugees and immigrants.

As Canadians we must decide whether we have responded appropriately to the security challenges. **WHERE DO YOU STAND?**

Does Immigration Make Canada Vulnerable to Terrorism?

There is no connection between immigration and terrorism. In fact, it is arguable that a positive immigration program actually discourages violence by promoting an open, diverse, dynamic and tolerant society, with opportunities for all.

Immigrants come to Canada seeking a better quality of life and a chance to bring up their families in freedom and peace. Like Canadians born here, immigrants want security for themselves and for their society.

Linking immigration with terrorism has a very damaging effect on immigrants, making them feel that they are always under suspicion. This is particularly the case with Muslim and Arab immigrants, who have been the chief targets of such suspicions.

Are Canada's Immigration Laws Too Lax?

Canada has in place a rigorous immigration system that gives priority to keeping out anyone who might be a security threat.

The *Immigration and Refugee Protection Act* contains a whole series of provisions making people inadmissible on criminality and security grounds. They cover every conceivable security threat. The problem is not that these provisions are too narrow, but rather that they are too wide and therefore penalize many innocent people. For example, all past and current members of Nelson Mandela's party, the African National Congress (ANC), are inadmissible to Canada on security grounds and can only enter Canada with a special waiver. In the summer of 2003, the immigration law allowed the arrest and public labelling of 23 South Asian immigrants as "suspected terrorists" based on the flimsiest of evidence (as part of "Operation Thread"). It soon became clear that the suspicions were unfounded and the accusations relating to terrorism were dropped. Yet many of their lives were drastically affected by being publicly associated with terrorism.

Canada's security processes for immigrants include the use of the "security certificate", a measure that many experts feel unnecessarily sacrifices the rights of individuals. The security certificate process allows the government to arrest, detain and deport immigrants on security grounds, without ever showing them, or their lawyers, the evidence against them.

Canada's immigration laws are very strict when it comes to excluding terrorists, but weak when it comes to protecting the rights of non-citizens.

Do We Need to Tighten Immigration Controls in Order to Promote Our Security?

It is a mistake to think that security can be achieved by building stronger walls. Canada is part of the global community and we must work collectively to promote global security. Since there is no connection between immigration and terrorism, tightening immigration controls will not improve security; all it will do is hurt immigrants.

There are serious security challenges facing the world. They need to be addressed by focusing on the real threats, not by letting ourselves get distracted by controlling immigration.

Are Canada's Immigration Controls Weaker Than Those in the U.S.?

Canada's immigration controls are in some ways actually tougher than those in the U.S. Many people report finding it easier to get into the U.S. than Canada. Often people refused a visa to Canada are granted one to the U.S. A significant proportion of refugee claimants who come to Canada arrive via the U.S. (72% of claims made at an airport or land border in 2003).

According to conservative estimates, there are some eight million people living without any status in the U.S., equivalent to a quarter of the population of Canada. The size of the undocumented population in the U.S. suggests that immigration controls in that country are not particularly tight. But the large undocumented population hasn't threatened the security of the U.S.: despite large scale arrests, detentions and registration programs, no terrorists have been found to be hiding among the undocumented in the U.S.

Some people in authority in the U.S. have accused Canada of having lax immigration controls. These accusations seem to be based either on misinformation or on a desire to find someone to blame.

Should We Harmonize our Immigration Policies With Those of the U.S.?

There is much in U.S. immigration policies of which Canadians should be very wary, including a wide range of draconian and unfair measures. These include the wide use of detention, often in degrading conditions. The U.S. also has discriminatory policies that apply special measures to immigrants based on their nationality, something that is unacceptable in Canada. Given the power relationship between the two countries, harmonization would inevitably mean Canada adopting U.S. policies, without Canadians' elected representatives being able to influence decisions.

The case of Maher Arar is a clear warning about the dangers in the U.S. system. Even though he is a Canadian citizen, Mr. Arar was deported by the U.S. to Syria where he was imprisoned without charge and tortured.

Should We Be Worried About Refugee Claimants Entering the Country Without Identity Documents?

It is far more difficult to enter Canada as a refugee than as a visitor. Refugee claimants make up only one-tenth of one percent of the visitors and immigrants entering Canada each year. Sophisticated wrong-doers are extremely unlikely to choose to go through the refugee claim system, which involves fingerprinting, photographing and interviews. Those involved in the September 11 attacks all seem to have entered the U.S. on visitor or business visas. They didn't make refugee claims.

Many refugees arrive without identity documents because it would be dangerous for them to carry identification while they are fleeing persecution, especially in ethnic conflicts. Others, such as Somalis, come from countries where there is no government left to issue documents. Since 1993 the Canadian government has required Convention refugees to produce identity documents in order to obtain permanent residence. The effect has been to put thousands of refugees in long-term legal limbo, without permanent residence status and all its accompanying rights. On the other hand, years later no one has been able to produce any evidence that criminals or security threats have been hiding in this group of refugees.

Would it Be Safer to Detain Refugee Claimants on Arrival?

Detaining refugee claimants is not a logical or effective way of fighting terrorism. It would be equivalent to arresting everyone found near the scene of a crime on the off-chance that one of them might be guilty of the crime.

In 1996 the U.S. adopted laws which resulted in massive increases in immigration detention, including detention of asylum seekers. These measures did nothing to protect the country from the September 11 attacks. On the other hand, many genuine refugees have had to spend months and even years in detention, often in appalling conditions.

Under current laws, Canadian immigration officials are fully empowered to detain any refugee claimant or other migrant who appears to present a security risk. Detaining refugee claimants *en masse* would do nothing to improve our security, but would be fundamentally unjust and a disgrace to Canada. Refugees are fleeing serious human rights abuses and seeking our protection: we owe it to them not to put them behind bars unless absolutely necessary.

Isn't It Easy to Be Accepted as a Refugee in Canada, Even if You Are a Terrorist?

It is not easy to be accepted as a refugee in Canada. Even before a refugee claim is considered, the claimant is screened for security and criminality. Having passed this initial screening, each claim is individually scrutinized and each claimant interviewed, often at great length, by the Immigration and Refugee Board (IRB). The Immigration and Refugee Board's documentation centre is highly regarded internationally and is probably one of the best in the world. Claimants have to respond to any contradictory evidence or apparent inconsistencies in their testimony.

While the refugee determination system is designed to find out who is a refugee, not who might be a terrorist, the Refugee Convention explicitly excludes from the refugee definition people who have committed serious crimes: where there is evidence of this, claimants will be denied refugee status.

Should We Be Concerned About the Large Numbers of Refused Claimants That Are Not Deported?

Refused refugee claimants do not in themselves represent a security threat, any more than a group of Canadian citizens does. Many of the refused claimants are families with children. Where an individual case raises security concerns, the Canadian government can use special measures to deal with it, including detaining the person. Many refused claimants come from countries at war or where there is unrest or public disorder (such as Afghanistan, Iraq or the Democratic Republic of Congo). Because sending people to these countries would cause significant hardship, the government does not immediately remove people there, unless there are criminality or security concerns.

Are Refugees Being Scapegoated in the Aftermath of September 11?

The attacks of September 11, 2001, had absolutely nothing to do with refugees, and yet since that date our refugee policies and refugees themselves have come under attack. We have to ask ourselves why. Is it because refugees are easy targets and because people's fears are ill-informed by stereotypes and racism? Refugees are among the most vulnerable people in our society: their own government was unable or unwilling to protect them and on arrival in Canada as claimants they have no status here. They depend on Canadians' sense of justice and hospitality to ensure that their basic rights are respected.

During the Second World War, mass internment of Japanese Canadians made them victims of Canadians' desire for security. Their mistreatment was clearly fed by racist prejudice. This chapter in our history is now recognized as one of the most shameful. How do we ensure that we don't make the same mistakes today in our treatment of refugees?

Source: "Questions about Refugees and Refugee Policies Post-September 11" at web.net/~ccr/sept11qs.htm

Review Questions

1. How is economic migration related to population growth and decline?
2. How are the United States and Canada alike in their immigration policies and how do they differ?
3. Why are immigration debates in North America generally not as politically charged as they are in Europe?
4. What are the differences in the ways in which Sweden and Switzerland have attempted to bolster their workforces with immigrant workers?

Discussion/Essay Questions

1. Consider the case of Pedro, the Mexican-American immigrant whose story was presented at the beginning of this module. In many ways, he is the model immigrant promoted by pro-immigration advocates. He is a law-abiding, hardworking family man who contributes to the American economy and makes no demands of the country. However, his undocumented status prevents him from expanding his business or paying income taxes. He seems like a poster case for relaxing immigration laws and/or providing amnesty to undocumented aliens already in the country.

   On the other hand, he is also the kind of person who is cited by opponents of immigration. If he were not around, his job at the hardware store probably would have gone to a local African-American. Furthermore, Pedro has *not* chosen to break his ties to Mexico and become a fully assimilated American. While this may in part be due to legal barriers, it appears as though Pedro and his family would remain Mexican first and American second even if his status were legalized and his family moved to the United States.

   Use the readings presented in this module to construct an argument either for or against granting amnesty to Pedro.
2. The European Union has recently been expanded to include more countries in Eastern and Central Europe and further expansion is being contemplated. Opponents fear a mass migration of individuals from the labor-surplus countries of Eastern and Central Europe to the wealthier countries of Western Europe. Supporters of expansion note that only through this migration could the industrial powerhouses of Western Europe be expected to compete with areas such as North America that always have been relatively open to the immigration of low-wage labor. Based on the

arguments presented in the readings, make an argument for or against expansion of the European Union and the freedom of individuals from member nations to migrate freely within the EU.

3. Fear of immigrants, whether or not it is well founded, seems to be behind much of the opposition to immigration. This includes fear of what immigrants will do to a country's culture, what they will do to its economy, and what they will do to its political stability. If you were the spokesperson for a business group supporting immigration (like the Essential Worker Immigration Coalition or the Employment Policy Foundation), what policy would you propose for reducing local residents' fears: the German "guestworker" model, the Swedish "assimilation" model, or something all together different?

List of Readings

1. Steve Brady, "Patterns of Prejudice," posted on the website of the National Front (revised version of an article that appeared in *Vanguard* magazine April 1987), *http://www.natfront.com/prejudic.html.*
2. Excerpts from "Program of the Austrian Freedom Party," adopted October 30, 1997, published on the website of the Freiheitliche Partei Österreichs, *http://194.96.168.106/fileadmin/pdfs/partieprogamm_eng.pdf.*
3. "Pim Fortuyn: Man of Paradox," May 9, 2002, posted on the website of the Cable News Network, *http://edition.cnn.com/2002/WORLD/europe/05/06/fortuyn.profile/?related.*
4. Patrick J. Buchanan, "To Reunite a Nation," transcript of a speech delivered at the Richard M. Nixon Library, Yorba Linda, California, January 18, 2000, posted on the website of VDARE, a project of the Center for American Unity, *http://www.vdare.com/pb/speech.htm.*
5. "What Is the Coalition for the Future American Worker?" posted on the website of the Coalition for the Future American Worker, *http://www.americanworker.org/whatis.html.*
6. "Documenting the Labor Shortage," report published by the Essential Worker Immigration Coalition, March 2002, posted on the website of the Essential Worker Immigration Coalition, *http://www.ewic.org/documents/EWIC_Documenting_Labor_Shortage_%203_02.pdf.*
7. "Immigrant Workers Are Critical to America's Future Growth," news release from the Employment Policy Foundation, June 11, 2001, posted on the website of the Employment Policy Foundation, *http://www.epf.org/news/nrelease.asp?nrid=77.*
8. "Building Understanding, Creating Change: What Union Members Should Know About the AFL-CIO Policy on Immigration," posted on the website of the American Federation of Labor–Congress of Industrial Organizations, *http://www.aflcio.org/issuespolitics/immigration/upload/AFLCIOPO.pdf.*
9. "Should the President Grant Amnesty to Mexicans Working Illegally in the U.S.?" transcript from CNN *Crossfire*, September 7, 2001, posted on the website of the Cable News Network, *http://www.cnn.com/TRANSCRIPTS/0109/07/cf.00.html.*
10. Herbert London, "Good Fences" in *American Outlook Today*, July 1, 2003, posted on the website of The Hudson Institute, *http://www.hudson.org/index.cfm?fuseaction=publication_details&id=2911.*
11. "Questions about Canada's Security Agenda. . . and It's Impact on Refugees & Immigrants," posted on the website of the Canadian Council for Refugees, *http://www. web.net/~ccr/security.htm.*

Websites for Additional Research

1. The website of the Migration Policy Institute, *http://www.migrationinformation.org/index.cfm*, offers a wealth of data about immigration around the world. You can read about migration-related stories in the news, search an archive for past news topics, or search the database for the number of individuals born in a particular country who are living in a certain U.S. state.
2. The British newspaper *The Guardian* maintains a website with links to the official pages of several European countries' far-right/anti-immigrant political parties at *http://www.guardian.co.uk/farright/subsection/0,11981,711842,00.html.* (Many of the linked party sites are not in English.)
3. One of the largest U.S. anti-immigration groups is the Federation for American Immigration Reform. Besides presenting the anti-immigration argument, their website, *http://www.fairus.org*, also contains many statistics on immigration to the United States and updates on legislative initiatives. For the other side of the story, see the website of the Essential Worker Immigration Coalition at *http://www.ewic.org.*

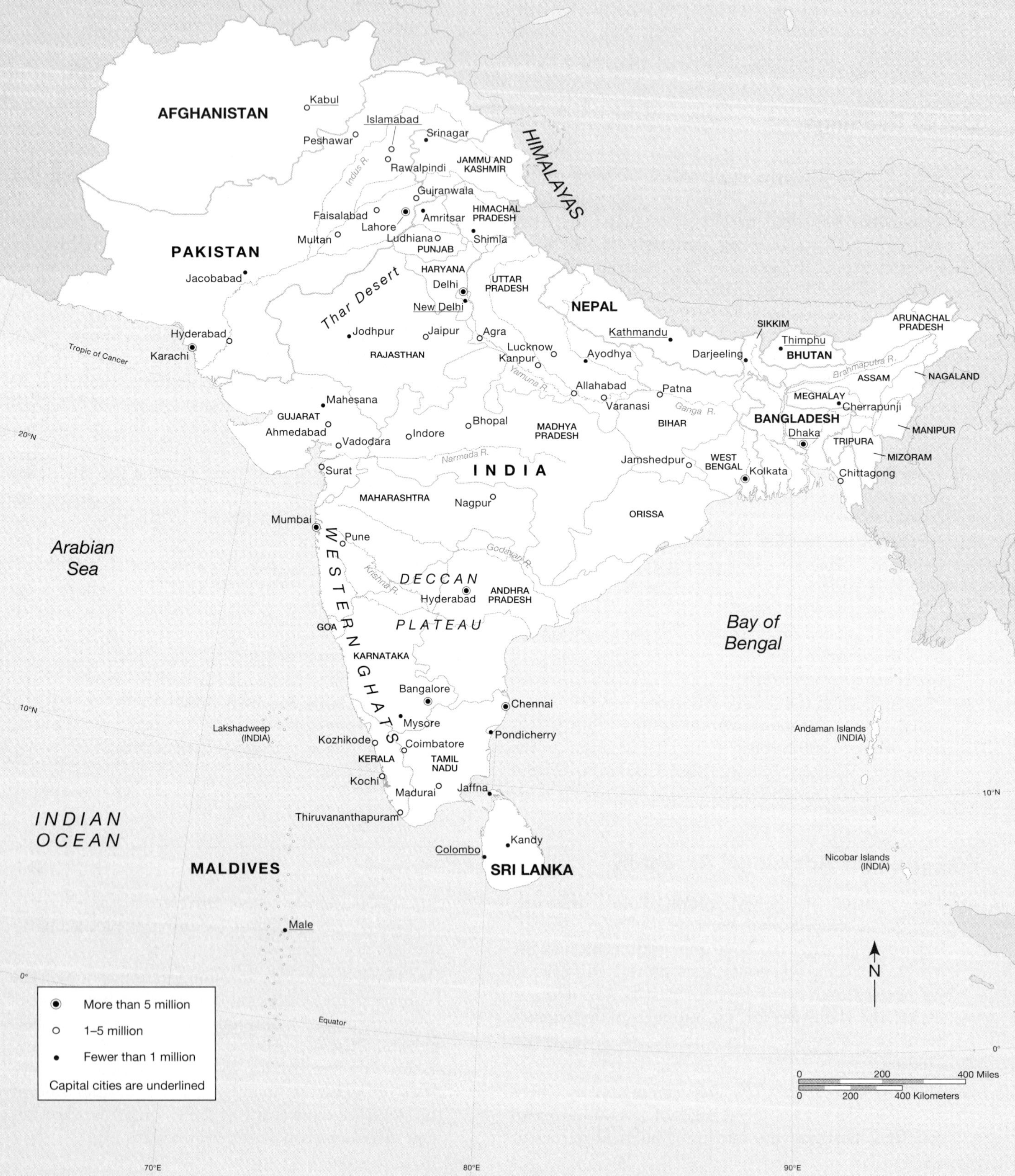

AFGHANISTAN
Kabul
Islamabad
Peshawar
Srinagar
Rawalpindi
JAMMU AND KASHMIR
HIMALAYAS
Indus R.
Gujranwala
Faisalabad
Lahore
Amritsar
HIMACHAL PRADESH
Multan
Ludhiana
Shimla
PUNJAB
PAKISTAN
Jacobabad
HARYANA
Delhi
UTTAR PRADESH
Thar Desert
New Delhi
NEPAL
Hyderabad
Jodhpur
Jaipur
Agra
Kathmandu
SIKKIM
ARUNACHAL PRADESH
Tropic of Cancer
Karachi
RAJASTHAN
Lucknow
Kanpur
Ayodhya
Darjeeling
Thimphu
BHUTAN
Brahmaputra R.
NAGALAND
Yamuna R.
Allahabad
Patna
ASSAM
Mahesana
Varanasi
MEGHALAY
Cherrapunji
GUJARAT
Ganga R.
BANGLADESH
Ahmedabad
Bhopal
MADHYA PRADESH
BIHAR
MANIPUR
20°N
Indore
Dhaka
Vadodara
TRIPURA
Narmada R.
MIZORAM
Jamshedpur
WEST BENGAL
Surat
INDIA
Kolkata
Chittagong
MAHARASHTRA
Nagpur
ORISSA
Mumbai
Pune
Arabian Sea
WESTERN GHATS
Godavari R.
Krishna R.
DECCAN
Hyderabad
ANDHRA PRADESH
Bay of Bengal
GOA
PLATEAU
KARNATAKA
Bangalore
Chennai
10°N
Mysore
Lakshadweep (INDIA)
Pondicherry
Andaman Islands (INDIA)
Kozhikode
Coimbatore
KERALA
TAMIL NADU
Kochi
Madurai
Jaffna
10°N
Thiruvananthapuram
INDIAN OCEAN
Kandy
Colombo
SRI LANKA
Nicobar Islands (INDIA)
MALDIVES
Male
N
0°
More than 5 million
1–5 million
Fewer than 1 million
Capital cities are underlined
Equator
0°
0
200
400 Miles
0
200
400 Kilometers
70°E
80°E
90°E

Civil Society, Social Movements, and Gender in South Asia

Companion to Chapter 4 Folk and Popular Culture

Manjamma has four daughters, ages 1 to 8. She and her husband worked in the fields all day. She had to leave the children in the care of the 8-year-old to go to work. She worried about their safety, but her family makes so little money she had no choice.

When she decided to attend a women's meeting in her village, her father-in-law, who lives with them, bellowed in rage that a woman's place is in the home. She had no business leaving her family just to meet with other women. Manjamma risked his wrath and went anyway.

The meeting led to the formation of an Opportunity Loan Group, and Manjamma soon had a $111 loan. She bought a milk cow so her children could have milk. She sells the rest for a steady monthly profit of $22. Her father-in-law still did not approve of the women's weekly meetings, but he enjoyed the milk. He grumbled, but he did not yell at her.

Community loan groups enable South Asian women to provide for their families and empower themselves. *Source:* Mark Edwards/© Still Pictures/Peter Arnold, Inc.

He might still be grumbling, but one day he doubled over in pain. Manjamma quickly sold some milk and gave him money for the doctor. It was the first time the family did not have to borrow from the local loan shark to meet an emergency.

Manjamma wants to give a milk cow to each of her daughters when they are young women so they will have some security. Her father-in-law thinks that is a fine idea.

Source: "A Woman's Place is at Her Meeting," from povertyfighters.com. ■

Culture, Civil Society, and Social Movements

As the textbook notes (in Chapter 1), culture is "the body of customary beliefs, social forms, and material traits that together constitute a group of people's distinct tradition." A culture thus may include a very specific combination of religion, diet, dress, language, rules of behavior, affiliations, and lifestyle, or only a single attribute such as allegiance to a sporting team. Essentially, culture is everything you have learned since you were born.

In recent decades, the effects of new technologies, increased global trade, and continuing migration have had a homogenizing effect on the world; many traditional cultures are being eroded. Although the transfer of ideas and goods is not new, it now is occurring on an unprecedented scale.

Some argue that we are now entering an era of "global culture," or "McWorld." Big Bird has invaded China; children everywhere sing along with Barney the purple dinosaur; if you travel wearing thick, round, black glasses you can expect children in the most remote corners of the world to point at you and shout "Harry Potter!" Foods provide a great example of this globalization of culture: "Mexican" salsa is the number one condiment in the United States (ahead, even, of ketchup); fish and chips has been replaced by "Indian" curries as the top take-away dish in the United Kingdom. At some point, salsa stops being a "Mexican" food and curry stops being an "Indian" food (many Americans no longer think of pizza as an "Italian" food; they expect it to be on the menu at "non-ethnic" restaurants and don't assume that it will be on the menu at Italian restaurants). Of course, in the process of crossing borders and becoming a new cultural symbol, the content of the cultural symbol itself changes, to affect a compromise with the culture that it is joining. Thus, for example, most processed salsa sold in the United States contains vinegar and sugar, two ingredients rarely found in fresh Mexican salsa (and, not coincidentally, as a result, most United States salsa is considerably more ketchuplike than the Mexican original).

As globalization exerts power on local and national societies to adhere to new norms and standards, culture frequently becomes a battleground of ideas. The domain of civil society—roughly, all aspects of society not directly associated with state power or economic processes—becomes a site of contestation as individuals and organizations attempt to create new (or maintain old) cultural bonds and norms that may unite a dynamic society. Social movements of civil society often are particularly strong in countries where Western capitalist modes of production and modern values and goals are permeating areas that still maintain a more "traditionalist" social system. Amid this disjuncture—when the practices of culture fail to support the realities of society, and vice versa—individuals form social movements to reshape culture on their own terms. To further investigate civil society and social movements as agents of cultural change and resistance, this module focuses on social movements that are challenging gender relations in South Asia and, in particular, India.

Civil Society and Cultural Politics

In some senses, the boundaries that we draw between the spheres of politics, economics, and culture are artificial. Cultural change often is advocated as a means toward changing state policy or economic conditions, and movements often seek political power so that they can use that power to change cultural norms. For instance, a group opposing welfare programs may focus its efforts on creating a cultural norm that values independent self-reliance and devalues reliance on "handouts." To achieve this change in cultural norms, however, the group may seek to change government policy (reduction of welfare payments). At the same time, as the group attempts to bring about change in the political and cultural arenas, it also may employ economic arguments (for example, the group may argue that a reduction in welfare payments will increase national productivity, which will increase national wealth, and that this will permit a reduction in tax rates, which will increase individual wealth). Should such a group be characterized as cultural, political, or economic? Clearly, it is all three. Although this group seeks political and economic change, we classify it as being *based* in civil society, because its roots are neither in an organization attempting to control state power (such as a political party) nor in a group seeking economic power (such as a corporation or a labor union).

Whether or not a social movement attempting to influence civil society attempts to change state policy is, in most cases, a matter of strategy. Some social movements operate in both the state and civil society arenas; for instance, many social movements are closely linked with political parties. Others engage elected officials and policymakers when necessary, but mostly operate on their own. Still others believe that the state will always be an enemy and they see no reason to waste scarce resources attempting to influence state policy or capture state power. Our point here is that, for many activist groups, the state is at best a useful political ally; for these groups, the ultimate goal is change in the sphere of civil society.

Some argue that the domain of civil society is particularly relevant today, as cultures around the world attempt to cope with outside influences and internally generated transformations. When societies and cultures undergo change, the state and market institutions that normally give a society much of its cohesion and cultural meaning lose some of their power. No matter how much a government wishes to legislate a new household ethic (such as suggesting that husbands share in the housekeeping because their wives are now working at paid jobs away from home), such changes in household behavior (which require substantial cultural change) will be implemented only if they emerge from the ground up. Likewise, certain necessary social practices—such as child rearing—are almost always money losers in industrial and postindustrial societies, so these societies require some organizing force beyond the market to build cultural norms that value family reproduction.

Looking at a society from the outside, we may be able to say that, for the society as a whole, cultural change is *necessary*. However, in almost all instances, some individuals and institutions will resist cultural change. Others will have divergent views regarding what kinds of change should be implemented. In short, efforts to define and transform culture nearly always are contested.

When individuals get together to influence and shape culture, they form a social movement. Because there is no such thing as a perfectly stable social structure there will always be social movements, whether fighting for the interests of women, animals, gun owners, political prisoners, or any other cause that enough people believe is worthy of their time and energy. In short, civil society is wrought with a cascading set of struggles that collectively can be characterized as cultural politics.

Social Movements and Gender

Gender equity (or women's rights) is among the issues frequently taken up by social movements. Although movements seeking to change the status of women in society may involve attempts to change state policy, at their heart these struggles are rooted in an effort to change cultural norms, and thus these struggles are based in the realm of civil society.

There are probably two reasons why women's struggles typically concentrate less on achieving state power and more on changing cultural norms. The first reason is that, while women traditionally have been shut out of most positions of power within the state (as well as within economic organizations), they have long held positions of responsibility and leadership in the less formalized world of civil society. For instance, the lay leadership in many religious institutions is female (if you grew up attending a religious institution, it's likely that your church, synagogue, or mosque could not have functioned without a group of dedicated, strong women, even if the formal spiritual leader of the congregation was male). At the household scale, too, mothers are the ones who usually are credited with "holding the family together." Some scholars, from both feminist and antifeminist perspectives, argue that women's tendency to nurture stems from their biological role in childbearing and feeding, while others hold that the association between women and nurturing has nothing to do with any natural difference between men and women.

Regardless of whether women are "naturally" suited for these roles, in practically all societies women are assigned primary responsibility for serving these valuable household-reproduction functions. Even if women are not rewarded monetarily or given opportunities to translate their household responsibilities into *state* power, women can translate these responsibilities into political power within civil society. Indeed, when making forays into the political arena, women often build on the authority that they already have in the household (for example, *Mothers* Against Drunk Driving). Women are particularly active in social movements that expand their home-maintenance role to quality-of-life issues in the broader "home-place" of the community. For instance, women are often the leaders in struggles to prevent the siting of hazardous waste facilities in residential communities where they could cause a health threat, especially to children.

A second reason why women's political efforts are concentrated in social movements outside the state arena is that much of the explicit violence and discrimination against women takes place within the home, an area where the state has little authority or power. If home-based violence against women is to be combated or if men are to be pressured to perform more child-care duties, the most effective route will probably be through cultural change, which is more likely to come about through the social movements of civil society than through government mandates.

South Asia as Cultural Battleground

South Asia is a region of many distinct, yet overlapping, cultural traditions. India, the second most populous country in the world, with almost 17 percent of the world's population, is the heartland of Hinduism and Sikhism and the birthplace of Buddhism. India, along with neighboring Nepal, are the world's two majority-Hindu countries. Pakistan, the seventh most populous country in the world, and Bangladesh, the eighth most populous, are the world's second- and third-largest predominantly Muslim countries (after Indonesia). Sri Lanka and Bhutan are predominantly Buddhist, although Sri Lanka has a significant Hindu minority, while the remaining two—Afghanistan and Maldives—are predominantly Muslim. The three large countries (India, Pakistan, and Bangladesh) alone account for more than 21 percent of the world's population. The Population Research Bureau predicts that by 2050 India will have surpassed China to become the world's most populous country, and Pakistan will have become the fourth most populous country in the world (behind India, China, and the United States). The region as a whole is predicted to have around 24 percent of the world's population by 2050.

Long before the arrival of armed European fleets at the end of the fifteenth century, South Asia was a well-integrated region. There was frequent trade with China and Southeast Asia to the east, and Arab traders brought Indian goods westward to Africa and Europe. In the centuries that followed their arrival, Europeans used their military might to take over these trade routes, and, following this commercial conquest, Europe began to establish its presence on land as well. By the middle of the nineteenth century, India (which included the current countries of Pakistan and Bangladesh, as well as what is now known as India), was formally under the control of the British government, which called India the "jewel in the crown" of the British Empire. Britain used Indian cotton to provide raw material for its textile industry (some of the product of which was then sold back to India at a profit); Indian soldiers staffed Britain's battle regiments; Indian workers produced sugar on plantations in British colonies in the Caribbean and the Pacific; and Indian merchants sold British goods in the British colonies of Africa. Although Indians were not formally enslaved, the British levied taxes that could only be paid in cash, and to earn cash Indians were forced to integrate their economic activities into the British-dominated world economy, whether by working on British-run plantations or by selling their crops (or manufactured products) to British traders who could largely set the price. (Similar strategies were adopted by European countries in other colonies as well, in the Caribbean, Africa, and in other parts of Asia.)

While cultural tensions certainly existed in the region prior to colonialism, these tensions often were exacerbated by the British. As rulers of a very large colony with a vastly different culture from that found in Britain, the British found it useful to implement what was called indirect rule. In practicing indirect rule, the British exploited pre-existing hierarchies and differences within Indian society, so that Indians would get other Indians to do what the British wanted. Indirect rule thus protected the British from the full wrath of their Indian subjects while also saving the empire from the cost of a full-scale British occupation. Whether intentionally or not, indirect rule as practiced by the British increased tensions among ethnicities, castes, and genders.

Tensions between ethnicities were increased by the common British practice of empowering leaders of minority ethnicities. In these instances, the minority rulers often would take advantage of British support to avenge what they perceived as centuries of oppression at the hands of the majority. The British understood that the minority rulers would realize that their power was due solely to the support that they received from their British overlords, and therefore the local rulers could be depended on to implement whatever policies the British wanted. Later, when the British finally left their colonies (including India) after World War II, it was payback time, and majority populations frequently reasserted their power over resident minorities. Thus, through their practice of empowering minority rulers, the British frequently took what had been a series of long-standing tensions between ethnicities throughout the region and exacerbated them into outright hostility. (This practice was not unique to the British in India; see Chapter 9 of the textbook.)

Recognizing the cauldron that they had created in India, the British divided India into two countries at independence. One region, made out of predominantly Hindu areas, became the present country of India. The other region, composed of predominantly Muslim villages, was called Pakistan and included two noncontiguous territories—West Pakistan and East Pakistan. In 1971, East Pakistan seceded from Pakistan and became the independent country of Bangladesh.

Although there may have been no better solution, given the ethnic tensions that existed, the Hindu-Muslim partition was by all accounts a tragedy. Because many Hindus lived on the Pakistan side of the line and many Muslims lived on the India side, and because these minorities feared living in what were about to become self-identified "Hindu" and "Muslim" states, huge numbers of South Asians were compelled to leave their homes. By most estimates, 12 to 14 million people were forced to move, and somewhere between 500,000 and 1.5 million died in the violence that ensued. Further problems were encountered by members of the many other ethnic and religious groups that were not given their own states under the terms of the partition. Today, governments in both Pakistan and India are forced to cope with (and, by some accounts, abet) ultranationalist Muslim and Hindu movements that would like to cleanse their countries of non-Muslim and non-Hindu elements. Thus, Pakistan recently has seen a number of attacks on its Hindu and Christian communities, while the Hindu nationalist party that ruled India until 2004 was often accused of failing to interfere with mob violence against Muslims and Sikhs. Almost since independence, primarily Buddhist Sri Lanka has been suffering from a civil war waged by its Hindu minority.

A second set of tensions exploited and exacerbated by the British were associated with differences in class status. Probably every society has class distinctions, but historically these have been particularly intense in Hindu India, with its system of castes. Every family belongs to a caste, and caste determines one's occupation and status in society (and one's potential marriage partner, since marriage between castes is frowned upon). Class systems in the West work in a similar way, but castes in India historically have been much more rigid, and with the social gap between the status of a high-caste person and a low-caste person much larger.

For the British colonial rulers, the caste system was ideal. There already was a system in place whereby the upper-caste people were keeping the rest of the populace in line, so all that the British had to do was win the allegiance of the upper caste. Working within this system, the British often exacerbated caste differences, as they armed the upper castes with unprecedented economic, legal, and military power. At the same time, the British destroyed some of the cultural norms and social structures (such as the need for upper-caste people to maintain the respect of their neighbors in closely knit rural villages) that had placed some checks and balances on abuses of caste power.

With independence, the new state of India outlawed the caste system, but it still persists in many forms. Indeed, the overlay of modern class differences (whereby higher classes have greater access to education, life in cities, emigration, and so on) with the caste system may be intensifying caste differences further. As urban India becomes an industrial and even postindustrial society (India is known for its motion picture and computer software industries, in particular), lower-caste people in rural areas remain impoverished, and they may be worse off than ever as the main current of Indian development passes them by.

A third tension that has been exacerbated in the colonial and postcolonial period has been the division between men and women. Like nearly all societies, traditional Indian society was dominated by men. But, as in most male-dominated societies, women had power within certain spheres of life. On balance, women and men (within the household and within the community) needed each other, and this usually placed a limit on male abuses of power.

This relative balance of power began to unravel under the British. The British emphasis on individual households generating cash income (as opposed to families and communities living off the land) generally gave men more importance in society, since both English and Indian standards of propriety dictated that men would have greater access to cash-earning activities. This imbalance increased after independence, as more opportunities arose for men to accumulate wealth in cities, far from the village-based social systems where women, if not dominant, at least had a substantial social role. Societies in the more developed countries had gone through similar transformations as they underwent industrialization and urbanization, but in most MDCs these transformations were accompanied by changes in cultural values that reflected new ideas about what forms of work were valuable and what cultural roles were appropriate. By now, in most communities in North America and Europe, it is expected that women work outside the home and increasingly it is acceptable for men to have a large role in raising children. In India (and much of the rest of the less developed world), by contrast, traditional cultural practices tend to prevail, especially in rural areas, but they are now distorted (or exaggerated) by the destruction of elements of the traditional society that had placed checks on men's goals of achieving wealth and power.

The result of this combination of traditional cultural practices with modern concepts of wealth has been a devaluation of women, not just as economic actors but as people. This can be illustrated by the example of the dowry. When an Indian woman gets married, she is considered to be joining the groom's family, and the bride's family provides the bride with a dowry (a gift) to take with her to her new household as a sign of respect. Historically, dowries were relatively insignificant, but in modern India the dowry has become a major payment and grooms' families at all caste-levels see dowry payments as an important source of income. Brides have come to be seen by grooms and their families as bonuses that accompany dowries, rather than the other way around.

Dowry inflation thus reflects a devaluation of women, since a bride's parents effectively ask the groom's: "How much do I have to pay you so that you will do me the favor of taking my daughter off my hands?" For parents having children, dowry inflation leads to further discrimination. Because the dowry imposes a significant burden on brides' families (and, conversely, families stand to "make money" if they can produce grooms), boy children often are valued more highly than girl children. In some instances, families practice female infanticide (killing girl children at birth); a survey in the 1990s estimated 10,000 cases of female infanticide annually in India. More often, families practice selective abortion (aborting a fetus when a sonogram shows that the baby will be a girl); one study at a Bombay (Mumbai) clinic found that 7,999 of 8,000 aborted fetuses were female. Even if the girl is allowed to live, she may be treated as an unwanted child, which can lead to a host of psychological problems (and physiological problems as well, if, for instance, she is not allowed to eat until other, more "valuable" members of the family are fed, even if there is not enough food to go around). The Hunger Project, a New York–based nongovernmental organization, reports that due to mistreatment and inequalities in hunger and nutrition, girls are 43 percent more likely than boys to die between ages 1 and 5 in India. As a result of these practices, the Indian population has approximately 50 million fewer females than one would predict based on the number of males.

Problems continue once a bride is "married off." Frequently the bride's parents cannot pay the entire dowry up front upon marriage, and so they promise that the remaining portion will be paid at a later date. If the bride's family cannot meet its obligation, the groom's family offers to return the bride but the bride's family generally refuses to accept her. The groom's family may then threaten to harm the bride as a means of blackmailing her family, or she may be cast out on the streets as "damaged goods." In some extreme cases, brides have been killed by their new family, either because they bring the family shame or as a means of following through on a blackmail threat. The Indian government has estimated that 6,000 women die each year as a result of dowry abuse, and, according to the Digital Freedom Network, an international human rights organization, this number may actually be closer to 25,000. Many more suffer from psychological and physical injuries.

Finally, even if the bride fares well under her husband and his family, she faces another set of hardships if her husband dies. Her original family often will not take her back and her new family may see no reason to support her. Left with few means of supporting herself as a single, widowed woman, she may be encouraged to kill herself as a way to avoid bringing shame to either household.

The Indian government outlawed dowries (except when provided voluntarily by the bride's family) in 1961, but the practice is still extremely common. Although state policy can work to eliminate dowries and the devaluation and abuse of women that often accompanies the practice, success will require a shift in cultural norms as well. This cultural shift likely can come about only from social movements working within civil society.

Social Movements in Indian Civil Society

Civil society has become a popular terrain of struggle throughout the world, as individuals and groups avert their focus from the traditional objects of social struggle—economic state power—and instead attempt to transform culture. This turn to cultural politics is particularly popular in the world's less developed countries, where neither states nor markets hold much promise of empowerment. Social movements use the terrain of civil society to fill the gaps where state and market cannot reach. These movements range from ad hoc grassroots efforts whose messages are directed at others in the community, to institutionalized nongovernmental organizations (NGOs) working to change the national political agenda, to groups whose messages are directed at intergovernmental agencies like the United Nations.

South Asia, and India in particular, has been a fertile ground for these social movements. In the first reading, published in the journal of the German Institute for International Development (DSE), Indian journalist Patralekha Chatterjee attempts to identify some of the reasons why groups based in—and trying to transform—civil society are so prominent in India. Chatterjee's analysis focuses mainly on the failings of the political arena: India has a strong democratic tradition and it is formally a democracy, but its government typically is unresponsive to people's demands. Social movements thus arise both to pressure the state and to serve as a substitute for the state.

The next four readings all report on social movements in India that are concerned with gender issues. Despite their common concern, these groups differ in their attitudes toward attempting change in state policy, their attitudes toward becoming institutionalized as permanent organizations, and the degree to which they connect gender oppression with other cultural, political, and economic issues. In the second reading, the Forum Against Oppression of Women (FAOW), a Bombay (Mumbai)-based organization, presents a statement of its beliefs and goals. An urban and apparently middle-class group (note the professions of its members), FAOW works through direct protest action, social-service provision, and lobbying for legislative change. Partly for political reasons (because it is a self-identified feminist organization that does not seek institutional power), FAOW remains outside the worlds of both the government and institutionalized NGOs.

The next reading comes from a civil-society organization that, in contrast to FAOW, is highly integrated within the world of politics. The Asian Legal Resource Centre (ALRC) is an organization of Asian lawyers that brings human rights issues to the attention of the international community. It has been granted official nongovernmental organization status by the United Nations, and this reading is a transcript of a statement that it submitted to the U.N. Commission on Human Rights regarding violence against women in Myanmar (Burma), India, and Pakistan.

The fourth reading is from *The Hindu*, one of India's major national daily newspapers. The article reports on a local organization that is encouraging low-caste women to empower themselves through a series of skills-building and confidence-building programs, ranging from literacy training to organizing collective resistance to sexual harassment by upper-caste men. At first glance, this appears to be a classic local civil-society-based social movement, building off the grassroots to change cultural practices. On closer reading, though, it emerges that the organization is government-funded, with the money originating in an overseas aid program sponsored by the Dutch government. Again, the division between politics and civil society, and between the global and the local, remains murky.

The fifth reading addresses how women taking action in civil society can improve not just the status of women, but society as a whole. This article (published in the journal of the New Delhi-based Institute for Peace & Conflict Studies) discusses the role that women must have if peace is ever to come to Kashmir, a disputed (and continually fought over) region on the border of India and Pakistan.

The final three readings move beyond South Asia to consider broader issues regarding the relationship between civil society, state, and economy. These readings all approach the question of whether there really is a distinct arena of civil society. They also consider the related question of whether it makes sense to focus one's efforts there if one wishes to change society, or if instead one should focus on the more familiar targets of state and economic power. In the first of these readings, the World Bank explains why the development of civil society (and its institutions) is a crucial component of the World Bank's overall goal of economic development. Although the document is generally positive about the role of civil society, in the final paragraphs it notes some of the problems (both for democracy and for economic development) that can occur if the institutions of a society are excessively centered on civil-society-based social movements. The next article, from the journal of Washington, D.C.–based Civitas International, argues that in a world in which states are increasingly unable to control the global economy, the only force emerging to counteract the nondemocratic globalizing power of corporations is a "global civil society" that is forming as NGOs around the world unite in loose networks. In the final reading, antiglobalization activist Aziz Choudry counters this assertion, arguing that if one wishes to counter the negative effects of economic globalization one must approach it head-on, rather than attempting to achieve progress in a separate world of civil society that often is idealized as being insulated from political and economic power.

Readings

1

from Die Deutsche Stiftung für Internationale Entwicklung (the German Institute for International Development)

Civil Society in India

A Necessary Corrective in a Representative Democracy

Patralekha Chatterjee
November/December 2001

India is the largest democracy in the world. But without its lively NGO scene, many ills in society would continue unchallenged. Civil society derives its strength from the Gandhian tradition of volunteerism, but today, it expresses itself in many different forms of activism.

It was dark, around midnight. The place—on the outskirts of the Indian capital, New Delhi. There was no house or car or road to be seen—only dark mounds that turned out to be excavated rock when one went near. Next to each mound was a near vertical pit—about 20 meters deep on the average. That was where the rock had been dug out from—the rock that went into making the big houses of the political leaders and bureaucrats and rich citizens of New Delhi.

Stepping carefully around the pits and mounds, four men approached the only source of light to be seen—a small kerosene lantern burning inside a mud and grass hut. There were four residents inside—a couple and their two children, ten and eight. They dug the pits, cut the rock and hauled it to the trucks, all without salary. They were bonded labourers, bonded for life and for generations to the owner of the pit because some ancestor sometime had borrowed money and had been unable to pay it back.

Two of the four men who visited the hut that night in 1985 were from a non-governmental organization called Bandhua Mukti Morcha (Bonded Labour Liberation Front). The other two were journalists brought by the NGO to prove that bonded labour—a form of slavery—did exist right in the nation's capital. After the visit, the men from the NGO went to the police station to lodge a complaint, because bonded labour is illegal in India, and so is child labour in a profession as hazardous as this.

The complaints, and the articles written by the journalists after the visit, were part of the NGO campaign to make the government implement the law.

Every day, different NGOs all over India are doing things like this. Sometime it may be taking a sample of water from a well that has been polluted by a nearby factory, getting the water analysed and then filing a "public interest petition" in a court to force the factory to follow anti-pollution laws. Another time, it may be a heated debate with a bureaucrat on why all citizens should have the right to be informed about all government decisions that affect their lives.

Though the term NGO became popular in India only in the 1980s, the voluntary sector has an older tradition. Since independence from the British in 1947, the voluntary sector had a lot of respect in the minds of people—first, because the father of the nation Mahatma Gandhi was an active participant; and second because India has always had the tradition of honouring those who have made some sacrifice to help others.

In independent India, the initial role played by the voluntary organizations started by Gandhi and his disciples was to fill in the gaps left by the government in the development process. The volunteers organized handloom weavers in villages to form co-operatives through which they could market their products directly in the cities, and thus get a better price. Similar cooperatives were later set up in areas like marketing of dairy products and fish. In almost all these cases, the volunteers helped in other areas of development—running literacy classes for adults at night, for example.

In the 1980s, however, the groups who were now known as NGOs became more specialized, and the voluntary movement was, in a way, fragmented into three major groups. There were those considered the traditional development NGOs, who went into a village or a group of villages and ran literacy programmes, crèches for children and clinics, encouraged farmers to experiment with new crops and livestock breeds that would bring more money, helped the weavers and other village artisans market their products and so on—in short, became almost a part of the community in their chosen area (usually in rural India) and tried to fill all the gaps left in the development process by the government. There are many examples of voluntary organizations of this kind running very successfully in India for the last five decades. Perhaps the most celebrated example would be the treatment centre for leprosy patients run by Baba Amte in central India.

The second group of NGOs were those who researched a particular subject in depth, and then lobbied with the government or with industry or petitioned the courts for improvements in the lives of the citizens, as far as that particular subject was concerned. A well-known example of an NGO of this type is the Centre for Science and Environment. It was a CSE member who picked up that sample of well water and then submitted the results of the chemical analysis to a court because the organization had not been able to get the factory to change its polluting practices in any other way.

In the third group were those volunteers who saw themselves more as activists than other NGOs did. Of course, all NGOs undertook a certain amount of activism to get their points across—they petitioned the bureaucrats, they alerted the media whenever they found something wrong and so on. But this third group of NGOs saw activism as their primary means of reaching their goals, because they did not believe they could get the authorities to move in any other way. Perhaps the best-known example of an NGO in this category is the Narmada Bachao Andolan (Save Narmada Campaign), an organisation that opposed the construction of a series of large dams in a large river valley of central India. The members of this NGO believe that large dams worsen water scarcity for the majority of the people in the long run rather than solve the problem, and they oppose the displacement it entails upstream of the dam. When the NBA found that it could not persuade the planners in India to agree to its point of view, the NBA members put up pickets, held demonstrations and tried every other way they could think of to oppose the construction of the first of the big dams. Most of the NBA members went to jail a number of times as a result. Right now, some of them—including celebrated novelist Arundhati Roy—face the prospect of being jailed again, because they criticized the Supreme Court of India when the court's decision on dam construction did not go in their favour.

There is no strict boundary between these three groups of NGOs—in fact, Baba Amte is now an important member of the Narmada Bachao Andolan. And whatever be the category a particular NGO falls into, all of them play an important role in modern India—they hold the politicians accountable to the people.

India is a representative rather than a participatory democracy. Once the elections are over, the politicians who run the federal and state governments do not really need to go back to the electorate for every major decision—there is no tradition of referendums in India, as there is in Switzerland or Denmark. So, in the five years between one election and another, the NGOs—and parts of the media, to some extent—are often the only means available to the citizens to voice their opinions on any decision taken by a government.

In a large developing country like India, there are numerous gaps left by the government in the development process—sometimes by intention, sometimes due to lack of funds, sometimes due to lack of awareness. These are the gaps that many NGOs try to fill in modern India. Some of them may work in areas that the government does not want to get into—like fighting discrimination on the basis of caste. Most Indian politicians do not really want to upset the existing caste hierarchy in his or her constituency, because the politician is dependent for votes on the dominant castes of that particular constituency. In the process, laws prohibiting discrimination on the basis of caste are often ignored unless there is an NGO working in the area that is willing to take up the cause of those being discriminated against.

Working in Health and Education

Then there are many NGOs who work in areas where the government effort proves inadequate. Two well-known examples are the areas of education and healthcare. In the area of education, there are often not enough government-run schools, especially in rural regions. Or there may be schools without adequate facilities, because a particular state government does not have the necessary money. There are many situations where the government runs a co-educational school, but the girls do not go there because their conservative parents (the overwhelming majority) refuse to send their daughters where they may meet boys. Then there are many cases where the government runs a largely-empty school, because most of the boys and girls are out working during school hours. NGOs have played an important role in all these cases—running special classes at night for children whose parents send them out to work, running special classes for girls and so on. By and large, governments have been supportive of such initiatives by NGOs, and the only problem is that there are not enough NGOs to educate all the uneducated people in India. The mammoth NGO called Kerala Sastra Sahitya Parishad is largely credited for the hundred percent literacy in that state in the south-western corner of India.

In the area of healthcare, too, NGOs play a stellar role in modern India—by supplementing the government effort to provide healthcare to citizens, and by raising awareness in society about issues like child and maternal malnutrition, which is perhaps more important than adding a few more clinics. Again, in modern India it is the NGOs who have battled social evils in the area of healthcare, like the neglect of the girl child, which can sometimes take the extreme form of female foeticide or infanticide. It is largely through the lobbying by NGOs through the media that many state governments have now passed laws banning sex-determination tests of foetuses, as such tests were often leading to the abortion of female foetuses.

In the last 20 years or so, a very large number of NGOs in India have been active in the area of environmental protection. They have been in the forefront of reforestation campaigns, they have lobbied against deforestation or overuse of pesticides in agriculture, and they have taken polluting industries to task. In this sustained campaign, the NGOs have often been helped by the judiciary whenever the government of the day has proved unresponsive. For this, NGOs in India have almost developed into a fine art a device called public interest litigation, by which any citizen can petition a court to intervene where (s)he feels it is in the public interest for the court to intervene.

Another field in which certain Indian NGOs have been active, especially in urban areas, is in trying to turn the right to shelter into a reality. This is an area where constructive work and activism have intermingled most often, as NGOs such as YUVA and SPARC in cities like Mumbai (Bombay) repeatedly oppose the demolition of hutments even as they try to improve the quality of life in the sprawling slum clusters.

The struggle by NGOs to make governments more accountable to citizens is an ongoing struggle in India. For years now, NGOs have been lobbying for the right to information to become a legal right, and it now appears that the federal Parliament may soon pass a bill to this effect.

The '80s were the hey days of activist journalism in India. NGOs became the media's key allies in exposing injustice and clear violations of rules. Today, human rights reportage has to fight for column space with a myriad other issues and NGOs have to speak louder to be heard by the public but their influence in public affairs is growing. But with that, has come corruption. Several NGOs have come under a cloud because of alleged misappropriation of public funds. The jet-set lifestyle of some NGO representatives in the country spurred one keen observer of the NGO in Delhi to quip that today there is a new category of NGOs—"airport NGOs"—who flit from one international airport to another, hopping from one cause to another, all in the name of the poor and grassroots activism.

Apart from this, there are many NGOs in India to represent special interest groups, ranging from the disabled to women to children to the aged to refugees and to people in specific professions. In the course of their work, almost all NGOs come up against an unfeeling or even hostile bureaucracy sometime or the other. It is part of the strength of Indian democracy that the state is by no means the winner in all these confrontations.

Source: Patralekha Chatterjee, "Civil Society in India: A Necessary Corrective in a Representative Democracy," *Development and Cooperation* (Frankfurt), Nov. 2001, p. 23–24. Reprinted with permission.

2

from the Forum Against Oppression of Women

Some Background Information

FAOW was formed in 1979 as a platform to respond to an extremely unjust judgement on a rape case. Soon from a Forum Against Rape as it was earlier called, it changed its name to the present one to encompass varied forms of women's oppression. FAOW is part of what in India is recognised as the Autonomous Women's groups which has played a crucial role in this third phase of the women's movement [late 70s onward].

Today it is mainly a campaign group consisting of members from varied background—students, housewives, professional women, lecturers, etc. The members meet regularly once in a week and everyone puts their voluntary time in the work that needs to be done. There are no paid staff.

In FAOW, some of the issues that have been taken up are rape, dowry, wife beating, sexual harassment, indecent portrayal and sex stereotyping of women in the media, sex determination and sex preselection techniques, against harmful long acting contraceptives, against the growing trend of communalism and the Development

Paradigm, especially the new economic policies and population control policies. These issues have been dealt with using different strategies. Direct action has been one of the important strategies used. Shaming the man and his family in case of harassment or murder of his wife, retrieval of distressed women's belongings from her matrimonial home, etc. Consciousness raising through plays, poster campaigns mainly in the ladies' compartments and special ladies' trains, discussions, and holding specific programmes like celebrating International Women's Day has been another strategy. FAOW has been active in networking with other women's organisations and has hosted the first two national conferences of [the] autonomous women's movement and has been active in organising the other three. The other important strategies [have] been to demand legal changes—lobbying for changes in laws, demanding new laws as well [as the] need for better implementation. One of the recent legal campaigns FAOW has been part of is to demand [a] ban on use of sex determination techniques like amniocentesis.

FAOW identifies itself as a feminist organisation and has consistently attempted to function as a collective, without 'leaders.' It has decided to function independently without any control either from the State or from the funding organisations. That is why it has consciously remained un-registered and also does not accept any institutional funding. All financial requirements for the programmes are collected by each member through donations. The office is housed in different members' houses. . . .

Source: Jagdish Parikh, "Forum Against Oppression of Women," April 5, 1994.

from the Asian Legal Resource Centre

3

Violation of the Human Rights of Women in Asia

Testimony Submitted to United Nations Commission on Human Rights December 29, 1999

1. The 1995 Beijing Platform for Action, which contained a comprehensive list of recommendations for the integration of the human rights of women and the gender perspective in all the concerns of the States Members of the United Nations, is currently under review by States as well as non-governmental organizations. The general impression of the progress made on implementation of the Platform for Action is that the gains which have been achieved, mostly by the tenacious efforts of NGOs, are minuscule compared to the persistence of discriminatory policies and violent practices committed by the State as well as private citizens or groups. Women are still impeded from participating in the society as equals, and are punished, either personally or as a group, when they defy or, through no fault of their own, are in contravention of the rules and controls which society imposes on them.
2. ALRC would like to highlight the following serious and systematic violations of the human rights of specific groups of women in some Asian countries. This is the life-threatening violence which is perpetrated on Asian women on a daily basis.
3. Myanmar: women in a war situation. The women and children of Myanmar bear the brunt of the suffering as the result of one of the longest civil wars of this century. Over 200,000 refugees, most of them women and children, have fled to Thailand. Women flee their country because under the SPDC (State Peace and Development Council) military dictatorship, women and children are forced by the

military to build roads and rebuild buildings. They are not paid for their work and are beaten if they do not work hard enough. In the war zones, ethnic minority women are forced by SPDC soldiers to carry heavy loads of ammunition and supplies through the jungle. At night these women and girls are raped by the soldiers. In some cases, women and children have been used as minesweepers or as human shields during fighting.

4. Poverty and the lack of democracy under military rule in Myanmar have denied women adequate food and access to decent health care and many children cannot attend school. Thousands of women and girls each year leave Myanmar to work in prostitution in Thailand, in order to escape forced labour by the military and in order to sustain themselves and their families.

5. India: Dalit women—violence on the bases of caste and gender. Fifty years after independence, successive Governments of India have not made any progress in the elimination of discrimination and the ensuing violation of the rights of the Dalit population. Dalits (also referred to as "untouchables"; Dalit = "Broken" people) are deemed the lowest caste of human beings in the highly caste-stratified/conscious Indian society. They may not enter the higher-caste sections of villages, may not use the same wells, may not wear shoes in the presence of upper castes, may not visit the same temples, may not drink from the same cups in tea stalls or lay claim to land that is legally theirs. Dalit children are frequently made to sit in the back of classrooms. Dalit villagers have been the victims of many brutal massacres in recent years. Since the start of a Dalit rights movement in 1990, violence against Dalits has increased proportionate to the growth of the movement.

6. The "Charter of Dalit Human Rights" drawn up by the National Campaign on Dalit Human Rights describes the numerous violations of the rights of Dalits. The Indian Government must be held accountable for the structural denial to 260 million Dalit men, women and children of their rights to have access to resources to maintain their livelihood, to education and adequate health care, and to participate in social, political and economic institutions. But, according to the Campaign Charter: "In India, the State and civil society go hand in glove in the denial of rights to the Dalits. While the State abets violations by the civil society it is forced to take sides with the dominant caste society in its favour."

7. Dalit women suffer threefold discrimination: on the basis of gender because they are women, on the basis of caste, because they are Dalits, and, as Dalit women, by their own menfolk (bases of gender and caste). In India caste and gender discrimination are perpetrated in their worst forms on Dalit women. The main areas of discrimination are with respect to the right to work and to just and favourable conditions of work, violations of article 23 of the Universal Declaration of Human Rights and articles 6 and 7 of the International Covenant on Civil and Political Rights.

8. Dalit women's labour is labelled unskilled and, therefore, is unrecognized, underpaid, and even unpaid. About 85 percent of Dalit women work in the agricultural sector, which is unorganized and does not have the social security benefits found in organized sectors, such as maternity benefits, medical support, etc. Dalit mothers have to bring their infant children with them to work in the fields, where there are no childcare facilities. Sometimes they are not allowed to do this, and lose their jobs in the agricultural sector.

9. In urban areas, Dalit women also work in the unorganized, self-employed sector as hawkers, scrap collectors, petty traders and house servants. Or they may earn wages in domestic work, construction or small-scale manufacturing. In some areas Dalit women work as nightsoil removers, without any considerations for hygiene, for as little as one roti per day. All these sectors of employment are characterized by low wages, irregular work and wages, absence of social security, sexual harassment, and dependency on the whims of middlemen and employers.

10. Almost all Dalit women workers enter the labour market before the age of 20; 31 percent of all girl children from Dalit communities are child workers. Girls' labour is needed, in agriculture and in household work, and poor people will choose not to spend money on the education of girls. Thus, there is a higher dropout rate for

Dalit girls at all levels, and over 83 percent drop out of school at the secondary stage. In addition, women are the ones who mainly take responsibility for cleaning, maintaining and running a household, and in fact, 70–75 percent of Dalit households are female-headed. Since, on an average, 70 percent of Dalit households have no electricity and more than 90 percent have no sanitation facilities, Dalit women (and girls) have to spend a great amount of energy doing household labour, walking long distances to collect food, fodder, fuel and water.

11. Violation of the right to life and security, and freedom from torture or cruel, inhuman or degrading treatment or punishment (Universal Declaration of Human Rights, arts. 4, 15, 12). Incidents are regularly reported in various newspapers, in different states of India, illustrating the systematic manner in which Dalit women are subjected to extreme inhumane treatment, as punishment for asserting their rights, or standing up to dominant castes. Often police are standing by, or have done nothing to prosecute the perpetrators of crimes such as the following:

(a) Amta, 30, of Randevi village, under the jurisdiction of the Nakud police station, had her face blackened and her bottom thrashed for accusing two of her neighbours of a theft at her house (Hindustan Times, 18 September 1997);

(b) Five teenage girls in Bihar were raped and mutilated in an attempt by landlords to reassert authority over increasingly vocal Dalits. All five girls were shot in the vagina and their breasts cut off. In addition to the organized massacres of the residents of entire Dalit villages, private armies in Bihar practise unlawful and dehumanizing programmes aimed at insulting members of the lowest castes and preventing their rise in society. One of the most heinous crimes [was] the Savarna Liberation Army's (SLA) mass rape campaign, conducted between March and July in Gaya and Jehanabad districts, when more than 200 Dalit women between the ages of 6 and 70 were raped. The perpetrators of these crimes publicized each of the incidents. Because of the stigma attached to rape victims, the operation was such that it broke the morale of Dalits in many villages (Frontline, 12 March 1999);

(c) In Andhra Pradesh, a Dalit woman was paraded naked by upper-caste people following a petty dispute over using water from a borewell at Malasamudram village in Anantapur district. Thirteen people were arrested in this connection. The Dalit woman had argued with upper-caste women about her right to use the borewell. Later, the upper-caste people went on a rampage, damaging houses of Dalits and beating up the women (The New Indian Express, 19 September 1999).

12. Pakistan: sexual harassment at work and on the streets. Women in Pakistan are constantly being harassed, at work and on the street. But they do not report these incidents for fear of being restricted in their movements, the only form of "protection" available. Women also fear retaliation, stigmatization, and the uncooperative and humiliating attitude of officials and law enforcers. "If we raise a voice against such harassment, we are told that we should not go out of our homes," one woman dejectedly says. "Once I reported to policemen at a checkpost that I was being chased and teased but they asked me instead why I wasn't wearing a veil," says another. Human rights and women's organizations, both non-governmental and governmental, have acknowledged the seriousness of the situation.

13. The Human Rights Commission of Pakistan (HRCP) said in its annual report for 1999 that, "It was not unusual for women to encounter remarks or experience physical push and shove in offices and shops, in houses and in other public places. These have ceased to offend only because of their recurrence." In Punjab Province, the HRCP documented 242 cases of crimes against women reported in the newspapers and magazines. Of these 113 were attempted rape incidents and 77 involved stripping and assault of women in public. However, only fewer than half of the cases were registered with the police and in only 23 of those cases were the accused taken into police custody. Although the Pakistan Penal Code prescribes punishments for sexual harassment offences, often policemen turn a blind eye even when they are approached by women.

14. The Islamabad-based Progressive Women's Association believes every second woman in Pakistan is a victim of a direct or indirect form of mental or physical violence. The most vulnerable are those who work in the informal sector, like domestic and brick-kiln workers.

15. An internal document assessing the impact of Islamic laws on women says: "In the past 15 years, discriminatory laws, along with exploitation of religion to control women's sexuality and productivity have been instrumental in increasing institutionalized violence in women's lives, both in the incidence of violence against women and in the number of women in prison."

16. The Government is committed to eliminating all forms of gender discrimination under the Convention on the Elimination of All Forms of Discrimination against Women. The Government's National Plan of Action (NPA), a follow-up to the 1995 Beijing Conference prepared in consultation with women's groups and rights organizations, admits to widespread sexual violence against women in the country, saying that it is rooted in the patriarchal system of male domination and female subordination. The NPA hoped to put in place redress mechanisms, where women could file complaints, by 2000. Also, management and labour inspectors were supposed to monitor sexual harassment in the workplace, but these measures have yet to materialize.

17. These are but a few situations, but they illustrate that deep-rooted prejudices against women still persist in Asian societies today. It is the responsibility of Asian Governments to take measures for radical change, in order for progress to be made in the protection and enhancement of the human rights of women, and so that all women can live lives free from discrimination and violence.

Source: "Violation of the Human Rights of Women in Asia," testimony submitted December 29, 1999 to the United National Commission on Human Rights posted at www.alrc.net/mainfile.php/56written/13.

4

from *The Hindu*

Breaking the Shackles

Kalpana Sharma
4 April 1999

While urban India is hurtling towards liberalisation, life in rural India still goes on, as it always has. Yet, something is changing, representing a virtual social revolution. KALPANA SHARMA highlights an instance in Bidar, where the women seem unaware of the import of what they are initiating.

"OUR young daughters had to sit on the laps of the Gowdas (upper caste men) in our village and remove the money the men were clenching between their teeth."

"An old man in our village raped an eight-year-old harijan girl because he had been told by the government doctor that he would be cured of a sexually transmitted disease in this way."

"The Gowdas would taunt us and say that even if you educate your daughters, who will give them jobs? Better send them to us."

These stories are not unfamiliar. You read about them, see programmes on television. Yet, each time they are reported, you are surprised that while urban India is hurtling towards integration with a global economy, large parts of rural India remain untouched. Life here goes on as it always did. Centuries of caste divisions refuse to be erased even if mud has given way to plastic and bright yellow and green plastic waterpots and lotas have replaced their mud and brass originals.

Yet, something is changing, albeit slowly. It is not one single process contributing to the change but probably a combination of several. Some of the changes are so dramatic that they represent a virtual social revolution. Yet, the people caught in the middle of it, particularly the women, seem blissfully unaware of the import of what they are initiating.

On March 8, International Women's Day, I sat talking to Lalita from Banmali village in Humnabad taluka of Karnataka's Bidar district. Lalita was one of over 400 women who had gathered on the grounds of the famous Mahilar Mallana Mandira in Khanapur, Bidar, for a "mela" organised by the women's empowerment and literacy programme, Mahila Samakhya.

"For years," said Lalita, "our community (Dalits) was forced to give our first daughter to the temple at Yellamma. We also had to send our young girls to the Gowdas. They would make them dance, take photographs, and make them sit on their laps. The girls would have to use their mouths to take the money the Gowdas clenched between their teeth."

Three years ago, the Panchasheela Mahila Sangha, of which Lalita is a part, decided to stop sending their daughters to the Gowdas. Instead of coming to their aid, the police first came and asked the women why they were refusing to send their daughters. The women also decided to break with the age old custom that required dalits to spread the news if any person died in a Gowda household and to perform certain rites connected with death.

Retaliation was inevitable. The Gowdas refused them water and access to the grain mill. Twelve of the Sangha women went to the police and complained; they managed to get the water restored and the mill opened. "If it hadn't been for the Sangha, we would have been doing the same thing even today," says Lalita.

Making women believe that they have the power to change the circumstances into which they were born is the first step to these changes. The programme that has facilitated the process, Mahila Samakhya, is different from other government programmes because it recognises the importance of this. Thus, it was designed to approach literacy, for instance, not as an end in itself, but as a tool which women would use as and when they felt the need for it. It began deliberately by providing women the space to meet and talk about their problems. Through these discussions, different needs were expressed and a variety of initiatives and programmes emerged.

Literacy was a by-product of this process. Balamma from Hirevenkelkunta in Raichur district acknowledged that she realised she needed to be literate after she stood and won a seat in the panchayat in the last elections. She was asked by three men to sign a meeting notice. As she did not know the contents of the document, she gave her thumb impression. Only later did she realise that she had signed withdrawal papers for the election for the chair of the panchayat. Instead of her, a man was elected.

Neelamma is another woman panchayat member who suffered because she did not know how to read and write. From Nilkunda village in Bidar district, Neelamma, who is a landless labourer, says that she realised she could not ask questions about the budget because she could not read. All the three women elected to the Panchayat are illiterate. Now she has realised the importance of literacy, has learned how to sign her name and has made sure that all her children are educated. "Before Panchayat Raj, we never got anything. We used to feel afraid to go to the gram panchayat. But now we are not afraid and we go and sit on the chair in the gram panchayat," she says with pride.

Kalavati from Netur B in Bidar district, a fiery young 22-year-old who heads the local Sangha, told the tale of the old man who had raped the 8-year-old Dalit girl because he had been advised by the government doctor to do this to cure himself of a sexually transmitted disease. The Sangha women took the girl to the same doctor who had given this advice as they did not realise the role he had played. When they found the old man and took him to the police, he confessed. The doctor was also arrested, but managed to get out on bail and is still practising in the village. Although Kalavati is incensed about what has happened, she feels helpless to do anything about the doctor. "He is a Brahmin and his father is a lawyer," she says by way of an explanation.

Kalavati is exceptional in that she is a younger woman. Many of the Sangha women are older, already grandmothers, who have decided to take on the upper castes in their villages. Most of the Sanghas are almost entirely made up of Dalit women. Neither women from other castes, or other creeds, are part of the Sanghas. For instance, there are hardly any Muslim women. When I ask Kalavati why she has not tried to enlist the Muslim women in her village, she says they do not want to come.

So while the Sanghas are tackling one set of rigid divisions, they are leaving others untouched. Perhaps that will be the next step, if they feel the need. Ms. Revathy Narayan, director of MS in Karnataka, says that programmes evolve according to the needs of the Sanghas. Because MS chose to begin with the most marginalised amongst the women, it has become predominantly a programme that works with Dalit women. It has made no real effort, except in one of the newer districts where it has just begun its work, to widen the caste base of the Sanghas.

Mahila Samakhya (MS) has now completed eight years. It began as a five-year programme, wholly funded by the Dutch government, in three states, Karnataka, U.P. and Gujarat. It has had several extensions and its present funding is expected to end in another three years. Will the programme be able to sustain itself?

Although it claims to work like a non-governmental programme, in fact it is a top-down programme but with a difference. It was conceived outside the villages in which it works, yet its basic structure ensures that what it does is dictated by the women at the bottom of the structure. Thus it varies crucially from other government programmes. It also differs from non-governmental programmes because it has a structure and has the funds to spread out and extend its reach. Many well-intentioned NGOs remain limited because of the absence of either or both these elements.

What is evident when you meet the women who have been part of the programme these last eight years or so in Karnataka, where the programme has been more successful than in some of the other States, is that the process of making women realise their power cannot be reversed too easily. Whether it is physically beating up drunken men, or tackling wife beating, or caste problems, the common theme that emerges is the fact that these women now feel strong, they have the courage to do things they would never have done in the past. And all of them recognise the importance of education, especially educating their daughters.

Chandramma from Chambal village in Bidar district says that her educated daughter came home after a bad marriage. In her village, when a Gowda passed by, the women had to cover their heads. Now they tie their paloo around their waists. And as for her daughter, she has come back now and is working and is not interested in marriage, saying, "My job is my husband. I don't want or need a husband." Chandramma never went to school, yet of her three daughters, one has graduated and the other two are in school.

In many parts of India, through a variety of programmes and processes like Mahila Samakhya, we are beginning to hear the voices of women like Lalita, Kalavati and Chandramma. They need support, of the kind the Sanghas can give them. They also need recognition, and participation in governance through the Panchayati Raj system. These are the stories that illustrate what is really meant by the over-used term "empowerment."

Source: Kalpana Sharma, "Breaking the Shackles," *The Hindu*, April 4, 1999.

from the Institute for Peace and Conflict Studies

Women in Conflict Resolution: The Road Ahead in Kashmir

Sumona Das Gupta and Ashima Kaul Bhatia
31 December 2001

Since the outbreak of violent conflict in Kashmir, the women's movement has undergone many twists and turns. Adapting to its changing character, the women of Kashmir have carved out their own place and have in many cases risen above the victimhood discourse, seeking their own ways of negotiating and coming to terms with the violence that has scarred their lives for more than a decade.

In the ultimate analysis the women of Kashmir have had to bear the end of the violence that has wracked the valley. It is they who as widows, half widows, rape victims, victims of religious dictates, and victims of displacement have to ensure that the pattern of life continues as normally as possible even when the times are abnormal. It is they who had to come to terms with violence from the guns of the militants or the security forces. It is the women who have had to provide solace to the household following the loss of a son, husband or brother in the crossfire apart from shouldering the additional responsibilities placed on them consequent to the loss. Shouldering these additional responsibilities has empowered them by compelling them to come out of the traditional confines of their perceived place in a male dominated society—but the price for this empowerment is heavy and laden with tragedy.

There has been an ominous silence when it comes to articulation of women's concerns in the midst of this exploding violence. Two reasons can be discerned for this. One, positing a question related to conflict resolution in terms of gender often helps to cut through political differences. Two, women's experiences of operating under multiple pressures and moving in and out of their multiple identities—as daughters, wives, mothers, peacemakers, and in many cases bread winners, probably places them at an advantage when it comes to conflict resolution and peace-building—a fact acknowledged in brainstorming sessions with a core group of Kashmiri women.

In the light of this scenario how can we chart the road ahead to sustainable peace? First by sensitizing those affected and those concerned about the Kashmir situation to the trauma, indignity and pain of the people of Kashmir in the last 11–12 years of violence that cuts across communal lines and has left its imprint on the psyche of the majority and the minority community, in different ways. Harrowing first hand testimonies of human rights abuses by the security forces are fused with equally horrifying accounts of militant abuses. Narratives of the agonies and indignities suffered by the Kashmiri Muslim population are interspersed with narratives of a traumatized community of displaced Kashmiri Pandits. These realities do not negate each other. They coexist. The road to conflict resolution requires that no one narrative is privileged over the other due to a vested political interest. A holistic, non sectarian approach must be adopted.

Second, we need to acknowledge that by definition, sustainable peace cannot be gender blind. There is no indication that there would be any attempt to include women's voices in the peace process although it is clear that women's different experiences can lead to an alternative formulation of peace. This is a lacuna that must be addressed. There is an urgent need to listen to the voices of women of Kashmir directly but ensure that there is a non sectarian inclusive representation of women who can cut across the political, religious and regional divide in the peace process.

A two pronged strategy is necessary. Economic reconstruction and rebuilding which would entail not just pumping in money but ensuring that it reaches the people for whom it was meant. Due to their control over resources the government of India and the state government would have to take primary responsibility for this though civil society would have a role to play. The other dimension would be reconciliation and dialogue—both between the different communities within Kashmir and Kashmir with the rest of India. The Indian government can help rebuild this trust only if it allows a transparent democratic process to operate in the state. Civil society both within Kashmir and the rest of India specially women's groups also have a vital role to play in this process of economic and social peacebuilding. It is the appropriate time now to reflect upon and identify the space for women in the reconstruction of civil society, establish the constituencies for peace in the valley and reinforce them where they exist.

Source: Sumona Das Gupta and Ashima Kaul Bhatia, "Women in Conflict Resolution: The Road Ahead in Kashmir," 12/31/01.

6 from **the World Bank**

What Is Civil Society?

Civil society consists of the groups and organizations, both formal and informal, which act independently of the state and market to promote diverse interests in society. Social capital, the informal relations and trust which bring people together to take action, is crucial to the success of any non-governmental organization because it provides opportunities for participation and gives voice to those who may be locked out of more formal avenues to affect change.

While individual groups form the building blocks of civil society, the concept's value lies in the extent and density of relations among groups as well as the synergy between civil society, state and market (Evans 1996, Renshaw 1994). Therefore, social capital is an integral part of civil society at the micro and macro levels. Civil society is not a constant, rather it is continuously evolving and its roles vary in different contexts and at different levels of economic development.

Civil Society and Social Capital

Social Capital within a Non-Governmental Organization (NGO)

Trust and willingness to cooperate allows people to form groups and associations, which facilitate the realization of shared goals.

- Grameen Bank (Bangladesh) lends money to rural poor, especially women, at a daily loan volume of $1.5 million and has a 98% repayment rate. The organization was started in 1976 by Muhammad Yunus, who lent money to 42 persons because he trusted they would repay it. Since then members have developed rules to maximize repayment of loans, but trust still plays a critical role in the organization's success, particularly in the absence of collateral (Uphoff, Esman and Krishna, 1997).

Social Capital and Civil Society Can Promote Welfare and Economic Development

When the state is weak or not interested, civil society and the social capital it engenders can be a crucial provider of informal social insurance and can facilitate economic development.

- Six-S (Burkina Faso) is a loose federation of rural organizations supported by more formal international NGOs. Social capital has helped Six-S to mobilize over 1 million villagers in 1,500 communities in West Africa in an effort to create better agriculture opportunities during the dry season. (Uphoff, Esman and Krishna, 1997).

Social Capital across Sectors

State, market and civil society can increase their effectiveness by contributing jointly to the provision of welfare and economic development. The success of this synergy is based on complementary rather than substitutable inputs, trust, freedom of choice and incentives of parties to cooperate (Evans 1996, Ostrom 1996).

- Plan Puebla (Mexico) is a joint project between farmer groups, university, government and private institutions to develop appropriate technology for farmers growing rain-fed maize. It has increased yields and incomes for almost 50,000 farmers, generated institutional changes and has been a model for other rain-fed agricultural programs (Uphoff, Esman and Krishna, 1997).

Social capital facilitates exchanges of resources and skills across sectors. Through such exchanges, civil society can serve as the beneficiary (via philanthropy, tax exemptions and management training) as well as the benefactor (offering political support, policy recommendations and service provision).

Social Capital and Civil Society Can Strengthen Democracy or Promote Change

A strong civil society has the potential to hold government and the private sector accountable. Civil society can be a crucial provider of government legitimacy.

- Putnam's seminal work *Making Democracy Work* (1993) shows that citizens who are active in local organizations, even non-political ones, tend to take a greater interest in public affairs. This interest, coupled with interpersonal social capital between government officials and other citizens which is fostered when both belong to the same groups and associations, renders the government more accountable.

Civil society gives a voice to the people, elicits participation and can pressure the state.

- Successive alliances between civic organizations and receptive state officials created increasingly greater tolerance for reforms in some of the rural states of Mexico and increased the "thickness" of civil society (Fox 1996).

Social Capital and Civil Society Impact Global Issues

Globalization offers unprecedented opportunities and risks. With increasing global integration, a new kind of inequality is emerging, marked by a widening divergence between the "haves" and the "have-nots." As those at the bottom feel increasingly disenfranchised from the policy decisions which affect them, the social fabric of a country is weakened. Civil society depends on participation; it is threatened when people and societies become excluded (Dahrendorf 1996).

Effective civil society unites those who may have little power individually and gives weight to their ideas and aspirations.

- Cross-national trade unions can unite workers to tackle common problems, strengthen their bargaining power, and avoid exclusion.
- Community-based organizations preserve the environment and local jobs when they band together to protect rain forests from destruction through teaching environmentally-friendly practices locally and publicizing cases of corporate irresponsibility globally.

Downside of Civil Society and Social Capital

Risk of Exclusion

The presence of a strong civil society does not automatically lead to a politically or socially healthier society. The strong ties that benefit members of groups and associations enable them to exclude outsiders (Portes and Landolt 1996). Internally cohesive groups which isolate themselves from the rest of society may use their social capital to pursue goals at odds with the public good.

- In some cases, tight-knit ethnically-based groups can lead to discrimination, violence, or even ethnic cleansing, as recently witnessed in the former Yugoslavia.

Risk of Decreased Autonomy

Another downside of social capital impacts individual members. Many people join community-based associations in order to pool resources and receive the immediate benefits of additional assets. However, as individuals become increasingly successful, the group may start to demand more in contributions than the individual feels he or she is receiving in benefits. In such cases, individual desires may have to be subjugated for the desires of the group. Strong social capital may make it difficult for individuals to opt out of a group to which they have belonged.

- The Mafia, drug gangs, political associations and immigrant merchant groups all have the capacity to serve as facilitators and impediments to an individual member's long-term success.

Source: "What Is Civil Society?" October 10, 2002, posted at www.worldbank.org/poverty/scaptial/sources/civil1.htm.

7

from Civitas International

What Is Global Civil Society?

David Callahan
January/February 1999

The Term Globalization Conjures Up a Panoply of Bleak Images

To many, the term globalization conjures up a panoply of bleak images: multinational corporations wielding more power than governments; dirt cheap labor replacing well-paying jobs; and unregulated international capital flows spawning economic bubbles that burst with dire consequences. Globalization, it is often claimed, is creating a world inhospitable to democratic values and economic justice—a world in which the proverbial little person is fast becoming smaller than ever.

But there is another, more benign face of globalization that has received less attention: the rise of a "global civil society" is linking together political activists and social reformers in different countries more tightly than at any time in history. Just as advances in communications technology have allowed global commerce to be conducted at a rising tempo, so, too, have they facilitated the growing transnational exchange of ideas and organizing tactics, creating a new sense of community among non-governmental organizations spread across the planet. "The global information revolution has transformed civil society before our very eyes," commented UN Secretary General Kofi Annan earlier this year.

The implications of global civil society are hazy, yet tantalizing. Will huge networks and coalitions of citizen activists come to rival international governmental organizations (IGOs) in the next century as leading vehicles of transnational cooperation? Will new democratic processes arise at the worldwide level that can offset the clout of global capital? And, will national public policy debates increasingly be influenced by social and economic norms that hold sway globally?

There Is Another, More Benign Face of Globalization

The concept of global civil society is hardly new, although the term has only come into widespread circulation in the 1990s. Over the past hundred years, efforts to strengthen cross-border links among nongovernmental organizations (NGOs) have run parallel to the far more visible crusade to create IGOs that could bring the rule of law to global affairs. At the dawn of this century, there was a proliferation of citizen associations in both the United States and Europe. This development, along with advances in transportation and communications in the early 1900s, produced early attempts to institutionalize global civil society. One notable example was the creation of the Central Office of International Associations, which was founded in 1907 by Henri La Fontaine (who would win the Nobel Peace Prize in 1913). Later renamed the Union of International Associations and still in existence today, this organization was founded with the goal of linking together nongovernmental groups in many different countries with the hope that such ties would help solve world problems.

The first half of the twentieth century also saw the creation of NGO coalitions to press specific agendas: international women's associations, labor union alliances, coalitions dedicated to disarmament and world peace, and associations aimed at strengthening international organizations. By 1939, there were an estimated 700 international NGOs. To many mid-century "One World" idealists, the creation of the United Nations in 1945 was but an interim step to true world government; and transnational links between citizens groups were seen as helping to pave the way to this dream. The World Federalist Movement, founded in 1947, is among the more well-known organizations dedicated to the goal of stronger world governance that emerged during this period.

The Implications of Global Civil Society Are Hazy, Yet Tantalizing

Decades of Cold War put a damper on the One World crusade. Today, however, the spirit of that movement can be seen in a range of new efforts to strengthen global civil society. While the current generation of transnational citizen activists do not talk so much of world government, they do still invoke other One World ideals—mainly the dream that some day a single set of political, social, and economic norms will bind all the planet's people together in a more just and democratic fashion. The rapid rise of global civil society during the 1990s can be seen in three main developments: the growing size and sophistication of international NGO coalitions and networks; the strengthening of ties between NGOs and IGOs; and the increasing frequency and effectiveness of *ad hoc* international campaigns by citizen activists.

In recent years, NGOs have become increasingly adept at banding together in common purpose. By pooling resources and coordinating their actions, they have strengthened their presence in international deliberations on a range of global issues. At a more formal level, the General Assembly of European NGOs has institutionalized efforts to build linkages among citizen groups in the European Union. Delegates to this organization are elected from each of the 15 EU countries and come together for regular conferences. At a less formal level, there now exist scores of international NGO networks that link NGOs at both a regional and global level. These include organizations like the Third World Network, a Malaysian-based organization that links up NGOs in the developing world that are working on environmental issues; and the International Federation of Human Rights, a Paris-based network of 89 human rights groups located in more than 70 countries.

In November 1995, over 80 NGO networks came together in Manila to discuss ways to leverage their growing presence in international affairs. Many of the participants at this meeting saw a stronger global NGO community as a critical counterweight to the rising power of transnational corporations. A paper written for the meeting stated: "In the long run, we have to invent the infrastructure so citizens can participate effectively in the democratic management of the global system. In the next decade, NGOs and their networks are one of the important precursors of an accountable global civil society. They are one of the few actors who try to articulate the global public interest."

The Concept of Global Civil Society Is Hardly New

In addition to organizing through networks, NGOs have also allied themselves across borders in committees or coalitions focused on particular issues. Prominent examples include the NGO Coalition for an International Criminal Court, the NGO Committee on Disarmament, and the NGO Initiatives on the UN Financial Crisis. Typically, these committees focus on influencing the United Nations or other international organizations. Some NGO groups endure for years; others fade away after a short time. Some are more mainstream and conservative; others operate at the fringe of accepted debate. One newcomer to this scene is the Conference of Peoples' Global Action Against Free Trade and the World Trade Organization. Holding its first international gathering in Geneva in May 1998, this coalition includes NGOs from Africa, Asia, and Latin America that are united in the view that economic globalization is having a highly destructive impact in developing countries.

The growth of global civil society has been manifested not just by new ties between NGOs, but by stronger links between NGOs and IGOs. When the UN was first founded in 1945, provisions were made to allow NGOs a very limited consultative status. Over time, the consultative rights of NGOs in regard to international organizations has expanded dramatically. NGOs now have a consultative role with a number of the UN's agencies, and the overall number of NGOs with consultative status at the UN has skyrocketed. Some 2,000 NGOs now have some of consultative status at the UN, with more than 70 large international NGOs enjoying the highest level of such status: some participation in UN proceedings. Kofi Annan has pledged to further strengthen these ties, saying in July 1998 that "a true partnership between NGOs and the United Nations is not an option; it is a necessity." In 1999, the United Nations University is sponsoring a major conference that will examine ways to build partnerships between civil society actors and the United Nations system. Cyril Ritchie, the main organizer of the event, explained that new institutional arrangements were needed to reflect the radically changed nature of power in the global system: "It is the people's participation in managing global problems and the increasingly complex role of civil society that distinguish our era from past ones."

Another focal point for NGOs seeking a greater formal role in global decision-making has been the major international conferences held during the 1990s. The Earth Summit in Rio in 1992 marked the emergence of NGOs as important players in such conferences. The 1,500 NGOs accredited to participate in the Rio conference often stole the show, putting forth a range of dramatic proposals and generally stirring up the proceedings at every turn. NGOs also played a prominent role in the Commission on Sustainable Development, an institution created as a result of Rio. At the 1994 world population conference at Cairo, NGO participation was even greater. And, in September 1995, the Fourth World Conference on Women attracted a stunning 35,000 NGO representatives to Beijing. The preparatory sessions for these international conferences, along with the new institutions that emerged in their wake, have been critical in strengthening cooperative ties among NGOs, as well as sharpening their advocacy skills within the global arena.

NGOs Have Become Increasingly Adept at Banding Together in Common Purpose

Perhaps the most dramatic display of rising NGO power during the 1990s has been the global campaigns for a ban on landmines and the creation of an International Criminal Court (ICC). The recent effort to stop the Multilateral Agreement on Investment (MAI) was also an impressive flexing of NGO muscle. In each case, NGOs demonstrated their ability to organize globally, exerting pressure on national governments and international organizations in a way that would have been unthinkable just a decade ago.

The International Campaign to Ban Landmines united a massive coalition of NGOs from all over the world to press successfully for a treaty, finally completed in 1997, to ban the production, stockpile, and export of landmines. From the beginning, this coalition was more virtual than actual. It was comprised of 1,000 NGOs in nearly 60 countries and was coordinated mainly over the Internet. Before the advent of e-mail, assembling and sustaining such a coalition would have been both enormously labor intensive and financially costly. The Internet allowed the campaign to have global reach on a shoestring budget. In awarding the Nobel Peace Prize to the Campaign and its lead coordinator, Jody Williams, the Nobel Committee cited the uniqueness of an effort that made it "possible to express and mediate a broad wave of popular commitment in an unprecedented way."

The advocacy by NGOs on behalf of an International Criminal Court is another striking example of what might be called "e-mail diplomacy." In the months leading up to an international conference on the ICC held in Rome in July 1998, a coalition of hundreds of NGOs swung into high gear. Operating again largely through the Internet, the NGO Coalition for an International Criminal Court mounted a global grassroots campaign to build support for the ICC, and also coordinated high-level lobbying for the ICC by international legal experts and other public policy leaders around the world.

A third notable example of this kind of global NGO organizing occurred earlier this year, when the Organization for Economic Cooperation and Development (OECD) were negotiating the Multilateral Agreement on Investment, a new set of rules to govern international investment among the OECD's 29 member states. NGOs organized to protest the MAI, coordinating their efforts over the Internet. During the campaign, new details of the emerging agreement were posted on the Internet within hours of becoming known, allowing a far-flung network of activists to respond rapidly to changing developments. The outpouring of global opposition led the OECD to postpone negotiations on the MAI in April.

It is premature to speculate about the long-term implications of the emerging global civil society. While it is abundantly clear that the dynamics of world politics have been dramatically altered in recent years, it remains to be seen whether this change presages real and lasting power shifts. Ultimately, cyber-savvy global activists may prove to be no match for transnational corporations and national governments intent on protecting their position and prerogatives. Moreover, while the UN is taking steps to collaborate more closely with NGOs, two of the most formidable power brokers on the international scene—the World Bank and the IMF—have showed few signs of following suit. Through the foreseeable future, the most likely scenario is that NGOs will exercise growing influence in the areas where they have already scored successes: namely, human rights, international security issues, and the environment. At the same time, the rise of global civil society is likely to have a far more limited impact on such core problems as the unequal distribution of global wealth and destabilizing flows of international capital—although this prediction could be confounded if backlash continues to mount against the negative aspects of globalization.

Certainly one thing is clear: the listservs and websites of global citizen activists are likely to be busier than ever in the next century.

Source: "What Is Global Civil Society?" by David Callahan, posted at www.civnet.org/journal/vol3no1/ftdcall.htm.

8

from the Centre for Research on Globalization

All This "Civil Society" Talk Takes Us Nowhere

Aziz Choudry
January 5, 2002

If there's one phrase I could do with hearing less of during 2002, it's "civil society." I'm not alone. Many of my friends, community activists and organisers in a number of countries, also cringe at the ritualized, ubiquitous usage of the phrase. We shudder at the thought that we might be mistaken for being part of it.

John Grimond, in The Economist's "The World In 2002," says of the phrase:

> It is universally talked about in tones that suggest it is a Great Good, but for some people it presents a problem: what on earth is it? Unless you know, how can you tell if you would want to join it?

I couldn't agree more. And if you found out, would you want to?

I still don't know what "civil society" is supposed to mean, let alone "international civil society."

Is it the name given to a group of representatives from various NGOs deliberating the latest sign-on statement or declaration at a meeting? It certainly seems to have become a kind of grandiose shorthand to describe groupings of NGOs which may or may not be connected with communities and broader peoples' movements in the countries in which they are based. Or is it something else?

Who gets to be in civil society and how? Are the people taking direct action on the streets against the authorities and being teargassed, pepper-sprayed, beaten and arrested part of civil society? Who gets to represent civil society and decide what, for whom, and on whose behalf?

There is no shortage of definitions of civil society; Gramsci, de Tocqueville, Putnam, Hegel, Marx and many others have written volumes on the subject. But other than general agreement that it spans all forms of organisations between the household and the state, the notion seems to mean all things to all people.

I cannot see how uncritical adoption and use of this term advances peoples' struggles for basic rights, for self-determination, liberation and decolonisation, and against imperialism and the neoliberal agenda in all their various guises. This is not a matter of a term emerging from peoples' struggles being coopted by the power elites and subverted—although the terminology lends itself entirely to the service of divide and rule strategies, and the marginalisation of social movements and more critical voices.

Civil society is a construct which allows politically and economically powerful institutions to decide who is in, and who is out, when and if it suits their interests. Furthermore, it has the added value of sounding broad and inclusive enough to add a gloss of legitimacy to any institution, programme, or system which can be shown to the public as somehow engaging with civil society, whoever that is.

Referring to the use of the phrase in the Philippines, after the recent mass mobilizations to oust President Estrada, Edmundo Santuario III observes, "it is now likewise being claimed by civic clubs, groups of rightist military elements, and politico-religious organisations, some legislators and others." Moreover, he argues, "civil society is actively bannered not necessarily as an antidote to poverty, corruption or as a vehicle for democratization, but to steer grassroots organisations away from the radical influence of political organisations calling for radical comprehensive revolutionary reforms."

In New Zealand, the Trade Liberalisation Network, a new business organization, entirely business led and driven, which was set up just prior to the Doha WTO Ministerial to promote "better public understanding and support for trade liberalisation" claims to be part of civil society. And of course, it is. So is the Mafia. So are feudal landlords, financial speculators, neoliberal thinktanks, corporate front groups, public relations spindoctors and CEOs of the most rapacious corporations. Yep, they're all part of the wondrous diversity of civil society, too.

James Petras and Henry Veltmeyer note in *Globalisation Unmasked* (Madhyam Books, New Delhi, 2001) that:

> Most of the greatest injustices against workers are committed by the wealthy bankers in civil society who squeeze out exorbitant interest payments on internal debt; by landlords who throw peasants off the land; and by industrial capitalists who exhaust workers on starvation wages in sweatshops. In their incisive chapter on 'NGOs in the Service of Imperialism,' they add: By talking about 'civil society,' NGOers obscure the profound class division, class exploitation and class struggle that polarizes contemporary 'civil society'.

At a time when the most powerful nation in the world has been pursuing the near obliteration of a country in a naked show of 21st century imperialism is this really a time to be "civil"? In confronting the power of global capital and exposing the role of neoliberal policies in inflicting misery and poverty in our communities and across the planet can we, should we, be "civil"?

There are now numerous foundations, institutions, courses, publications—and NGOs dedicated to studying, strengthening or building civil society. Over the past few years, astronomical sums of funding seem to have been sloshing around in the NGO world on "civil society"—related projects.

The Bretton Woods institutions, the "baby banks," many governments and big business are full of civil society rhetoric. The World Bank says it "welcomes the opportunity to work with civil society." According to its official documents, The Inter-American Development Bank's "work with civil society takes on many forms. At the operational level the Bank and its borrowers consult with civil society organisations (CSOs) and affected populations during the course of project preparation and implementation."

The Asian Development Bank seeks to "strengthen cooperation with civil society actors and to respond to their concerns." WTO Director-General Mike Moore says he welcomes scrutiny from civil society and that their engagement with the WTO "informs us and encourages us to do better."

How very nice. While their policies are helping to deepen poverty, inequality, desperation, injustice, and environmental destruction we can all feel so much better because these agencies will be engaging in "constructive dialogue" or consulting with carefully selected "representatives" of civil society.

There is no shortage of exhortations to "civil society" to form "partnerships" with business, government, and international institutions in order to supposedly eradicate poverty, save the environment or work towards some other noble-sounding goal. And plenty of takers in the NGO world where the term "civil society" seems well and truly entrenched and many seem willing to walk through fire to earn the right to mingle and meet with those in power.

Maori lawyer Moana Jackson says: "One of the difficulties we confront in the struggle against globalisation and the terrible things that neoliberalism, capitalism do to us is that we give away, we forfeit the right to name our world. We seek to demolish or deconstruct globalisation with the naming tools and ideologies of the capitalists who have constructed it. We seek to oppose the weapons of colonization with the language and words of colonization."

We need to have some clear starting points for how we define ourselves. And some clear values. As someone who has long described myself—for want of a better phrase—as part of "uncivil society," I won't be queueing to join civil society. And as a Filipina activist said to me recently, I think we need less NGOs, and more mass movements.

If we are to successfully dismantle the oppressive economic world order and create real alternatives to the neoliberal agenda, we need to stop defining ourselves in half-baked terms. We need much more precision. And we must develop strategies and campaign plans that are not driven by the frameworks of the institutions and processes which we oppose—urgently.

Source: Aziz Choudry, "All This 'Civil Society' Talk Takes Us Nowhere," January 5, 2002 from www.zmag.org/sustainers/content/2002-01/09 choudry.cfm.

Review Questions

1. Define *culture, civil society*, and *social movements*. How are the three related to each other? How are they distinct from each other?
2. What are some reasons why social movements attempting to change gender relations are so prominent in India?
3. How did British rule exacerbate cultural tensions in colonial India?
4. How is the dowry system connected to the problem of "missing women" in the population of India?
5. Why do many feminists focus their energies on the arena of civil society, rather than attempting to obtain political power (by gaining control of governmental policymaking) or economic power (by gaining control of corporte decision making)?

Discussion/Essay Questions

1. Consider the story of Manjamma, the woman profiled at the beginning of this module. Manjamma's father-in-law told her that "a woman's place is in the home" and that she had no business leaving her family just to meet with other women. Manjamma's father-in-law might have added that, in his opinion, people should gather outside the home only to conduct business and discuss politics, and neither of these are appropriate things for women to engage in. How might Manjamma respond to this in light of what you know about both the status of women and the role of civil society in India today? How might her father-in-law respond to Manjamma's reply?
2. Although the social movements profiled in the second through fifth readings all support working through civil society in order to transform culture, each movement has a different idea about the role that governments should play in this process. Briefly discuss each group's position on the role of the state in transforming culture (and/or of the role of cultural change in transforming government policy), and discuss which group you think will be most successful in its mission. Could this group serve as a model for a similar movement in an MDC? Conversely, which of the four groups do you think is least likely to succeed? Could this group serve as an example for what movements MDCs should avoid?
3. In the seventh reading, David Callahan asserts that civil society organizations can put a brake on the negative aspects of economic globalization. In the eighth reading, Aziz Choudry disagrees, arguing that institutions of civil society in practice are often used to subvert the political and economic movements that might actually create obstacles. Based on what you have read in the other six readings, do you agree with Choudry or Callahan?

List of Readings

1. Patralekha Chatterjee, "Civil Society in India: A Necessary Corrective in a Representative Democracy," in *D + C: Development and Cooperation*, November/December, 2001, pp. 23–24, posted on the website of Deutsche Stiftung für Internationale Entwicklung, *http://www.dse.de/zeitschr/de601-9.htm.*
2. Jagdish Parikh, "Forum Against Oppression of Women: Some Background Information," posted April 5, 1994 to the *misc.activism.progressive* newsgroup.
3. "Violation of the Human Rights of Women in Asia," testimony submitted December 29, 1999, by the Asian Legal Resource Centre to the United Nations Commission on Human Rights, U.N. document E/CN.4/2000/NGO/65 (February 8, 2000), posted on the website of the Asian Human Rights Commission,

http://www.ahrchk.net/hrsolid/mainfile.php/2000vol10no04/398.

4. Kalpana Sharma, "Breaking the Shackles," *The Hindu*, April 4, 1999, posted on the website of India Mediawatch, *http://www.media-watch.org/articles/0499/76.html.*
5. Sumona Das Gupta and Ashima Kaul Bhatia, "Women in Conflict Resolution: The Road Ahead in Kashmir," December 31, 2001, posted as Article No. 671 on the website of the Institute for Peace and Conflict Studies, *http://www.ipcs.org/ipcs/issueIndex2.jsp?action=showView&kValue=222&issue=1012&status=article&mod=b.*
6. "What Is Civil Society?" and "Social Capital and Civil Society," both revised October 10, 2002, posted on the website of the World Bank, *http://www.worldbank.org/poverty/scapital/sources/civil1.htm* and *http://www.worldbank.org/poverty/scapital/sources/civil2.htm.*
7. David Callahan, "What Is Global Civil Society?" in *Journal*, January–February 1999, posted on the website of Civitas International, *http://www.civnet.org/journal/vol3no1/ftdcall.htm.*
8. Aziz Choudry, "All This 'Civil Society' Talk Takes Us Nowhere," posted on the website of the Centre for Research on Globalization, January 5, 2002, *http://www.globalresearch.ca/articles/AZI201A.html.*

Websites for Additional Research

1. The South Asian Women's Network (SAWNET) provides a forum for discussion of South Asian women's issues as well as descriptions of (and links to) numerous South Asian women's NGOs, both in Asia and abroad, at *http://www.umiacs.umd.edu/users/sawweb/sawnet.*
2. The World Bank's PovertyNet website includes a library of online (or partially online) documents on topics with which it is involved. While some of these documents are generated by the World Bank, most come from other sources, such as newspapers, magazines, and academic journals. The library can be found at *http://poverty.worldbank.org/library* and the list of articles on civil society can be found at *http://poverty.worldbank.org/library/topic/4294/5026.*
3. If a global civil society is forming, the World Wide Web certainly is one of the network's main media. Many groups that are part of this global civil society are linked through the Association for Progressive Communications (APC), a network of networks that uses information and communication technologies to support nongovernmental organizations. APC (and its various member organizations) can be found on the web at *http://www.apc.org.* Dutch sociologist Albert Benschop maintains a list of links to social movements (and links to online articles about social movements) at *http://www2.fmg.uva.nl/sociosite/topics/activism.html.*

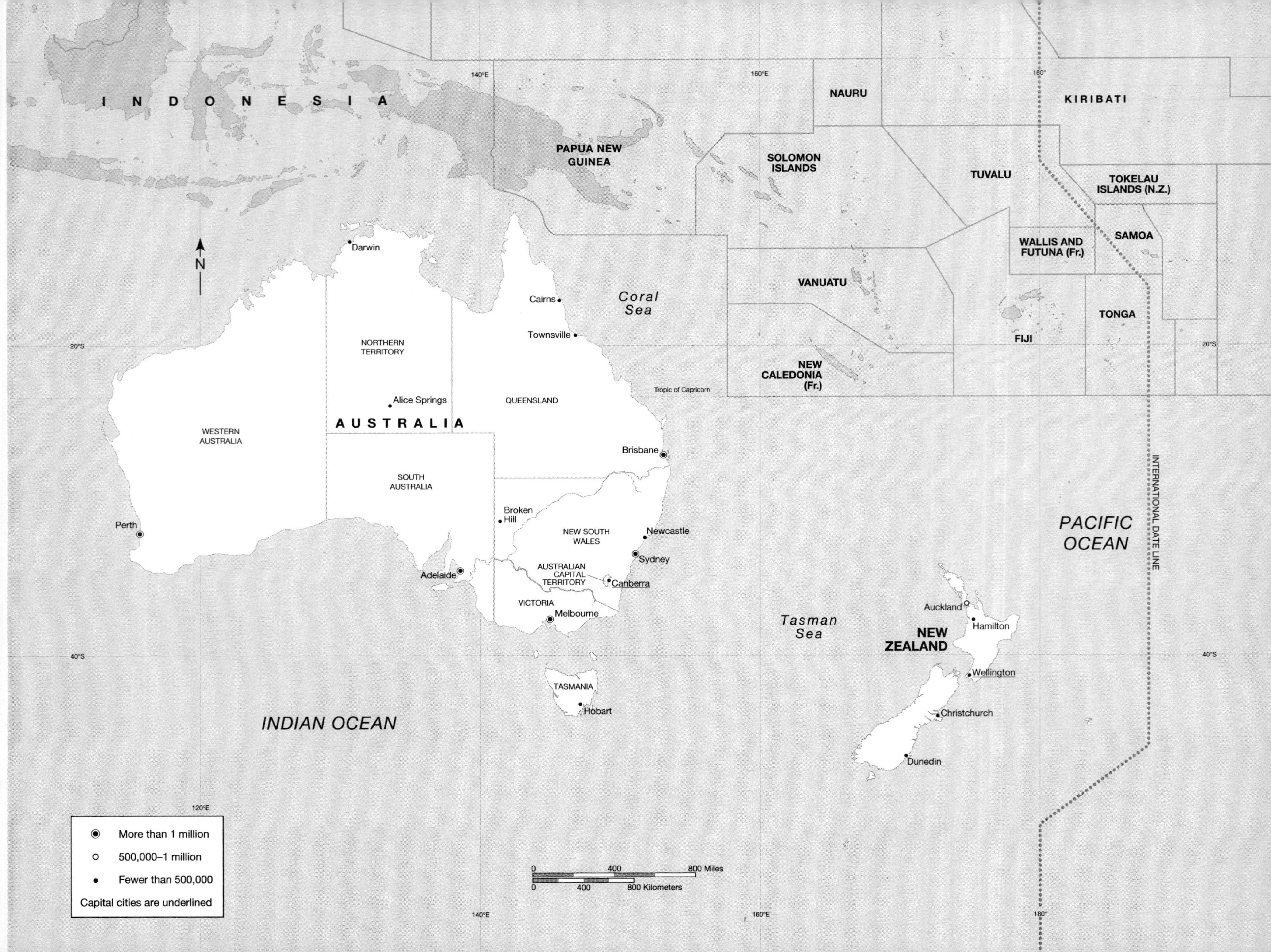

INDONESIA
PAPUA NEW GUINEA
NAURU
KIRIBATI
SOLOMON ISLANDS
TUVALU
TOKELAU ISLANDS (N.Z.)
SAMOA
WALLIS AND FUTUNA (Fr.)
VANUATU
FIJI
TONGA
NEW CALEDONIA (Fr.)
Coral Sea
Tropic of Capricorn
AUSTRALIA
Darwin
Cairns
Townsville
NORTHERN TERRITORY
Alice Springs
QUEENSLAND
WESTERN AUSTRALIA
SOUTH AUSTRALIA
Brisbane
Broken Hill
NEW SOUTH WALES
Newcastle
Sydney
AUSTRALIAN CAPITAL TERRITORY
Canberra
Perth
Adelaide
VICTORIA
Melbourne
TASMANIA
Hobart
INDIAN OCEAN
Tasman Sea
PACIFIC OCEAN
INTERNATIONAL DATE LINE
NEW ZEALAND
Auckland
Hamilton
Wellington
Christchurch
Dunedin
140°E
160°E
180°
120°E
20°S
40°S
N
More than 1 million
500,000–1 million
Fewer than 500,000
Capital cities are underlined
0
400
800 Miles
800 Kilometers

Language and Education Policy in Australia and New Zealand

Companion to Chapter 5 Language

When **Jennifer Wang** arrived in Australia from China a simple "hello" greeting in English was way beyond her capabilities. Now, as payroll officer and human resources assistant for a large financial company, her communication skills are daily put to the test, as she writes letters of offer to prospective employees. Not bad for someone who not so long ago "couldn't string four words together!"

"When I first arrived, my husband took me to a shopping centre and I couldn't understand anything—not even 'hello, how are you?'," said Jennifer. "My ambition when I arrived was to resume the career I had in China, practising in Chinese medicine, but I knew if I was going to get anywhere, I had to begin by learning the language."

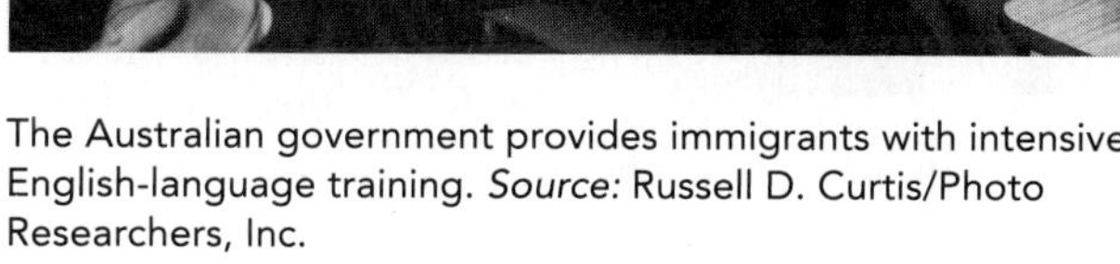

The Australian government provides immigrants with intensive English-language training. *Source:* Russell D. Curtis/Photo Researchers, Inc.

And she was very glad to be able to do so with the help of the Australian government's Adult Migrant English Program (AMEP). "The class teachers dealt with very basic English, which was just what I needed. The first sentence I remember learning was 'How long have you been here?'," she said.

"My home tutor was a wonderful teacher. She was very patient and went out of her way to tailor the lessons for my needs," she said. "When I got a job as a receptionist with an electrical business, I had problems with the names of things and pronunciation, so my teacher would especially prepare my lessons using the words I needed for work—socket, lead, and so on. Because of her, my English improved very quickly, and I was given the job permanently and got a salary raise."

With her confidence and her language skills growing, Jennifer felt it was time to move on to "bigger and better" things, so she applied for and won her current job. Describing it as "quite a jump," she nonetheless feels relaxed about handling the greater responsibility that this entails—and with her new life in general. "I'm really happy with how things have turned out," she said. "I have a great job and I'm making lots of friends who don't mind me practising my English on them! When I feel ready, I'll practise in Chinese medicine. Then I will have achieved my ultimate goal, which is to help Australian people benefit from the natural treatments that Chinese medicine offers.

Adapted from: "Success Story: Jennifer Wang Prescribes English For a Happy and Healthy Future," posted on the website of the Australian Department of Immigration and Multicultural & Indigenous Affairs' Adult Migrant English Program, http://www.immi.gov.au/amep/success/wang.htm.

Map Source: Map adapted from pp. 572–573 in *World Regions in Global Context* by Marston et al., © 2002. Reprinted by permission of Pearson Education, Inc., Upper Saddle River, NJ. ■

The Personal Politics of Language

On first consideration, language appears to be among the least political aspects of a country's culture. Language is something that we use and reproduce without consciously thinking about it. One could almost say that language comes to us "naturally."

And yet, if you are an English speaker who grew up in a predominantly English speaking country, your command of the English language is anything but natural. You probably were continually enrolled in "English" classes from your first year in school—where you learned to read and write in English—through the end of high school, if not longer. When your teacher spoke to you only in English and demanded responses in English as well, she effectively was telling you that you needed to speak English if you wanted to get by. You may have been given the message that other languages (for instance, the non-English language that you may have spoken at home or perhaps in your neighborhood) might be acceptable in specific settings, but in most cases your school sent you a consistent message that English was to be the language of public social interactions.

School is not the only place where one receives information about linguistic norms and expectations. As one grows older and moves away from one's home place, one receives further messages, not only about English being the standard language of interaction but also about how one should speak the language. A woman from the southern United States who moves to the northeast finds that when she speaks with her slow southern drawl people assume that she is lazy and stupid. She quickly learns to speak more rapidly and authoritatively when she goes to a job interview. A northerner who moves to the south finds that when people hear his quick, straightforward speech they assume that he is aggressive and arrogant. He quickly learns to slow down his speech and engage in pleasantries before ordering food at a restaurant.

In short, when we use a language we make some very *personal* political decisions, not unlike many of the other personal political decisions that we make as we go through life. With every word that one speaks one makes a decision about the extent to which one is going to compromise to fit in with society and the extent to which one will retain one's individuality. Clearly, some compromise is necessary; if everyone spoke an "individual language" no one would understand anyone else. At the same time, though, individuals and subnational communities have very good reasons for wishing to retain their distinct languages (or their distinct ways of speaking a language). When the southern woman shortens her vowels so that northerners will not think she is lazy, is she just making a pragmatic decision so that people will take her more seriously, or is she at some level declaring that the northern way of speech is somehow "better"? Has she changed her pronunciation because she is just trying to fit in, or has she changed it because she has internalized the northerners' suspicion of her southern roots and is trying to stop them from appearing in her language? How can she defend herself when she visits back home and her relatives, noting that she is "talking like a Yankee," accuse her of being ashamed of her heritage?

The Politics of Language Policy

While the act of speaking (or not speaking) a language is a personal decision, the politics behind this decision is wrapped up in broader arenas of politics. To continue with the examples from the previous section, in the United States there is a difference between the southerner who starts talking like a northerner and the northerner who starts talking like a southerner. When the northerner who has begun to use southern speech patterns returns home, he may be thought of as strange, but, unlike the southerner who starts talking like a northerner, it is unlikely that he will be accused of "selling out" or of being ashamed of his heritage. The reason for this is simply that the cultural history of the United States is laden with examples (whether justified or not) of northerners urging southerners to adjust their behavior so that they will fit better into the (northern-defined) national norm. The southern response typically has not been that northerners should start behaving more like southerners, but rather that northerners should simply leave them alone. These dynamics, in turn, reflect an imbalance in economic power that historically has characterized relations between the north and the south. Thus, the southerner who starts talking like a northerner likely will appear to some as having acquiesced to a long-standing campaign to obliterate the distinct cultural practices of the southern United States, while the northerner who starts talking like a southerner at worst will be teased for having "gone native."

Not only are individual decisions about whether to preserve a regional language or dialect wrapped up in broader power relations between the majority and the minority, they also reflect governmental decisions about how important (or desirable) it is to integrate subcultures into the national identity. Some national governments, like Switzerland or Belgium, for instance, recognize that the country consists of several distinct national-linguistic groups, and each group (and its language) is accorded equal prominence in constructing the Swiss or Belgian identity. Other national governments, such as the United Kingdom, recognize the majority language as the defining language (and culture) of the nation and it is expected (or in some cases mandated) that all official business at the national scale will be conducted in that language. At the same time, however, these governments recognize the legitimacy of regional minority languages (such as Welsh, in the case of the United Kingdom) and instruction in these languages is encouraged and even subsidized by the national government. In contrast, there are other countries with single dominant languages in which the retention of minority languages (whether by indigenous cultures or by immigrant communities) is seen as a threat and those who persist in speaking minority languages are viewed as unpatriotic. Prior to the mid-twentieth century this was the attitude of the British government toward Welsh and other regional languages, and it remains the attitude of "English Only" proponents in the United States.

Language policy can be expressed through a wide variety of laws and mandates. For instance, some municipalities in the United States prohibit the publication of official documents in any language other than English, and many countries have official languages. Most often, however, language policy debates center on education. Education is where future citizens are molded and instructed in what it means to be a citizen of their nation. If one defines being American as speaking English (or, to be more precise, speaking "American English"), then a school whose language curriculum is guided by any goal other than instructing all students to use standard American English as their primary language is falling short of its goal. Spanish may be acceptable as a "foreign language" subject and Victorian English novels may be acceptable as "literature," but these subjects are to be taught simply as subjects to be mastered, whereas instruction in the (American) English language is to be taught so that it shapes and defines the student as a member of American society.

Others, who view the relationship between language and identity as more flexible, may see a greater role for non-English languages (or non-standard-American forms of English) in the classroom. Recognizing the growing number of Americans whose primary language is Spanish, some educators encourage schools to adopt language curricula where students who speak English at home are immersed in Spanish and those who speak Spanish at home are immersed in English. Such an educational program creates bilingual adults and effectively constructs (or acknowledges the reality of) a bilingual America that is unique in its blend of English and Spanish linguistic and cultural elements.

Dueling Policies in Language Education

In practice, language education policies are usually determined with reference to new members of society (whether children or immigrants) who do not speak the majority language. Few people deny that there are benefits to multilinguality, and almost no one is against school programs that instruct speakers of the majority language in a second, foreign language. Like the person from the northern United States who learns to speak like a southerner, his identity as a member of the majority is secure, and any skill that he gains in communicating with others in their language (or dialect) generally is viewed as a positive development.

Issues of language instruction, however, become much more politically loaded when the student entering the classroom speaks a language (or dialect) different from the officially or unofficially sanctioned majority language. Expanding on the debate as it has played out in the United States, two paradigms generally prevail toward education of non-English-speaking school children and immigrants. Proponents of English as a Second Language (ESL) note that English is the standard language of business and all other social interactions in the United States and, if a member of an immigrant or minority community hopes to "make it" in the United States, she or he must become culturally and linguistically fluent in the dominant, English-language society. Thus, the typical educational program advocated by ESL proponents for immigrant school children is to immerse them for a few years in an ESL program and then, once their English language skills are up to speed, mainstream them into regular classes that are taught in the English language, so that these children can receive the same education as their native-English-speaking counterparts.

Proponents of the other paradigm, bilingual education, acknowledge that success in the United States requires mastery of English, but they feel that command of English should not be the *only* goal of education for non-English-speaking students. For bilingual education proponents, English-language skills should be considered, like mathematics, as an essential skill for students to master so as to prepare them for adulthood. However, they argue, the ESL paradigm treats the student's prior language skills as a liability to be unlearned, when instead these skills should be used as the foundation for teaching students about other subjects (including the important skill of mastering English). In addition, bilingual education proponents charge that ESL programs do more than simply tell students that their language is not the language of America. ESL programs also (whether intentionally or unintentionally) give the message to students that their culture and outlook are not sufficiently American and that the purpose of school is to "Americanize" them. Given the low rates of performance and low confidence among students in predominantly minority and immigrant schools, bilingual education advocates argue that ESL sends precisely the *wrong* message by telling students that they must ignore the cultural and linguistic skills that they have acquired at home and instead learn the English language and American social skills. Furthermore, bilingual advocates assert that, under an ESL curriculum, by the time a student is mainstreamed into, say, an English-language math class, he or she is several grades behind, and this experience of being "left back" further reduces the student's confidence.

As an alternative to ESL, bilingual education advocates typically favor curricula where the bulk of classes (including a required English-language class) are taught in the student's native language. Over time, as the student's English skills improve, the student may take more and more classes taught in English. At the same time, to encourage integration of the immigrant students into the school's mainstream and to emphasize that all of the cultures and languages within the school are legitimate American cultures, native-English-speaking students may be encouraged (or required) to take classes where they learn the language and culture of the school's major immigrant or minority group.

Although ESL programs typically are advocated by representatives of majority communities and bilingual education programs are more often favored by immigrants and minorities, the divisions are not always so clear. The textbook mentions the controversy that arose when the Board of Education in Oakland, California, instituted a bilingual "Ebonics" school curriculum that was devoted to teaching students standard American English alongside instruction in the African-American dialect that was said to be the first language of most Oakland students. Some of the opposition to the Ebonics curriculum came from cultural conservatives who argued that the purpose of education should be to teach students how to integrate into the American mainstream and that therefore the "improper" grammatical constructions of African-American English should be corrected rather than affirmed. Other opponents, however, included a number of African-American parents. These parents, like parents everywhere and especially in poor communities, wanted a better life for their children. They perceived that the only way that their children could achieve that better life was through acquiring the skills that would allow them to assimilate into mainstream (white-dominated) America. They argued that by affirming cultural (and linguistic) patterns that whites perceive as indicating ignorance or provincialism, the Ebonics curriculum would only hinder that assimilation. In contrast to these parents, other African-Americans argued that Oakland's Ebonics program—by recognizing African-American language as a legitimate dialect of American English—was a step toward recognition of African-American culture as a legitimate American subculture, and they argued that it therefore warranted the support of the African-American community.

The politics of language education becomes even more complicated in countries where multiple languages prevail. In South Africa, for instance several of the riots and strikes that paved the way for the resistance movement that eventually toppled the white supremacist *apartheid* regime were led by black youths fighting *for* the teaching of English in the country's segregated blacks-only schools. Why were Africans fighting for the right to be taught in a colonial language? The answer lies in the specifics of South African colonial politics and history (described in greater detail in Chapter 7 of your textbook). The South African white population is split between the Afrikaners, descendents of Dutch and French Protestants who arrived in the seventeenth and eighteenth centuries, and the English, who came in the nineteenth and twentieth centuries. During the first half of the twentieth century, the English dominated the country economically and politically, and the Afrikaners, who typically were poor farmers, were only a cut above the black Africans in the country's hierarchy. After World War II, although the English maintained economic power, the Afrikaners gained control of the government. Because they could not wrest economic power from the English, they used their political power to further exploit and segregate black Africans. The English language posed a threat to the Afrikaners because it provided a vehicle for black and English South Africans to unite. Thus, the Afrikaner-dominated government instituted a policy where only indigenous African languages and Afrikaans (the Afrikaner language, which is similar to Dutch) could be taught in black African schools, while Afrikaner schools instructed in Afrikaans (often with English as a second language) and English schools instructed in English (often with Afrikaans as a second language).

Canada also has experienced some unusual language politics. Some of the most pronounced opposition to mandatory bilingual education (in English and French) has come from indigenous Canadians and from immigrants whose native language is neither English nor French. Normally, one would expect these communities to support multilingual education, as it would indicate a general tolerance toward a multiplicity of Canadian subcultures (including immigrant and native subcultures). For indigenous groups, however, the French-Canadian drive to define Canada as a *bi*cultural country often is seen as working against their desire to construct a Canada tolerant of all cultures that recognizes the special place of native-Canadians in the Canadian ethnic mosaic. For immigrants who are seeking to integrate into and move up within Canadian society, the requirement that their children learn French as well as English is seen as an unnecessary diversion when they recognize that English is the more important language for achieving success in global (or even Canadian) society.

Language Policy in Australia and New Zealand

As in South Africa and Canada, historic and ongoing language policy debates in Australia and New Zealand reflect the competing interests of English settlers, indigenous populations, and non-English settlers. Both countries' policies have changed several times in the past hundred years, and the two countries now have quite different policies, in part due to their very different populations.

Turning first to Australia, British immigration did not begin until the late eighteenth century. With the help of European-borne diseases and violence, the European population quickly surpassed the native—or Aboriginal—population. Soon, however, the prospect of large-scale migration from Asia emerged as a potential threat to European settler dominance. Although very far from England, the north coast of Australia lies just across the Torres Strait from what is now Indonesia and Papua New Guinea, and China and the rest of Asia lie not far beyond. The prospect of Asian migration has always been a threat to white Australians seeking to build a British commonwealth in the South Pacific.

In response, the British developed the "White Australia" immigration policy, which prohibited non-Europeans from immigrating. The origins and eventual demise of the "White Australia" policy is detailed in the first reading, a fact sheet on the country's immigration history published by Australia's Department of Immigration and Multicultural & Indigenous Affairs. As the fact sheet demonstrates, Australia's immigration policy was, from the beginning, interconnected with its language policy: One of the means used to exclude non-Europeans was a clause requiring every prospective migrant to "write out dictation and sign in the presence of an officer a passage of 50 words in a European language directed by the officer." After World War II, the "White Australia" policy began to be dismantled. The dictation test was eliminated in 1957 and by 1978 Australia had developed an immigration policy that welcomed non-Europeans. Between 1978 and the present the Australian government has issued a number of reports asserting that Australia is a multicultural country, and today one fourth of all Australians are immigrants, in many cases from non-European countries.

Multicultural, however, does not necessarily mean multilingual. According to the group that produced the second reading, the Australian Alliance for Languages (AAL), in 1987 Australia produced a National Policy on Languages that celebrated the diversity of languages spoken in a multicultural Australia, but since then the government has changed its emphasis. Although still advocating multiculturalism and seeking immigrants from around the world, now the English language is being promoted as a tool for uniting the various cultures that constitute Australia into one nation. According to the AAL, educational initiatives (for young Aboriginals and for adult immigrants and their school-age children) now are oriented toward teaching English rather than maintaining a multiplicity of languages. This policy of fusing multiculturalism with monolingualism is expressed in the third reading, excerpts from the Australian government's *New Agenda for Multicultural Australia*, published in 1999. The first part of the document advocates multiculturalism, but the recommendation on language policy (Recommendation 28) is restricted to English language instruction. The next recommendation (no. 29) is about instruction in non-English languages, but, while this paragraph advocates Australians learning non-English languages as *second* languages, it ignores any desire that native non-English speakers might have to maintain fluency in their ancestral tongue.

Having made its decision to build a multicultural society through universal use of the English language, the Australian government has recognized the need to assist immigrants in learning English. To this end, the government has established the Adult Migrant English Program, which is discussed in the personal profile that introduced this chapter and also in the fourth reading.

Consistent with the Australian government's move toward promoting monolingualism over multilingualism, the government of the Northern Territory (the most remote of Australia's eight states and territories, consisting primarily of barren "outback," much of which is reserved Aboriginal land) announced in 1998 that it would eliminate its bilingual education programs in Aboriginal schools, ostensibly as a cost-savings measure. In the fifth reading, the Northern Territory government announces this change as part of an overall education restructuring package. The sixth reading is a letter from the president of the Australian Linguistic Society, opposing this change in the Northern Territory language policy. The author, Peter Austin, a professor of linguistics at the University of Melbourne, expands on some of the themes introduced earlier in this module as he argues for bilingual education instead of ESL, in part by making the point that English can be taught within a bilingual-bicultural education program and that students learn better in a bilingual environment that reinforces and affirms their distinct identity. In the next reading, Aboriginal activist and scholar Lester Irabinna Rigney addresses some of these points while stressing the role that preservation of language plays in preserving endangered indigenous cultures, which must be included in the new, multicultural Australia.

The remaining four readings in this module are from New Zealand. Like Australia, New Zealand, was colonized by British settlers around the beginning of the nineteenth century and, like Australia, these settlers and their descendents until recently were dedicated to maintaining the country's "British" character. Those similarities notwithstanding, there are substantial differences between the two countries, which are reflected in their divergent language policies. Some 1,200 miles (2,000 kilometers) east of Australia, New Zealand is far from any large land mass or island. So, unlike Australia, New Zealand has not (until recently) attracted much immigration by Asians or non-British Europeans. Even the native Polynesians of New Zealand, the Māori, arrived there only a little over 1,000 years ago. On the other hand, while there are many fewer non-British settlers in New Zealand, the indigenous population is much more significant than in Australia. While Australia's population is only about 1 percent Aboriginal, New Zealand's population is between 10 and 13 percent Māori, which gives an entirely different climate to language policy.

As in Australia, until recently much of language policy in New Zealand was devoted to affirming the centrality of the English language. While in Australia this policy was used primarily to keep out non-European immigrants, in New Zealand it primarily was used to oppress the Māori. As in Australia, after World War II these policies began to undergo revision. However, the revision went in a different direction in New Zealand. While Australia redefined itself as a multicultural land of immigrants, all of whom were united by the English language, New Zealand redefined itself as a bilingual society built on the fusion of English and Māori culture and language. To this end, in 1987 the New Zealand Government passed the Māori Language Act, which named Mā ori and English as the country's two official languages and established the Māori Language Commission to promote te Reo Māori, the Māori language. The eighth reading outlines the objectives of the Commission and the ninth reading is a brief statement by the Commission about the language's current status in New Zealand. Note the relatively few New Zealanders who regularly speak Māori. A 1990 survey by researchers Rangi Nicholson and Ron Garland revealed that, at most, 3 percent of New Zealanders classify themselves as fluent in te Reo Māori (and the actual percentage probably is considerably lower). Of those who classified themselves as fluent, all were of Māori heritage and 80 percent were over age 50.

The last two readings show the difficulties encountered when one attempts to revive a minority language and reclaim it as a major national language. Bilingual education requires highly skilled teachers and, in cases where few speak the language, teachers may be hard to find. The tenth article reports on a school in New Zealand that has been having trouble finding a teacher to fill a bilingual education position. The final reading demonstrates that, even as its path has diverged from Australia's, New Zealand is still in many ways an English-language society. Practically all New Zealanders, including Māori who also speak te Reo Māori, are fluent in English. Even while the country's educational policy is committed to bilingualism, its immigration policy, like that of Australia, is increasingly built around a recognition that English is and will remain the country's one, universal language. In fact, the linguistic element of New Zealand's immigration policy is considerably stricter than Australia's; while the Australian government only requires that immigrants learn English (and it helps them by providing free English lessons), New Zealand requires that immigrants enter the country with enough English-language skills to study for a doctoral degree.

Readings

from the Australian Department of Immigration and Multicultural & Indigenous Affairs

Abolition of the "White Australia" Policy

Updated 13 February 2003

The abolition of the "White Australia" policy was a gradual process that took place over a period of 25 years.

The first step towards a less discriminatory migration policy was taken by Immigration Minister Harold Holt, with bipartisan support from the Australian Labor Party.

Following the election of a coalition of the Liberal and Country parties in 1949, Mr. Holt allowed 800 non-European refugees to remain in Australia and Japanese war brides to enter Australia.

Over the next 24 years, the government gradually removed most of the remaining discrimination, with the final vestiges being removed in 1973 by the new Labor Government.

The History

The origins of the "White Australia" policy can be traced back to the 1850s. White miners' resentment towards industrious Chinese diggers culminated in violence on the Buckland

River in Victoria, and at Lambing Flat (now Young) in New South Wales. The governments of these two colonies introduced restrictions on Chinese immigration.

Later, it was the turn of hard-working kanakas (indentured labourers from Pacific islands) in Northern Queensland. Factory workers in the south became vehemently opposed to all forms of immigration, which might threaten their jobs—particularly by non-white people who they thought would accept a lower standard of living and work for lower wages.

Some influential Queenslanders felt that the colony would be excluded from the forthcoming Federation if the kanaka trade did not cease. Leading NSW and Victorian politicians warned there would be no place for "Asiatics" or "coloureds" in the Australia of the future.

In 1901, the new Federal Government passed an Act ending the employment of Pacific Islanders. The new *Immigration Restriction Act 1901* received Royal Assent on 23 December 1901. It was described as an Act "to place certain restrictions on immigration and to provide for the removal from the Commonwealth of prohibited immigrants."

Among those it prohibited from immigration were the insane, anyone likely to become a charge upon the public or upon any public or charitable institution, and any person suffering from an infectious or contagious disease "of a loathsome or dangerous character."

It also prohibited prostitutes, criminals, and anyone under a contract or agreement to perform manual labour within the Commonwealth (with some limited exceptions). Other restrictions included a dictation test, used to exclude certain applicants by requiring them to pass a written test in a specific language, with which they were not necessarily familiar.

The Act stated the migrant had to "write out dictation and sign in the presence of an officer, a passage of 50 words in a European language directed by the officer."

Regardless of these severe measures, the implementation of the White Australia policy was warmly applauded in most sections of the community. In 1919 the Prime Minister, William Morris Hughes, hailed it as "the greatest thing we have achieved."

Second World War

After the outbreak of hostilities with Japan, Prime Minister John Curtin reinforced the philosophy of the White Australia policy, saying, "this country shall remain forever the home of the descendants of those people who came here in peace in order to establish in the South Seas an outpost of the British race."

During World War II, many non-white refugees entered Australia. Most left voluntarily at the end of the war, but many had married Australians and wanted to stay. Arthur Calwell, the first Immigration Minister, sought to deport them, arousing much protest.

Minister Holt's decision in 1949 to allow 800 non-European refugees to stay, and Japanese war brides to be admitted, was the first step towards a non-discriminatory immigration policy.

The Next Major Step

The next major step was in 1957 when non-Europeans with 15 years residence in Australia were allowed to become Australian citizens.

The revised *Migration Act of 1958* introduced a simpler system of entry permits and abolished the controversial dictation test. The revised Act avoided references to questions of race. Indeed, it was in this context that the Immigration Minister, Sir Alexander Downer, stated that, "distinguished and highly qualified Asians" might immigrate.

After a review of the non-European policy in March 1966, Immigration Minister Hubert Opperman announced applications for migration would be accepted from well-qualified people on the basis of their suitability as settlers, their ability to integrate readily and their possession of qualifications positively useful to Australia.

At the same time, the Government decided a number of non-Europeans, who had been initially admitted as "temporary residents", but who were not to be required to leave Australia, could become residents and citizens after five years (the same as for Europeans), instead of 15 years as had earlier been required.

There was also an easing of restrictions on non-European migrants. The criteria of "distinguished and highly qualified" was replaced by the criteria of "well qualified" non-Europeans and the number of non-Europeans allowed to immigrate would be "somewhat greater than previously."

A Watershed

The March 1966 announcement began a period of steady expansion of non-European migration and was the watershed in abolishing the "White Australia" policy. Yearly non-European settler arrivals rose from 746 in 1966, to 2696 in 1971, while yearly part-European settler arrivals rose from 1498 to 6054.

In 1973 the Whitlam (Labor) Government took three further steps in the gradual process to remove race as a factor in Australia's immigration policies. These were to:

- legislate to make all migrants, of whatever origin, eligible to obtain citizenship after three years of permanent residence;
- issue policy instructions to overseas posts to totally disregard race as a factor in the selection of migrants; and
- ratify all international agreements relating to immigration and race.

Because the Whitlam Government reduced the overall immigration intake, the reform steps that it took had very little impact on the number of migrants from non-European countries. An increase in the number and percentage of migrants from non-European countries did not take place until after the Fraser Government came into office in 1975.

Continuing the Trend

In 1978 there was a comprehensive review of immigration in Australia. Far-reaching new policies and programs were adopted as a framework for Australia's population development. The main points adopted were three-year rolling programs to replace the annual immigration targets of the past, a renewed commitment to apply immigration policy without discrimination, a more consistent and structured approach to migrant selection and an emphasis on attracting people who would represent a positive gain to Australia.

An average net population gain of 70,000 was forecast for the first triennium 1978–81 and this was estimated to assist in raising Australia's population to about 19 million in 2001.

The Present

Today, almost one in four of Australia's 19 million people were born overseas. In 2001/2002 the number of arriving settlers by country of birth totalled 88,900 and they came from more than 150 countries. Most came from New Zealand (17.6 percent), the United Kingdom (9.8 percent), China (7.5 percent) South Africa (6.4 percent) India (5.7 percent) and Indonesia (4.7 percent).

Australian Multiculturalism

The *New Agenda for Multicultural Australia* was tabled in Parliament on 9 December 1999. It contains a framework which aims at making multiculturalism relevant to all Australians, ensuring that the social, cultural, and economic benefits of our diversity are fully maximised in the national interest, and encouraging harmonious relationships between people or organisations of different cultural backgrounds.

Australian multiculturalism recognises and celebrates our cultural diversity. It accepts and respects the right of all Australians to express and share their individual cultural heritage within an overriding commitment to Australia and the basic structures and values of Australian democracy. It also refers specifically to the strategies, policies and programs that are designed to:

- make our administrative, social and economic infrastructure more responsive to the rights, obligations and needs of our culturally diverse population;
- promote social harmony among the different cultural groups in our society; and
- optimise the benefits of our cultural diversity for all Australians.

Source: "Fact Sheet No. 8: Abolition of the 'White Australia' Policy," revised February 13, 2003 at immi.gov.au/facts/08abolition.htm.

from the Australian Alliance for Languages

2

Statement of Needs and Priorities for Language Policy for Australia

Preface

2001, the year of the Centenary of Australia's political federation, represents an excellent historical moment in which Australian cultural democracy can reflect and be supported by the public policy processes and institutions of Australian life. In the report of the Centenary of Federation Committee to the Council of Australian Governments (August, 1994) the Committee described Australia's language policies as one of the significant achievements of Australian federation:

> In 1987 the federal Government adopted a National Policy on Languages, becoming the first English speaking country to have such a policy and the first in the world to have a multilingual languages policy.

Australia is uniquely well placed among the nations of the world to set in place public policies for languages and literacies which are enlightened, progressive, socially just, and economically wise as well as practical. During the 1990s the scope and range of the policies enunciated by Federal authorities have shrunk so that at present there is no overall set of principles that guide policy, resulting in fragmented provision, and contradictory tendencies between states and within the Federal sphere.

This is characterised by the following retreats from and omissions in policy:

- During the 1990s, all federal governments have retreated from a commitment to issues of cultural and linguistic diversity, and have failed to recognise the benefit to Australian society and the centrality in all policies of social justice.
- Government and policy rhetoric often deliberately omit a commitment to linguistic and cultural diversity with the consequence that there has been a failure to promote cultural and linguistic diversity as a means of fostering social cohesion.
- There has been a re-categorisation of linguistic and cultural diversity as being linked to disadvantage, rather than seeing it as resource for the entire Australian community.

Australian identity is well served by a pluralistic and inclusive multicultural language policy based on the rights of all of Australia's language and cultural groups to develop their unique differences within the context of a united and harmonious nation. At a time of rapid and intense economic globalisation, the preservation of the diversity and skills of Australia's population will assist the nation in its efforts to cultivate its skills and knowledge assets to meet the challenges of a competitive trading environment.

The Alliance views language and literacy policy as central to the promotion of Australia's interests and much wider than education, encompassing the media, law, health, the arts, all levels of education, and public affairs. We urge that Australian public authorities make the concept of a tolerant, diverse yet united Australia a central part of the entire mode of governance so that political forces that seek to sow the seeds of division and conflict can be marginalised.

Source: "Preface to the Statement of Needs and Priorities for Language Policy for Australia," revised October 5, 2001 on latrobe.edu.au/rett/als/aa/p.html.

3

from the Government of the Commonwealth of Australia

A New Agenda for Multicultural Australia

December 1999

The Commonwealth Government remains committed to multicultural Australia. It has been central to our social, political, cultural and economic growth as a nation over the past fifty years, and is vital to our further development in the new millennium and beyond.

Australia is comprised of people who were born in this country and who have migrated here. Together, we have witnessed many changes in our nation. Our many shared experiences have produced a complex, cosmopolitan society, but together we have met and overcome challenges and striven for harmonious relationships between Australians from all backgrounds.

For its part, the Commonwealth Government has worked to ensure that our cultural diversity and all its implications are appropriately addressed through the development of policies and principles based on tolerance, humanity and mutual respect. A particular commitment by the Government has been to ensure that all Australians have the opportunity to be active and equal participants in Australian society, free to live their lives and maintain their cultural traditions.

But the democratic foundations of our society contain a balance of rights and obligations. The freedom of all Australians in practice is dependent on their abiding by mutual civic obligations. Thus, all Australians are expected to have an overriding commitment to Australia and the basic structures and principles common to Australian society. These are the Constitution, Parliamentary democracy, freedom of speech and religion, English as the national language, the rule of law, tolerance, and equality—including equality of the sexes.

Within this broad framework, each individual and group is welcome to make a contribution to the common good. We do not seek to impose a sameness on all our people. Nor do we seek to discourage the further evolution of the Australian culture, which already includes the heritage of Indigenous Australians, our British and Irish settlers, our Australian-grown customs, and those of our more recently-arrived migrant groups. We are, in reality as well as by definition, a multicultural nation.

The term **Australian multiculturalism** summarises the way we address the challenges and opportunities of our cultural diversity. It is a term which recognises and celebrates Australia's cultural diversity. It accepts and respects the right of all Australians to express and share their individual cultural heritage within an overriding commitment to Australia and the basic structures and values of Australian democracy. It also refers specifically to the strategies, policies and programs that are designed to:

- make our administrative, social and economic infrastructure more responsive to the rights, obligations and needs of our culturally diverse population;
- promote social harmony among the different cultural groups in our society; and
- optimise the benefits of our cultural diversity for all Australians.

The values of Australian multiculturalism form one dimension of the values which make up Australian citizenship which is built on a set of common civic values, rights and obligations that can unify Australians. These include: respect for institutional structures, participation in support of democracy and its institutions, respect for the law, respect for and tolerance of others' beliefs and practices, individual freedom of association and prime loyalty to Australia's interests.

The shared civic values that underpin Australia's civic identity and citizenship must necessarily reflect the cultural diversity of Australian society. Indeed Australian citizenship has played an important unifying role in the development of Australia's nationhood and the modern multicultural society which has evolved from it.

A strong multicultural framework can provide a firm basis for citizenship, and vice versa. Although Australians are very different as individuals we are unified by our common citizenship.

Multicultural policy also has its roots in Australian culture and contributes to its continuous enhancement and projection on the world stage. Australian culture includes Indigenous Australians, our British and Irish heritage, our Australian-grown customs, and those of our more recently arrived migrant groups as part of a dynamic and interacting set of life patterns. Australians all share a common culture, but every group and individual makes its own unique contribution to it.

For multiculturalism to be a unifying force in our developing nationhood and identity, it needs to be inclusive. It is about and for all Australians; it is not concerned mainly with immigration and minority ethnic communities. Multicultural Australia emphasises the things that unite us as a people—our common membership of the Australian community, our desire for social harmony, the benefits of our diversity and our evolving national character and identity.

In this context, the multicultural framework of our society has broadly benefited Indigenous peoples through its promotion of the integrity of diverse cultures and their harmonious intermingling. Aboriginal and Torres Strait Islander peoples are Australia's original inhabitants—its "First Peoples." While, therefore, they provide a foundation for the cultural diversity of the nation, it is appropriate that their distinct needs and rights be reaffirmed and accorded separate consideration.

Australian society has gained breadth and depth through the many benefits of our multicultural population. Our non-discriminatory immigration policy has drawn people from all around the world rather than from any one particular region. We thus have a reservoir of talent, energy, skills and knowledge which facilitates the way we do business with the rest of the world, especially given the reality of modern life and the "global village."

Modern communication facilities allow businesses located here to deal with markets and exchanges the world over as well as with customers in the towns and cities of Australia. Likewise, the growth of global media systems allows business arrangements to be conducted from local urban centres by way of the "electronic super highway," using satellite, cable and other technology.

Products may thus be developed in Australia specifically for markets which may have quite different cultures to our own and marketed through international outlets. In such an environment, businesses which are able to communicate in the languages of their customers and appreciate their cultural preferences will obviously have a distinct competitive advantage over those that do not.

The Government's Productive Diversity strategy aims to capitalise on the linguistic and cultural skills, business networks and market knowledge of individuals in our diverse population and to remove any impediments to their effective contribution in the workforce. This is to the advantage of all Australians and is yet another example of the benefits of our multicultural policy.

The Government will continue actively to seek opportunities to work in partnership with the private sector to help maximise the economic and social benefits of our diversity. Among other things, the Government has initiated a Productive Diversity Partnership Program, a cooperative venture between the Commonwealth, a group of Australia's foremost business schools and the private sector, including some of Australia's largest and most prominent corporations. Its purpose is to develop curriculum material for business education in both the university and TAFE sectors.

The ultimate goal is to achieve widespread appreciation of the fact that productivity and performance improvements are achievable through diversity management strategies, and that diversity planning should be viewed as an integral part of an organisation's business planning process.

In summary, we have built a social infrastructure of institutions, traditions and processes on our democratic foundation. Cultural diversity is one of our great social, cultural and economic resources. Australian unity in this diversity has been built on such moral values as respect for difference, tolerance and a common commitment to freedom, and an overriding commitment to Australia's national interests. For multicultural Australia to continue to flourish for the good of all Australians, multicultural policies and programs should be built on the foundation of our democratic system, using the following principles:

- **Civic Duty**, which obliges all Australians to support those basic structures and principles of Australian society which guarantee us our freedom and equality and enable diversity in our society to flourish;
- **Cultural Respect**, which, subject to the law, gives all Australians the right to express their own culture and beliefs and obliges them to accept the right of others to do the same;
- **Social Equity**, which entitles all Australians to equality of treatment and opportunity so that they are able to contribute to the social, political and economic life of Australia, free from discrimination, including on the grounds of race, culture, religion, language, location, gender or place of birth; and
- **Productive Diversity**, which maximises for all Australians the significant cultural, social and economic dividends arising from the diversity of our population.

Australia's cultural diversity will be a continuing and fundamental strength of our society if the community and institutions base their action and measure their achievements on these four principles.

The Government also believes these principles form a sound basis for workplace diversity planning strategies, both within government and the private sector. Indeed, the Commonwealth Government has taken a leadership role in implementing workplace diversity. For example, "respecting and valuing the diversity of the workforce by helping to prevent and eliminate discrimination on the basis of race, colour, sex, sexual preference, age, physical or mental disability, marital status, family responsibilities, pregnancy, religion, political opinion, national extraction or social origin" is a principle object of the *Workplace Relations Act 1996*.

The private sector faces the same diversity issues in relation to its workforce and clients as the public sector. The Government is itself committed to encouraging a holistic approach to the management of diversity and is keen to work with business to address the common issues.

Such involvement is consistent with the overall aim of the Government's multicultural policy: the achievement of enhanced community harmony and maximum benefits from our diversity, in the national interest. It is also consistent with a Plan of Action the Government has formulated for implementation in the years ahead to ensure that our cultural diversity is indeed a unifying force for Australia.

Recommendation 28

English is Australia's national language. Because it is a significant unifying influence and the ability to speak English is fundamental to full participation in Australian society, there would appear to be virtually no disagreement in the community about the importance of English language skills. The importance of English language proficiency has recently increased significantly because English has become the defacto standard for business and Internet communications throughout the world. Accordingly, the Council fully supports, and strongly **recommends** the continuation of, the high priority that has been given for many years to English language tuition for adult migrants.

Government Response

Supported. The Government will continue the high priority that has been given for many years to English language tuition for adult migrants.

For instance, the Adult Migrant English Program gives migrants and humanitarian entrants free access to 510 hours of English language training soon after their arrival in Australia. The Workplace English Language and Literacy Program provides funding to employers to support the training of workers in vocational skills integrated with English language and literacy skills, that are sufficient to enable them to meet the demands of their current and future employment and training needs. The Advanced English for Migrants Program provides advanced level English language assistance to help job seekers to gain employment or enter vocational courses or other post-secondary institutions.

Recommendation 29

In a multicultural society such as ours, proficiency in a language other than English is more than desirable; it can be a business or social imperative. If we are to engage the global marketplace and derive maximum benefit from it, Australia must maintain expertise in languages other than English, particularly the major languages of our region and the world. It is therefore very important that teaching languages other than English continues to be a priority and that the value of a multilingual community be better appreciated. The Commonwealth's specific priorities for funding languages other than English include the National Asian Languages and Studies in Australian Schools (NALSAS) Strategy and two language elements—Community Languages and Priority Languages. The Council fully endorses all these language programs, and **recommends** their continuation.

Government Response

Supported. The Government will continue the high priority that has been given for many years to teaching languages other than English. The Commonwealth funds three elements of the school languages program—the National Asian Languages and Studies in Australian Schools (NALSAS), Priority Languages and Community Languages.

The NALSAS Strategy is a collaborative initiative between the Commonwealth, States and Territories. Through NALSAS the Commonwealth funds school education jurisdictions to enhance and expand the provision of four Asian languages—Chinese (Mandarin), Indonesian, Japanese and Korean—and studies of Asia across the curriculum. The Strategy aims to improve Australia's capacity and preparedness to interact internationally, in particular, with key Asian economies.

In the 1999-2000 Budget the Commonwealth announced a further $90 million in funding for NALSAS to the end of 2002. Funding for Priority Languages and Community Languages is an ongoing commitment.

It is also noted that the Special Broadcasting Service (SBS) has extensive radio and television broadcasts in languages other than English. SBS Radio broadcasts 650 hours of programming each week in 68 languages. SBS television policy is that half of scheduled programming will be in languages other than English, which requires hundreds of hours of subtitling in some sixty languages.

The Government recognises the economic opportunities created by the rich pool of language resources that already exist in Australia's multicultural society and will continue to promote their full utilisation.

Source: "A New Agenda for Multicultural Australia," December 1999, posted immi.gov/au/multicultural/__inc/pdf_doc/agenda/agenda.doc

4

from the Australian Department of Immigration and Multicultural & Indigenous Affairs

What Is the Adult Migrant English Program?

Updated 14 March 2002

The Adult Migrant English Program (AMEP) provides up to 510 hours of basic English language tuition to migrants and refugees from non-English speaking backgrounds.

The program is provided by the Commonwealth Government through the Department of Immigration and Multicultural Affairs (DIMA) to help newly-arrived migrants and refugees settle successfully in Australia. Around six million hours of adult English language tuition are provided each year from an annual budget of approximately $92 million.

It is 1949. Magda and her husband are displaced persons from Poland. They are living in the Bonegilla Reception Centre near Albury. On this scorching Australian summer's day it is too hot to stay in the airless ex army hut that serves as their usual classroom. So they and their 20 or so classmates go outside to sit on wooden benches under a tree, while their teacher writes on a blackboard propped on an old easel.

"It is 1977. Tri is 19 years old. He and his brothers and sisters escaped from Vietnam on a small boat. Now they are living in Sydney in a migrant hostel. Every day Tri goes to English classes and in the evenings he studies hard. He hopes that if he can improve his English he'll be able to get a job in the factory where his older brother works.

"It is 1998. Mirsad was a lawyer in his native Bosnia. Now he has settled in Melbourne with his young family. He has been lucky to find full time work in an office but realises that he must continue to improve his English if he wants to study in an Australian university. So he attends English classes two nights a week and also practises his English using the Virtual Independent Learning Centre that he can access via the Internet.

"These three stories exemplify how more than a million new arrivals since 1948 have begun their new life in Australia. In order to be able to participate fully in the social, economic and cultural life of their new country, they need to speak its national language, English.

"The Commonwealth government has recognised the vital importance of learning English ever since the early days of large scale migration to Australia. As well as providing formal classroom tuition in a range of settings, the Adult Migrant English Program (AMEP) has, over the decades offered new arrivals to Australia the opportunity to learn English in the way most convenient and appropriate to them: via correspondence or distance learning courses, with the help of a Home Tutor, in small groups in community settings, via television or radio programs, and in more recent years, using computers and the Internet.

"But the AMEP is proud to be more than just a language program. It is a major settlement tool, enabling students to avoid the isolation which comes from being unable to communicate. You only have to visit an AMEP classroom to understand what an important role it plays in easing recently arrived migrants into their new environment—the practical advice and information provided by teachers, the lively multicultural atmosphere where tolerance is both necessary and appreciated, the opportunities for friendship during what can be a very lonely and bewildering period in a person's life, and of course the chance to learn and practise new linguistic and cultural skills in an encouraging and non-threatening environment.

"This history of the AMEP (is) a fascinating record of the people and policies that have made the AMEP what it is today—a world leader in adult language learning."

Philip Ruddock
Minister for Immigration and Multicultural Affairs
Minister Assisting the Prime Minister for Reconciliation

Source: "What is the Adult Migrant English Program?" revised March 14, 2002 on immi.gov/au/amep/what/what.htm AND immi.gov/au/amep/what/what1.htm.

from the Northern Territory Department of Education

Schools Our Focus—Information Package

Background

Our core business is the development of healthy, balanced and self-sufficient young adults who are able to take responsibility for their own lives.

The 1998 Education Review was about whether our frontline educators were getting what they needed to achieve this goal. Equally importantly, if there were any shortfalls, what we could do about them?

In the Territory, the business of teaching has to overcome the same fundamental problems that every Territory enterprise has to face. There is the tyranny of distance, the smaller number of clients spread over a much wider area, and the difficulties of finding and retaining skilled staff.

But, even against the challenges of this immutable strategic background, your feedback made us ask fundamental questions about how our Department was conducting its business.

For example, was our people power in the right places? Were the curriculum modules delivering the right educational outcomes in the classroom? Could we improve our handling of resources? Were our administrative procedures making unnecessary work for our front-line educators?

As Territorians we are all used to paying slightly more than other Australians for our goods and services. For the education of our children, we are paying considerably more.

Across Australia in 1997, it cost an average of some $5,365 a year to educate a child. Here in the Territory, it cost $7,923—just over 47% more.

Given that our taxpayers are already paying more than the average Australian, it is only fair that we make sure they are getting the best value for their education dollars.

Seventeen key areas were identified for improvement. The process of change has already begun.

The purpose of this information pack is to update every single member of the Department of Education team on the progress of those changes to date, and on what is still to come.

No. 10: Phase Out the Bilingual Language Program

Progressively Withdraw the Bilingual Education Program, Allowing the Schools to Share in the Savings and Better Resource English Language Programs.

The Bilingual program dates back to the days of Commonwealth responsibility for education in the Northern Territory but has continued with funding exclusively provided by the NT Government. Of the 91 schools in remote Aboriginal communities, 20 schools in the Territory (including 4 non-government schools) have additional resources for bilingual education programs.

There is no evidence to show that children in these schools are performing better in English literacy than children in other schools, which do not have extra resources for bilingual education. In fact, on average, children in schools with funded bilingual programs are performing slightly worse in English literacy and in numeracy.

The Government has decided to examine how these resources can be redirected more equitably to provide for improvements in literacy.

The Government has made it clear that the positions of Indigenous teachers and non-teaching support staff employed with bilingual education funds are protected. These teachers will be able to continue teaching in the schools where they are presently employed.

A detailed consultation will take place, over the next six months, with all schools and communities affected by this decision, to explore how it is to be implemented most constructively, so that resources are best allocated for improvements in literacy.

Source: "Schools Our Focus—Information Package" revised May 18, 2003 from ntde.nt/gov/au/prod/NTEDWeb/PRUWeb.nsf/pages/announce_review.

6 from **Peter K. Austin**

8th December 1998
Hon. Peter Adamson,
Minister for Education,
Parliament House,
GPO Box 3146,
Darwin, NT, 0801

Dear Minister,

I am writing to you in my capacity as President of the Australian Linguistic Society. The Australian Linguistics Society wishes to express its concern at the Northern Territory Government's proposal to dismantle the bilingual-bicultural education programmes in Aboriginal schools. A number of our members have worked in these programmes, or served them, or given support, often unpaid, to them. These programmes are a shining light in Northern Territory education, and have acted as models for programmes elsewhere in the world.

We do not understand the reason given for the dismantling, namely that this will enable better teaching of English as a Second Language (ESL). It suggests a misunderstanding of the nature of bilingual education programmes. ESL teaching is a fundamental part of bilingual programmes. We do not understand why sensible, staged, ESL teaching cannot be carried out in bilingual schools. The premise of bilingual education in the Northern Territory, that indigenous and materially disadvantaged children learn better in their first language, is backed by the findings of a wide range of international second language acquisition research. We therefore do not understand why the Government believes that Aboriginal children will acquire English better in English-only schools.

We recognise the right of indigenous communities to decide the medium of instruction for their children. However, we believe this decision rests at the level of the individual community. Some communities may wish for English-only instruction for their children. That is their right. But those communities who want to continue bilingual instruction should be able to do so.

Three side benefits of the NT bilingual education programmes deserve mentioning.

1. the bilingual education programmes have allowed the entry of Aboriginal people into the schools in a way that truly values their knowledge and skills, and accords them dignity. It is hard to see how the English-only programmes can provide such opportunities;
2. the indigenous languages of Australia and the cultural heritage and knowledge of indigenous people are part of Australia's heritage. The bilingual programmes

have encouraged collaborative work between indigenous people and researchers in documenting indigenous knowledge of the natural world which is of great scientific value, and in the case of plants, potentially of great economic value.

3. the loss of the language diversity that the world is currently undergoing has been compared to the tragedy of the loss of biological diversity. Bilingual education programmes such as those in the Northern Territory have been one of the major instruments for halting language and cultural loss.

We respectfully request you to reconsider the decision and to allow each affected Aboriginal community to decide whether they prefer bilingual-bicultural programmes or English-only programmes.

Yours sincerely,
Prof. Peter K. Austin
Foundation Professor in Linguistics
Head of Department
University of Melbourne
President, Australian Linguistic Society
cc. Hon. Tim Baldwin, Minister for Aboriginal Development and Community Government; Hon. Shane Stone, Chief Minister, Northern Territory

Source: Letter from the President of the Australian Linguistic Society to Peter Adamason, Dec. 8, 1998 at dnathan.com/VI/alert_2.htm#1

from Lester Irabinna Rigney

7

(Brief introduction in the Indigenous Australian language of KAURNA):

Ngankinnna meyunna! Na marni purrutye? Ngai narri Irabinna Kudnuitya Rigney. Ngai Bukkiyana ungko worni. Ngai yaitya meu, Narungga/Kaurna/Ngarrindjeri birkounungko. Martuitya Kaurna meyunna Ngai wanggandi

Thank you Madame Chair

Today I bring you greetings from the Narungga, Kaurna and Ngarrindjeri peoples of Australia. As is our cultural custom I acknowledge all Indigenous brothers sisters here today and thank them for sharing their struggle. I also acknowledge the chair and organisers of this the 19th session, of the Working Group on Indigenous Populations. I also thank ATSIC for their financial sponsorship for me to attend this session.

Madame Chair, the inherent rights of Indigenous peoples form the basis of any development strategy or definition. Development means improving the quality of our lives. However, the concept of development in relation to colonised peoples cannot be meaningfully discussed outside the context of Education and Language. In other words, the ideals of Indigenous rights to development cannot be realised if the house of education that trains the future architects of Indigenous culture is in decay. Let us not be misled by the mischievous assumption that the right to development can occur in absence of quality Indigenous education. Therefore, to compliment the discussion on Indigenous rights to development I wish to speak on education and language rights.

I begin with education first. Schooling is one of the largest industries within member states. The Education Departments of member states are one of the largest units of the State. Millions of dollars are spent on education that is rivalled only by private enterprise and corporate business. Equally, education is a major cultural force with equal power to mass media and popular culture. Education in countries like Australia has local and global markets leading to the exporting of educational methods, curriculums and technology to other countries. Education is simply big business for member states.

We Indigenous peoples realise the role western education has played in our colonisation with its alignment to the policies and priorities of member states. However, we also realise that the twentieth century has seen a shift toward the integration of Indigenous education. More recently, Indigenous peoples have embraced non-Indigenous education as a tool for social and economic mobility, although with some reservation. Education is fundamental to prepare Indigenous peoples with the necessary skills not only to promote, protect and nurture Indigenous cultures but also the preparation of the next generation for the ever-changing modern society. Therefore, the right to Indigenous development can be realised both by education and through education. Schools are knowledge producing factories that can equip colonised peoples with tools to contribute to Indigenous development. Some examples are:

Back to Basics:	Reading, Writing, Arithmetic (three "R's") in Indigenous and non-Indigenous languages
Cultural Education:	Promote, protect and maintain Indigenous cultures
Work Education:	Experience/education for employment
Technology Education:	Internet/Email
Sex Education:	Health/Environment/Aids/contraception
Driver Training:	Safety/eliminate drunk drivers/speed kills
Lifestyle Morals:	Decent citizens and the role in society
Immunisation:	Preventable diseases
Economy:	School follows the trends and need of society to be competitive on local/global markets.

Such skills are important for Indigenous peoples engaging in development and assisting Indigenous peoples in decision-making affecting them. What we know from research and experience is that if member states education systems fail our peoples, Indigenous unemployment rates rise contributing to extreme poverty and reducing the potential for Indigenous development. The employment/population rates in Australia are substantially lower for Indigenous peoples at all ages than for the general population. Literacy and Numeracy skills of Indigenous Australian students are well below those of the non-Indigenous students in primary and secondary schools. Educational outcomes for Indigenous peoples are much worse in some regions of Australia than others, particularly in rural and remote areas in the interior of Australia. Indeed, many Indigenous peoples in remote communities continue to have no direct access to secondary school education.

Factors such as racism, poor health, crowded housing and extreme poverty contributes to poor educational outcomes for Indigenous Australians. Equally, discriminatory structures implemented by member states that are unjust and impede Indigenous peoples progress in education are also prevalent. To illustrate this point, I will use the recent cuts to Indigenous Bi-Lingual education in the Northern territory, Australia.

The Northern Territory's bilingual education programs, in which local Indigenous Australian languages and English were used side by side in a minority of Aboriginal primary schools in remote northern Australia, came into being in 1973 under the broader policy language of "self-determination" for Indigenous Australians. Twenty-five years later, in December 1998, the Northern Territory Government dismantled these unique programs in a top-down decision, without consultation with the affected communities. Axing bilingual education programs in Aboriginal schools in the NT was replaced by ESL (English as a Second Language) which could be interpreted as the return to English only education. The programs' closure was effected on the recommendation of a Review Panel made up of four government appointees from outside of those communities, and against the stated and often-expressed wishes and cries of protest of the Indigenous community members. Since the official closure of the programs, some of the schools that were formerly bilingual have been trying to "go it alone" in order to keep their bilingual education programs alive and operational.

Finally, Madame Chair, I wish to briefly draw your attention to the state of endangered Indigenous languages in Australia, as Language rights are essential to Indigenous development. The world's Indigenous languages are in crisis. The way things are going, only a few hundred languages, amongst the world's 6,000 or so, look like surviving in the long term. The rate of extinction of languages and cultures far exceeds that of fauna and flora. And Australia has one of the worst records. Indeed, Indigenous activists argue that if our languages were like animals under threat of extinction there would be global outcry. The silence is deafening. It is no crime (as far as the law is concerned) to speak an Indigenous language in Australia in public. Sadly, this is not the case in other countries (e.g., the Kurdish people in Turkey, whose language is banned outright). However, in some areas of Australia, there are racist attitudes that persist which make people ashamed to speak their language in public. It took in Australia until 1994 before an Australian Indigenous language was offered for accredited study at senior secondary school level. Prior to this, some 34 languages, including languages such as Estonian with very few students, had already been accredited. Similarly, another issue that needs urgent attention is in regard to the statistics held by governments. Neither the 1986 nor the 1991 census data collection, data on the numbers of speakers of specific Australian languages to accurately reflect the crisis at hand. What was recorded only was how many people spoke an Aboriginal or Torres Strait Islander language. Therefore, those Indigenous groups whose remaining language speakers total a number under five could not be identified. Moreover, such communities whose languages are under the threat of extinction could not be recognised statistically in government records under the current census system. At the same time, speakers of other small languages like Nauruan or Tetum were counted. In the 1996 census, some 48 specific Australian languages were counted (in SA only Pitjantjatjara, Yankunytjatjara, Kukatha, Adnyamanthanha and Arabana). No NSW, Victorian or Tasmanian languages were counted.

Australia's Indigenous languages remain outside the official language status of the country and therefore receive little financial resources compared with international economic languages of French, Japanese and German. Nor are they recognised as national languages. ATSIC's financial contributions to Indigenous Languages Centres to reclaim and maintain languages are well under resourced. More funds are urgently needed to begin to arrest the extinction of Indigenous Languages.

Other issues can be recognised in the following. As a result of no official language status recognition, government naming policies and practices in relation to newly formed parks and public spaces disregard and ignore Indigenous naming practices. This continues to over-ride Indigenous cultural values and the naming of land and seascapes.

Therefore, Madame Chair, I conclude by suggesting that Indigenous Rights to development draw on the educational and Language Rights enshrined in UNESCO declarations.

When our children engage in the journey of education that does not do violence to their culture, it teaches them to dream of possibilities and not to be a prisoner of certainty. It teaches our children to be the best they can be. Education that welcomes Indigenous identities reinforces Indigenous cultural views of the world. Let me be clear. Education is a sacred activity and must be done with extreme care. There can be no development without human beings or the education of human beings. Bright futures are only possible from strong pasts. For being strong is what it means to be Indigenous.

Ngaito Yungadalya Yakkandalya (thank you)

Source: "Intervention Titled: Building Stronger Communities: Indigenous Australian Rights in Education and Language" submission to the United Nations Commission on Human Rights, July 2001 at fatsil.org/papers/research/rigney1.htm.

from the Māori Language Commission 8

About Us

What Are the Goals of the Commission?

Mana Māori

To empower iwi Māori to maintain and generate reo development amongst their communities

Excellence
To maintain and improve the quality of te Reo Māori
Learners
Increase the number of people using te Reo Māori by increasing the opportunities to learn
Regeneration
The maintenance and dissemination of Māori language resources and corpus
Innovation
To increase the rate of language development so that te Reo Māori can keep pace with technological development
Good Will
To foster positive attitudes towards te Reo Māori by all New Zealanders
Language Domains
To increase the number of situations where Māori is used

What Is the Māori Language Commission?

The Māori Language Commission was set up under the Māori Language Act 1987 to promote the use of Māori as a living language and as an ordinary means of communication.

What Does the Māori Language Act 1987 Do?

The Act came into force in August 1987. It does three things:

- It declares the Māori language to be an official language of New Zealand.
- In Courts of Law, Commissions of Inquiry and Tribunals, it confers the right to speak Māori upon any member of the Court, any party, witness or counsel.
- It establishes Te Taura Whiri i te Reo Māori (Māori Language Commission).

Government Māori Language Objectives

- To increase the number of people who know the Māori language by increasing the opportunities to learn Māori.
- To improve the proficiency levels of Māori in speaking, listening, reading and writing Māori.
- To increase the opportunities to use Māori by increasing the number of situations where Māori can be used.
- To increase the rate at which the Māori language develops so that it can be used for the full range of modern activities.
- To foster among Māori and non-Māori positive attitudes, accurate beliefs and positive values about the Māori language so that Māori-English bilingualism becomes a valued part of New Zealand.

What Does the Commission Do?

The Commission's work includes:

- Promoting and raising awareness of the Māori language and Māori language issues;
- Promote quality standards of written and spoken Māori through initiatives such as advanced immersion courses, a Māori language checking service, the Commission's quarterly newsletter He Muka and public service attestations;
- Administering examinations for candidates seeking formal certification as translators and interpreters;

- Researching and formulating policy related to the promotion, maintenance and progression of the Māori language;
- Lexical expansion work including the production of glossaries.

How Is the Commission Organised?

The Commission meets at least six times a year. The secretariat is headed by a Chief Executive which carries out research, policy advice, translation checking work, promotional activities, and tasks assigned by Commission members.

Source: "About Us" at tetaurawhiri.govt.nz/english/about_e/index.shtml

from **the Māori Language Commission** 9

Te Reo

The Māori Language Act 1987 declared Māori to be an "official" language and created a right to use Māori in court proceedings. This article provides a perspective of the history, current use, and likely future of te reo Māori in the light of the Māori Language Act 1987.

Māori is the foundation language of New Zealand, the ancestral language of the tangata whenua and one of the taonga guaranteed protection under the Treaty of Waitangi. It also provides this country with a unique language identity in the rest of the world, as this is the only place where Māori is spoken widely. In more tangible terms, the Māori language is a powerful social force for the reconstruction of a damaged and deteriorated self-image among Māori youth, a vehicle of contribution to society, and therefore a means of regaining dignity. Finally, human freedom is dependent at all levels on choice and diversity; linguistic pluralism can be nothing other than a guardian of individual freedom and identity against the forces of conformism.

According to the latest figures on Māori language, it is estimated that some 10,000 New Zealanders, almost all of Māori descent, are in the very high, medium high, and high range of speakers of Māori, while perhaps a further 153,000–163,000 speak or understand the language to some extent. The ability to converse in Māori was highest among those aged 60 years and over and lowest among those aged 25–29 years (21 percent). Figures from the National Māori language survey taken in 1995 state that 41% of Māori adults do not speak any Māori and of the 53% that do, 14% hold conversations in their home every day.

In terms of absolute numbers, Auckland has the lion's share of Māori speakers, accounting for almost a third of North Island figures. But areas of concentration are also to be found in the secondary urban centres and rural communities of Northland, Waikato, Bay of Plenty and East Cape.

Source: "Te Reo" at tetaurawhiri.govt.nz/english/issues_e/reo/index.shtml

from ***The Nelson [New Zealand] Mail*** 10

Teacher Sought for New Bilingual Unit

by Paula Batters
September 3, 2002

The possibility of a bilingual Māori education unit at Nelson Central School is under threat due to a shortage of suitably qualified teachers.

Principal Paul Potaka plans to hold a public meeting at the school tomorrow afternoon to discuss the problem.

Mr. Potaka said while the future of the school's Māori immersion unit was in no doubt, he was "having difficulties" in finding teachers to set up a bilingual unit. "Finding people who meet the criteria is the hard part, as we set the bar that little bit higher because we want to do the best we can for the children," he said.

Mr. Potaka, who has been principal at the 450-pupil school for six years, said it boasted the only 100 percent immersion unit in Nelson city. This unit's 39 pupils speak only Māori during classtime.

Mr. Potaka said he had been approached by some parents wanting an alternative provided by a bilingual unit.

"We want to make sure there are also options for English-speaking New Zealanders wanting to learn Māori and about Māori culture," he said.

The current immersion unit teacher, Marama Ataera, is leaving the school but Mr. Potaka said it was likely that he had found a replacement.

But he said finding someone suitable for a bilingual unit would be harder.

"It takes longer to fill a position like this as there just aren't many people out there who will meet all our needs," he said.

Mr. Potaka said suitable candidates would need to be fluent speakers of the Māori language, hold a teaching degree and have classroom experience. "We want to preserve the language; it's important for Māori children and their parents."

Mr. Potaka said the number of pupils enrolled in the unit had remained "steady" since it opened in the mid-1980s.

While children in the immersion unit spoke only Māori in class, they spoke English with their peers in the playground. "But they might talk in Māori among themselves," he said.

The meeting will be held at 3.15pm at the school.

Source: Paula Batters, "Teacher Sought for New Bilingual Unit," *The Nelson Mail,* September 3, 2002 at asu.edu/educ/epsl/LPRU/newsarchive/Art1319.txt

11

from **Agence France Presse**

New Zealand Toughens English Language Rules for Migrants

November 19, 2002

New Zealand Tuesday significantly toughened its language requirements for people planning to emigrate here.

Immigration Minister Lianne Dalziel said with immediate effect, applicants under a general skills category will have to pass a test showing they have enough English knowledge to cope with study for a doctorate at a university. Previously applicants had to show enough skills to handle the equivalent of secondary school study.

Immigration officials predicted about 9,000 people who gained residency in the category last year would not be able to pass the new test.

The move came in the midst of political debate here over whether the country is taking too many migrants while at the same time large numbers of fee paying students from China are studying here.

Dalziel said the changes would protect both migrants and New Zealand's economic interests.

"It is unfair for people to be able to come here believing their English skills equip them fully for life in New Zealand, and then find that they cannot communicate well enough to get a good job," she said.

The government aims for around 50,000 immigrants to enter the country each year. A point system is used in the general skills category to assess skills and other factors. At the moment, applicants must have at least 30 points to qualify and Dalziel expected this "high" number to come down in order to meet the government's quota.

Figures showed that the top migrant providing nations in the last year were from China with 17 percent of all migrants, India 16 percent, Britain 12 percent, South Africa 8 percent and Samoa, South Korea and Malaysia with 4 percent each.

Source: "New Zealand Toughens English Language Rules for Migrants," published by Agence France Presse, Nov. 19, 2002 found at asu.edu/educ/epsl/LPRU/newsarchive/Art1393.txt.

Review Questions

1. What are the differences between the English as a Second Language (ESL) and bilingual approaches toward educating non-English-speaking students?
2. What was the "White Australia" policy, and why was it abandoned after World War II?
3. What are some of the key differences between the Australian and New Zealand governments' positions toward their indigenous populations? What are some of the reasons for these differences?
4. What was the Māori Language Act of 1987?
5. What positions were taken on multiculturalism and multilingualism in the Australian government's *New Agenda for Multicultural Australia?*

Discussion/Essay Questions

1. Jennifer Wang, the Chinese immigrant to Australia profiled at the beginning of this module, is a strong supporter of the Australian government's Adult Migrant Education Program (AMEP). But the resources of the Australian government, like those of all governments, are limited. Suppose that a new Australian government proposes eliminating the AMEP and directing these resources to bilingual education. Now, Australia's Department of Immigration and Multicultural & Indigenous Affairs is holding hearings at which the public has been invited to comment on this proposal. Peter Austin (the author of the sixth reading) and Lester Irabinna Rigney (the author of the seventh reading) have just given testimony in favor of the reallocation of funds, making many of the same arguments in favor of bilingual education and a multilingual Australia that they made in the readings. Wang has been recruited by ESL advocates to speak against the change. She had intended to retell her life story, pointing to herself as a success story of someone who, thanks to ESL, is contributing to a new, multicultural Australia. Now, however, she realizes that she also must work with (or refute) the points just made by Austin and Rigney. Write her testimony.
2. The dispute over the Northern Territory's decision to eliminate bilingual education in Aboriginal schools is the focus of the fifth and sixth readings and is referred to in the seventh reading. Imagine a case in which a coalition of indigenous rights groups have filed suit against the Northern Territory government, arguing that the elimination of bilingual education in Aboriginal schools is inconsistent with the government's stated commitment to building a multicultural Australia. Write the Northern Territory government's response. Make your case by discussing how the Northern Territory policy is consistent with the general Australian approach toward multiculturalism.
3. Both countries—Australia and New Zealand—are often accused of practicing racism and ethnocentrism in their past policies toward immigration, linguistic rights, and the position of indigenous peoples. Although both countries have reformed their policies, they still encounter frequent criticism in these areas. On balance, which country—Australia or New Zealand—do you feel provides a better environment for people who are not of British heritage to realize their personal potential and improve their lives? Which country is more tolerant of

people who are not of British descent retaining their cultural distinctiveness and linguistic heritage? Can the first objective (providing an environment for personal development of non-British-heritage citizens) be met without also meeting the second objective (fostering the continuity of non-British cultural identities)? Can the second objective be met without meeting the first?

List of Readings

1. "Fact Sheet no. 8: Abolition of the 'White Australia' Policy," revised February 13, 2003, posted on the website of the Commonwealth of Australia's Department of Immigration and Multicultural & Indigenous Affairs, *http://www.immi.gov.au/facts/08abolition.htm.*
2. "Preface to the Statement of Needs and Priorities for Language Policy for Australia," revised October 5, 2001, posted on the website of the Australian Alliance for Languages, *http://www.latrobe.edu.au/rclt/als/aalp.html.*
3. "A New Agenda for Multicultural Australia," December 1999, posted on the website of the Commonwealth of Australia's Department of Immigration and Multicultural & Indigenous Affairs, *http://www.immi.gov.au/multicultural/_inc/pdf_doc/agenda/agenda.doc.*
4. "What Is the Adult Migrant English Program?" revised March 14, 2002, posted on the website of the Commonwealth of Australia's Department of Immigration and Multicultural & Indigenous Affairs, *http://www.immi.gov.au/amep/what/index.htm* and *http://www.immi.gov.au/amep/what/what1. htm.*
5. "Schools Our Focus—Information Package," revised May 18, 2003 and "Fact Sheet no. 10: Phase Out the Bilingual Language Program," revised April 2, 2003, posted on the website of the Northern Territory Government's Department of Employment, Education and Training, *http://www.ntde.nt.gov.au/prod/NTEDWeb/PRUWeb.nsf/pages/announce_review* and *http://www.ntde.nt.gov.au/prod/NTEDWeb/PRUWeb.nsf/4432de3b22c170878525631600 7db7a0/ea916ca84ca5fa8169256706000ac55f?OpenDocument.*
6. Peter K. Austin, Letter from the President of the Australian Linguistic Society to Peter Adamson, Northern Territory Government Minister for Education, December 8, 1998, posted on the website of the Internet Library for Australian Languages, *http://www.dnathan.com/VL/alert_2.htm#1.*
7. Lester Irabinna Rigney, "Intervention Title: Building Stronger Communities: Indigenous Australian Rights in Education and Language," submission to the United Nations Commission on Human Rights' Sub-Commission on the Promotion and Protection of Human Rights' Working Group on Indigenous Populations, July 2001, posted on the website of the Federation of Aboriginal and Torres Strait Islander Languages, *http://www.fatsil.org/papers/research/rigney-1.htm.*
8. "About Us," posted on the website of the New Zealand Government's Māori Language Commission, *http://www.tetaurawhiri.govt.nz/english/ about_e/index.shtml.*
9. "Te Reo," posted on the website of the New Zealand Government's Māori Language Commission, *http://www.tetaurawhiri.govt.nz/english/issues_e/reo/index.shtml.*
10. Paula Batters, "Teacher Sought for New Bilingual Unit," *The Nelson Mail*, September 3, 2002, posted on the website of Language Policy Research Unit Press Archives, *http://www.asu.edu/educ/epsl/LPRU/newsarchive/Art1319.txt.txt.*
11. "New Zealand Toughens English Language Rules for Migrants," published by Agence France Presse, November 19, 2002, posted on the website of the Language Policy Research Unit Press Archives, *http://www.asu.edu/educ/epsl/LPRU/newsarchive/Art1393.txt.*

Websites for Additional Research

1. For information on Australian Aboriginal languages, including policy debates over bilingual education, sound clips of spoken languages, and links to advocacy groups, see the Aboriginal Languages of Australia Virtual Library webpage maintained by David Nathan, at *http://www.dnathan.com/VL/austLang.htm*. The website has annotated links to 180 resources for about 60 of Australia's 200 indigenous languages.
2. The New Zealand Government's Māori Language Commission maintains a website of issues relating to the history and preservation of the Māori language in New Zealand, at *http://www.tetaurawhiri.govt.nz*.
3. Since 2001, the Language Policy Research Unit of Arizona State University's Education Policy Studies Laboratory has maintained Language Policy in the News, a full-text database of articles from newspapers around the world concerning language policy issues. The fully searchable database is available at *http://www.asu.edu/educ/epsl/LPRU/newsarchive*.

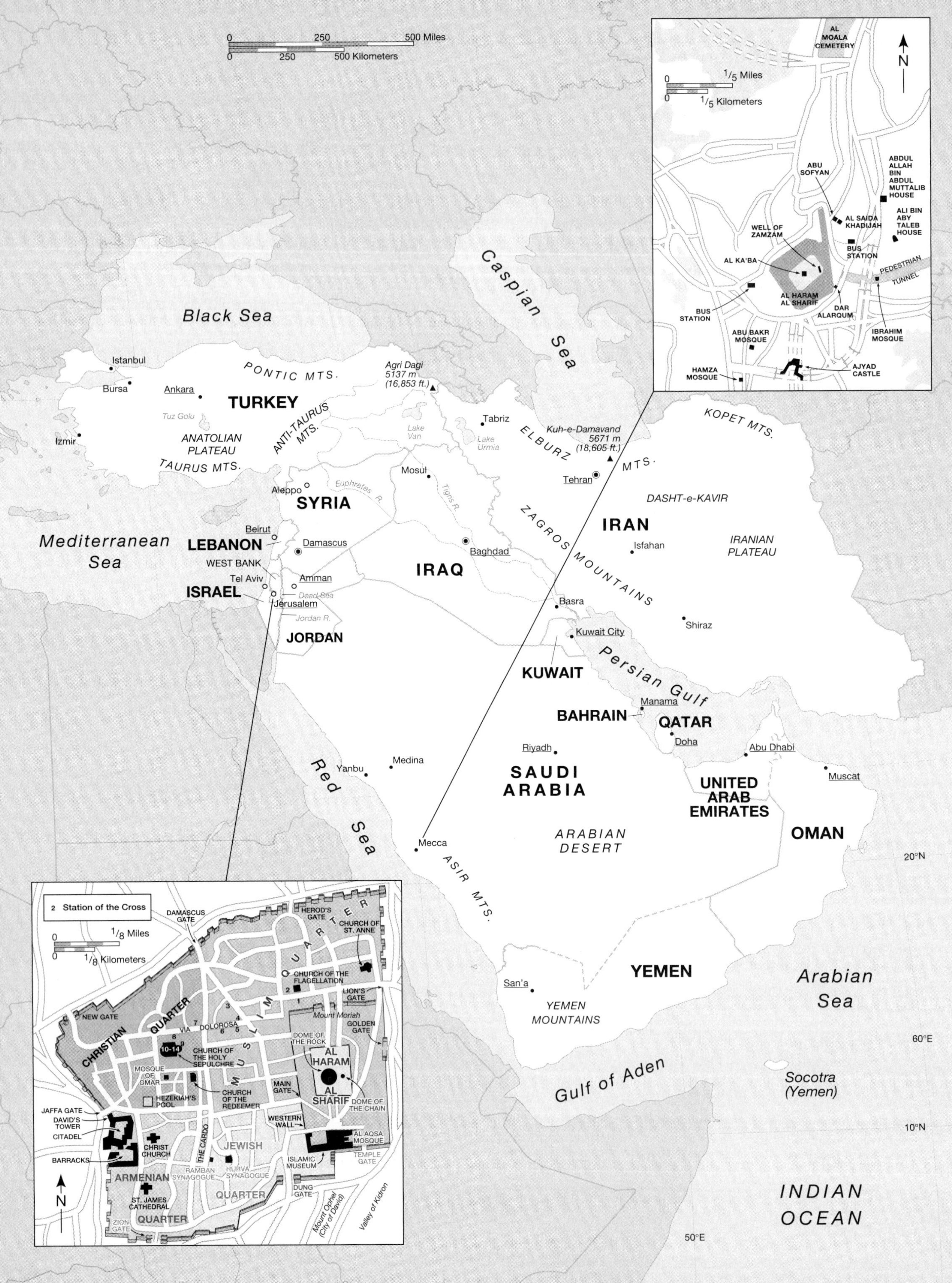

0 250 500 Miles
0 250 500 Kilometers
Black Sea
Caspian Sea
Mediterranean Sea
Red Sea
Persian Gulf
Gulf of Aden
Arabian Sea
INDIAN OCEAN
TURKEY
SYRIA
LEBANON
WEST BANK
ISRAEL
JORDAN
IRAQ
IRAN
KUWAIT
BAHRAIN
QATAR
SAUDI ARABIA
UNITED ARAB EMIRATES
OMAN
YEMEN
Istanbul
Bursa
Ankara
Izmir
Tuz Golu
ANATOLIAN PLATEAU
PONTIC MTS.
ANTI-TAURUS MTS.
TAURUS MTS.
Agri Dagi 5137 m (16,853 ft.)
Lake Van
Lake Urmia
Tabriz
ELBURZ MTS.
Kuh-e-Damavand 5671 m (18,605 ft.)
Tehran
KOPET MTS.
DASHT-e-KAVIR
IRANIAN PLATEAU
Isfahan
Shiraz
ZAGROS MOUNTAINS
Mosul
Tigris R.
Euphrates R.
Aleppo
Beirut
Damascus
Baghdad
Tel Aviv
Amman
Dead Sea
Jerusalem
Jordan R.
Basra
Kuwait City
Manama
Doha
Riyadh
Abu Dhabi
Muscat
Yanbu
Medina
Mecca
ASIR MTS.
ARABIAN DESERT
San'a
YEMEN MOUNTAINS
Socotra (Yemen)
20°N
10°N
60°E
50°E
AL MOALA CEMETERY
N
0 1/5 Miles
0 1/5 Kilometers
ABU SOFYAN
ABDUL ALLAH BIN ABDUL MUTTALIB HOUSE
AL SAIDA KHADIJAH
ALI BIN ABY TALEB HOUSE
WELL OF ZAMZAM
BUS STATION
AL KA'BA
PEDESTRIAN TUNNEL
AL HARAM AL SHARIF
DAR ALARQUM
IBRAHIM MOSQUE
BUS STATION
ABU BAKR MOSQUE
HAMZA MOSQUE
AJYAD CASTLE
2 Station of the Cross
0 1/8 Miles
0 1/8 Kilometers
DAMASCUS GATE
HEROD'S GATE
CHURCH OF ST. ANNE
MUSLIM QUARTER
CHURCH OF THE FLAGELLATION
LION'S GATE
Mount Moriah
GOLDEN GATE
NEW GATE
CHRISTIAN QUARTER
VIA DOLOROSA
CHURCH OF THE HOLY SEPULCHRE
10-14
DOME OF THE ROCK
AL HARAM AL SHARIF
DOME OF THE CHAIN
MOSQUE OF OMAR
HEZEKIAH'S POOL
CHURCH OF THE REDEEMER
MAIN GATE
JAFFA GATE
DAVID'S TOWER
CITADEL
WESTERN WALL
AL AQSA MOSQUE
CHRIST CHURCH
THE CARDO
JEWISH QUARTER
BARRACKS
RAMBAN SYNAGOGUE
HURVA SYNAGOGUE
ISLAMIC MUSEUM
TEMPLE GATE
ARMENIAN QUARTER
ST. JAMES CATHEDRAL
DUNG GATE
ZION GATE
Mount Ophel (City of David)
Valley of Kidron

The Politics of Pilgrimages in the Sacred Spaces of Makkah and Jerusalem

Companion to Chapter 6 Religion

My name is **Umar Abdul Salam**. I was born and raised in Harlem, New York City. I became a Muslim when I was 19 years old. I find myself living halfway across the world. Saudi Arabia for me is home, sweet home. I am on my way to the Hajj.

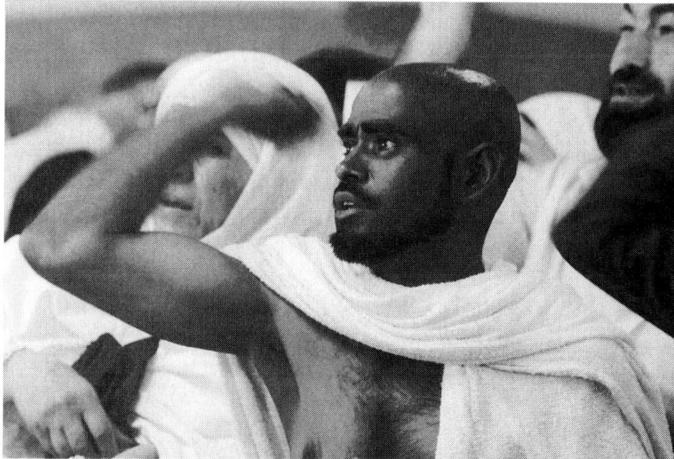

Every Year, millions of Muslims from around the world journey to Makkan for the Hajj. *Source:* AP/Wide World Photos

All of the markets in Jeddah are alive and packed with shoppers buying new outfits for Eid. The offices and businesses are closed. Everybody has three days off. Celebrations are going on everywhere. On Mecca Highway, a feeling comes over me. There is no name for it. Here I am, making the same journey that millions of Muslims have made for centuries. There are hundreds of buses and cars. Yet the traffic moves smoothly.

As I near Mecca, I make my intention to perform Hajj. Reflecting on my life's accomplishments, they are nothing next to this one moment. Being blessed to be invited to perform Hajj is the greatest feeling. It humbles me when I look around seeing people from countries where the yearly income is a mere fraction of the average American's. Yet they worked, they saved and are now right next to me.

Knowing that I worked hard to be here is a comforting thought. Yet there were sacrifices that others made that I will never know. Pilgrims from war-torn countries walk humbly by with the smile of satisfaction and peace on their faces. We made it. More than two million of us.

The faces are the same—satisfied, happy, peaceful, and focused. Same, and yet different. Pilgrims performing their required rituals with an energy that revives you. All my thoughts in these moments are of becoming a better person, a better father, a better husband, and a better friend. I pray for peace in the countries of the pilgrims that I'm walking with.

The atmosphere of peace and caring is overwhelming. Strangers are trying to give pilgrims water and food every step of the way. Nowhere else on Earth can you watch more than two million people from different countries speaking different languages, caring, sharing, and helping each other while they happily perform their obligations to their Creator. Only on the Hajj performed in Mecca and Medina can you find 2.5 million people praying to God at the same time in the same manner.

There is no way to explain the feeling of belonging to a community that encourages right for all, and stands against wrong for all. That is real globalization.

Adapted from: "For American, Hajj is Exhilarating, Humbling," by Umar Abdul Salam.

Map Source: Figure 6-10 and 6-14, p. 203 and p. 215, from *The Cultural Landscape* by James Rubenstein, ■

Sacred Spaces and Pilgrimages

Perhaps every religion seeks a middle ground between the idea that God (or the gods) is everywhere (including within oneself) and the idea that to get close to God (or the gods) one must venture to specific places, practice specific rituals, and/or interact with specific objects. On the one hand, religious practice involves making a personal commitment. At some level, this commitment must be individualized as each believer develops her or his own faith in the deity(s). On the other hand, every religion, to a greater or lesser extent, has a public component, as believers demonstrate and reaffirm their faith through the public performance of rituals.

Shared and standardized rituals are particularly important in universalizing religions, because these rituals often are the only source of connections among believers, whose cultures otherwise may vary widely. In contrast to ethnic religions, universalizing religions do not claim that there is anything "natural" that unites, say, one Christian with other Christians around the world or one Muslim with other Muslims around the world. Therefore, special importance is attributed to the community that Christians and Muslims reproduce on regular occasions in the church or mosque. Equally powerful for universalizing religions are the imagined communities that are reproduced when a Christian thinks of the millions of other Christians around the world who are celebrating Jesus' resurrection through participation in the Easter Mass or when a Muslim thinks of the millions of other Muslims around the world who are professing their faith by bowing toward Makkah (Mecca) and praying to Allah. These connections that are forged through shared rituals in turn can reinforce a believer's commitment to the tenets of the religion.

These rituals that bind the community of religious practitioners together need not necessarily be tied to specific, public places. For instance, Muslims around the world are unified by the practice of stopping what they are doing and praying five times a day. Although this ritual is centered on a reference to a place (all Muslims face toward Makkah when they pray) the power of this ritual in part derives from the fact that it can be (and is) practiced anywhere. Likewise, some of the most powerful and important rituals in Judaism are practiced in the private space of the home.

Nonetheless, some of the most profound affirmations of religious community occur when adherents of a religion travel to special places, whether that special place is the local house of worship that one visits weekly or the distant shrine that one tries to visit once in one's lifetime. Shrines emerge as important places for a variety of reasons. Often these sites have historical significance: A shrine may be established at a place where the religion's founder or one of its historic holy persons witnessed or performed a miracle, where a sacred object is housed, or where the religion's first worshippers lived. For religions that at one time had political control of a region, a shrine may be the former capital of the old political-religious empire. For animist religions that recognize deities in specific objects, a shrine may be established at a specific place where a god lives.

The rituals and timing surrounding visits to shrines (pilgrimages) also vary from religion to religion. Buddhism places little emphasis on the group experience of religion, and its shrines typically are visited year-round with little formal ceremony. Shrines honoring Catholic saints may be visited year-round, but devotees most typically visit a saint's shrine on the day of the year reserved for honoring that particular saint. Likewise, pilgrimages to holy cities like Rome and Jerusalem may be made at any time and always have religious significance, but for maximum spiritual effect these visits often are made on major holidays. The Islamic pilgrimage to Makkah—the Hajj—is always timed so as to conclude on Eid al-Adha, the feast that commemorates the day when Ibrahim (Abraham) offered to sacrifice his son to Allah (God). In ancient Judaism, access to sacred spaces was restricted based on one's status in society as well as the calendar. Only Judaism's high priest could enter the innermost sanctum of the Temple in Jerusalem, and he could enter it on just one day of the year. Today, the remnants of the Temple remain a focus of Jewish worship and Jerusalem continues to attract Jewish pilgrims, ranging from those who briefly visit so as to experience the religion's historic center to those who choose to live in this city so that they can be surrounded by like-minded Jews who are seeking to live according to Jewish law.

Political Control of Religious Sites

Throughout history, many religions have sought to couple their leader's religious authority with political power. In many cases, the two have been inseparable, as when a king declares that his authority to rule comes directly from God, and that therefore the people under his domain should respect both his political authority and his method of worship. Countries in which a large majority of people belong to one religion often have had difficulty accepting governments that forsake religious identification, and in recent decades several secular governments gradually or abruptly have added a religious dimension to their claim to authority, often by melding religious tenets into their legal system. While this trend has been most prominent in the Muslim world, other countries, including India recently have been moving in this direction. Some have argued that the United States should also move in this direction, as, for instance, with the movement to post the Ten Commandments inside courtrooms.

Although there are numerous instances of political leaders successfully using religion to increase their authority, religious leaders generally have had a more difficult time claiming political authority, particularly over large areas where political power (and, in some cases, religious practices) had been fragmented. For instance, a common theme in the history of Europe from around 400 to 1800 was an ongoing power struggle between the Papal hierarchy in Rome and regional princes and kings. Although many of these regional political rulers were Catholic and professed *religious* allegiance to Rome, they placed limits on their *political* loyalty to the Church. In the end, the Church lost this struggle and the modern state system emerged independent of the Catholic hierarchy (see Chapters 7 and 8 in the textbook and the accompanying modules in this workbook for more on the origins of the modern state system). Although, according to the *Annuario Pontifico* (Pontifical Yearbook), there were 1.06 billion baptized Catholics in the world in 2001, the Church retains formal political control only over Vatican City, a .44 square kilometer (.17 square mile) area with a population of 900, surrounded entirely by Rome.

Because of the problems involved in extending religious belief systems over large swaths of territory, many empires, including the Roman and Ottoman empires, chose to allow conquered lands with unique religious practices to retain their religious systems so long as these did not directly challenge the authority of the new political rulers. (Rome had reached an agreement along these lines with the Jewish leaders in Jerusalem: They would respect Roman political rule and pay taxes to Rome in return for Rome allowing them to govern most religious and political affairs in their community. Jesus was perceived as a threat and ultimately was executed by the Romans because they perceived that he was advocating a strain of Judaism that would threaten this accommodation).

This uneasy fit between religious authority and political territory becomes an issue in the administration of religious pilgrimage sites because, with the exception of Vatican City, all pilgrimage sites are within politically organized territories. The problem thus emerges, particularly in the case of religions whose followers are dispersed globally: Does the pilgrimage site belong to the country in which it is located or does it belong to believers, scattered around the world, who view the site as a part of their cultural and religious heritage and whose beliefs, in some case, require pilgrimage to the site? Can the rulers of the country within which a holy site is located be trusted to maintain the site for all members of the religious community that flock there, even if the government that controls the site is at war with countries from which many of the pilgrims come and even if the government's rulers are of a different religion? Is it enough to promise global access to specific holy places, or must entire cities be opened to the world? In the remainder of this module, we delve deeper into these questions by focusing on two particularly problematic holy cities: Makkah (also know as Mecca) and Jerusalem (also known as Yerushalayim in Hebrew or Al-Quds in Arabic).

Makkah

Makkah, the birthplace of the Prophet Muhammad, is the holiest site in Islam. The city's ancient mosque, the Ka'ba, is said to have been built by Ibrahim (Abraham, the patriarch of Judaism and Christianity as well as Islam) with the help of his son Ismail, and restored by Muhammad when he founded Islam. In the Quran (or Koran, the Islamic holy book), Muhammad commands every Muslim to make a pilgrimage to Makkah and perform the rituals of the Hajj once in his or her lifetime unless prevented by poor health or financial hardship. In modern times, with relatively easy international travel, over two million people participate annually in the Hajj. Many individuals coming to Makkah for the Hajj also participate in the rituals of Umrah, a less significant pilgrimage ritual that can be undertaken at any time of the year but is often coupled with the Hajj.

The first reading, from *Aramco World*, a magazine published by the Saudi national petroleum corporation that introduces Arab culture to the West, provides a comprehensive introduction to the rituals of the Hajj, its religious significance, and the emotions experienced by Hajj pilgrims. Many Hajj pilgrims, like Umar Abdul Salam, the American Muslim profiled at the beginning of this module, view the Hajj as a fundamentally *global* experience. If one of the functions of public religious rituals is to demonstrate that one is not alone in one's faith, then the experience of being around millions of people, all speaking different languages but united by one faith, is a tremendous affirmation of one's sense of community that, in turn, is likely to renew one's own belief system. This is why Riz Khan, the author of the second reading, describes the Hajj more as a "celebration" than a sacred or solemn affair. This sense of the Hajj as a *global* celebration and of Makkah being temporarily turned into a *global* Islamic city is also expressed in the third reading, a report by Saudi teen Yousef Saleh Ba-Isa. For Ba-Isa, the global nature of the Hajj is particularly striking because, as a Saudi going to a religious site within Saudi Arabia, he had been expecting a more recognizably Saudi, or at least Arab, experience. Ba-Isa is surprised to see such a diverse collection of Muslims, many of whom do not even understand the Arabic of the prayers that they recite, but united by a belief in Allah and a commitment to the five pillars of Islam.

Although the Hajj is for many a uniquely global experience, it also is a trip to a distinct city, Makkah, within a distinct country, Saudi Arabia. Although Saudi Arabia is wealthy from petroleum sales, it nonetheless has many characteristics of a less developed country. The fourth reading, "Hajj and Umrah Tips & Tricks" is from Jannah.org, a website run by a U.S.-based Muslim and oriented toward American Muslim youth. This website advises readers that, although they will be experiencing the Hajj as a global event, they also will be traveling to a distinctly local place, and one that is very different from the town or city in the United States that they will be leaving behind. Thus, along with practical religious advice ("Write all the Duas [required prayers] on index cards"), the website is filled with practical advice similar to what might be given to any American college student preparing to visit an LDC for the first time (beware of beggars, avoid malls because things will cost more there than in the United States and you'll be wasting time when you could be experiencing the local culture, avoid being heard speaking English in shops because merchants will raise the prices if they think that you're a rich American tourist).

Because Makkah is indeed a *place* and not just a global gathering, it falls within the territory of a government, in this case the Kingdom of Saudi Arabia. The status of Makkah, and the importance of the Hajj within Islam (which is also the official religion of Saudi Arabia), gives the Kingdom a unique responsibility. The Saudi government has assumed the responsibility for maintaining a piece of its territory on behalf of global Islam and for making this piece of its territory accessible to Muslims from around the world. To this end, as the fifth reading details, the Saudi king has formally taken on the title "Custodian of the Two Holy Mosques" (the second holy mosque is in the city of Madinah [Medina], where Muhammad preached and was buried), and the Saudi government has built infrastructure to house and transport the millions of pilgrims who come to these mosques each year. Along with spending vast sums of money on facilities to ease pilgrims' visits, the Saudi government has also established a Ministry of Pilgrimage and instituted a special category of Hajj visas, which, unlike other visas for visiting Saudi Arabia, are free of charge but are available only to pilgrims.

Despite these efforts at providing for pilgrims, the Saudi government has opened itself to criticism by taking on this added, *global* responsibility. Some Western politicians have charged that in order to maintain its legitimacy as spiritual guardian of the Islamic world, the Saudi royalty has been forced to restrain itself when criticizing Islamic fundamentalists, and that this restraint has prevented the Saudis from being a reliable ally in the war against Islamic-inspired terrorism. The Saudi government's legitimacy as protector of global Islamic territory has also been criticized by Osama bin Laden. Although bin Laden is best known for his struggle against the United States, this struggle in large part is an outgrowth of his long-standing war against the Saudi monarchy which, he charges, is not fit to serve as custodian for Islam's two most holy places. For bin Laden, a key indication of the monarchy's unfitness to serve as stewards of Islam was its decision to allow American troops onto Saudi soil to defend the Saudi nation. Thus, just as Americans worry that the Saudi monarchy's global responsibilities (tied to its religious sites) compromise its dependability as a friendly national government, bin Laden worries that the position of the monarchy as a national government committed to defending its territory compromises its global responsibilities as guardian of Islam. All of these concerns probably are far from the minds of most pilgrims who come to Makkah for the Hajj, but as tensions in the region increase they have the potential to disrupt the Hajj experience, and such a disruption would have the effect of confirming bin Laden's claim that the king is unfit to hold the title "Custodian of the Two Holy Mosques."

Jerusalem

The political tension surrounding the governance of Makkah pales in comparison to the conflict surrounding the governance of Jerusalem and its many holy sites. While Makkah is a holy city for one religion—Islam—Jerusalem is a holy city for three religions—Judaism, Christianity, and Islam, as is introduced in the sixth reading, from *Time* magazine.

For Jews, Jerusalem is the site of the first and second Temples—the focal point of Jewish worship—whose final restoration will occur with the coming of the Messiah. The pilgrimages to the Temple that took place in biblical times, described in the *Time* article, resembled the modern Hajj experience in their sense of celebration. Today, the site of the Temple attracts countless informal Jewish pilgrims. For Muslims, Jerusalem is the third holiest city, after Makkah and Madinah, as it was the place from which Muhammad arose from Earth for a mystical night flight to the heavens. For Christians, Jerusalem and surrounding towns (especially Bethlehem) were the places where Jesus walked, preached, and ultimately was betrayed and crucified.

The modern history of Jerusalem is intertwined with the complex and ongoing Israeli-Palestinian conflict, which is discussed in the accompanying chapter of the textbook (Chapter 6, on religion) and also in the module in this workbook that accompanies Chapter 7 (the module on political geographic conflict in southwest Asia). In brief, Palestine (which refers to the territory of the modern country of Israel plus the lands that Israel presently occupies) was controlled by the British between the First and Second World Wars. During that period, and especially immediately after World War II, there was considerable Jewish migration from Europe to this historic Jewish homeland, and after World War II the world Jewish community demanded an independent Jewish state in Palestine so as to protect itself from a repeat of the Holocaust that had just occurred in Europe. The people who had resided in Palestine prior to the immigration of European

Jews—primarily Muslims, along with small numbers of Eastern Orthodox Christians—opposed this plan.

In 1947, the United Nations attempted to resolve this conflict through General Assembly Resolution 181, which called for Palestine to be divided into two countries, a Jewish state and an Arab state. The border between the two countries was to run through the middle of Jerusalem. The western half of the city, which was newer and predominantly Jewish, was to go to the Jewish state, while the predominantly Arab eastern half, which included the Old City where most of the three religions' holy sites are located, was to go to the Arab state. Overlaying this division was the creation of a *corpus separatum* (separate body) for the entirety of Jerusalem plus a ring of neighboring towns that was purposely extended to include the Christian holy town of Bethlehem. The *corpus separatum* was to be ruled as a special international zone, administered by the United Nations. The portion of Resolution 181 concerning Jerusalem is reprinted as the seventh reading.

As it turned out, Resolution 181 was never fully implemented. Neighboring Arab states had voted against the resolution in the U.N. and, the day after the State of Israel declared its independence in 1948, they attacked Israel in an effort to reclaim a united, Arab Palestine. By the end of the war, Israel had acquired somewhat more land than it had been given under Resolution 181, there was no Arab Palestinian state, and the portions of land that had been designated for the Arab state that had not been seized by Israel were under the control of Jordan and Egypt. Jerusalem remained split down the middle, but with no *corpus separatum*. Israel controlled the newer, western part, while Jordan controlled the historically Arab east, including the holy Old City. Jordan allowed Jews from Israel to visit the holy Jewish sites in the Old City one day a year, but, according to the Israelis, Jordan did little to maintain these sites or facilitate access.

In 1967, the neighboring Arab states once again attacked Israel and Israel once again successfully repelled the invasion. By the end of that war, Israel had seized the remaining territory in what had been the British Mandate of Palestine (the areas that are now generally referred to as the West Bank and the Gaza Strip), as well as small adjacent pieces of Syria and Egypt (the Golan Heights and the Sinai Desert, respectively, the latter of which was given back to Egypt in 1978). Israel thus reunified Jerusalem. It proclaimed that it would allow people of all religions to visit their holy sites in the city, but it also asserted that it would maintain political control of the entire city, including the predominantly Arab East. In fact, Israel declared that the newly unified city of Jerusalem was to be its capital, although most countries refuse to recognize this, instead maintaining their embassies in the pre-1967 capital city of Tel Aviv, which is solidly located in the part of Palestine that was allocated to the Jewish state by Resolution 181. See Figure 6–15 in the textbook for maps of Israel's boundary changes from 1947 through the present.

Since 1967, Jerusalem has been perhaps the greatest sticking point in attempts to broker a peace in Israel/Palestine. Indeed, most of the provisional accords signed over the past decades have consciously avoided the issue of Jerusalem because all parties recognize that if they were to concentrate on Jerusalem they would not be able to reach agreement on other, somewhat less contentious issues. In the eighth and ninth readings, two individuals from opposing sides of the Jerusalem debate state their cases. First, Moshe Sasson, an Israeli government official and former diplomat, argues that the religious issue of Jerusalem should be separated from the political issue. Politically, he argues that a unified Jerusalem should be the capital of Israel because it historically had that status in biblical times and it has no historical political importance for Muslims or Christians. That said, however, he is happy to let the Muslim and Christian communities administer their holy sites, within the context of Israeli sovereignty, much as foreign countries administer embassies that are located on another country's soil. In the next reading, Islamic History and Culture professor Abdel-Haleem E'weis takes an opposing view. E'weis argues that of the three monotheistic religions, Islam is uniquely suited to care for all of the region's holy sites because Islamic theology incorporates the holy sites and prophets of the Jewish and Christian religions whereas Judaism, in particular, is intolerant of the other two traditions. Thus, he argues, Islam should be granted political authority over all of Jerusalem so as to ensure that all of the city's holy sites are honored and preserved. In the tenth reading, religious scholar Karen Armstrong takes a position halfway between Sasson and E'weis: In contrast to Sasson, she argues that Jerusalem is indeed an important city to Muslims, but she argues that its importance lies in its being a historic site of Muslim tolerance rather than Muslim claims to religious superiority.

The eleventh reading, a statement by the Vatican's Minister of Foreign Affairs, Monsignor Jean-Louis Tauran, directly addresses the issue of Jerusalem's political status, and he, like Armstrong, takes a position between that of E'weis and Sasson. On the one hand, Tauran rejects E'weis' position by stating that Jerusalem is a universal city that should be enjoyed by all. "It is the cultural heritage of everybody," he says, "including those who visit it simply as tourists." On the other hand, he also differentiates himself from Sasson's position. He rejects the Israeli argument that political sovereignty can be considered independently from religious control. For Tauran, Jerusalem as a holy *city* is greater than the sum of its sacred sites, and for the city to be preserved for all it must be governed in a manner that shares sovereignty among the three religions. Tauran makes it clear that he is not satisfied with the current arrangement whereby Israel is sovereign but allows

Christian and Muslim communities to maintain and facilitate access to their specific holy sites:

> The simple "extraterritoriality" of the Holy Places, with the assurance that pilgrims would be able to visit them without hindrance, would not suffice. The identity of the City includes a sacred character which belongs not just to the individual sites or monuments, as if these could be separated from one another or isolated from the respective communities. The sacred character includes Jerusalem in its entirety.

Tauran is vague about what precise legal mechanism he favors for protecting Jerusalem as "a special internationally guaranteed Statute," but one can imagine an arrangement similar to the *corpus separatum* advocated fifty years earlier by the United Nations. A similar solution is advocated in the final reading, a sermon given on Rosh Hashanah (the Jewish New Year) in 2000 by American Rabbi Arthur Rulnick. Rulnick advocates a return to Resolution 181, with the Jewish and Arab states controlling their respective areas of Jerusalem, and with the United Nations (or some other international or binational entity) governing the areas where different religions' holy sites are so close to each other that control by one country or the other is not possible.

Readings

1

from *Aramco World Magazine*

Hajj: The Journey of a Lifetime

by Ni'mah Isma'il Nawwab
July-August 1992

One fifth of humankind shares a single aspiration: to complete, at least once in a lifetime, the spiritual journey called the Hajj.

The hajj, or pilgrimage to Makkah, a central duty of Islam whose origins date back to the Prophet Abraham, brings together Muslims of all races and tongues for one of life's most moving spiritual experiences.

For 14 centuries, countless millions of Muslims, men and women from the four corners of the earth, have made the pilgrimage to Makkah, the birthplace of Islam. In carrying out this obligation, they fulfill one of the five "pillars" of Islam, or central religious duties of the believer.

Muslims trace the recorded origins of the divinely prescribed pilgrimage to the Prophet Abraham, or Ibrahim, as he is called in Arabic. According to the Qur'an, it was Abraham who, together with Ishmael (Isma'il), built the Ka'bah, "the House of God," the focal point toward which Muslims turn in their worship five times each day. It was Abraham, too—known as Khalil Allah, "the friend of God"—who established the rituals of the hajj, which recall events or practices in his life and that of Hagar (Hajar) and their son Ishmael.

In the chapter entitled "The Pilgrimage," the Qur'an speaks of the divine command to perform the hajj and prophesies the permanence of this institution:

> And when We assigned for Abraham the place of the House, saying "Do not associate Anything with Me, and purify My House for those who go around it and for those who stand and bow and prostrate themselves in worship. And proclaim the Pilgrimage among humankind: They will come to you on foot and on every camel made lean By traveling deep, distant ravines."

By the time the Prophet Muhammad received the divine call, however, pagan practices had come to muddy some of the original observances of the hajj. The Prophet, as ordained by God, continued the Abrahamic hajj after restoring its rituals to their original purity.

Furthermore, Muhammad himself instructed the believers in the rituals of the hajj. He did this in two ways: by his own practice, or by approving the practices of his Companions. This added some complexity to the rituals, but also provided increased flexibility in carrying them out, much to the benefit of pilgrims ever since. It is lawful, for instance, to have some variation in the order in which the several rites are carried out, because the Prophet himself is recorded as having approved such actions. Thus, the rites of the hajj are elaborate, numerous and varied; aspects of some of them are highlighted below.

The hajj to Makkah is a once-in-a-lifetime obligation upon male and female adults whose health and means permit it, or, in the words of the Qur'an, upon "those who can make their way there." It is not an obligation on children, though some children do accompany their parents on this journey.

Before setting out, a pilgrim should redress all wrongs, pay all debts, plan to have enough funds for his own journey and for the maintenance of his family while he is away, and prepare himself for good conduct throughout the hajj.

When pilgrims undertake the hajj journey, they follow in the footsteps of millions before them. Nowadays hundreds of thousands of believers from over 70 nations arrive in the Kingdom of Saudi Arabia by road, sea and air every year, completing a journey now much shorter and in some ways less arduous than it often was in the past.

Till the 19th century, traveling the long distance to Makkah usually meant being part of a caravan. There were three main caravans: the Egyptian one, which formed in Cairo; the Iraqi one, which set out from Baghdad; and the Syrian, which, after 1453, started at Istanbul, gathered pilgrims along the way, and proceeded to Makkah from Damascus.

As the hajj journey took months if all went well, pilgrims carried with them the provisions they needed to sustain them on their trip. The caravans were elaborately supplied with amenities and security if the persons traveling were rich, but the poor often ran out of provisions and had to interrupt their journey in order to work, save up their earnings, and then go on their way. This resulted in long journeys which, in some cases, spanned ten years or more. Travel in earlier days was filled with adventure. The roads were often unsafe due to bandit raids. The terrain the pilgrims passed through was also dangerous, and natural hazards and diseases often claimed many lives along the way. Thus, the successful return of pilgrims to their families was the occasion of joyous celebration and thanksgiving for their safe arrival.

Lured by the mystique of Makkah and Madinah, many Westerners have visited these two holy cities, on which the pilgrims converge, since the 15th century. Some of them disguised themselves as Muslims; others, who had genuinely converted, came to fulfill their duty. But all seem to have been moved by their experience, and many recorded their impressions of the journey and the rituals of the hajj in fascinating accounts. Many hajj travelogues exist, written in languages as diverse as the pilgrims themselves.

The pilgrimage takes place each year between the eighth and the 13th days of Dhu al-Hijjah, the 12th month of the Muslim lunar calendar. Its first rite is the donning of the ihram.

The ihram, worn by men, is a white seamless garment made up of two pieces of cloth or toweling; one covers the body from waist to ankle and the other is thrown over the shoulder. This garb was worn by both Abraham and Muhammad. Women generally wear a simple white dress and a headcovering, but not a veil. Men's heads must be uncovered; both men and women may use an umbrella.

The ihram is a symbol of purity and of the renunciation of evil and mundane matters. It also indicates the equality of all people in the eyes of God. When the pilgrim wears his white apparel, he or she enters into a state of purity that prohibits quarreling, committing violence to man or animal and having conjugal relations. Once he puts on his hajj clothes the pilgrim cannot shave, cut his nails or wear any jewelry, and he will keep his unsown garment on till he completes the pilgrimage.

A pilgrim who is already in Makkah starts his hajj from the moment he puts on the ihram. Some pilgrims coming from a distance may have entered Makkah earlier with their ihram on and may still be wearing it. The donning of the ihram is accompanied by the primary invocation of the hajj, the talbiyah:

> Here I am, O God, at Thy Command! Here I am at Thy Command! Thou art without associate; Here I am at Thy Command! Thine are praise and grace and dominion! Thou art without associate.

The thunderous, melodious chants of the talbiyah ring out not only in Makkah but also at other nearby sacred locations connected with the hajj.

On the first day of the hajj, pilgrims sweep out of Makkah toward Mina, a small uninhabited village east of the city. As their throngs spread through Mina, the pilgrims generally spend their time meditating and praying, as the Prophet did on his pilgrimage.

During the second day, the 9th of Dhu al-Hijjah, pilgrims leave Mina for the plain of 'Arafat for the wuquf, "the standing," the central rite of the hajj. As they congregate there, the pilgrims' stance and gathering reminds them of the Day of Judgment. Some of them gather at the Mount of Mercy, where the Prophet delivered his unforgettable Farewell Sermon, enunciating far-reaching religious, economic, social and political reforms. These are emotionally charged hours, which the pilgrims spend in worship and supplication. Many shed tears as they ask God to forgive them. On this sacred spot, they reach the culmination of their religious lives as they feel the presence and closeness of a merciful God.

The first Englishwoman to perform the hajj, Lady Evelyn Cobbold, described in 1934 the feelings pilgrims experience during the wuquf at 'Arafat. "It would require a master pen to describe the scene, poignant in its intensity, of that great concourse of humanity of which I was one small unit, completely lost to their surroundings in a fervor of religious enthusiasm. Many of the pilgrims had tears streaming down their cheeks; others raised their faces to the starlit sky that had witnessed this drama so often in the past centuries. The shining eyes, the passionate appeals, the pitiful hands outstretched in prayer moved me in a way that nothing had ever done before, and I felt caught up in a strong wave of spiritual exaltation. I was one with the rest of the pilgrims in a sublime act of complete surrender to the Supreme Will which is Islam."

She goes on to describe the closeness pilgrims feel to the Prophet while standing in 'Arafat: "... as I stand beside the granite pillar, I feel I am on Sacred ground. I see with my mind's eye the Prophet delivering that last address, over thirteen hundred years ago, to the weeping multitudes. I visualize the many preachers who have spoken to countless millions who have assembled on the vast plain below; for this is the culminating scene of the Great Pilgrimage."

The Prophet is reported to have asked God to pardon the sins of pilgrims who "stood" at 'Arafat, and was granted his wish. Thus, the hopeful pilgrims prepare to leave this plain joyfully, feeling reborn without sin and intending to turn over a new leaf.

Just after sunset, the mass of pilgrims proceeds to Muzdalifah, an open plain about halfway between 'Arafat and Mina. There they first pray and then collect a fixed number of chickpea-sized pebbles to use on the following days.

Before daybreak on the third day, pilgrims move en masse from Muzdalifah to Mina. There they cast at white pillars the pebbles they have previously collected. According to some traditions, this practice is associated with the Prophet Abraham. As pilgrims throw seven pebbles at each of these pillars, they remember the story of Satan's attempt to persuade Abraham to disregard God's command to sacrifice his son.

Throwing the pebbles is symbolic of humans' attempt to cast away evil and vice, not once but seven times—the number seven symbolizing infinity.

Following the casting of the pebbles, most pilgrims sacrifice a goat, sheep or some other animal. They give the meat to the poor after, in some cases, keeping a small portion for themselves.

This rite is associated with Abraham's readiness to sacrifice his son in accordance with God's wish. It symbolizes the Muslim's willingness to part with what is precious to him, and reminds us of the spirit of Islam, in which submission to God's will plays a leading role. This act also reminds the pilgrim to share worldly goods with those who are less fortunate, and serves as an offer of thanksgiving to God.

As the pilgrims have, at this stage, finished a major part of the hajj, they are now allowed to shed their ihram and put on everyday clothes. On this day Muslims around the world share the happiness the pilgrims feel and join them by performing identical, individual sacrifices in a worldwide celebration of 'Id al-Adha, "the Festival of Sacrifice." Men either shave their heads or clip their hair, and women cut off a symbolic lock, to mark their partial deconsecration. This is done as a symbol of humility. All proscriptions, save the one of conjugal relations, are now lifted.

Still sojourning in Mina, pilgrims visit Makkah to perform another essential rite of the hajj: the tawaf, the seven-fold circling of the Ka'bah, with a prayer recited during each circuit. Their circumambulation of the Ka'bah, the symbol of God's oneness, implies that all human activity must have God at its center. It also symbolizes the unity of God and man.

Thomas Abercrombie, a convert to Islam and a writer and photographer for National Geographic Magazine, performed the hajj in the 1970's and described the sense of unity and harmony pilgrims feel during the circling: "Seven times we circled the shrine," he wrote, "repeating the ritual devotions in Arabic: 'Lord God, from such a distant land I have come unto Thee .... Grant me shelter under Thy throne.' Caught up in the whirling scene, lifted by the poetry of the prayers, we orbited God's house in accord with the atoms, in harmony with the planets."

While making their circuits pilgrims may kiss or touch the Black Stone. This oval stone, first mounted in a silver frame late in the seventh century, has a special place in the hearts of Muslims as, according to some traditions, it is the sole remnant of the original structure built by Abraham and Ishmael. But perhaps the single most important reason for kissing the stone is that the Prophet did so.

No devotional significance whatsoever is attached to the stone, for it is not, nor has ever been, an object of worship. The second caliph, 'Umar ibn al-Khattab, made this crystal clear when, on kissing the stone himself in emulation of the Prophet, he proclaimed: "I know that you are but a stone, incapable of doing good or harm. Had I not seen the Messenger of God kiss you—may God's blessing and peace be upon him—I would not kiss you."

After completing the tawaf, pilgrims pray, preferably at the Station of Abraham, the site where Abraham stood while he built the Ka'bah. Then they drink of the water of Zamzam.

Another, and sometimes final, rite is the sa'y, or "the running." This is a reenactment of a memorable episode in the life of Hagar, who was taken into what the Qur'an calls the "uncultivable valley" of Makkah, with her infant son Ishmael, to settle there.

The sa'y commemorates Hagar's frantic search for water to quench Ishmael's thirst. She ran back and forth seven times between two rocky hillocks, al-Safa and al-Marwah, until she found the sacred water known as Zamzam. This water, which sprang forth miraculously under Ishmael's tiny feet, is now enclosed in a marble chamber the Ka'bah.

These rites performed, the pilgrims are completely deconsecrated: They may resume all normal activities. According to the social customs of some countries, pilgrims can henceforth proudly claim the title of al-Hajj or Hajji.

They now return to Mina, where they stay up to the 12th or 13th day of Dhu al-Hijjah. There they throw their remaining pebbles at each of the pillars in the manner either practiced or approved by the Prophet. They then take leave of the friends they have made during the Hajj. Before leaving Makkah, however, pilgrims usually make a final tawaf round the Ka'bah to bid farewell to the Holy City.

Usually pilgrims either precede or follow the hajj, "the greater pilgrimage," with the 'umrah, "the lesser pilgrimage," which is sanctioned by the Qur'an and was performed by the Prophet. The 'umrah, unlike the hajj, takes place only in Makkah itself and can be performed at any time of the year. The ihram, talbiyah and the restrictions required by the state of consecration are equally essential in the 'umrah, which also shares three other rituals with the hajj: the tawaf, sa'y and shaving or clipping the hair. The observance of the 'umrah by pilgrims and visitors symbolizes veneration for the unique sanctity of Makkah.

Before or after going to Makkah, pilgrims also avail themselves of the opportunity provided by the hajj or the 'umrah to visit the Prophet's Mosque in Madinah, the second holiest city in Islam. Here, the Prophet lies buried in a simple grave under the green dome of the mosque. The visit to Madinah is not obligatory, as it is not part of the hajj or 'umrah, but the city—which welcomed Muhammad when he migrated there from Makkah—is rich in moving memories and historical sites that are evocative of him as a prophet and statesman.

In this city, loved by Muslims for centuries, people still feel the presence of the Prophet's spirit. Muhammad Asad, an Austrian Jew who converted to Islam in 1926 and made five pilgrimages between 1927 and 1932, comments on this aspect of the city: "Even after thirteen centuries [the Prophet's] spiritual presence is almost as alive here as it was then. It was only because of him that the scattered group of villages once called Yathrib became a city and has been loved by all Muslims down to this day as no city anywhere else in the world has ever been loved. It has not even a name of its own: for more than thirteen hundred years it has been called Madinat an-Nabi, 'the City of the Prophet.' For more than thirteen hundred years, so much love has converged here that all shapes and movements have acquired a kind of family resemblance, and all differences of appearance find a tonal transition into a common harmony."

As pilgrims of diverse races and tongues return to their homes, they carry with them cherished memories of Abraham, Ishmael, Hagar, and Muhammad. They will always remember that universal concourse, where poor and rich, black and white, young and old, met on equal footing.

They return with a sense of awe and serenity: awe for their experience at 'Arafat, when they felt closest to God as they stood on the site where the Prophet delivered his sermon during his first and last pilgrimage; serenity for having shed their sins on that plain, and being thus relieved of such a heavy burden. They also return with a better understanding of the conditions of their brothers in Islam. Thus is born a spirit of caring for others and an understanding of their own rich heritage that will last throughout their lives.

The pilgrims go back radiant with hope and joy, for they have fulfilled God's ancient injunction to humankind to undertake the pilgrimage. Above all, they return with a prayer on their lips: May it please God, they pray, to find their hajj acceptable, and may what the Prophet said be true of their own individual journey: "There is no reward for a pious pilgrimage but Paradise."

Source: Ni'mah Isma'il Nawwab, "Hajj: The Journey of a Lifetime" *Aramco World Magazine*, July-August 1992.

2

from the Cable News Network

For Pilgrims, Destination Is More Spiritual than Geographical

by Riz Khan
March 2001

The Hajj is misunderstood.

Not only by non-Muslims, but also by many of those who practice the faith. To some degree, the misunderstanding comes from the fact that it's not until a Muslim actually goes on this pilgrimage that he or she starts to get some idea of what it's all about and what it really involves spiritually.

Usually, any religious obligation or duty seems to gain an air of sombre importance. Well, for a Muslim, it *is* an important event—but not necessarily quite so sombre. The Hajj is one of the Five Pillars of Islam: At least once in a lifetime, any Muslim who is able, financially and physically, to complete this journey must do so.

Most people might imagine the Hajj to be a complicated and perhaps demanding version of a Catholic confession—a visit to God's holiest site for Muslims to plead forgiveness for sins. (Perhaps with a fear that one's sins might be so bad that they can't be overlooked.) In fact, the Hajj is a very happy affair. Essentially, it's a gathering of more than 2 million people who have achieved a "mission of a lifetime." They've arrived at the geographical and physical heart of their religion, so pilgrims are usually elated to have made it to Mecca—or Makkah, to use the old name that has been reinstated. Mecca marks the spot where, according to tradition, the prophet Abraham first built a shrine to worship God. It was a caravan crossroads through rocky outcroppings in the desert, which grew into a modern, noisy, bustling center.

During the Hajj, the atmosphere is more one of celebration in and around the city, which is open only to Muslims. In a way, that exclusion of non-Muslims is a pity. It prevents much of the world from seeing what a mixed community Islam embraces. There are the colorfully dressed, and often loud and cheerful, Africans—many from Nigeria. Small, closely huddled groups of women from Indonesia and the Philippines often wind their way through the huge crowds that take over Mecca. A chorus of Indians, Pakistanis, Bangladeshis, Bosnians, Arabs . . . it's truly a melting pot. Men, women and children are all equally in awe of this place they've known only through pictures, and perhaps television.

Add to that jam-packed streets and a mix of languages, and it's amazing that hardly a voice is raised in anger, or confrontation witnessed. Everyone really tries to help each other through this annual miracle of logistics and mass movement of people. And that's what the Hajj mostly is—a movement through the desert. The pilgrims flood into Mecca by air, sea, motor vehicles and sometimes still the odd camel caravan, which is how it was done in the "old" days.

Many pilgrims arrive having made a trip to the huge Prophet's Mosque in Medina—a city that first became home to Mohammed, when he and his early followers were driven out of Mecca. It's considered a great blessing to pray at this remarkable mosque.

Once in Mecca, the first goal of the pilgrims is to get over to the Great Mosque that surrounds the best known structure symbolizing Islam—the Kaaba. The Kaaba is a cubic stone structure the size of a modest house, but it's clad in a black silk cloth, with Arabic embroidered onto it in gold. Pilgrims rush to see this sight to remind themselves that they've actually arrived at the heart of Islam—the center of their religion. The Kaaba is the location that Muslims turn to pray toward daily. Blessings for prayers in the holy Great Mosque are considered to be multiplied thousands of times.

But it comes as a surprise to most that the core purpose of the Hajj is not to look out on the wonders of the Kaaba and the beautiful mosque, but to look within—to discover oneself. The pilgrimage involves a trek through the desert to the tent city at Mina, and from there, on to the plain of Arafat. It was here on a small hill that Mohammed preached his last sermon, declaring that Muslims must conduct the pilgrimage in this particular manner. It was a ritual dating back to the time of Abraham and follows mostly the events he experienced in affirming his faith to God. That final sermon by Mohammed set in stone the pilgrimage duty for those who were to follow Islam for centuries on.

The Hajj pilgrims stay in Arafat until sunset before weaving their way back to Mecca, again, via Mina. But it's those crucial hours in the desert where Muslims are supposed to discover what the Hajj is all about. That time is meant to be the most honest in a person's life—an honest reflection on all that a person has done right and wrong. Then comes the request to God for forgiveness, and the chance to make a fresh start.

In a way, it's not what you do during the Hajj that matters.

It's what you do after, and perhaps there comes the understanding.

Source: Riz Khan, "For Pilgrims, Destination is More Spiritual than Geographical," March 2001.

3

from the Cable News Network

Saudi Teen Amazed at How People Differ in Their Devotion to Allah

by Yousef Saleh Ba-Isa
March 4, 2001

I am a 17-year-old, living in Dhahran, Saudi Arabia, on the Kingdom's East Coast. I am an 11th grader at Dhahran Ahliyyah School. All my friends were very happy that I would be performing Hajj. They all asked me to pray for them in Mecca and at Arafat.

I am performing Hajj with my father, aunt and cousins. When my father was a teenager, he made Hajj with *his* father. They walked all the way from the Grand Mosque in Mecca to the plain of Arafat and back. My journey will be much easier. Because my cousins are from Mecca, I am staying at their home every night. So I am not sleeping out in a tent like most of the pilgrims.

I left my home Saturday morning at 9 a.m., caught a flight, and by 1 p.m. I was in Jeddah on the Kingdom's West Coast. After visiting my grandmother we moved on by SUV to Mecca. Three hours later I was drinking tea with my cousins. The crowds in Mecca were already diminishing when we arrived, because the pilgrims were moving to Mina where they would stay overnight before moving toward Arafat in the morning.

We waited until after midnight to go make Tawaf at the Grand Mosque. The circumambulation of the Kaaba was very easy because the Grand Mosque was about half empty. By 3 a.m. we had finished, but I wanted to kiss the Black Stone. My father wasn't too happy about this because the area is very crowded, but he gave me one hour to accomplish my goal.

I took half a turn around the Kaaba, so I could approach the Black Stone easily. There were many people pushing and shoving. I am tall and strong so I just gently moved forward without pushing. I was a little bit afraid mostly because some of the people around me were weeping. They were crying because they thought that they wouldn't get to touch the Stone. I held the crowd back so that two of the weaker pilgrims could slip in before me and then my turn came. The space was very small and I quickly kissed the stone and wiped it and then fell back so my place could be taken by another Muslim. Kissing the Black Stone is a Sunnah of Hajj and an act of obedience to Allah.

And speaking about Allah, I was surprised to see the number of people who could not understand Arabic in general except for the word Allah. These pilgrims who came from other lands would have guides with them and they would recite the prayers after their guides, line by line. They didn't know exactly what they were reciting but they believed that their prayers for forgiveness and blessings would be accepted nevertheless.

After finishing the rituals at the Grand Mosque I returned to my cousin's home and slept for about six hours. Then it was time to move to Arafat. The police did an excellent job of keeping traffic flowing smoothly and there were no major slowdowns.

We arrived in time for the prayer, and recited the combined Dhur and Asr prayers. I have been amazed at how people differ in their devotions to Allah. Some pilgrims are very quiet and sit or stand practically motionless while reciting the Koran. But I have also seen some astonishing outbursts. When we were at the Grand Mosque, I saw a man fall to the ground weeping and begging for Allah to forgive his sins. Here at Arafat, I have seen many people crying and making supplications to Allah for blessings and forgiveness. I pray that all their prayers will be answered.

Today it is very beautiful here at Arafat. The sky is overcast and there have been a few small rain showers. It is crowded but there is still some space available. The pilgrims are better behaved here than they were at the Grand Mosque. There is no pushing or shoving and everyone is being very gentle. I am troubled by the amount of litter and I hate it when I see pilgrims smoking cigarettes. Somehow smoking doesn't seem to have anything to do with obedience to Allah and that is what Hajj is supposed to be all about.

I want people to know that Hajj is truly a matter of faith. Yes, it has its hardships and inconveniences, but praying together as a community before Allah is very moving. I have actually felt that I am standing in front of Allah and feeling His strength. Being involved in such an experience results in a deep peace of mind that I have never felt before in my life. Some people unfortunately have very bad ideas about Arabs and Muslims. I wish they would watch us as we move together in prayer, begging for forgiveness and asking for blessings from Allah. Islam is a religion of peace.

Source: Yousef Saleh Ba-Isa, "Saudi Teen Amazed at How People Differ in Their Derotion to Allah," March 4, 2001.

from **Jannah.org**

4

Hajj and Umrah Tips & Tricks

- EXTREMELY IMPORTANT: DO study the fiqh of Hajj & Umrah before you go, well in advance. (I mean really advanced, not the weeks before it'll be so hectic you won't be able to do it at your leisure. Think 2-3 months in advance.) Get some good books, tapes and videos. Look online and print everything out. Write all the Duas out on index cards. Try to memorize as many as you can BEFORE you go. Make flash cards whatever it takes!

  DO NOT go there and expect to learn how to do everything from your guides/ppl with you. This is too important not to know for yourself. Remember if you do anything incorrect your umrah or hajj WILL NOT BE ACCEPTED. There are many stories of the tour people taking people to do things too early or too late. Also, the crowd there is so enormous it is incredibly difficult when millions of people are all trying to do the same thing at the same time. This is where your fiqh of hajj knowledge comes in. It makes it much easier if you know things like you can pray your 2 rakats of Maqam Ibrahim all the way back in the Haram, or on the different levels or even anywhere in the Haram, or that you can do your Saiyy on the second level or that you can throw your stones from after sunrise to anytime at night. If you know things like this you can avoid the crowds of people who will try to stick to only one thing because they don't know what else is allowed. Also if something happens you will know what you should do to compensate for it and you may save your hajj or umrah.

  A NOTE about all the books and info. Everyone will tell you something a little different. Try to piece everything together so that you at least know what the required conditions you must fulfill are and what things would break your hajj or umrah absolutely or what things will necessitate compensation. Ask your Imam or local knowledgable people to help you understand the Fiqh and answer any questions you have.
- Also VERY IMPORTANT: Learn the fiqh of praying a Janaza (funeral prayer). You will be doing it after every single prayer, no joke. (There was only one prayer in our entire 2 week stay that we didn't.) Learn the dua for it too. Praying at someone's janaza is a very good thing and you have ample opportunity of participating so take them. Also learn the different ways to send Salams on the prophet (saw) for the time you are in Madina. Another good suggestion is to draw a chart of the Kaaba (looking from above) and along each of its walls (and along each of the special points), write the relevant duas.
- The best times to visit the Haram of Makkah, do tawaf and saiyy, ibadah and to try to touch the black stone etc., is from 1AM until Fajr and from about an hour after Fajr until 9 or so. A good schedule would be to sleep from Fajr until Dhuhr, pray Dhuhr in the Haram then go back to the hotel and eat and rest. (After Dhuhr about noon to 2 it is HOT.) Go back for Asr. Then go back for Maghrib and STAY there for Isha/Taraweeh. After you can take a few hours to go shopping if needed. Otherwise stay at the haram, alternating doing different things like Tawaf, reading Quran, praying until Qiyam which is about 1AMish Then stay until Tahajjud which is about an hour before Fajr. Then stay for Fajr. If you are going to Ziyara or doing multiple

Umrahs and can't sleep after Fajr, take a few hours at night before Fajr to sleep. (You might say to yourself... gee doesn't look like there is any time for sleep there! That's true... don't waste your time sleeping... you can always sleep when you come home :) But know when your body is tired and can't take it anymore. Don't make yourself sick. Naps of 3–4 hours worked for me.) The best spots in the Haram are right in front of the Kaaba on the marble surrounding it or on the second level at the front. If you want to be on the haram floor you have to go early and stake out a spot. Sisters should head towards the Zamzam area and sit right near there, otherwise you'll get kicked out of any other area after awhile. A really good spot I found for sisters is upstairs all the way in the front of the sisters section. You have an incredible birdseye view of the entire haram and can still pray with the Kaaba in sight. Remember to bring your prayer rug to sit on. This is so that people don't come and sit right in front of you or back of you so you'll have room to pray later. Two things that I wish I knew beforehand about this: You can pray anywhere in the Haram without a sutrah i.e., people can walk in front of you without breaking your prayer. (This was extremely annoying to me at the beginning when people would walk right through my sutrah every time! But you should try your best not to do this to others). And you can also look at the Kaaba while you pray in the Haram.

- PREPARE yourself mentally for a lot of frustration, mental anguish, annoyance and anger. To obtain a Hajj Mabrur (one that is accepted and forgives all your sins) you must not let these things interfere with your goal.
- PREPARE yourself physically for a lot of walking and physical exercise. You may be walking back and forth from your hotel to the Haram at least 10 times a day, not to mention logging Tawafs and Sai'ys that come out to miles in the end.
- DO NOT bring a lot of clothes. All you need really is 3–4 jilbabs—mostly black for umrah time (lighter colors for hajj and summer season) and a few clothes underneath that you can mix and match. Bring enough socks and underthings. You can always wash clothes if you have to. But bringing a lot of clothes just wastes your time and suitcase room. A regular travelling rule of thumb is to never pack more than you can carry for a trip!

  It goes without saying that you should bring some comfy shoes/sneakers for when you go on Ziyara or shopping. Also bring some good chappals/shibshib for when you are going to the haram. Don't overpack your suitcase. You won't be able to get it all back home especially if you buy things. You can always give away some of your clothes/shoes while there or try this well-known travelling tip: bring an extra empty suitcase/bag with you or buy one there.
- A NOTE for Madina. The weather there is very different from Makkah. Madina is much cooler. We were in 80–90 degrees everyday in Makkah and then in Madina we were at 60–70s and it was extremely COLD around Fajr, so bring a sweatshirt/sweater and a few warm clothes so you don't get sick.
- DO NOT bother bringing anything expensive, like jewelry or electronic equipment. If it's lost or stolen you'll never see it again.
- DO NOT change money before you go. There are tons of exchanges there that you can go to on the streets and you'll get a better rate than the usual bank rate too. Bring enough cash for what you want to buy. Don't rely on using your credit card. There are finance charges of at least 2–3% and the exchange rate of Visa sux.
- DO bring alot of medicine—Tylenol, asprin, Theraflu, Tums, Pepto Bismol, Midol, Chapstick/Vaseline. Believe me, you'll be glad you did and if by some miracle you don't get sick other people in your group will.
- DO bring some snack food like granola bars, goldfish, dried fruit. You'll be glad when your stomach starts doing loops or you are starving in the haram. Not to mention sometimes getting stuck for hours on a bus during heavy traffic or waiting for clearance.
- DO bring a cheap plastic bag/canvas bag to put your shoes and janamaz (prayer rug) and quran in when you go into the haram.

  Don't bring a janamaz from home. Just buy a nice one there. You can also bring a small bag/purse you can wear under your jilbab for your money/passport. Don't bother with big handbags—you'll get searched every time you enter the haram and will have to watch it all the time.

- DO NOT waste your time shopping. You're only in Makkah once in your life (anything could happen and due to life circumstances, being busy, illness etc., you may never return again).

  Make a list of what gifts you have to give to whom beforehand and write down what you need to buy and quantities. That way you won't waste alot of time trying to decide what to get for who.

  Some common gifts include:

  - Dates (yes they have chocolate covered dates :))
  - Zamzam water
  - Tasbeeh Beads
  - Hijabs & Jilbabs
  - Prayer Rugs (called sajadahs there)
  - Jewelry (gold, fake and everything in between)
  - Metal work (tea sets etc.)
  - Incense Perfumes & Perfume bottles
  - Thobes
  - Kufis
  - Cards
  - Desi (IndoPak) dresses
  - Books
  - Watches

- DO NOT bother going [to] any malls. They're usually WAY too expensive, but if you want to waste your time window shopping or looking at American and European designer stuff go ahead.

  A NOTE about shopping, every shop owner and I mean EVERY (even those who are very nice and islamic and kind to you) will try to get every penny (riyaal) from you if they can. For your first time out, just walk around and ask for prices of different things until you get a handle of how much things go for. (Brothers please ask your mothers/wives what a good price for gold/gram is before you try to buy anything there!)
- DO NOT speak English or any foreign language in stores, just urduor Arabic. They will totally rip you off. Even if all you know is "Kam Hatha" and they know you don't speak Arabic as your native language because of your accent for some reason if you speak English they up the price 200% automatically. You must bargain for everything you buy. Everything is marked up as par for the course. A good trick is to offer half then work your way up. Remember to go to many stores to get an idea of price range. Sometimes they will try to bully you into buying something by grabbing it and stuffing it into a bag. Don't let them . . . take your time, get what you want and check it. If you feel that the guy is ripping you off but really want/need the stuff, buy it anyway because you'll regret it after. Anything there is cheaper than anywhere here. Don't let pride stand in your way :)
- DO learn all the Arabic numbers (1 to 20, and every 5 after that till a hundred and you'll be set) and phrases like that "Kam Hatha?"—How much is that?, "BiKam?" How much?, "Shuayyeh" Move over, and "Laa . . . " No . . .

  Don't tell them you're from America unless you're window shopping. Don't dress expensively, shopkeepers there are extremely sharp. You can even wait a few days until everyone else has found the best places to buy things cheap—find out where they are and THEN save time and just get what you want.

  SISTERS be careful about going shopping alone (as in, DON'T) always go with a bro (hate to say it but you're safer from harassment) or with a group of sisters.
- DO be careful about taking pictures and bringing cameras.They don't let cameras into either of the Harams and they'll check your bags too.
- DO agree on prices for cab rides, camel rides and any other service in between beforehand. A note about the camel rides. They will immediately go and take polaroids of you on the camel/horse and then force you to buy them. Tell them beforehand NO PICTURES please. (One guy cursed our group and made dua that Allah never answer our duas b/c someone refused to buy the pics !!)
- BEGGARS—There are beggars everywhere. On the streets of Makkah, at historical sites like Uhud, Arafat etc. They'll even come up to you (to sisters too!) and tell you stories about how they lost all their luggage and possessions and need to get

back to Jeddah/Makkah/Madinah. It's hard to figure out who is faking and who isn't. I would suggest deciding on an amount you would like to give as Sadaqah beforehand and then give that to a reputable Islamic relief agency or relatives you have in poor countries or even friends/ppl traveling back to their home countries.They would know better who is truly in need.

- MAKE SURE to go to Ziyara. Ziyara is visiting the local historical sites in Makkah or Madina. It's extremely important to visit those sites, give your salams to the dead, to really get a feeling and understanding of how Islam began in this place. I would say it should be almost as important to you as making sure you do your Umrah. (But please don't think it's fard or anything else. :)) Don't go there thinking it's part of any worship practices. Many people do go to those sites and commit strange innovative practices (you'll see all the Saudi signs in Arabic, English, French, Urdu and Turkish!! telling ppl not to). This is where learning some Seerah (study of the life of Muhammad (saw) and the early Muslims) is extremely important before you go so you can actually feel where you are visiting. Just take one day after fajr and have some cab driver take you out to the different places. Get a map of Jannat al-Baqi' and make sure you say salam to all the sahaba/saliheen buried there. For both Ziyara's go immediately AFTER Fajr at the haram, come back BEFORE dhuhr, pray dhuhr at the haram and then rest at the hotel for a while.
- DO NOT forget what gate you enter the Haram from and the hotel you are staying at.
- DO make multiple Umrahs. It's not hard to take the short cab ride to TAN'IM sometime after Fajr. (The place you'll need to go to state your intention and reenter Makkah in Ihram). Make ghusl and change into your ihram before you go. Then at Tan'im you just pray 2 rakats and make your intention for umrah again. Come back before Dhuhr and perform your second (or third or fourth) umrah... Remember you can also make umrah on behalf of another person, a dead relative or someone back home who could not make it.
- DO NOT look at pictures of the Kaaba before you go. From now until you see it with your own two eyes It RUINS the impact it has on your heart.
- REMEMBER you are there to worship. Shopping, eating, etc., are all just a waste of time. Food: it will be tempting to eat all this food there, but remember this much: the more you eat, the more tired you will get and in Makkah you will NEED your energy.

  Take all this advice with a grain of salt. Every situation/time is different. Most of the information here is suitable for umrah. Hajj may be a completely different experience!
- HOTEL—Choose one within 5 mins walking distance of Haram if you can. When the crowds descend it will take you 20–30 mins to get to the Mosque.
- JUMAH—Go EARLY!!!! Our first Jumah in Makkah the Adhaan was 12:05 or something and we got to Masjid Al Haram at 11AM. BIG mistake. The concourse outside the Masjid was full let alone trying to get in there. We sat in the beating sun for over an hour and prayed. The next week we got there at 9 AM!!!
- TAWAF—There are large groups who perform Tawaf together and link hands. If you see them approach you, then let them pass or step aside and give them way. Also watch out for the elderly who are being carried around the Kabah and people taking their relatives around in wheelchairs. As a rule the nearer the Kabah you perform your Tawaf the more squashed you will get and I mean squashed. This is especially true when going around Maqam Ibrahim because of the people trying to pray.
- ARAFAT—Make the most of this day as possible. Don't eat too much after Zuhur. The last thing you want to do is feel sleepy from having overeaten.
- MUZDALIFAH—You have to pray Fajr on the morning of Eid before you go back to Mina. Make sure you make a note of what time Fajr was when you were in Makkah. Some people make the Adhaan too early in an attempt to get to Mina early. Don't let them make you pray your Fajr before time.
- MINA—For those of you who haven't been, there are three Jamarat all in a line with each other separated by 200 meters or so. The Saudi authorities have created what I can only describe as a flyover so that you can perform your stoning from the above tier as well as the lower tier. There is also a one way system on the top tier so that everyone starts from one end and moves to the other. Each Jamarat is sur-

rounded by a circular wall and it is this circle that your stones have to enter after you've thrown. We were very fortunate enough to be on the side of the mountain and we could see how the Hajjis were performing the stoning. On the top tier most Hajjis walk in a straight line and reach the front of the circle. Consequently, there is a HUGE crowd at the front of the circle as people wait for those in front to finish. Because of this you should walk at the sides hugging the fence and avoid the front entirely. Walk PAST the Jamarat and then double back on yourself and throw from the back of the circle. Alhamdulillah we were able to place our hands on the circle wall and throw from there using this piece of advice.

- SHOPPING—Check out the many bookshops. Lots of good stuff out there.
- PATIENCE—You'll understand the meaning of that word when you encounter the crowds, the queues, the long waits etc., etc., etc.

Source: "Hajj and Umrah Tips & Tricks," jannah.org/hajj/tipsandtricks.html

from the Saudi Arabian Embassy, London

5

Guardian of the Holy Places

Saudi Arabia is the home of two of Islam's holy sanctuaries: Makkah the Blessed and Madinah the Radiant. The Al-Aqsa Mosque in Jerusalem, enclosing the place from which the Prophet Muhammad ascended to heaven, completes the trio of venerated shrines in the Islamic world.

To Saudi Arabia, caring for the holy cities of Makkah, the birthplace of Islam and the Prophet Muhammad, and Madinah, the Prophet's burial place, is a sacred trust exercised on behalf of all Muslims. Recognising the unique and historic tradition these holy sites represent, King Fahd Bin Abdul Aziz adopted the official title of the Custodian of the Two Holy Mosques as an expression of his deep sense of responsibility toward Islam.

Saudi Arabia's dedication to Islam is demonstrated by its superb maintenance and expansion of the holy sites, enabling greater numbers of Muslim pilgrims to perform the Hajj. This has always been an essential priority for the Kingdom. Every year funds from the annual budget are allocated exclusively for this purpose. The Ministry of Pilgrimage, in conjunction with other government agencies, oversees the annual logistical challenge of preparing for the Hajj and supports projects in the Kingdom and abroad that promote Islam's role in the community.

Beginning with the late King Abdul Aziz, Saudi leaders have directed a series of ongoing projects aimed at improving the quality of accommodation, health care and other services for the pilgrims. An essential component of this policy has been the expansion of the Holy Mosque in Makkah and the Prophet's Mosque in Madinah. At the time of Saudi Arabia's unification in 1932, the Holy Mosque could accommodate 48,000 worshipers and the Prophet's Mosque 17,000. A series of expansion plans, the latest of which was completed in 1992, increased the capacity of the two holy mosques to more than one million and over half a million, respectively. Under the personal direction of King Fahd, the work also improved the infrastructure and services necessary to enable the millions of pilgrims to carry out their religious observances in comfort and safety. This has included new airport and port facilities in Jeddah and other points of entry for pilgrims, roads to Makkah and Madinah, comfortable accommodations and an extensive health care network.

As part of the Kingdom's efforts to better serve Muslim pilgrims, whose numbers are expected to continue to increase, in 1993 King Fahd restructured the Ministry of Pilgrimage and Endowments into two separate organisations—the Ministry of Pilgrimage, which deals exclusively with the Hajj, and the Ministry of Islamic Affairs, Endowments, Call and Guidance.

The vast financial and human resources Saudi Arabia has committed to the Hajj reflect the dedication of the leadership and citizens of the Kingdom to the service of Islam and the holy sites and to preserving them as a haven of peace for all Muslims.

With the help of God, we will continue to honour this great trust, holding fast to our Islamic creed and upholding its teachings. We will make every effort to strengthen our relations with our brothers in Muslim and Arab countries and we will do our utmost for the Muslim nation.
—The Custodian of the Two Holy Mosques King Fahd Bin Abdul Aziz.

Source: "Guardian of the Holy Places", from saudiembassy.org.uk/profile-of-saudiarabia/islam/guardian-of-the-holy-places.htm

6

from *Time Magazine*

Jerusalem at the Time of Jesus

by David van Biema
April 16, 2001

It is impossible today to hear the word Jerusalem without thinking about the violence that is again bedeviling the Holy Land. Almost 400 Palestinians and 65 Israelis have died since last fall, when peace negotiations imploded over the question of Jerusalem's status.

The current agony is not atypical of the locale's holy, bloody history. Over the centuries, each of the West's great faiths has coveted the city; each alternately has controlled it, and each has constructed around it a separate sacred history. As the myths have collided, the result has been a play of extremes: physical splendor alternating with utter destruction; moments of pious exultation oscillating with the grossest carnage. Or sometimes carnage and exultation at once. "Men rode in blood up to their knees and bridle reins," wrote an 11th century Crusader fresh from a massacre of Muslims on the Temple Mount. He added, "Indeed, it was a just and splendid judgment of God."

The years from A.D. 1 to A.D. 33 happened to be a high point for the holy city. It was, says Eric Meyers, professor of Judaic studies at Duke University, "a great, great metropolitan area" and home to the lavishly restored Jewish Temple, a world-renowned wonder. It was prosperous and cosmopolitan. And it was also, unknowingly, the cradle for something else, a way of believing, of seeing, that would change the West and the rest of history. It is worth revisiting Jerusalem during this period not so much in celebration as in curiosity—to know the metropolis that shaped Jesus' last ministry and so wove itself into his great story, and to note, cautiously, the ways in which its vexations foreshadow those of Jerusalem today.

It is the Gospel of Luke that describes Jesus' childhood visit to Jerusalem. Though he had been there before—Luke says his family was visiting "as usual" for Passover—the 12-year-old from Nazareth, 60 miles to the north, must still have been agog walking south down the grand new Roman street toward the Temple's lower entrance. A stretch of that road is visible today, just below the Western Wall, majestically wide but piled high on one side with huge blocks of stone that rained from above during one of the city's many destructions.

There is a debate regarding exactly how citified the young Jesus would have been. Excavations of the city of Sepphoris, near Nazareth, reveal a bustling town, suggesting that he may have been less of a country lad than previous scholarship posited. But his native Galilee certainly had nothing to compare with this. Jerusalem was one of the biggest cities between Alexandria and Damascus, with a permanent population of some 80,000. During Passover, Succoth and Shavuoth, the great festivals during which Jews were obligated to make sacrifices at the Temple, between 100,000 and 250,000 visitors (historians differ) would stream down the long city thoroughfare.

The city was in a renaissance. Its initial splendor had been snuffed out by Babylonia in 586 B.C. Within 50 years, Jews had begun rebuilding, but full glory awaited the rule, from 37 B.C. to 4 B.C., of Herod the Great. Herod is one of ancient history's extraordinary figures. Ten times married, a serious drinker and a half-Jew who was half-trusted by his subjects, he played the superpower politics of his day consummately. In 63 B.C., Rome became Judea's ruler, succeeding Babylonia, Persia, Greece and the Jews themselves. Herod, who hailed from the neighboring province of Idumea (which included part of today's West Bank), won and maintained his position as the empire's proxy King of the Jews by allying himself successively with Julius Caesar, Mark Antony and Emperor Augustus, a dance involving very tricky pirouettes.

Herod killed thousands of Jerusalemites in the streets while taking power. But he was also a local who understood Judea's needs and its hard-won privilege of being governed under Jewish law. A builder king, he ordered up huge forts, palaces and indeed whole cities throughout Judea, and he created an artificial harbor at Caesarea Maritima that lasted 600 years.

But it was in Jerusalem, says Meyers, that Herod "undertook to make one of the major wonders of the ancient world." He rebuilt the existing meandering streets on a paved grid and created a moat-ringed palace featuring—in a moisture-starved region—picturesque water gardens. He added an amphitheatre and a hippodrome. But the jewel in the crown, the spiritual, economic and social center of Judea and an icon to Jews throughout the region, was the Temple. It was his bid to rival Solomon, biblical builder of the Jews' first great house of worship, which had been razed by the Babylonians some 570 years earlier.

Physical remains of Herod's masterpiece are scarce. But they tend to support descriptions in the four surviving written sources from approximately the same period: the Gospels and the biblical book of Acts; the part of the Jewish Talmud called the Mishnah; and the histories of Flavius Josephus, a Jewish priest and commander turned Roman military aide who lived in the years A.D. 30 to A.D. 100. For instance, a stone found later near the Temple's likely site was inscribed with the words *to the place of trumpeting,* which corroborate Josephus' description of the signal for the beginning of the Sabbath.

Tradition forbade the Temple's enlargement beyond Solomon's original dimensions. So Herod expressed his egomania by adding a 35-acre platform—"the greatest ever heard of," writes Josephus—on which the Temple could sit. The Western Wall where Jews pray today is a small slice of the platform's 16-ft.-thick western side. Some of the stones are 30 ft. long and weigh up to 50 tons. ("Look, Teacher, what large stones and what large buildings!" exclaims a disciple in the Gospel of Mark.) As Herod built out over the adjacent valleys, the outline of the mountain on which the compound sat gradually disappeared. The great stone featured in the Dome of the Rock, the Muslim shrine that now occupies Herod's immense pedestal, may be the mountain's peak.

At the time, the platform (Jews call it the Temple Mount) had up to seven entrances. Most experts believe the remains of an expansive, carved-stone stairway on the south side of the mount, perpendicular to the Roman street, were once the main entry for common pilgrims. At the foot of the stairs are the ruins of a series of baths, for ritual purification, and small shops, some of which still have hitches for animals.

Temple worship revolved around sacrifice: a lamb for Passover, a bull for Yom Kippur, two doves—"the poor woman's sacrifice"—to celebrate a child's birth. Before buying an animal, visitors changed their Roman denarii (the dollar of the day) for shekels, or Temple coins, that had no portraits on them and so did not violate the Jewish prohibition of graven images. Herod appears to have allowed the money changers onto the Temple platform, which may have spurred Jesus' scourging of them in "my father's house." Joshua Schwartz, a professor of historical geography at Israel's Bar Ilan University, styles the stairway as a Judean version of London's Hyde Park Corner. There would be "beggars and upper-class Jews and Gentiles from all over," he says. "Scholars would be teaching, and would-be prophets would be preaching. The steps were *the* experience in Jerusalem."

After the Temple itself, perhaps. Scholars have hypothesized that the southern steps led pilgrims into a tunnel under an administrative building and out again amid a series of courtyards. The outermost was open to curious Gentiles. The remaining enclosures were for Jews only, as indicated by another of the Temple's remaining relics—a sign, in Greek, warning that any non-Jew passing farther "is answerable himself for his ensuing death."

Next came the Court of Women, followed by the Court of Israelites, the Court of the Priests and, above all, the massive sacrificial altar. The Temple's innermost shrine, featuring the holy room that the Bible said had been occupied by the Ark of the Covenant in Solomon's Temple, loomed 80 ft. high, a glistening tower.

The scene must have been spectacular. Whether that spectacle is understood as deeply felt or empty depends on later interpretation. "The place was as vast as a small city. There were literally thousands of priests, attendants, temple soldiers and minions," writes historian Paul Johnson. "Dignity was quite lost amid the smoke of the pyres, the bellows of terrified beasts, the sluices of blood, the abattoir stench, the unconcealed and unconcealable machinery of tribal religion inflated by modern wealth to an industrial scale."

Bruce Chilton, a religion professor at Bard College whose book *Rabbi Jesus* was published in October, says recent scholarship finds a great deal more meaning and joy in the proceedings. Pilgrimages were festive occasions, with families or friends traveling together and camping overnight in the hills around the city and singing cheerful sacred songs outside the Temple. Although parts of the sacrifice would be immolated for the Lord or consumed by the priests, others would be cooked and shared by the pilgrims, who ate little meat the rest of the year. "Not only would they offer this very scarce protein to the deity," says Chilton, "but actually share a meal of meat with the Lord of Israel. The sense was one of wealth and celebration."

Jerusalem was a monoculture, comparable to Washington or Redmond, Wash. (It remains so today, although it is now tourism rather than religion that is the city's dominant business.) Unlike many company towns, however, the city in Jesus' time had a cosmopolitan feel. Its material needs drew caravans from Samaria, Syria, Egypt, Nabatea, Arabia and Persia. Two-thirds of its population were Jews (roughly the same percentage as today), practicing a religion that counted millions of adherents in the Roman Empire and a large group of "God fearers," Gentiles who observed some key precepts without full conversion. At the same time, the city was in its 15th generation of Greco-Roman influence (since the conquests of Alexander the Great in 332 B.C.). Parents gave their children Greek names; intellectuals were conversant in classical philosophy. Greek had become along with Hebrew and Aramaic one of the area's main languages, and one of the most commonly used versions of the Torah was in Greek. (Jesus presumably spoke all three languages.) The interaction of Jewish and classical thought would lend the Christian Bible much of its strength.

This Greco-Roman "modernism" was conflicted, however. A building full of soldiers loomed over the Temple courtyards like a watchtower over a prison. As Jesus and the other pilgrims performed the most sacred rites of their faith, they would never be beyond surveillance. After Herod's initial rise, the Roman yoke was relatively light, consisting mostly of tribute. But the Jews had been independent for a century before the imperial conquest, and many hoped to return to that state. In recognition of this, above the Temple's northwestern corner stood the city's great Roman garrison, the Antonia, named after Herod's patron Mark Antony and housing between 2,000 and 3,000 soldiers.

Their presence in the city's very soul posed a painful conundrum. Beneath its prosperous surface, says Neil Asher Silberman, director of the Ename Center for Public Archaeology in Brussels, Jerusalem was actually "extremely turbulent." To some, "the beautiful Temple of Herod was a horrible betrayal of Israelite tradition. Herod obliterated the original Temple and replaced it with a Roman one." Even the most prosperous citizens must have had some major identity issues.

This led, Silberman suggests, to "movements of desperation where people harked back to a purity of faith and looked for signs of messianic redemption." The city's dominant religious authorities, skewered in the Gospels, were the Sadducees, who made up most of the Temple élite, and the Pharisees, respected for their ongoing explorations of the correct interpretation of religious law. But the city also played host to groups like the Zealots, a militant nationalist group, and the Essenes. The Essenes detested the Temple priests, lived in monastic communities and may have been authors of the Dead Sea Scrolls, the treasure trove of texts uncovered in the Judean desert in 1947. Josephus assigns the Essenes a membership of 4,000, only 2,000 fewer than his count of Pharisees.

And then there were radical free-lancers like Jesus. Herod had managed to keep a lid on anti-Roman sentiment for most of his reign. But starting with his fatal illness in 4 B.C. and continuing over the careers of several less effective successors, a series of bloodily suppressed revolts erupted.

In 4 B.C., angry Jews, protesting the execution of students who had tried to remove a Roman eagle from the Temple decorations, threw stones down on their occupiers from the mount's porches and set off a citywide riot; eventually 2,000 rebels were crucified. In A.D. 26, the Roman governor provocatively ordered his troops to raise flags with Caesar's face within a few hundred feet of the central shrine. A mob marched to his house in Caesarea. His soldiers drew their swords. The Jews, in an extraordinary act of passive resistance, laid bare their necks and said they would rather die than see their religious laws flouted. The governor, a normally hot-tempered newcomer named Pontius Pilate, recalled the flags.

The situations now and then are not analogous. Israel's current Jewish government, unlike the Roman Empire, is not alien to Jerusalem. The Palestinians are not as defenseless as the ancient Jews. And Israeli opposition leader (now Prime Minister) Ariel Sharon's unwelcome stroll last September around the two Islamic shrines that now occupy the Temple platform—a provocation that may have sparked the Holy Land's current strife when Muslims responded by throwing rocks down on Jews at prayer below—has no precise 1st century cognate. Still, the intertwined dynamic of military occupation and religious clash is shockingly familiar.

Two thousand years ago, the man in the middle of this potentially deadly tug-of-war was the high priest. The position, ritually paramount at the Temple, had been politically hobbled by Herod. Nonetheless, as head of the Sanhedrin, a Jewish religious and civic body, and a key participant at city council meetings, the officeholder still had great power and responsibility.

The actions of Caiaphas, high priest from A.D. 18 to A.D. 36, are traditionally attributed to rage over Jesus' challenges to his class's power and his personal standing. But historians have begun to argue for a more nuanced appreciation. Caiaphas knew better than anyone that the doomed Jewish revolts inevitably started at the Temple, frequently during Passover, as keyed-up pilgrims celebrated Israel's liberation from an earlier oppressor. He knew Pilate as a ruler, says Richard Horsley of the University of Massachusetts, Boston, who "shot first and asked questions later." Personal pride notwithstanding, the high priest had reason to act against a Jew who had disrupted the Temple and may have been plotting another grand entrance on the second day of the feast. To Caiaphas, says Lee Levine, professor of Jewish history at Jerusalem's Hebrew University, "Jesus and others like him were just a bad idea. Bad for the Temple, bad for the Romans and bad for the Jews."

The path of the Via Dolorosa, the Stations of the Cross, through the Old City of Jerusalem is almost certainly inaccurate. It follows a 14th century grid of the city rather than a 1st century plan, and probably reflects the desire of 14th century merchants along the way to get pilgrims' business. But the hill of Golgotha (a.k.a. Calvary) and Jesus' burial cave, both located by tradition in Jerusalem's Church of the Holy Sepulcher, are a different matter.

In Jesus' day executions and burials took place outside the city. Today the church is tucked within the Old City's Christian Quarter, but at the time, the area would have been safely outside town walls. The niche-style grave is consistent with 1st century custom. Written attestations to its authenticity—and that of the Calvary rock a few yards away—date back more than 1,800 years. Tellingly, early rulers who might have been tempted to "adjust" the site's location did not do so. Says Dan Bahat, for many years Jerusalem's district archaeologist: "There's nothing to prove that this is not the site of the Crucifixion." If this sounds weak to a believer, coming from an archaeologist, it carries significant weight.

When the unnamed disciple remarked on the size of the Temple stones, Jesus replied that "not one stone will be left upon another; all will be thrown down." He was right. After one last rebellion, in A.D. 135, the Romans leveled Jerusalem, leaving only the bald platform behind. The city, of course, rose again and fell again, was conquered and reconquered ....

Yousef Abu Ghannam's family holds the key (and the souvenir concession) for the Mosque of the Ascension on the Mount of Olives; it was a Christian shrine until Saladin took Jerusalem back from the Crusaders. Abu Ghannam reports sadly that business is down. "We used to get 700 to 800 people a day," he says. "Now we're lucky to get 150. People are afraid." The few visitors who brave Jerusalem today encounter a metropolis again edgy and turbulent. In the sanctuary of the city's churches, mosques and synagogues, pilgrims can find momentary tranquillity. But the streets bear new pocks from the bullets that flew here late last year. Herod's ancient platform had been closed since last fall to all but Islamic worshippers to avoid further confrontation: Sharon's directive last week to reopen it to non-Muslims may make it a flashpoint again. Travel in the area is the riskiest in a decade, and a U.S. State Department warning against it remains in effect.

At the Ascension Mosque there is at least one optimist, albeit with a long view. The Rev. Frank Booke of Anniston, Ala., has led his tour group to the small off-white domed tower. On its floor is an indentation that pilgrims have thought for centuries is the imprint made by Jesus' right foot as he ascended to heaven.

Booke is an ebullient Pentecostal Christian in an orange Stetson. Consistent with his faith, he takes solace in Christ's expected return to earth and his re-establishment of God's kingdom here, regardless of humankind's errors. Like many, Booke believes his Saviour will arrive at precisely the point from which he left. "This is the place," he says. His flock responds with an explosive "Hallelujah!" and a rendition of The Old Rugged Cross.

But Booke will not simply leave it at that. His joy over the eschatological future does not render him blind to the scandalous present. "We love the Jewish people," he says, then glances at the Muslim gatekeepers and adds, "These are all God's people. When everybody else is afraid, we come to support this land. To support the souvenir sellers. We pray for this land. We pray for the peace of Jerusalem."

Source: David Van Biema, "Jerusalem at the Time of Jesus," *Time*, 4/16/01 from time.com/time/2001/jerusalem/cover.html

7 from **the United Nations**

General Assembly Resolution 181

November 29, 1947

Part III—City of Jerusalem

A. Special Regime

The City of Jerusalem shall be established as a corpus separatum under a special international regime and shall be administered by the United Nations. The Trusteeship Council shall be designated to discharge the responsibilities of the Administering Authority on behalf of the United Nations.

B. Boundaries of the City

The City of Jerusalem shall include the present municipality of Jerusalem plus the surrounding villages and towns, the most eastern of which shall be Abu Dis; the most southern, Bethlehem; the most western, 'Ein Karim (including also the built-up area of Motsa); and the most northern Shu'fat, as indicated on the attached sketch-map (annex B).

C. Statute of the City

The Trusteeship Council shall, within five months of the approval of the present plan, elaborate and approve a detailed statute of the City which shall contain, inter alia, the substance of the following provisions:

Government Machinery; Special Objectives

The Administering Authority in discharging its administrative obligations shall pursue the following special objectives:

> To protect and to preserve the unique spiritual and religious interests located in the city of the three great monotheistic faiths throughout the world, Christian, Jewish and Moslem; to this end to ensure that order and peace, and especially religious peace, reign in Jerusalem;
>
> To foster cooperation among all the inhabitants of the city in their own interests as well as in order to encourage and support the peaceful development of the mutual relations between the two Palestinian peoples throughout the Holy Land; to promote the security, well-being and any constructive measures of development of the residents having regard to the special circumstances and customs of the various peoples and communities.

Governor and Administrative Staff

A Governor of the City of Jerusalem shall be appointed by the Trusteeship Council and shall be responsible to it. He shall be selected on the basis of special qualifications and without regard to nationality. He shall not, however, be a citizen of either State in Palestine.

The Governor shall represent the United Nations in the City and shall exercise on their behalf all powers of administration, including the conduct of external affairs. He shall be assisted by an administrative staff classed as international officers in the meaning of Article 100 of the Charter and chosen whenever practicable from the residents of the city and of the rest of Palestine on a non-discriminatory basis. A detailed plan for the organization of the administration of the city shall be submitted by the Governor to the Trusteeship Council and duly approved by it.

Local Autonomy The existing local autonomous units in the territory of the city (villages, townships and municipalities) shall enjoy wide powers of local government and administration.

The Governor shall study and submit for the consideration and decision of the Trusteeship Council a plan for the establishment of special town units consisting, respectively, of the Jewish and Arab sections of new Jerusalem. The new town units shall continue to form part the present municipality of Jerusalem.

Security Measures

The City of Jerusalem shall be demilitarized; neutrality shall be declared and preserved, and no para-military formations, exercises or activities shall be permitted within its borders.

Should the administration of the City of Jerusalem be seriously obstructed or prevented by the non-cooperation or interference of one or more sections of the population the Governor shall have authority to take such measures as may be necessary to restore the effective functioning of administration.

To assist in the maintenance of internal law and order, especially for the protection of the Holy Places and religious buildings and sites in the city, the Governor shall organize a special police force of adequate strength, the members of which shall be recruited outside of Palestine. The Governor shall be empowered to direct such budgetary provision as may be necessary for the maintenance of this force.

Legislative Organization

A Legislative Council, elected by adult residents of the city irrespective of nationality on the basis of universal and secret suffrage and proportional representation, shall have powers of legislation and taxation. No legislative measures shall, however, conflict or interfere with the provisions which will be set forth in the Statute of the City, nor shall any law, regulation, or official action prevail over them. The Statute shall grant to the Governor a right of vetoing bills inconsistent with the provisions referred to in the preceding sentence. It shall also empower him to promulgate temporary ordinances in case the Council fails to adopt in time a bill deemed essential to the normal functioning of the administration.

Administration of Justice

The Statute shall provide for the establishment of an independent judiciary system, including a court of appeal. All the inhabitants of the city shall be subject to it.

Economic Union and Economic Regime

The City of Jerusalem shall be included in the Economic Union of Palestine and be bound by all stipulations of the undertaking and of any treaties issued therefrom, as well as by the decisions of the Joint Economic Board. The headquarters of the Economic Board shall be established in the territory City. The Statute shall provide for the regulation of economic matters not falling within the regime of the Economic Union, on the basis of equal treatment and non-discrimination for all members of the United Nations and their nationals.

Freedom of Transit and Visit: Control of residents

Subject to considerations of security, and of economic welfare as determined by the Governor under the directions of the Trusteeship Council, freedom of entry into, and residence within the borders of the City shall be guaranteed for the residents or citizens of the Arab and Jewish States. Immigration into, and residence within, the borders of the city for nationals of other States shall be controlled by the Governor under the directions of the Trusteeship Council.

Relations with Arab and Jewish States

Representatives of the Arab and Jewish States shall be accredited to the Governor of the City and charged with the protection of the interests of their States and nationals in connection with the international administration of the City.

Official Languages

Arabic and Hebrew shall be the official languages of the city. This will not preclude the adoption of one or more additional working languages, as may be required.

Citizenship

All the residents shall become ipso facto citizens of the City of Jerusalem unless they opt for citizenship of the State of which they have been citizens or, if Arabs or Jews, have filed notice of intention to become citizens of the Arab or Jewish State respectively, according to Part 1, section B, paragraph 9, of this Plan.

The Trusteeship Council shall make arrangements for consular protection of the citizens of the City outside its territory.

Freedoms of Citizens

Subject only to the requirements of public order and morals, the inhabitants of the City shall be ensured the enjoyment of human rights and fundamental freedoms, including freedom of conscience, religion and worship, language, education, speech and press, assembly and association, and petition.

No discrimination of any kind shall be made between the inhabitants on the grounds of race, religion, language or sex.

All persons within the City shall be entitled to equal protection of the laws.

The family law and personal status of the various persons and communities and their religious interests, including endowments, shall be respected.

Except as may be required for the maintenance of public order and good government, no measure shall be taken to obstruct or interfere with the enterprise of religious or charitable bodies of all faiths or to discriminate against any representative or member of these bodies on the ground of his religion or nationality.

The City shall ensure adequate primary and secondary education for the Arab and Jewish communities respectively, in their own languages and in accordance with their cultural traditions. The right of each community to maintain its own schools for the education of its own members in its own language, while conforming to such educational requirements of a general nature as the City may impose, shall not be denied or impaired. Foreign educational establishments shall continue their activity on the basis of their existing rights.

No restriction shall be imposed on the free use by any inhabitant of the City of any language in private intercourse, in commerce, in religion, in the Press or in publications of any kind, or at public meetings.

Holy Places

Existing rights in respect of Holy Places and religious buildings or sites shall not be denied or impaired.

Free access to the Holy Places and religious buildings or sites and the free exercise of worship shall be secured in conformity with existing rights and subject to the requirements of public order and decorum.

Holy Places and religious buildings or sites shall be preserved. No act shall be permitted which may in any way impair their sacred character. If at any time it appears to the Governor that any particular Holy Place, religious building or site is in need of urgent repair, the Governor may call upon the community or communities concerned to carry out such repair. The Governor may carry it out himself at the expense of the community or communities concerned if no action is taken within a reasonable time.

No taxation shall be levied in respect of any Holy Place, religious building or site which was exempt from taxation on the date of the creation of the City. No change in the incidence of such taxation shall be made which would either discriminate between the owners or occupiers of Holy Places, religious buildings or sites or would place such owners or occupiers in a position less favourable in relation to the general incidence of taxation than existed at the time of the adoption of the Assembly's recommendations.

Special Powers of the Governor in Respect of the Holy Places, Religious Buildings and Sites in the City and in any Part of Palestine.

The protection of the Holy Places, religious buildings and sites located in the City of Jerusalem shall be a special concern of the Governor.

With relation to such places, buildings and sites in Palestine outside the city, the Governor shall determine, on the ground of powers granted to him by the Constitution of both States, whether the provisions of the Constitution of the Arab and Jewish States in Palestine dealing therewith and the religious rights appertaining thereto are being properly applied and respected.

The Governor shall also be empowered to make decisions on the basis of existing rights in cases of disputes which may arise between the different religious communities or the rites of a religious community in respect of the Holy Places, religious buildings and sites in any part of Palestine.

In this task he may be assisted by a consultative council of representatives of different denominations acting in an advisory capacity.

D. Duration of the Special Regime

The Statute elaborated by the Trusteeship Council the aforementioned principles shall come into force not later than 1 October 1948. It shall remain in force in the first instance for a period of ten years, unless the Trusteeship Council finds it necessary to undertake a re-examination of these provisions at an earlier date. After the expiration of this period the whole scheme shall be subject to examination by the Trusteeship

Council in the light of experience acquired with its functioning. The residents the City shall be then free to express by means of a referendum their wishes as to possible modifications of regime of the City.

Source: United Nations General Assembly, Resolution 181, 11/29/47, found mideastweb.org/181.htm

8

from *Ma'ariv*

Jerusalem: The Battle for Sovereignty

by Moshe Sasson
8 July 1994

A. Neither the city Mecca (where the 'Ka'aba' is found—the holiest place in Islam) nor Medina (the second most important city in Islam) nor the city Najaf or the city Karbala (holiest to Shia Islam) are the capitals of Saudi Arabia or Iraq. The governments of these two Islamic states chose other cities (Riyadh and Baghdad) for the seat of their capital. In other words: there does not exist in Islam, neither Sunni nor Shia, not in the Koran and not in any more, not according to the traditional writings nor in practice, any instruction or decision that cities which contain the most holy of Islamic sites should be made the capitals of these countries.

B. Jerusalem is home to holy Islamic places (al-Aksa Mosque and the Dome of the Rock, also known as the Mosque of Omar). The al-Aksa mosque is, according to Islamic tradition, the first place that Muslims turned towards during prayer and the third in importance. Yet, in accordance with Muslim law and especially in accordance with precedents set by Saudi Arabia and Iraq, there is no religious requirement that the city of Jerusalem, wholly or in part, be a Muslim capital. The claim (by in large Palestinian) that as a result of the existence of these two very important mosques, the city must become, "the capital of Arab Palestine," has no backing either from the Muslim religion or from political Islamic or Arab custom.

C. Indeed, the Arab Palestinian leadership claims that east Jerusalem should be the capital of Arab Palestine. This claim, by nature, has no Islamic religious basis and is not widely endorsed by the Muslim or Arab countries.

In other words: the disagreements on the issue of Jerusalem are not between the Jewish and Muslim religions, not between Israel and the Arab world, and by no means between Israel and the Muslim world. Such a description is mistaken and incorrectly reflects the nature of the conflict over this issue. The problem which requires Israel's attention, in this context, (in answer to the Muslim nations and some Arab nations's stance), is not a problem of the political status of the eastern part of the city, but the existence of the holy sites in the city, the rights of their "owners," their status, their method of management, freedom of access to them and the guarantees that Israel should give in this context.

D. The religious (not political) importance of the al-Aksa mosque has roots in the Koranic verse which attributes the place to a specific saying of the prophet Muhammad, as it appears in the first verse of the "Night Journey" sura (Chapter 17 in the Koran), which was delivered in Mecca and which reads:

> Glory be to Him who made His servants go by night from the Sacred Mosque to the farther (al-Aksa) Mosque whose surroundings We have blessed, that We might show him some of Our signs.

The prophet Muhammad's reference to the al-Aksa mosque (which did not even exist at the time when Muhammad said these words) has no political Islamic meaning, and of course no religious meaning which would have implications for granting sovereignty over the city to any one religion. This saying had and still has, clear religious meaning, but nothing more (which does require attention from Israel, but not with respect to sovereignty over the city).

E. In addition to the two holy Islamic sites, there are also, as is well- known, holy Jewish places (the Western Wall and the Temple Mount) and ten important holy Christian sites, the three most important being: Church of the Holy Sepulchre, the Cenaculum, and the Church of Gethsemane which all have religious (and not political) meaning of the highest degree.

Israel should relate to the status of all of these places equally, should be responsible for them and make sure that freedom of access to them is retained.

F. In an annex to the Lateran Agreement signed in 1929 between the Italian government and the Vatican, it was determined that, among other things, the five largest chapels in the city outside of the Vatican walls (for example the "San Giovani in Latarano") which belong to the Vatican, would receive ex-territorial status (similar to the status of foreign embassies all over the world) and would be run exclusively by the holy throne.

We also have before us a precedent and a model which provide a solution for the problem regarding the status of the two Muslim mosques located on the Temple Mount, the other holy Christian sites in Jerusalem and maybe even Christian sites in other areas in Israel. Recognized and legally protected extra-territorial status should be granted to all of the sites, which would be an appropriate legal guarantee.

Once the principle issue is agreed upon, Israel will be responsible for conducting negotiations with the Vatican, the Greek Orthodox and Armenian Patriarchs, as well as with one (or various) Muslim group(s), in an effort to reach agreements regarding the details of the rights to be granted the ex-territorial sites, holy and important to Christianity and Islam.

G. It is not up to Israel to pick its negotiating partner for discussions regarding the status of holy Islamic places in Jerusalem. This is, in my opinion, the Muslim community's responsibility—and their's alone. This way, for example, if Arab and Muslim nations come to the conclusion that the ex-territorial rights to manage the two mosques in Jerusalem should be awarded to the King of Saudi-Arabia (who carries the title "Guardian of the Holy Places") or to the King of Jordan or the King of Morocco or the Palestinians or any other Arab or Muslim nation, or to the Arab League, or the Islamic Congress it is not Israel's place to deny them ex-territorial status (including having the flag of this nation wave from the two Muslim mosques). There is no doubt that the Palestinians and King Hussein have valid reasons for wanting control, solely or jointly, since they have been the actual managers of these sites for numerous years.

H. There is another very important point which must be considered: both Jews and Muslims view the Temple Mount as a holy site. Naturally, in the era of the political conflict between us and the Jordanians and Palestinians, each side emphasized its exclusive right to the Mount. However, in an era of peace and coexistence, it is important to understand that this site is holy to both sides, and it is therefore desirable that understanding, tolerance and cooperation replace conflict and struggle.

The "Tomb of the Patriarchs," as we call it, and the "Holy Ibrahimi Mosque," as Muslims call it, is a prominent and clear example of "partnership in holiness" (as is "Rachel's Tomb"). I have no doubt that the two sides now attempting to build the foundations of peace and coexistence between them, can reach agreement on "administration through agreement, cooperation and coordination" of these places considered holy by both. Any unilateral, coercive decisions regarding this subject is out of place.

Following the massacre at the Tomb of the Patriarchs, Israel is now unilaterally determining prayer and security arrangements there. Such a "military" approach is fundamentally wrong, and could lead to negative results. What still remains to be accomplished (especially on the eve of Hebron's inclusion into the Palestinian Autonomy), are negotiations leading to an agreement determining the Tomb of the Patriarchs' status, manner of administration and both sides' rights. This is required both for immediate implementation and also to set an important precedent regarding the Temple Mount (not regarding the mosques which, as we said, will be run by the designated Muslim group).

In my opinion, the Palestinians will be quite interested in quickly reaching an agreement with us regarding the Tomb of the Patriarchs and Rachel's Tomb, if only to acquire for themselves preferential standing in the Muslim and Arab world as the controlling power of the Temple Mount and the two important mosques located there.

As for us, united Jerusalem is the capital of Israel. In this framework, we must provide solutions and reach agreements regarding the status of the holy places in the city, which have nothing whatsoever to do with the question of sovereignty over the city.

Source: Moshe Sasson, "Jerusalem: The Battle for Sovereignty," in *Ma'ariv*, July 8, 1994 found at israel-mfa.gov.il/mfa/go.asp?MFAH082n0

9

from The Palestinian Information Center

The Status of Jerusalem in Islam

Muslims—Not Jews—Have the Right to Defend Jerusalem

Muslims strongly believe that Jerusalem and its surrounding neighbourhood is holy land which can never be given up because it is a part of their faith. Muslims are the only people who believe in all the Prophets (peace be upon them all) from the time of Adam through to Prophet Muhammed (peace be upon him). There is not even one statement in either the Qur'an or Sunna that attributes any sins or misdeeds committed by any of them. In addition, a Muslim's faith will not be perfect unless they believe in all of the Prophets (Surat Al-Baqarah [The Cow], verses 136 and 285, as well as tens of verses in the Holy Qur'an which honour all of them because they have all brought the same message to mankind). Thus, all Muslims are obligated to defend all religious places wherever they are because Muslims are the best and most honourable nation ever created for Mankind (Surat Al-Imran [The Family of Imran], verse 110; and Surat Al-Hajj [The Pilgrimage], verse 78). Actually, this a great responsibility compared with the Jews' attitude toward the Prophets throughout history; they have been described in the Holy Qur'an, as well as in both the Old and New Testaments, as the murderers of Prophets, the Sons of Serpents, the Stray or Blind, and the Cursed and Condemned because of their disbelief in many of the Prophets. In this light, it is obvious that they would be unable to carry out this mission.

God said: "I will destroy Jerusalem and the Yahuda, and I will give them up for their enemy because they had bad conduct," and "God has made this people blind to truth." His guides went astray and became rude to the poor and orphans. It has been evidenced in the Bible that the Jews murdered several Prophets, namely Hezeqial, Esh'ia Bin Amous, Ermia, Zachariah, and his son Yahya (Jonah). They tried to murder both Essa (Jesus) and Mohammed (peace be upon them both). Moreover, it is stated in the Bible that the Prophet Jacob (peace be upon him) decided to wipe out the Canaanites even though they believed in Judaism, but this is untrue because this would never be done by a Prophet. Then it has also been stated that all people are dogs and servants for the Jews who say that they did not occupy any land because it is their legacy which had been usurped by others.

The Talmud states that all nations are cursed and condemned except the Jews, who are honourable and have the right to capture whatever the non-Jews have because they deserve to be murdered. As a result, how can such people be trusted to defend and protect the holy places and the human legacy (Surat Al-Baqarah, verse 61), and who has more right to do so, Muslims or Jews?

Muslims have always viewed Jerusalem as a holy place which must be defended because it is similar to Mecca in its holiness and has been for more than 14 centuries. These places must be protected given that Abraham, the Father of all Prophets (peace be upon them all), had built the Kaaba in Mecca and thereafter moved to Palestine where he passed away and was buried in Hebron near Jerusalem. Muslims will never forget that they used to pray toward Jerusalem in the early stages of Islam before Allah ordered it to be changed to the Holy Shrine in Mecca. There is a mosque in Medina that still has the two directions (one pointing toward Jerusalem and one towards Mecca), namely Al-Qiblatain Mosque, as real evidence for this intimate connection between Jerusalem and Mecca. On the other hand, Muslims still consider the places where God spoke to Moses; where David and Soloman repented to God and where the mountains and birds had been put in their service; where Issac asked his sons to bury him; and where Christ was born, spoke in the cradle, the banquet was descended from the Heavens, where he was raised to the Heavens and where Mariam passed away, as holy places to be guarded. This is the real and sincere attitude of all Muslims toward Jerusalem, which shows their appreciation and respect for all Prophets and their holy places as an immense historical and religious responsibility. On the contrary to this honourable attitude, the Jews have been rude and aggressive toward all Prophets (peace be upon them all). They have changed and deformed all the real teachings and texts of the Bible. Muslims all over the world feel this responsibility because Muslims are ordered to defend and protect all of the Prophets' legacy and heritage, and they must fight any Jewish attempt to deform and forge the truth and facts about these human issues.

Source: Abdel-Haleem E'weis, "The Status of Jerusalem in Islam," February 6, 2003 on palestine.co.uk/am/publish/article_26.shtml

from *Time Magazine* 10

Islam's Stake: Why Jerusalem Was Central to Muhammad

by Karen Armstrong
April 16, 2001

Jerusalem was central to the spiritual identity of Muslims from the very beginning of their faith. When the Prophet Muhammad first began to preach in Mecca in about 612, according to the earliest biographies, which are our primary source of information about him, he had his converts prostrate themselves in prayer in the direction of Jerusalem. They were symbolically reaching out toward the Jewish and Christian God, whom they were committed to worshipping, and turning their back on the paganism of Arabia. Muhammad never believed that he was founding a new religion that canceled out the previous faiths. He was convinced that he was simply bringing the old religion of the One God to the Arabs, who had never been sent a prophet before.

Consequently, the Koran, the inspired scripture that Muhammad brought to the Arabs, venerates the great prophets of the Judeo-Christian tradition. It speaks of Solomon's "great place of prayer" in Jerusalem, which the first Muslims called City of the Temple. Only after the Jews of Medina rejected Muhammad did he switch orientation and instruct his adherents to pray facing Mecca, whose ancient shrine, the Kabah, was thought by locals to have been built by Abraham and his son Ishmael, the father of the Arabs.

The centrality of Jerusalem in Muslim spirituality is apparent in the story of Muhammad's mystical Night Journey to Jerusalem. Muslim texts make it clear that this was not a physical experience but a visionary one (not dissimilar to the heavenly visions of the Jewish Throne Mystics at this time). One night Muhammad was conveyed miraculously from the Kabah to Jerusalem's Temple Mount. There he was welcomed by all the great prophets of the past before ascending through the seven heavens. On his way up he sought the advice of Moses, Aaron, Enoch, Jesus, John the Baptist and Abraham before entering the presence of God. The story shows the yearning of the Muslims to come from far-off Arabia right into the heart of the monotheistic family, symbolized by Jerusalem.

Respect for other faiths was manifest in Islamic Jerusalem. When Caliph Umar, one of Muhammad's successors, conquered the Jerusalem of the Christian Byzantines in 638, he insisted that the three faiths of Abraham coexist. He refused to pray in the Church of the Holy Sepulcher when he was escorted around the city by the Greek Orthodox Patriarch. Had he done so, he explained, the Muslims would have wanted to build a mosque there to commemorate the first Islamic prayer in Jerusalem.

The Jews found their new Muslim rulers far more congenial than the Byzantines. The Christians had never allowed the Jews to reside permanently in the city, whereas Umar invited 70 Jewish families back. The Byzantines had left the Jewish Temple in ruins and had even begun to use the Temple Mount as a garbage dump.

Umar, according to a variety of accounts, was horrified to see this desecration. He helped clear it with his own hands, reconsecrated the platform and built a simple wooden mosque on the southern end, site of al-Aqsa Mosque today.

Jerusalem's Dome of the Rock, built by Caliph Abd al-Malik in 691, was the first great building to be constructed in the Islamic world. It symbolizes the ascent that all Muslims must make to God, whose perfection and eternity are represented by the circle of the great golden dome. Other Islamic shrines on the Temple Mount, which Muslims call al-Haram al-Sharif, the Most Noble Sanctuary, were devoted to David, Solomon and Jesus.

After the bloodbath of the Crusades, when Saladin reconquered Jerusalem for Islam in 1187, the Jews (barred from the city by the Crusaders) were invited to return, and even the Western Christians, who had supported the crusading atrocities, were allowed back. In the 16th century, Ottoman Sultan Suleiman the Magnificent permitted the Jews to make the Western Wall their official holy place and had his court architect Sinan build an oratory for them there.

So why the rejectionism that some Muslims in Jerusalem display today? In history, a holy city has always become more precious to a people after they have lost it. In the struggle for survival, the more compassionate traditions tend to get lost. As Muslims the world over feel that Jerusalem is slipping from their grasp, some espouse an intolerance that is far from the Koranic spirit. In an age in which religious atrocity occurs in nearly all faiths, it would be tragic if the Muslim tradition of inclusion and respect were lost to the world.

Source: Karen Armstrong, "Islam's Stake," *Time*, 4/16/01 also on time.com/time/2001/jerusalem/israel.html

11

from Monsignor Jean-Louis Tauran

Statement on Relations Between the Holy See and Jerusalem

October 26, 1998

The Holy See and Jerusalem

It is Jerusalem that has brought us together.

It is Jerusalem that urges us to look to the future.

And Jerusalem, yet again, wishes to impart its secret, the secret which the Prophet Ezekiel disclosed for all time: "And the name of the city henceforth shall be, The Lord is there" (Ez. 48:35).

On behalf of us all, I think it is right that I should thank His Beatitude Patriarch Michel Sabbah for the warm welcome extended to us, as well as for the spiritual joy he has brought us by gathering us together for the sake of the Holy City.

This cause of the Holy City has long been at the centre of the Holy See's concerns and one of its top priorities for international action, ever since the Jerusalem question existed.

I. The Jerusalem Question

Indeed, there is a conflict, or rather there are conflicts, because of and within Jerusalem—all related to its universally accepted uniqueness. It is unique in itself, and consequently it is also unique in its conflicts. It is different from any other city. The introduction to a book published in 1994 by a number of important Israeli academics begins thus: "At least in three respects Jerusalem differs from most other places: the City is holy to the adherents of three religions, it is the subject of conflicting national claims by two peoples, and its population is heterogeneous to a considerable degree.[1] Let us remember what Pope John Paul II wrote in his Apostolic Letter 'Redemptionis Anno' of 20 April 1984: '....Jews ardently love (Jerusalem) and in every age venerate her memory, abundant as she is in many remains and monuments from the time of David who chose her as the capital, and of Solomon who built the Temple there. Therefore, they turn their minds to her daily, one may say, and point to her as the sign of their nation.'

"Christians honour her with a religious and intent concern because there the words of Christ so often resounded, there the great events of the Redemption were accomplished: the Passion, Death and Resurrection of the Lord. In the City of Jerusalem the first Christian community sprang up and remained throughout the centuries a continual ecclesial presence despite difficulties."

"Muslims also call Jerusalem 'holy,' with a profound attachment that goes back to the origins of Islam and springs from the fact that they have there many special places of pilgrimage and for more than a thousand years have dwelt there, almost without interruption."

II

I think it is important to clarify from the very start that when we speak of Jerusalem the distinction often made between "the question of the Holy Places and the question of Jerusalem" is unacceptable to the Holy See. It is obvious that the Holy Places derive their meaning and their cultic and cultural uses from their intimate connection with the surrounding environment, to be understood not merely in terms of geography but also and most especially in its urban, architectural and above all human community and institutional dimensions.

In papal documents there certainly exist emphases and nuances, and they are seen more clearly the greater the span of time under consideration, for example, in a book edited by Archbishop Edmond Farhat,[2] in which he gathers papal documents from 1887 to 1986 (one hundred years), dividing this span of time into three periods:

1. from 1887 to 1947 (the first war between Arabs and Israelis), when the Popes spoke of the Holy Land in general and of Jerusalem, insisting primarily on the need to protect the physical integrity of the Holy Places and on the needs of the local Catholics;
2. from 1947 to 1964 (Pope Paul VI's pilgrimage): here the stress is on safeguarding the Holy Places, on freedom of access for all the faithful of the three religions and the right of each of the three religions to have control of its own holy sites;
3. from 1964 to the present day, a period during which the emphasis moves to Jerusalem in a global context and to the preservation of its identity and vocation: the Holy Places; the areas surrounding them; guarantees for everybody of their own cultural and religious identity; freedom of religion and conscience for the inhabitants and the pilgrims; the cultural dimension.

III

From the references to historical events, particularly those of the last fifty years, there emerges what is commonly referred to as the "political dimension" of Jerusalem in a complex of situations which have arisen regarding territorial control and the actions carried out

---

[1] Ruth Lapidoth and Moshe Hirsh, *The Jerusalem Question and its Resolutions: Selected Documents*, Dordecht–Boston, London 1994.

[2] *Gerusalemme nei Documenti Pontifici*, Libreria Editrice Vaticana, 1982.

to gain such control. The concern expressed in the interventions of the Popes and in other documents of the Holy See could not and cannot overlook this aspect. It is ever present, first, in order to prevent the Holy City becoming a battlefield and later to ensure that it does not become, as is the situation today, a case of manifest international injustice. The situation today has been brought about and is maintained by force. The Holy See has spoken out on this and will continue to speak out clearly, without mincing words and consistently adhering to the position of the majority within the international community, as expressed above all in the pertinent United Nations Resolutions. Since 1967, a part of the City has been occupied militarily and subsequently annexed. In that part of the City are to be found most of the Holy Places of the three monotheistic Religions. East Jerusalem is illegally occupied. It is therefore wrong to claim that the Holy See is only interested in the religious aspect or aspects of the City and overlooks the political and territorial aspect. The Holy See is indeed interested in this aspect and has the right and duty to be, especially insofar as the matter remains unresolved and is the cause of conflict, injustice, human rights violations, restrictions of religious freedom and conscience, fear and personal insecurity. Obviously, the Holy See's immediate and practical concern is with religious questions, while in other matters—political, economic, etc.—it interests itself inasmuch as they have a moral dimension. If the Holy See has no competence to enter into territorial disputes between Nations, to take sides, to seek to impose detailed solutions, on the contrary it has the right and duty of reminding the Parties of the obligation to resolve controversies peacefully, in accordance with the principles of justice and equity within the international legal framework.

In the case of Jerusalem, both aspects, the religious and the political and territorial, are closely linked, even though they are different in their constitutive elements, in the proper means of dealing with them and in finding a solution to them.

IV. What Is the Holy See Requesting for Jerusalem?

1. First of all, it asks that Jerusalem be respected for what it is in itself or rather what it should be, compared with what it actually is. That is what I defined a short while ago as the vocation or identity of the Holy City. Jerusalem is a treasure of the whole of humanity. In view of a situation of evident conflict and considering the rapid transformation of the Holy City, any unilateral solution or one brought about by force is not and cannot be a solution at all.

It is the view of the Holy See that every exclusive claim—be it religious or political—is contrary to the logic proper to the very City itself. I must insist: every citizen of Jerusalem and every person who visits Jerusalem should embody the message of dialogue, coexistence and respect evoked by the City. Exclusive claims cannot be backed up by numerical or historical criteria.

Having said that, I must add that there is nothing to prevent Jerusalem, in its unity and uniqueness, becoming the symbol and the national centre of both the Peoples that claim it as their Capital. But if Jerusalem is sacred to Jews, Christians and Muslims, it is also sacred to many people from every part of the world who look to it as their spiritual capital or travel there on pilgrimage, to pray and to meet their brethren in faith. It is the cultural heritage of everybody, including those who visit it simply as tourists.

2. Consequently, the Holy See believes that there is an obligation to find a realistic solution to the problems of Jerusalem, to all of them, according to their particular characteristics.

(a) There is a political problem concerning Jerusalem for Israelis and Palestinians first of all which is very practical. The Madrid Conference of 1991 and what followed gave birth to hopes of a peaceful future. Hopes founded on a willingness to talk, to negotiate and to seek to compromise. Hopes which appeared well-founded also by reason of the commitment and efforts of a large section of the international community, and in particular of the United States of America, as the events which took place at Wye Plantation in the last few days have demonstrated. Let us hope that the aspirations for dialogue and peace will contribute to the implementation of what has been agreed upon.

In this context, which is certainly both complex and delicate, the Jerusalem question has been placed at the bottom of the agenda. It is understandable that the difficulty and delicacy of the question of Jerusalem have meant that it has been left till last. But we all know, and the Israelis and the Palestinians are the first in this, that peace and coexistence in the Holy Land and Middle East have no future, unless an answer is found to the political question of Jerusalem. Allow me to quote once again from "Redemptionis Anno" of 1984, in which His Holiness Pope John Paul II wrote: "I am convinced that the failure to find an adequate solution to the question of Jerusalem, and the resigned postponement of the problem, only compromise further the longed-for peaceful and just settlement of the crisis of the whole Middle East."

What does the Holy See mean by an "adequate solution"? It means recognizing that the situation today is one of conflict. It means that Israelis and Palestinians, with the collaboration of all who can help them, have to reach an agreement which corresponds in some way to their particular legitimate and reasonable aspirations, and respects the principles of justice.

(b) As far as the Holy See is concerned, however, the solution of a territorial dispute alone is not enough for Jerusalem, precisely because Jerusalem is an unparalleled reality: it is part of the patrimony of the whole world. And the whole world has shown that it is fully aware of this when, for example, through resolutions of the United Nations it has sought to defend that patrimony.

Looking to Jerusalem, the Holy See continues to ask that it be protected by "a special internationally guaranteed Statute." What is meant by this? In the Holy See's view:

- the historical and material characteristics of the City, as well as its religious and cultural characteristics, must be preserved, and perhaps today it is necessary to speak of restoring and safeguarding those still existing;
- there must be equality of rights and treatment for those belonging to the communities of the three religions found in the City, in the context of the freedom of spiritual, cultural, civic and economic activities;
- the Holy Places situated in the City must be preserved, and the rights of freedom of religion and worship, and of access, for residents and pilgrims alike, whether from the Holy Land itself or from other parts of the world, must be safeguarded.

At stake is the basic question of preserving and protecting the identity of the Holy City in its entirety, in every aspect. For example, the simple "extraterritoriality" of the Holy Places, with the assurance that pilgrims would be able to visit them without hindrance, would not suffice. The identity of the City includes a sacred character which belongs not just to the individual sites or monuments, as if these could be separated from one another or isolated from the respective communities. The sacred character involves Jerusalem in its entirety, its holy places and its communities with their schools, hospitals, cultural, social and economic activities.

Israelis and Palestinians, in the desired search for a political settlement of their conflict over Jerusalem, cannot overlook the fact that the City has aspects which go far beyond their legitimate national interests. They, therefore, have to take these aspects into consideration in looking for and in reaching a lasting political and territorial solution. In the same way, they will not be able to avoid giving due consideration to the efforts and demands of all legitimately interested parties. In this, Israelis and Palestinians must not feel in any way restricted, but rather honoured and reassured.

V

It is essential that the parties to the negotiations take fair and appropriate account of the sacred and universal character of the City. This requires that any possible solution should have the support of the three monotheistic Religions, both at the local level and at the international level. Besides, as they are being proposed, the negotiations are expected to include the participation of the sponsors of the Peace

Process and other parties could also be invited to contribute. The Holy See believes in the importance of extending representation at the negotiating table: in order to be sure that no aspect of the problems is overlooked and to affirm that the whole International Community is responsible for the uniqueness and sacredness of this incomparable City.

Conclusion

In these coming days we shall listen to various other presentations and reflections. I would like to end my own intervention by expressing two feelings which I have experienced with great intensity:

(a) Sometimes I have felt great sadness and almost a sense of helplessness: the way forward to peace for the Holy Land and Jerusalem appears very precarious, alternating between progress and hesitation or failure. One has the impression that anything could happen: be it good or bad. Thinking also about the Year 2000, I wish to quote a few words which Pope John Paul II addressed to the Diplomatic Corps on 11 January 1992: "What a blessing it would be if this Holy Land, where God spoke and Jesus walked, could become a special place of encounter and prayer for peoples, if this Holy City of Jerusalem could be a sign and instrument of peace and reconciliation! It is here that believers have a mission of primary importance to accomplish. Forgetting the past and looking to the future, they are called to repentance, to re-examine their behaviour and to realize once again that they are brothers and sisters by reason of the one God who loves them and invites them to cooperate in his plan for humanity."

(b) And the second of my feelings: Episcopates of important Nations of the world are represented here. The Bishops are in communion and solidarity with each other, and the initiative of His Beatitude Patriarch Michel Sabbah is founded on this certainty. In the name of the Holy Father and together with the Patriarch I say to you all: let us remember Jerusalem, let us recall its essential nature, its vocation and the love which people have for it, let us help the world and those who wield power in it to remember Jerusalem and to understand that for its sake it should not be impossible to make it definitively a place of meeting, of harmony and of peace. It is my earnest hope that the Episcopates of the world will become Jerusalem's "Ambassadors" within the local Churches, to your respective Nations and societies and to the institutions and Authorities thereof. "Let my tongue cleave to the roof of my mouth, if I do not remember you, if I do not set Jerusalem above my highest joy!" (Ps. 137:6).

Source: Jean-Louis Tauran, "Discorsodi S. E. Mons. Jean-Louis Tauran sui Rapportio tra la Santa Sede e Geruslaemme" October 26, 1998 posted on the Holy See Press Office Site.

from Rabbi Arthur D. Rulnick

Jerusalem

October 1, 2000

There are two mountains that loom large in Jewish thought. There is Mount Sinai where the Torah was given to the Jewish people and there is Jerusalem's Mount Moriah where Abraham prepared to sacrifice Isaac, his beloved son, an event described in this morning's Torah reading. Which one do you think is more holy—Mount Sinai or Mount Moriah? Those of you who said to yourselves Mount Moriah are right. The reason, our rabbis tell us, is that because Abraham was willing to make the ultimate sacrifice for his faith there, that mountain transcended in holiness even the site where the Jewish people received the torah. Many generations after Abraham and Isaac, King Solomon built the holy Temple on that mountain.

This morning I would like to speak about the Temple Mount and Jerusalem. They have been very much in the news in the last several days because of the violent disturbances there. They stand at a fateful and historical crossroad as Israel seeks to resolve the conflict with the Palestinians. And it might be that the key to resolving the conflict is the paradigm of Abraham's willingness to sacrifice for his faith on the very spot that today is the main source of contention. For the Israeli–Palestinian conflict to be resolved, both sides will have to compromise where they have not been prepared to compromise previously. They are going to have to swallow some national pride, reign in nationalist strivings and accept only partial fulfilment of their dreams.

Let me tell you what has to happen at this crucial juncture in Jerusalem's history. And let me emphasize it is my opinion, although it is shared by many in Israel. Its chief advocate is Israel's Prime Minister, Ehud Barak, formerly the Chief of Staff of the Israeli army. But others of you may feel quite differently, just as there are many Israelis—and I spoke to a few of them on this subject during my visit to Israel in July—who disagree with my thoughts.

Let me tell you how I arrived at my opinion. It involves a little history because history tells best what Jerusalem means not only to Jews but also to Moslems and Christians. Only by understanding the attachment of three faiths to the Holy City, will you fully appreciate why issues regarding Jerusalem are not a simple matter of black and white.

For Jews, Jerusalem is ir ha'kodesh, the Holy City. The Bible tells us it was founded by King David to be the capital of the first Israelite Commonwealth. The core of its holiness is the Temple Mount where King Solomon built the first Temple to the God of Israel. Ever since, Jerusalem has been the spiritual center of Judaism. Throughout the millennia, when an independent Jewish State was only a dream, Jews have ended their Yom Kippur service with the phrase l'shana ha'ba'ah biyrushalayim, "Next year in his Jerusalem," just as they have concluded their seders each year with those same words and that same hope. The great Spanish Jewish poet, Yehuda Ha'Levi, expressed the eternal love of the Jewish people for Jerusalem in his immortal words, Libi bmizrach va'ani b'sof maarav. "My heart is in the East though I am in the furthest reaches of the West."

Jerusalem is the only capital our nation has ever had. Equally significant is the fact that no foreign occupier of the land of Israel has ever made Jerusalem the capital of its newly conquered territory. Jerusalem is too out of the way and strategically unimportant to be of interest to other peoples. Not even during the many centuries that Palestine was ruled over by the Moslem Turks, was Jerusalem considered the capital. But for the Jews, it didn't matter that Jerusalem was a provincial, backwater town. In the eyes of the Jewish people, Jerusalem is the center of the earth, the place where God's presence is felt most intimately.

For Christians, Jerusalem is also sacred ground. Jesus taught there and was crucified there. But most important of all, the central event of Christian belief, Jesus's resurrection, occurred in the Old City of Jerusalem, allegedly, where the Church of the Holy Sepulcher now stands. Today, because the number of Christians in Jerusalem is small and growing even smaller, Christians are not making any claims for power. They are no longer major players in the political process. The Vatican, which was once a vocal advocate to internationalize the entire city now is insisting only on international guarantees that its holy sites be safeguarded and that Christians be entitled to freely practice their faith.

Moslems venerate Jerusalem as their third holiest city—even though Jerusalem is not once mentioned in the Koran. The city first appears in a story told by one of Muhammad's early biographers. According to this story, one night Mohammad mounted his horse for a miraculous midnight journey from Mecca to the Temple Mount. From there, Mohammed ascended through the seven heavens to God's throne. Moslems believe that the Dome of the Rock in the Old City marks the very spot from which Mohammad made his ascent.

Mohammad's journey enhanced Jerusalem's holiness for Moslems in two ways. It linked Jerusalem to Mecca, and it made the Temple Mount the sacred launching pad for the prophet's journey to the heavens.

Now it is easy to dismiss Moslem and Christian claims to Jerusalem as being based on myths, bubba meises. Do I, personally, believe that Muhammad miraculously traveled on his magical steed from Mecca to Jerusalem and then ascended to Heaven? No. Do I believe in Jesus' resurrection? No. But please keep in mind that the biblical stories about Abraham, Moses and King David are also unprovable. Other than the Bible's word, no other evidence for their existence has ever been found. Since one person's religious truths may appear to others as mere myths, it doesn't behoove any of us to cast aspersions on other people's beliefs. We must treat them with respect as we would want our beliefs to be respected.

What history shows is this: Jews do not hold a monopoly on venerating Jerusalem. There is a joke surfing through the Internet that goes like this. A trilateral meeting is taking place between President Clinton, Prime Minister Barak and Chairman Arafat. Barak begins with a history lesson. "Thousands of years ago when the Temple stood in Jerusalem, the Cohayn Gadol, the High Priest, was preparing for the Yom Kippur service. As the torah requires, he took off his robes, went into the mikveh to purify himself before entering the Holy of Holies. When he emerged from the mikveh, to his astonishment, his golden robes had been stolen. And we Jews think a Palestinian stole them." At this point, Chairman Arafat rises from his seat livid at this accusation. "This is a Zionist lie," He shouts. "There were no Palestinians in those days." "Aha," Mr. Barak, exclaims. "Now that we have agreed on that point we can start the negotiations."

This is a cute story and makes an important historical point. However, the truth is that regardless of who got there first, and whether we like it or not, Jerusalem is holy to three faiths. Moreover, one of those faiths, Islam, happens to have 200,000 believers living in Jerusalem, and neither their presence there nor their spiritual attachment to Jerusalem can be safely ignored as the last several days of violence demonstrate.

So what can be done about Jerusalem? A solution begins to take form if we put unnecessary passion aside and think realistically. In order to reach a compromise, which is the only hope for peace, let the Palestinian authority be in charge in areas of East Jerusalem, including the Arab quarter of the Old City and nearby Arab neighborhoods, which are currently 99.9 percent Moslem and Christian. According to an interview with Prime Minister Barak in Friday's Jerusalem Post, if a final peace treaty is signed, this area will be called Al Quds, which means the holy city in Arabic. It will be the capital of a future Palestinian entity.

Dividing Jerusalem so that the Palestinians have control over the areas where Arabs live and pray, and Israel has control over the sacred places that resonate in the Jewish heart and soul and that we have prayed for for 2,000 years, is the only path to peace. And doing so won't even be creating a new reality. Jerusalem is already, virtually, two entities. West Jerusalem, where most Jews live, is a modern, western city; whereas Arab East Jerusalem has the feel of the third world. In Arab sections of Jerusalem, Palestinians run their own bus system, they have their own hospitals that operate independently of the Israel health system. They get electricity from their own Palestinian controlled electric company. Their schools follow a Palestinian curriculum.

So an invisible border already exists. Few Arabs venture into Western Jerusalem, and few Jews enter the Christian or Moslem areas of East Jerusalem because each feels unwelcome on the other's turf. The one place where Palestinians and Israelis used to mingle—the shuk or market place in the Arab Quarter—no longer is common ground. Israelis don't go there because it is potentially dangerous, and because they can get better prices on Ben Yehudah Street or at the local mall. In reality, Jerusalem is already divided because neither the Jewish nor the Arab populations aspire to a shared community.

As for the Temple Mount area, where the First and Second Temples once stood and which today is home to the Dome of the Rock and the El Aqsa Mosque, sovereignty there will have to be shared. And the truth is that it is already shared. Do you know who at this very moment daily exercises control over the Temple Mount? The Moslems! How is this possible? Well, when Israel conquered the Old City in 1967, Moshe Dayan, then the Israeli Chief of Staff, kept the Moslem religious council, called the Waqf, as the ruling authority over the El Aqsa Mosque and the Dome of the Rock—even though Israel technically has sovereignty over the area.

This verbal arrangement has basically worked for all the last 33 years. The problem arises today because of the need to put jurisdictional authority down on paper as part of a final peace agreement. To overcome the obstacles, several ideas have been suggested in recent weeks. Sovereignty would be turned over to the Security Council of the United Nations which would then delegate control of the Temple Mount to the Palestinians and control of the Western Wall to the Israelis. An old idea being revived is similar but leaves the UN out. It places the Temple Mount under Palestinian sovereignty, the Western Wall under Israeli sovereignty, whereas the earth under the El Aqsa Mosque and the Dome of the Rock, which presumably holds the remains of the ancient Israelite Temples, will be governed by a yet to be determined special arrangement. Whatever the particulars, it all boils down to one principle—sharing "God's holy half-kilometer."

What has become clearer than ever in the past few months, and especially in the last few days, is this: if the Jews and Moslems in Israel and in the world cannot envision sharing Jerusalem in some equitable way, there will never be peace in Israel or in the Middle East. For the Israel-Palestinian conflict to be resolved, the Arabs must accept that for Jews, Jerusalem is the paramount symbol of Jewish religious and national identity. They must accept not merely the existence—but the legitimacy—of a Jewish Zionist State. They must acknowledge that Jews are in the Holy Land not just because the Arabs couldn't defeat them, but because the Jewish people have a deep historical connection to this place.

Nor will the conflict end unless Israel recognizes that although the Moslems have the holy cities of Mecca and Medina, Jerusalem is also integral to Islamic history and faith. Only by acknowledging and respecting the others' claims will Jerusalem live up to and become what its lofty name states—Ir Shalom, the "City of Peace."

Shin Shalom, an Israeli poet wrote in 1968:

Ishmael, my brother
How long will we fight each other.
My brother from time bygone,
My brother—Hagar's son . . .
The heat of the desert has narrowed our mind,
Our common grazing ground we cannot find . . .
Time is running out, put hatred to sleep.
Shoulder to shoulder, let us water our sheep.

In Psalms, Kings David urges us, Sha'alu Shalom Yerushalayim, "Pray for the peace of Jerusalem." Together with Jews around the world, let us pray for the peace of Jerusalem, for Jerusalem belongs to all Jews, and its fate affects us all.

With God's help and with a renewed spirit of compromise and good will among Palestinians and Israelis, may this new year see the inhabitants of Jerusalem—Jew, Moslem and Christian—living together, shoulder to shoulder in peace and in harmony

V'chen yehi ratzon. May this be God's will.

Source: Arthur D. Rulnick, "Jerusalem" sermon delivered October 1, 2000. Used by permission of the author.

Review Questions

1. Why do specific places become designated as shrines by religions?
2. How does the ritual of a pilgrimage strengthen both the communal aspects of a religion and the faith of its adherents?
3. What are the arguments for and against the importance of Jerusalem to Muslims?
4. What is a *corpus separatum* and how would Jerusalem have been administered if one had been established there?
5. What is a key difference between the way that Muslims visit their holiest site and the way in which Jews and Christians visit theirs?

Discussion/Essay Questions

1. In the profile that began this module, American Muslim Umar Abdul Salam describes the Hajj as "real globalization" and implies that the rest of the world could learn from the example of the Hajj where diverse people from around the world peacefully work together for a common purpose. Given that allegiance to a religion involves private commitments of faith as well as public acts of participation, could Salam's vision actually be applied to the world at large?
2. In the tenth reading, Karen Armstrong writes, "In history, a holy city has always become more precious to a people after they have lost it." Considering what you know about how Muslims, Christians, and Jews view Jerusalem (a city that all of them at one time or another has "lost") and what you know about how Muslims view Makkah (a city that Muslims never really "lost"), would you agree with this statement?
3. In the eleventh reading, Monsignor Tauran writes that the holy city of Jerusalem must be preserved not just for practitioners of the three religions that consider it holy but also for "those who visit it simply as tourists." Given the tense situation that already exists between the three religions that lay claim to Jerusalem, does it make sense to also consider the desire of secular tourists? Is consideration of secular tourists likely to just complicate things further, or could it begin to open paths to a solution?

List of Readings

1. Ni'mah Isma'il Nawwab, "Hajj: The Journey of a Lifetime," in *Aramco World Magazine*, July–August 1992, posted on the website of Zawaj.com, *http://www.zawaj.com/events/hajj2001/journey_lifetime.html.*
2. Riz Khan, "For Pilgrims, Destination Is More Spiritual than Geographical," March 2001, posted on the website of the Cable News Network, *http://www.cnn.com/SPECIALS/2001/hajj/stories/overview/.*
3. Yousef Saleh Ba-Isa, "Saudi Teen Amazed at How People Differ in Their Devotion to Allah," March 4, 2001, posted on the website of the Cable News Network, *http://www.cnn.com/SPECIALS/2001/hajj/stories/yousef.notebook/.*
4. "Hajj and Umrah Tips & Tricks," posted on the website of Jannah.org, *http://www.jannah.org/hajj/tipsandtricks.html.*
5. "Guardian of the Holy Places," posted on the website of the Royal Embassy of Saudi Arabia, London, *http://www.saudiembassy.org.uk/profile-of-saudia-arabia/islam/guardian-of-the-holy-places.htm.*
6. David Van Biema, "Jerusalem At the Time of Jesus," in *Time Magazine*, April 16, 2001, posted on the website of *Time Magazine*, *http://www.time.com/time/2001/jerusalem/cover.html.*
7. United Nations General Assembly, Resolution 181, November 29, 1947, posted on the website of the MideastWeb for Coexistence, *http://www.mideastweb.org/181.htm.*
8. Moshe Sasson, "Jerusalem: The Battle for Sovereignty," in *Ma'ariv*, July 8, 1994, posted on the website of the Israeli Ministry of Foreign Affairs, *http://www.mfa.gov.il/MFA/Archive/Articles/1994/JERUSALEM+--+THE+BATTLE+FOR+SOVEREIGNTY+-+08-Jul-9.htm.*
9. Abdel-Haleem E'weis, "The Status of Jerusalem in Islam," February 6, 2003, posted on the website of the Palestinian Information Center, *http://www.palestine-info.co.uk/am/publish/article_26.shtml.*
10. Karen Armstrong, "Islam's Stake," in *Time Magazine*, April 16, 2001, posted on the website of *Time Magazine*, *http://www.time.com/time/2001/jerusalem/islam.html.*
11. Jean-Louis Tauran, "Discorso di S.E. Mons. Jean-Louis Tauran sui Rapportio tra la Santa Sede e Geruslaemme (Statement of S.E. Mons. Jean-Louis Tauran on Relations Between the Holy See and Jerusalem)," October 26, 1998, posted on the website of the Holy See Press Office, *http://www.vatican.va/news_services/bulletin/news/3824.php?index=3824&po_date=26.10.1998%20%20&lang=en.*
12. Arthur D. Rulnick, "Jerusalem," sermon delivered October 1, 2000 (Rosh Hashanah Day 2) at the Woodbury Jewish Center, posted on the website of the Woodbury Jewish Center, *http://www.thewjc.org/sermons/jerusalem.htm.*

Websites for Additional Research

1. An excellent starting point for more information on the Hajj (and other aspects of Islam) is the "Web Resources" page that accompanied CNN's coverage of the 2001 Hajj, at *http://www.cnn.com/SPECIALS/2001/hajj/stories/web.resources/index.html*. This page contains links to several annotated English-language links pages that are maintained by Islamic organizations. Especially useful is the "Islamic Gateway" page (*http://www.ummah.net*), which is referred to from the CNN page, and its Hajj section, *http://www.ummah.net/hajj/*.
2. To obtain balanced information on Jerusalem one needs to peruse a number of websites. The "Resource Center" page accompanying *Time Magazine*'s special report on "Jerusalem As Jesus Saw It" *http://www.time.com/time/2001/jerusalem/resource.html* contains many links for information about specific holy shrines, but it steers clear of links to any groups that directly address the political controversy that surrounds control of the city. For the hard-line Jewish perspective that an undivided Jerusalem should be the capital of Israel, see the website of One Jerusalem, at *http://www.onejerusalem.org*, and especially the "About Jerusalem" area of the website, at *http://www.onejerusalem.org/AboutJerusalem.asp*. For a comprehensive set of links to websites presenting the Palestinian perspective (and, occasionally the Israeli perspective as well) on Jerusalem and, more generally, on the Israel/Palestine conflict, see the Palestine section of the Al Mashriq website, at *http://almashriq.hiof.no/base/palestine.html*. Also, see the websites listed at the end of the next module in this workbook.
3. For general information on the world's religions, see the non-sectarian BeliefNet website at *http://beliefnet.com*. This website contains links to information on dozens of world religions, from Baha'i to Zoroastrianism. To find out what religion most closely matches your personal spiritual beliefs and moral positions, check out the Belief-O-Matic, a 20-question interactive quiz that compares your answers with the creeds of 27 world religions, at *http://beliefnet.com/story/76/story_7665_1.html*.

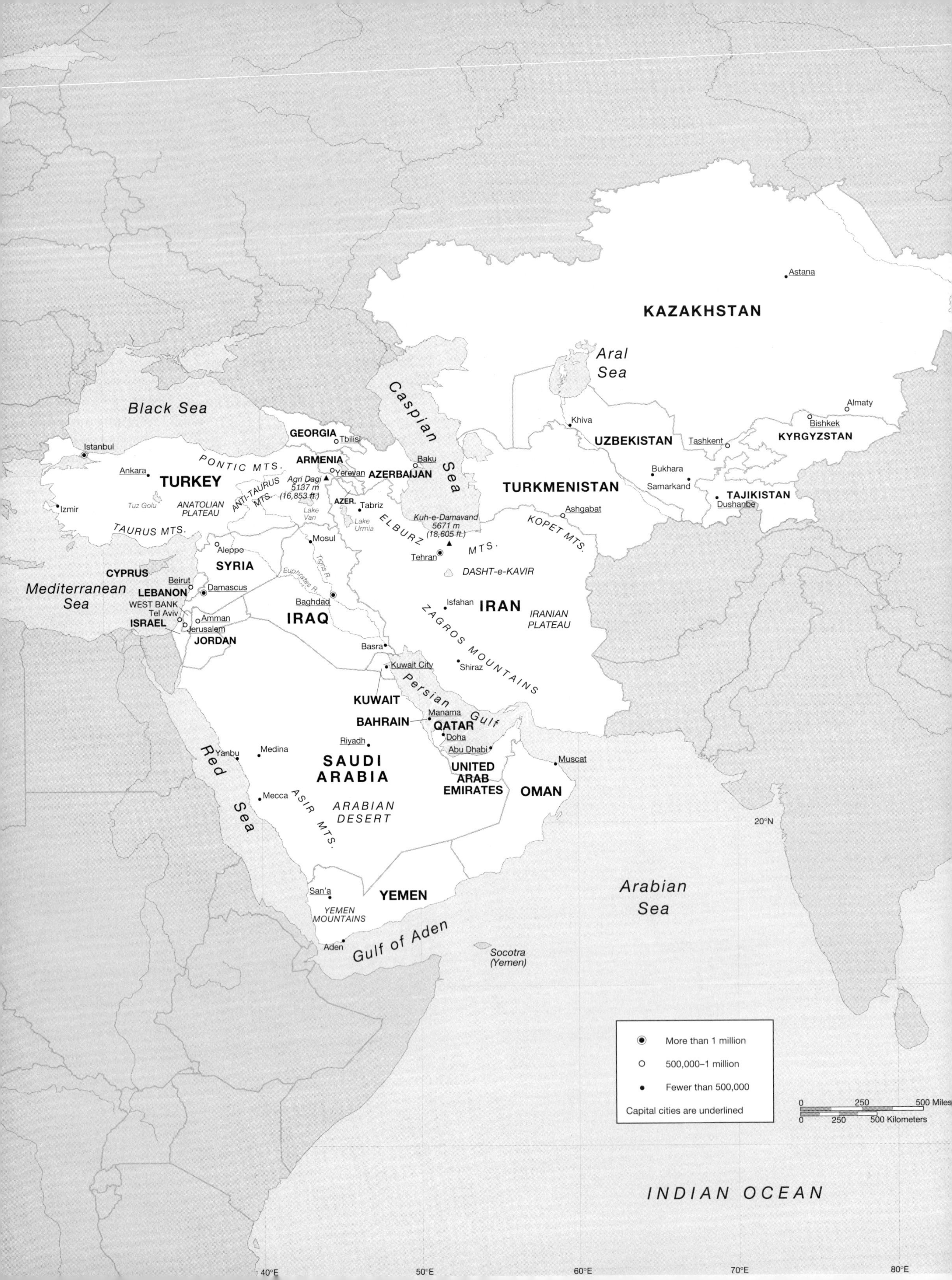

KAZAKHSTAN
Astana
Aral Sea
Almaty
Bishkek
KYRGYZSTAN
Khiva
UZBEKISTAN
Tashkent
Bukhara
Samarkand
TAJIKISTAN
Dushanbe
TURKMENISTAN
Ashgabat
KOPET MTS.
Caspian Sea
Baku
Black Sea
GEORGIA
Tbilisi
Istanbul
ARMENIA
Yerevan
AZERBAIJAN
AZER.
Ankara
TURKEY
PONTIC MTS.
ANTI-TAURUS MTS.
Agri Dagi 5137 m (16,853 ft.)
Izmir
Tuz Golu
ANATOLIAN PLATEAU
TAURUS MTS.
Lake Van
Tabriz
Lake Urmia
ELBURZ MTS.
Kuh-e-Damavand 5671 m (18,605 ft.)
Tehran
DASHT-e-KAVIR
Mosul
Aleppo
SYRIA
CYPRUS
Beirut
Damascus
Mediterranean Sea
LEBANON
WEST BANK
Tel Aviv
ISRAEL
Amman
Jerusalem
JORDAN
Euphrates R.
Tigris R.
Baghdad
IRAQ
Isfahan
IRAN
IRANIAN PLATEAU
ZAGROS MOUNTAINS
Basra
Kuwait City
Shiraz
Persian Gulf
KUWAIT
Manama
BAHRAIN
QATAR
Doha
Abu Dhabi
Riyadh
Medina
Yanbu
Red Sea
SAUDI ARABIA
UNITED ARAB EMIRATES
Muscat
OMAN
Mecca
ASIR MTS.
ARABIAN DESERT
20°N
San'a
YEMEN
YEMEN MOUNTAINS
Arabian Sea
Aden
Gulf of Aden
Socotra (Yemen)
More than 1 million
500,000–1 million
Fewer than 500,000
Capital cities are underlined
0 250 500 Miles
0 250 500 Kilometers
INDIAN OCEAN
40°E
50°E
60°E
70°E
80°E

Nationalism and Self-Determination in Southwest and Central Asia

Companion to Chapter 7 Ethnicity

It is March 21, Kurdish New Year. **Nesibe Yilmaz**, secretary of HADEP Women's Committee, and the people at HADEP (The Kurdish People's Democracy Party) are celebrating. According to Kurdish history, Newroz (as it is called in Kurdish, which means "New Day") dates back to 612 B.C., when a Kurdish smith called Kawa led a successful people's revolt against a tyrannical ruler. But the festival has been banned in Turkey since the 1920s. It has become a symbol for Kurds of freedom, independence, and unity for the Kurdish nation.

In the last decade, the Turkish government has resorted to violent means to suppress it. 109 Kurdish civilians were killed while celebrating Newroz in 1992. In 1995, then Prime Minister Tansu Çiller attempted a new strategy. Turkey claimed Newroz as its own festival. Despite this, all Newroz celebrations have to receive official permission and are monitored by the Turkish police.

Kurds celebrate Newroz, the Kurdish New Year. *Source:* © AFP/Agence France Presse/Getty Images

Source: Adapted from "New Day Yet to Dawn: The Kurds," *New Internationalist*, January–February 1998. Copyright © 1998 Wayne Ellwood. Reprinted by permission. ■

Nationalism and Identity

Every nation is, at root, a cultural entity. As the textbook notes, a nation is "a group of people tied together to a particular place through legal status and cultural tradition." This cultural tradition could include sharing a common history or the sharing of a certain belief system by which one group of people distinguishes itself from another group.

Although the words *nation* and *state* are often used interchangeably, a state, unlike a nation, is fundamentally a *political* entity. That said, for the past two hundred years the concepts of nation and state have been linked in the ideal of the *nation-state*. A nation-state is a state in which the government bases its legitimacy on a claim to represent the people of a nation, or, as the textbook states, a nation-state is "a state whose territory corresponds to that occupied by a particular ethnicity that has been transformed into a nationality." For a nation-state to function, there should be a direct correspondence between the social boundaries of the nation (where people perceive themselves to be unified by a similar history and culture) and the territorial boundaries of the state. This is the principle of nationalism: the right of every nation to its own state and the ideal that every state should contain only one nation of people. (For further definition and discussion of the concept of the *state*, see Chapter 8 in the textbook.)

A broader definition of a nation expands the term to include any group of people who share a common identity. For instance, some observers of and participants in pop culture have identified the existence of a "hip-hop nation." In this example, a claim to nationhood ("We are the hip-hop nation") combines an assertion of shared identity ("We are people who have something in common—a love of hip-hop music") with a statement that the bonds among the nation's citizens go beyond the specific experiences that tangibly unite the group ("We are characterized by an outlook on life that we share even when we're not listening to hip-hop"). Like other nations, the hip-hop nation is also unified by a common language and a common cultural heartland (see Chapter 4 of the textbook for more on culture and, in particular, hip-hop culture).

Once a nation is declared to exist based on a shared identity, the nation then is associated with certain prevailing attitudes and norms, and these frequently are used to guide policymaking, especially when the nation has its own state within which to implement these policies. The following statement serves as an example of this use of national identity to guide public policy: "Because we are Americans, we cherish the right of each individual to defend himself or herself. Therefore, we must oppose efforts to register or control the sale of firearms." A statement of national identity ("We are Americans") is associated with a cultural mindset that is said to permeate the American nation ("Americans cherish the right of each individual to defend himself or herself"), and this cultural mindset is extended to the public policy that logically follows ("Americans should be free from registration or control of firearm sales").

Besides resting on a contestable argument about policy (one could argue that although there is an American ethic of self-defense it would not be violated by gun registration), this argument also rests on a contestable statement about Americans' identity. The statement assumes that we know who Americans are and what they believe, and that these cultural attributes do not vary over time, space, or individuals. Perhaps, however, Americans felt strongly about their right to defend themselves back when the Constitution was drafted in 1789 but they no longer feel it, or perhaps southerners, rural people, and men feel strongly about this right but northerners, urbanites, and women don't.

The lesson here is that any political statement based on the singular identity of a nation is complicated by the fact that individuals have various identifications. These may be based on gender, ethnicity, location, beliefs, or any other factor that might define a common interest. Sometimes these identities contradict each other. Consider this alternate statement of identity: "Because we are mothers and givers and nurturers of life, we oppose all instruments that can be used to extinguish life, and therefore, we must support efforts to register or control the sale of firearms." How is an American mother to decide where her identity lies and what cultural mindset (and position on gun control) is associated with that identity? Is she more like mothers around the world who, although speaking different languages and living very different lives, all nurture their children? Or is she more like the all-male framers of the Constitution who, more than two hundred years ago, placed a high value on the right to self-defense?

To take another example, imagine an African-American Floridian attending Florida State University, where the school team is called the Seminoles, a reference to a Native American tribe that inhabited much of Florida prior to the state's colonization. As a Floridian, the student has pride in things Floridian, including the Gators of the University of Florida. However, as a proud Seminole fan, the student has little sympathy for the arch-rival Gators. Finally, as an African-American, the student also may be influenced by the fact that the NAACP, the nation's largest African-American advocacy group, has opposed the practice of using Native American symbols as team mascots, arguing that these symbols are degrading and potentially can create an environment of hostility toward all ethnic minority groups.

Fortunately, one can identify oneself with many nations. The principle of sovereignty dictates that there cannot be a state within a state, but no such principle applies to nations. In many ways, Texans (particularly white, Anglo Texans) treat Texas as a distinct nation. In comparison with any other state in the United States, the Texan flag and depictions of the state's shape are everywhere, and Texans frequently stress the distinctiveness of their "Texan" culture. But Texans do not portray their culture as opposed to American culture; if any thing, Texans (and outsiders as well) view Texan culture as an *extreme* form of American culture.

These examples all point to the complex ways in which individuals identify themselves with groups and the complex ways in which groups identify their core beliefs and values. The membership and values of any nation are continually shifting and overlapping. And yet the ideal of the nation-state fuses each nation to a state, which, by definition, is permanent and mutually exclusive. Not only does the complexity of national affiliation create mixed allegiances for individuals who feel that they belong to multiple nations; it also causes problems for residents and governments of countries that have minority populations, where some residents' national affiliations are different from that with which the state is primarily identified. As the textbook notes, most states that purport to be nation-states actually are "multinational states." This uneasy fit between nation and state may go a long way toward explaining why nation-states have been the focus of so much conflict in the twentieth and twenty-first centuries.

Nationalism and Self-Determination

Closely linked with, but distinct from the ideal of nationalism is that of self-determination. As the previous section demonstrated, the ideal of nationalism—the belief that every nation deserves a state and that every state should contain only one nation—rests on shaky ground. There are no clear standards for defining just what is a nation and, even if one could define a nation, it would be unlikely that the residents of that nation would be the sole occupants of a contiguous swath of territory that could then be transformed into a nation-state.

In contrast with nationalism, the doctrine of self-determination does not say that every nation must have its own sovereign state. Rather, it says that every nation should be allowed to determine its own destiny. The distinction is subtle, but important. When a nation seeks self-determination, it often achieves political power through control over territory, but that is not necessarily the case. Even in instances when a nation does assert its power through control over territory, that power may be consolidated in ways other than through formation of a sovereign nation-state.

The history of the African-American movement for self-determination demonstrates the many ways in which self-determination can be realized. For at least the past hundred years, many African-American intellectuals and activists have been asserting that African-Americans constitute a distinct nation, with a culture that is distinct, both from mainstream (white) American culture and from any culture found in Africa. Yet relatively few African-American nationalists (at least since the early twentieth century) have called for the establishment of an African-American state, either in Africa or in North America. Instead, African-American nationalists have tended to advocate a combination of individual rights (an end to discrimination against African-Americans within the United States) and national rights (respect for African-American culture as a distinct and legitimate national culture within the United States). In Canada, the Québecois (or French-Canadians) appear to be seeking (and perhaps achieving) self-determination along similar lines. The Québecois are a minority that historically has faced discrimination, both inside and outside of the province of Québec (where they are the majority). After years of political struggle, the Canadian government now permits a high level of French-language-only self-government within Québec and mandates bilingual administration nationally.

Thus, nationalism represents only one of the many routes to self-determination. In fact, according to international relations expert Tom Barry, the principle of self-determination has long been enshrined in international treaties, usually through clauses that limit the ability of states to withhold basic rights from their minority nations. It was only in the twentieth century, according to Barry, that this ideal of basic rights for *nations* became transformed into the ideal of basic rights for *states*. Once this shift occurred, leaders of nations that sought the right to control their own destiny were left with no option but to advocate statehood. Thus the doctrine of nationalism replaced self-determination as the fundamental driving force behind national empowerment movements.

The Roots of Nationalism in Southwest and Central Asia

Perhaps no region of the world has suffered more from nationalist conflict in the twentieth and twenty-first centuries than Southwest Asia. Most of Southwest Asia (or the Middle East, as it is also known) was at some point under the rule of the Ottoman Empire, which reached its height in the mid-sixteenth century when it controlled most of Southwest Asia and also much of North Africa and southeastern Europe. Over the next several centuries, European powers—first Russia and Austria-Hungary, and later Britain, France, and Italy—slowly chipped away at the empire, until, at the end of World War I, it was dissolved, with its core area becoming present-day Turkey.

The Ottoman Empire was known for its pluralism, tolerating the presence of minority cultures and communities, as long as these minorities politically affiliated themselves with the Ottoman rulers. In comparison with the Christian rulers in Europe during the same era, the Muslim rulers of the Ottoman Empire were notable for their ideas about religious freedom, as non-Muslim minorities (including Christians and Jews) were allowed to form their own communities within the empire. Non-Muslim communities were forced to contribute taxes and soldiers to the empire like everyone else, but otherwise they basically were allowed to govern themselves, continue their non-Muslim practices, and maintain their identities as distinct, non-Ottoman communities. In short, the Ottoman Empire and its administrative regions were organized according to a principle very different than that of nationalism.

Ironically, the lack of nationalism in the Ottoman Empire may have contributed to the ferocity of the nationalist struggles that now characterize the region. When the European powers, especially after World War I, drew lines around areas of territory, called them countries, and identified national leaders, these countries typically had little sense of a "national culture." As is also true in much of Africa, many of the countries in Southwest Asia are European creations. Typically, the people who emerged to lead these new "nations" were individuals (or families) who previously had authority over just one of the many tribes or clans within the country's borders but who had earned the support of the European powers. Ruling over a fictitious nation with something less than a popular mandate, the national leader typically responded to this situation in two ways. On the one hand, he would proclaim a new national culture, frequently tracing the new nation's lineage back to a historic empire in the same region, while, at the same time associating the new nation with progress and modernization. On the other hand, he would appoint members of his own tribe or clan to positions of authority within the government, a move that would tend to undermine any attempts to associate his administration with the entire new "nation." Not surprisingly, members of other ethnic groups (who previously had been left alone by the Ottomans to do as they pleased) did not take kindly to this process of centralization without democratization. As a result, movements for self-determination grew among religious and ethnic minorities (or, in some cases, disempowered majorities).

Throughout the world, most struggles for self-determination have taken a nationalist form, as when a group of people located inside one country (or as a colony of another country) assert that they are a distinct nation and that they therefore deserve their own state (for example, the Hindu Tamils, who have been fighting for the independence of their homeland, which is the northeast portion of majority-Buddhist Sri Lanka). Another typical form of nationalist self-determination movement occurs when a group that spans the border of two (or more) existing countries declares itself a nation (such as the Basques, whose region spans the border of France and Spain).

In Southwest Asia, movements like these rarely have had much success, in large part because the core powers have intervened to preserve the countries that they created out of the ruins of the Ottoman Empire. Although there are various explanations for the core countries' high level of interest in the region's political stability, a popular explanation is that core countries are concerned that the growth of separatist movements could lead to an overall unraveling of the region's political system, which could limit their access to the region's strategically important petroleum supplies. Whatever the motivation, the United States (with the support of its allies) frequently has signaled that it would not permit boundary changes in the region. For instance, this attitude was expressed during the 1991 Gulf War, both when it intervened militarily to prevent Iraq's annexation of Kuwait, and again just after the war when it refused to aid Kurdish and Shi'ite rebel groups in northern and southern Iraq, respectively, who were attempting to achieve national independence.

A form of self-determination struggle unique to Southwest Asia is regional nationalism, known as pan-Arabism. Pan-Arabism was advocated by Gamal Abdel-Nasser, who ruled Egypt from 1954 to 1967. Nasser's version of nationalism was not that different from what was being practiced elsewhere in the world (he favored a strong state that, while looking backward to a common heritage, also was strongly oriented toward modernization), but he proposed redefining the scope of the nation so that it covered roughly the entire territory of the former Ottoman Empire. In the end, only one other country, Syria, joined with Egypt to form Nasser's United Arab Republic (U.A.R.). In most countries, national leaders had only recently come to power, installed by European forces to govern their newly created countries. They saw little reason to cede this power to Nasser for the U.A.R. After a few years, Syria left the U.A.R. and this experiment in pan-Arabism was ended.

The Central Asia and Caucasus region also has a long history of nations struggling for self determination. The region has long been subject to incursions by various groups including the Persians, the Mongol Empire of Genghis Khan, the Timurid Empire, and the Kazakh, Uzbek, Tatar and Russian empires. Finally, the region came under the political control of the Soviet Union. The region's present state boundaries are a legacy of Soviet administrative divisions.

The Soviet Union combined the region's nationalities into various administrative units based on the size of a nation's population. While the Soviet constitution on paper appeared to give equal treatment, including autonomy, to all the nationalities that made up the Soviet Union, treatment was at times far from equal, and local authority was always superceded by central authority. Distinctions were also made between larger minority groups and smaller groups. While there were 15 Soviet Socialist Republics (SSRs), each of which was associated with a majority nationality, the territories of smaller ethnic groups were marked off as Autonomous Soviet Socialist Republics (ASSRs), administrative entities that lacked some of the rights of the SSRs. Other, even smaller units included oblasts, krays, and rayons.

In the early history of the USSR, there were ideological differences among the leaders on how to deal with the various nationalities. Karl Marx's views are briefly mentioned in the textbook in Chapter 7. He believed that nationalism would cease to exist after the global workers' revolution, because workers—the proletariat—would identify with other workers rather than with members of their nation. In *The Communist Manifesto*, Marx and Frederick Engels wrote:

> The Communists are distinguished from the other working class parties by this only: 1. In the national struggles of the proletarians of the different countries, they point out and bring to the front the common interests of the entire proletariat, independently of all nationality. 2. In the various stages of development which the struggle of the working class against the bourgeoisie has to pass through, they always and everywhere represent the interests of the movement as a whole.
>
> The Communists are further reproached with desiring to abolish countries and nationality.
>
> The working men have no country. We cannot take from them what they have not got. Since the proletariat must first of all acquire political supremacy, must rise to be the leading class of the nation, must constitute itself the nation, it is, so far, itself national, though not in the bourgeois sense of the word.
>
> National differences and antagonisms between peoples are daily more and more vanishing, owing to the development of the bourgeoisie, to freedom of commerce, to the world market, to uniformity in the mode of production and in the conditions of life corresponding thereto.

Lenin, while agreeing with Marx and Engels on the place of national differences in the communist state, believed also that the rights of ethnic minorities should be protected, and that they should be granted some degree of flexibility in determining their fate. In *The Declaration of the Rights of the Working and Exploited People*, Lenin wrote:

> Supporting Soviet power and the decrees of the Council of People's Commissars, the Constituent Assembly considers that its own task is confined to establishing the fundamental principles of the socialist reconstruction of society.
>
> At the same time, endeavouring to create a really free and voluntary, and therefore all the more firm and stable, union of the working classes of all the nations of Russia, the Constituent Assembly confines its own task to setting up the fundamental principles of a federation of Soviet Republics of Russia, while leaving it to the workers and peasants of each nation to decide independently at their own authoritative Congress of Soviets whether they wish to participate in the federal government and in the other federal Soviet institutions, and on what terms.

Lenin thus shared the goal of Marx to abolish distinctions between nationalities but thought this goal must be achieved voluntarily. This philosophy was not shared by Lenin's successor, Josef Stalin. Although Stalin himself was from Georgia, one of the republics in the Caucuses, and gave lip service to preserving the autonomy of the non-Russian republics and the identity of non-Russian ethnicities, his policies were based on centralization, using whatever means were necessary. The republics were "russified" by suppressing local languages in schools in favor of teaching Russian, and local government officials were replaced with Russian officials.

Contemporary Struggles for Self-Determination in Southwest and Central Asia

The best-known struggle for self-determination in the Southwest/Central Asia region is that between Palestinians and Israelis, which is described in Chapter 6 of the textbook and in the accompanying module in this workbook. Palestinians and Israelis have both waged what could be called classic nationalist struggles, wherein they each have asserted that they are a distinct nation, with a historic tie to a certain swath of land, and that therefore they each have a right to establish their own state on that land. (The problem, of course, is that they're talking about the same land.) In the first reading, this vision is articulated by Theodor Herzl, considered the father of modern political Zionism, in an excerpt from his 1896 book *Der Judenstaat (The Jewish State)*. A similar vision is found in the second reading, the Declaration of Independence issued by the Palestinian National Authority in 1988.

More than 105 years after Herzl issued his statement, and more than 55 years after Britain partitioned Palestine into what were intended to be separate Jewish and Arab nation-states, the land of Israel-Palestine remains wrought by nationalist conflict. The apparent failure of a nationalist solution has led some to argue that it is time to abandon the rigid concept of nationalism in favor of a more flexible form of self-determination. In the third reading, Edward Said, a Palestinian intellectual who lived in the United States from the 1960s until his death in 2003, makes this case. Said urges that the current situation of two opposing nationalist forces at best will lead to a tense two-state solution but more likely will just lead to a continued stalemate. Instead, he urges that Jews and Palestinians work toward the creation of a tolerant binational state wherein both groups can achieve self-determination.

In the fourth reading, Shibley Telhami, another Palestinian living in the United States, makes just the opposite argument. Telhami states that nationalism is built on the idea that your nation is one among many, which means that nations must allow other nations to exist. Therefore, if Israelis and Palestinians could ever find a way to divide the territory in a way that satisfied both groups' nationalist aspirations, peace would be at hand. By contrast, if the two sides were to abandon their nationalist ideals and turn the current political conflict into an ethnic-religious conflict, it would become even more violent as each side would use ideologies of ethnic superiority to try to wipe out the other.

Similar debates regarding what self-determination is and how to achieve it are taking place among the Kurds, an ethnic group with some 25 million people, whose situation is discussed in Chapter 7 of the textbook. Since it effectively stopped the Kurds from achieving statehood in 1923, the government of Turkey, where Kurds constitute 15 to 20 percent of the population, has held that Kurds are not a national minority. This sentiment is expressed in the fifth reading, a statement from the Turkish Ministry of Foreign Affairs. Turkey recognizes that some people in eastern Turkey speak a strange language and have some unique customs, but, according to the Turkish government, there is no distinct Kurdish nation. Whatever characteristics of "Kurdishness" exist can be assimilated into the expansive Turkish national character. To that end, the Turkish government has had a long-standing policy of encouraging migration of Kurds to non-Kurdish regions and their integration into mainstream Turkish culture. Expressions of Kurdish identity that challenge the official ideology—that Kurds are one of the ethnic groups that constitute the Turkish nation—were until recently met with repression and are still discouraged.

To a large extent, this strategy has worked. Today there are many Turks of part-Kurdish descent, and more Kurds live in cities in non-Kurdish regions than in the remote mountains of Kurdistan. As with the example of African-Americans, nationalism is no longer a simple project for Turkish Kurds to undertake. The Turkish-Kurd (or Kurdish-Turk) identity, while still unique, is at the same time wrapped within one's identity as a member of the Turkish nation. It is unclear how many urban Turkish-Kurds would leave Istanbul or Ankara to "return" to a "homeland" that in many cases has never been their home, just as not many African-Americans voluntarily would go "back" to Africa.

Self-determination for "nested" nations like Turkish-Kurds, African-Americans, or French-Canadians can be achieved when the smaller nation finds a way to live within the larger nation and the larger nation accepts the integrity of the smaller nation. However, because states insist on sovereignty (under which there is no provision for "nested" states), a conflict emerges when a "nested" nation adopts a nationalist program and advocates independent statehood. Differing approaches to this dilemma are reflected in the sixth, seventh, and eighth readings, which present three different visions for Kurdish self-determination: one from the PKK, a nationalist organization active in Turkey; one from the KDP, an Iraqi-based organization that favors regional autonomy and cultural rights closer to that of the Québecois model; and one from AKIN, a Washington, D.C.-based group that seeks to facilitate communication among Kurds and create awareness of the Kurdish situation among Americans and U.S. policymakers.

The last group of readings turns the focus to the former Soviet republics in Central Asia and the Caucasus. First, a summary of the present situation in the region is provided by Mustafa Adyin, Associate Professor of International Relations at Ankara University in Turkey. Adyin discusses how the structure of the Soviet Union essentially stoked the inter-ethnic tensions that are at the root of today's nationalist conflicts in Central Asia and the Caucuses. The final two readings delve deeper into the region's present conflicts, first to Uzbekistan, where the plight of a small ethnic group—the Luli—is profiled, and then to Tajikistan, where ethnonationalism is fueling an ongoing civil war.

Readings

1

from Theodor Herzl

Der Judenstaat (The Jewish State)

Originally published in 1896

The idea I have developed in this pamphlet is an ancient one: It is the restoration of the Jewish State.... The decisive factor is our propelling force. And what is that force? The plight of the Jews.... I am profoundly convinced that I am right, though I doubt whether I shall live to see myself proved so. Those who today inaugurate this movement are unlikely to live to see its glorious culmination. But the very inauguration is enough to inspire in them a high pride and the joy of an inner liberation of their existence....

The plan would seem mad enough if a single individual were to undertake it; but if many Jews simultaneously agree on it, it is entirely reasonable, and its achievement presents no difficulties worth mentioning. The idea depends only on the number of its adherents. Perhaps our ambitious young men, to whom every road of advancement is now closed, and for whom the Jewish state throws open a bright prospect of freedom, happiness, and honor perhaps they will see to it that this idea is spread....

It depends on the Jews themselves whether this political document remains for the present a political romance. If this generation is too dull to understand it rightly, a future, finer, more advanced generation will arise to comprehend it. The Jews who will try it shall achieve their State; and they will deserve it....

I consider the Jewish question neither a social nor a religious one, even though it sometimes takes these and other forms. It is a national question, and to solve it we must first of all establish it as an international political problem to be discussed and settled by the civilized nations of the world in council.

We are a people—one people.

We have sincerely tried everywhere to merge with the national communities in which we live, seeking only to preserve the faith of our fathers. It is not permitted us. In vain are we loyal patriots, sometimes superloyal; in vain do we make the same sacrifices of life and property as our fellow citizens; in vain do we strive to enhance the fame of our native lands in the arts and sciences, or her wealth by trade and commerce. In our native lands where we have lived for centuries we are still decried as aliens, often by men whose ancestors had not yet come at a time when Jewish sighs had long been heard in the country....

Oppression and persecution cannot exterminate us. No nation on earth has endured such struggles and sufferings as we have. Jew-baiting has merely winnowed out our weaklings; the strong among us defiantly return to their own whenever persecution breaks out.... Wherever we remain politically secure for any length of time, we assimilate. I think this is not praiseworthy....

Palestine is our unforgettable historic homeland.... Let me repeat once more my opening words: The Jews who will it shall achieve their State. We shall live at last as free men on our own soil, and in our own homes peacefully die. The world will be liberated by our freedom, enriched by our wealth, magnified by our greatness. And whatever we attempt there for our own benefit will redound mightily and beneficially to the good of all mankind.

Source: Excerpt from Theodor Herzl, *Der Judenstaat*, trans. Sylvie D'Avigdor. Published by Dover Publications.

from the Palestinian National Authority

2

Declaration of Independence

November 15, 1988

In the name of God, the Compassionate, the Merciful:

Palestine, the land of the three monotheistic faiths, is where the Palestinian Arab people was born, on which it grew, developed and excelled. Thus the Palestinian Arab people ensured for itself an everlasting union between itself, its land, and its history.

Resolute throughout that history, the Palestinian Arab people forged its national identity, rising even to unimagined levels in its defense, as invasion, the design of others, and the appeal special to Palestine's ancient and luminous place on the eminence where powers and civilizations are joined. All this intervened thereby to deprive the people of its political independence. Yet the undying connection between Palestine and its people secured for the land its character, and for the people its national genius.

Nourished by an unfolding series of civilizations and cultures, inspired by a heritage rich in variety and kind, the Palestinian Arab people added to its stature by consolidating a union between itself and its patrimonial Land. The call went out from Temple, Church, and Mosque that to praise the Creator, to celebrate compassion and peace was indeed the message of Palestine. And in generation after generation, the Palestinian Arab people gave of itself unsparingly in the valiant battle for liberation and homeland. For what has been the unbroken chain of our people's rebellions but the heroic embodiment of our will for national independence. And so the people was sustained in the struggle to stay and to prevail.

When in the course of modern times a new order of values was declared with norms and values fair for all, it was the Palestinian Arab people that had been excluded from the destiny of all other peoples by a hostile array of local and foreign powers. Yet again had unaided justice been revealed as insufficient to drive the world's history along its preferred course.

And it was the Palestinian people, already wounded in its body, that was submitted to yet another type of occupation over which floated that falsehood that "Palestine was a land without people." This notion was foisted upon some in the world, whereas in Article 22 of the Covenant of the League of Nations (1919) and in the Treaty of Lausanne (1923), the community of nations had recognized that all the Arab territories, including Palestine, of the formerly Ottoman provinces, were to have granted to them their freedom as provisionally independent nations.

Despite the historical injustice inflicted on the Palestinian Arab people resulting in their dispersion and depriving them of their right to self-determination, following upon U.N. General Assembly Resolution 181 (1947), which partitioned Palestine into two states, one Arab, one Jewish, yet it is this Resolution that still provides those conditions of international legitimacy that ensure the right of the Palestinian Arab people to sovereignty.

By stages, the occupation of Palestine and parts of other Arab territories by Israeli forces, the willed dispossession and expulsion from their ancestral homes of the majority of Palestine's civilian inhabitants, was achieved by organized terror; those Palestinians who remained, as a vestige subjugated in its homeland, were persecuted and forced to endure the destruction of their national life.

Thus were principles of international legitimacy violated. Thus were the Charter of the United Nations and its Resolutions disfigured, for they had recognized the Palestinian Arab people's national rights, including the right of Return, the right to independence, the right to sovereignty over territory and homeland.

In Palestine and on its perimeters, in exile distant and near, the Palestinian Arab people never faltered and never abandoned its conviction in its rights of Return and independence. Occupation, massacres and dispersion achieved no gain in the unabated Palestinian consciousness of self and political identity, as Palestinians went forward with their destiny, undeterred and unbowed. And from out of the long years of trial in ever-mounting struggle, the Palestinian political identity emerged further consolidated and confirmed. And the collective Palestinian national will forged for itself a political embodiment, the Palestine Liberation Organization, its sole, legitimate representative recognized by the world community as a whole, as well as by related regional and international institutions. Standing on the very rock of conviction in the Palestinian people's inalienable rights, and on the ground of Arab national consensus and of international legitimacy, the PLO led the campaigns of its great people, molded into unity and powerful resolve, one and indivisible in its triumphs, even as it suffered massacres and confinement within and without its home. And so Palestinian resistance was clarified and raised into the forefront of Arab and world awareness, as the struggle of the Palestinian Arab people achieved unique prominence among the world's liberation movements in the modern era.

The massive national uprising, the intifada, now intensifying in cumulative scope and power on occupied Palestinian territories, as well as the unflinching resistance of the refugee camps outside the homeland, have elevated awareness of the Palestinian truth and right into still higher realms of comprehension and actuality. Now at last the curtain has been dropped around a whole epoch of prevarication and negation. The intifada has set siege to the mind of official Israel, which has for too long relied exclusively upon myth and terror to deny Palestinian existence altogether. Because of the intifada and its revolutionary irreversible impulse, the history of Palestine has therefore arrived at a decisive juncture.

Whereas the Palestinian people reaffirms most definitively its inalienable rights in the land of its patrimony:

> Now by virtue of natural, historical and legal rights, and the sacrifices of successive generations who gave of themselves in defense of the freedom and independence of their homeland;
>
> In pursuance of Resolutions adopted by Arab Summit Conferences and relying on the authority bestowed by international legitimacy as embodied in the Resolutions of the United Nations Organization since 1947;
>
> And in exercise by the Palestinian Arab people of its rights to self-determination, political independence and sovereignty over its territory, The Palestine National Council, in the name of God, and in the name of the Palestinian Arab people, hereby proclaims the establishment of the State of Palestine on our Palestinian territory with its capital Jerusalem (Al-Quds Ash-Sharif).

The State of Palestine is the state of Palestinians wherever they may be. The state is for them to enjoy in it their collective national and cultural identity, theirs to pursue in it a complete equality of rights. In it will be safeguarded their political and religious convictions and their human dignity by means of a parliamentary democratic system of governance, itself based on freedom of expression and the freedom to form parties. The rights of minorities will duly be respected by the majority, as minorities must abide by decisions of the majority. Governance will be based on principles of social justice, equality and non-discrimination in public rights of men or women, on grounds of race, religion, color or sex, and the aegis of a constitution which ensures the rule of law and an independent judiciary. Thus shall these principles allow no departure from Palestine's age-old spiritual and civilizational heritage of tolerance and religious coexistence.

The State of Palestine is an Arab state, an integral and indivisible part of the Arab nation, at one with that nation in heritage and civilization, with it also in its aspiration for liberation, progress, democracy and unity. The State of Palestine affirms its obligation to abide by the Charter of the League of Arab States, whereby the coordination of the Arab states with each other shall be strengthened. It calls upon Arab compatriots to consolidate and enhance the emein reality of state, to mobilize potential, and to intensify efforts whose goal is to end Israeli occupation.

The State of Palestine proclaims its commitment to the principles and purposes of the United Nations, and to the Universal Declaration of Human Rights. It proclaims its commitment as well to the principles and policies of the Non-Aligned Movement.

It further announces itself to be a peace-loving State, in adherence to the principles of peaceful co-existence. It will join with all states and peoples in order to assure a permanent peace based upon justice and the respect of rights so that humanity's potential for well-being may be assured, an earnest competition for excellence may be maintained, and in which confidence in the future will eliminate fear for those who are just and for whom justice is the only recourse.

In the context of its struggle for peace in the land of Love and Peace, the State of Palestine calls upon the United Nations to bear special responsibility for the Palestinian Arab people and its homeland. It calls upon all peace and freedom-loving peoples and states to assist it in the attainment of its objectives, to provide it with security, to alleviate the tragedy of its people, and to help it terminate Israel's occupation of the Palestinian territories.

The State of Palestine herewith declares that it believes in the settlement of regional and international disputes by peaceful means, in accordance with the U.N. Charter and resolutions. With prejudice to its natural right to defend its territorial integrity and independence, it therefore rejects the threat or use of force, violence and terrorism against its territorial integrity or political independence, as it also rejects their use against territorial integrity of other states.

Therefore, on this day unlike all others, November 15, 1988, as we stand at the threshold of a new dawn, in all honor and modesty we humbly bow to the sacred spirits of our fallen ones, Palestinian and Arab, by the purity of whose sacrifice for the homeland our sky has been illuminated and our Land given life. Our hearts are lifted up and irradiated by the light emanating from the much blessed intifada, from those who have endured and have fought the fight of the camps, of dispersion, of exile, from those who have borne the standard for freedom, our children, our aged, our youth, our prisoners, detainees and wounded, all those ties to our sacred soil are confirmed in camp, village, and town. We render special tribute to that brave Palestinian Woman, guardian of sustenance and Life, keeper of our people's perennial flame. To the souls of our sainted martyrs, the whole of our Palestinian Arab people that our struggle shall be continued until the occupation ends, and the foundation of our sovereignty and independence shall be fortified accordingly.

Therefore, we call upon our great people to rally to the banner of Palestine, to cherish and defend it, so that it may forever be the symbol of our freedom and dignity in that homeland, which is a homeland for the free, now and always.

In the name of God, the Compassionate, the Merciful:

Say: 'O God, Master of the Kingdom,
Thou givest the Kingdom to whom Thou wilt,
and seizes the Kingdom from whom Thou wilt,
Thou exalted whom Thou wilt, and Thou
abasest whom Thou wilt; in Thy hand
is the good; Thou are powerful over everything.

Source: "Declaration of Independence" posted at http://www.pna.gov.ps/subject_details2.asp?DocId=98.

3

from *The New York Times Magazine*

The One-State Solution

By Edward Said
January 10, 1999

Given the collapse of the Netanyahu Government over the Wye peace agreement, it is time to question whether the entire process begun in Oslo in 1993 is the right instrument for bringing peace between Palestinians and Israelis. It is my view that the peace process has in fact put off the real reconciliation that must occur if the hundred-year war between Zionism and the Palestinian people is to end. Oslo set the stage for separation, but real peace can come only with a binational Israeli-Palestinian state.

This is not easy to imagine. The Zionist-Israeli official narrative and the Palestinian one are irreconcilable. Israelis say they waged a war of liberation and so achieved independence; Palestinians say their society was destroyed, most of the population evicted. And, in fact, this irreconcilability was already quite obvious to several generations of early Zionist leaders and thinkers, as of course it was to all Palestinians.

"Zionism was not blind to the presence of Arabs in Palestine," writes the distinguished Israeli historian Zeev Sternhell in his recent book, *The Founding Myths of Israel*. "Even Zionist figures who had never visited the country knew that it was not devoid of inhabitants. At the same time, neither the Zionist movement abroad nor the pioneers who were beginning to settle the country could frame a policy toward the Palestinian national movement. The real reason for this was not a lack of understanding of the problem but a clear recognition of the insurmountable contradiction between the basic objectives of the two sides. If Zionist intellectuals and leaders ignored the Arab dilemma, it was chiefly because they knew that this problem had no solution within the Zionist way of thinking."

David Ben-Gurion, for instance, was always clear. "There is no example in history," he said in 1944, "of a people saying we agree to renounce our country, let another people come and settle here and outnumber us." Another Zionist leader, Berl Katznelson, likewise had no illusions that the opposition between Zionist and Palestinian aims could be surmounted. And binationalists like Martin Buber, Judah Magnes and Hannah Arendt were fully aware of what the clash would be like, if it came to fruition, as of course it did.

Vastly outnumbering the Jews, Palestinian Arabs during the period after the 1917 Balfour Declaration and the British Mandate always refused anything that would compromise their dominance. It's unfair to berate the Palestinians retrospectively for not accepting partition in 1947. Until 1948, Jews held only about 7 percent of the land. Why, the Arabs said when the partition resolution was proposed, should we concede 55 percent of Palestine to the Jews, who were a minority in Palestine? Neither the Balfour Declaration nor the mandate ever specifically conceded that Palestinians had political, as opposed to civil and religious, rights in Palestine. The idea of inequality between Jews and Arabs was therefore built into British, and subsequently Israeli and United States, policy from the start.

The conflict appears intractable because it is a contest over the same land by two peoples who always believed they had valid title to it and who hoped that the other side would in time give up or go away. One side won the war, the other lost, but the contest is as alive as ever. We Palestinians ask why a Jew born in Warsaw or New York has the right to settle here (according to Israel's Law of Return), whereas we, the people who lived here for centuries, cannot. After 1967, the conflict between us was exacerbated. Years of military occupation have created in the weaker party anger, humiliation and hostility.

To its discredit, Oslo did little to change the situation. Arafat and his dwindling number of supporters were turned into enforcers of Israeli security, while Palestinians were made to endure the humiliation of dreadful and noncontiguous "homelands" that make up about 10 percent of the West Bank and 60 percent of Gaza. Oslo required us to forget and renounce our history of loss, dispossessed by the very people who taught everyone the importance of not forgetting the past. Thus we are the victims of the victims, the refugees of the refugees.

Israel's raison d'etre as a state has always been that there should be a separate country, a refuge, exclusively for Jews. Oslo itself was based on the principle of separation between Jews and others, as Yitzhak Rabin tirelessly repeated. Yet over the past 50 years, especially since Israeli settlements were first implanted on the occupied territories in 1967, the lives of Jews have become more and more enmeshed with those of non-Jews.

The effort to separate has occurred simultaneously and paradoxically with the effort to take more and more land, which has in turn meant that Israel has acquired more and more Palestinians. In Israel proper, Palestinians number about one million, almost 20 percent of the population. Among Gaza, East Jerusalem and the West Bank, which is where settlements are the thickest, there are almost 2.5 million Palestinians. Israel has built an entire system of "bypassing" roads, designed to go around Palestinian towns and villages, connecting settlements and avoiding Arabs. But so tiny is the land area of historical Palestine, so closely intertwined are Israelis and Palestinians, despite their inequality and antipathy, that clean separation simply won't, can't really, occur or work. It is estimated that by 2010 there will be demographic parity. What then?

Clearly, a system of privileging Israeli Jews will satisfy neither those who want an entirely homogenous Jewish state nor those who live there but are not Jewish. For the former, Palestinians are an obstacle to be disposed of somehow; for the latter, being Palestinian in a Jewish polity means forever chafing at inferior status. But Israeli Palestinians don't want to move; they say they are already in their country and refuse any talk of joining a separate Palestinian state, should one come into being. Meanwhile, the impoverishing conditions imposed on Arafat are making it difficult for him to subdue the highly politicized inhabitants of Gaza and the West Bank. These Palestinians have aspirations for self-determination that, contrary to Israeli calculations, show no sign of withering away. It is also evident that as an Arab people—and, given the despondently cold peace treaties between Israel and Egypt and Israel and Jordan, this fact is important—Palestinians want at all costs to preserve their Arab identity as part of the surrounding Arab and Islamic world.

For all this, the problem is that Palestinian self-determination in a separate state is unworkable, just as unworkable as the principle of separation between a demographically mixed, irreversibly connected Arab population without sovereignty and a Jewish population with it. The question, I believe, is not how to devise means for persisting in trying to separate them but to see whether it is possible for them to live together as fairly and peacefully as possible.

What exists now is a disheartening, not to say, bloody, impasse. Zionists in and outside Israel will not give up on their wish for a separate Jewish state; Palestinians want the same thing for themselves, despite having accepted much less from Oslo. Yet in both instances, the idea of a state for "ourselves" simply flies in the face of the facts: short of ethnic cleansing or "mass transfer," as in 1948, there is no way for Israel to get rid of the Palestinians or for Palestinians to wish Israelis away. Neither side has a viable military option against the other, which, I am sorry to say, is why both opted for a peace that so patently tries to accomplish what war couldn't.

The more that current patterns of Israeli settlement and Palestinian confinement and resistance persist, the less likely it is that there will be real security for either side. It was always patently absurd for Netanyahu's obsession with security to be couched only in terms of Palestinian compliance with his demands. On the one hand, he and Ariel Sharon crowded Palestinians more and more with their shrill urgings to the settlers to grab what they could. On the other hand, Netanyahu expected such methods to bludgeon Palestinians into accepting everything Israel did, with no reciprocal Israeli measures.

Arafat, backed by Washington, is daily more repressive. Improbably citing the 1936 British Emergency Defense Regulations against Palestinians, he has recently decreed, for example, that it is a crime not only to incite violence, racial and religious strife but also to criticize the peace process. There is no Palestinian constitution or basic law: Arafat simply refuses to accept limitations on his power in light of American and Israeli support for him. Who actually thinks all this can bring Israel security and permanent Palestinian submission?

Violence, hatred and intolerance are bred out of injustice, poverty and a thwarted sense of political fulfillment. Last fall, hundreds of acres of Palestinian land were expropriated by the Israeli Army from the village of Umm al-Fahm, which isn't in the West Bank but inside Israel. This drove home the fact that, even as Israeli citizens, Palestinians are treated as inferior, as basically a sort of underclass existing in a condition of apartheid.

At the same time, because Israel does not have a constitution either, and because the ultra-Orthodox parties are acquiring more and more political power, there are Israeli Jewish groups and individuals who have begun to organize around the notion of a full secular democracy for all Israeli citizens. The charismatic Azmi Bishara, an Arab member of the Knesset, has also been speaking about enlarging the concept of citizenship as a way to get beyond ethnic and religious criteria that now make Israel in effect an undemocratic state for 20 percent of its population.

In the West Bank, Jerusalem and Gaza, the situation is deeply unstable and exploitative. Protected by the army, Israeli settlers (almost 350,000 of them) live as extraterritorial, privileged people with rights that resident Palestinians do not have. (For example, West Bank Palestinians cannot go to Jerusalem and in 70 percent of the territory are still subject to Israeli military law, with their land available for confiscation.) Israel controls Palestinian water resources and security, as well as exits and entrances. Even the new Gaza airport is under Israeli security control. You don't need to be an expert to see that this is a prescription for extending, not limiting, conflict. Here the truth must be faced, not avoided or denied.

There are Israeli Jews today who speak candidly about "post-Zionism," insofar as after 50 years of Israeli history, classic Zionism has neither provided a solution to the Palestinian presence, nor an exclusively Jewish presence. I see no other way than to begin now to speak about sharing the land that has thrust us together, sharing it in a truly democratic way, with equal rights for each citizen. There can be no reconciliation unless both peoples, two communities of suffering, resolve that their existence is a secular fact, and that it has to be dealt with as such.

This does not mean a diminishing of Jewish life as Jewish life or a surrendering of Palestinian Arab aspirations and political existence. On the contrary, it means self-determination for both peoples. But it does mean being willing to soften, lessen and finally give up special status for one people at the expense of the other. The Law of Return for Jews and the right of return for Palestinian refugees have to be considered and trimmed together. Both the notions of Greater Israel as the land of the Jewish people given to them by God and of Palestine as an Arab land that cannot be alienated from the Arab homeland need to be reduced in scale and exclusivity.

Interestingly, the millennia-long history of Palestine provides at least two precedents for thinking in such secular and modest terms. First, Palestine is and has always been a land of many histories; it is a radical simplification to think of it as principally or exclusively Jewish or Arab. While the Jewish presence is longstanding, it is by no means the main one. Other tenants have included Canaanites, Moabites, Jebusites and Philistines in ancient times, and Romans, Ottomans, Byzantines and Crusaders in the modern ages. Palestine is multicultural, multiethnic, multireligious. There is as little historical justification for homogeneity as there is for notions of national or ethnic and religious purity today.

Second, during the interwar period, a small but important group of Jewish thinkers (Judah Magnes, Buber, Arendt and others) argued and agitated for a binational state. The logic of Zionism naturally overwhelmed their efforts, but the idea is alive today here and there among Jewish and Arab individuals frustrated with the evident insufficiencies and depredations of the present. The essence of their vision is coexistence and sharing in ways that require an innovative, daring and theoretical willingness to get beyond the arid stalemate of assertion and rejection. Once the initial acknowledgment of the other as an equal is made, I believe the way forward becomes not only possible but also attractive.

The initial step, however, is a very difficult one to take. Israeli Jews are insulated from the Palestinian reality; most of them say that it does not really concern them. I remember the first time I drove from Ramallah into Israel, thinking it was like going straight from Bangladesh into Southern California. Yet reality is never that neat.

My generation of Palestinians, still reeling from the shock of losing everything in 1948, find it nearly impossible to accept that their homes and farms were taken over by another people. I see no way of evading the fact that in 1948 one people displaced another, thereby committing a grave injustice. Reading Palestinian and Jewish history together not only gives the tragedies of the Holocaust and of what subsequently happened to the Palestinians their full force, but also reveals how in the course of interrelated Israeli and Palestinian life since 1948, one people, the Palestinians, has borne a disproportional share of the pain and loss.

Religious and right-wing Israelis and their supporters have no problem with such a formulation. Yes, they say, we won, but that's how it should be. This land is the land of Israel, not of anyone else. I heard those words from an Israeli soldier guarding a bulldozer that was destroying a West Bank Palestinian's field (its owner helplessly watching) to expand a bypass road.

But they are not the only Israelis. For others, who want peace as a result of reconciliation, there is dissatisfaction with the religious parties' increasing hold on Israeli life and Oslo's unfairness and frustrations. Many such Israelis demonstrate against their Government's Palestinian land expropriations and house demolitions. So you sense a healthy willingness to look elsewhere for peace than in land-grabbing and suicide bombs.

For some Palestinians, because they are the weaker party, the losers, giving up on a full restoration of Arab Palestine is giving up on their own history. Most others, however, especially my children's generation, are skeptical of their elders and look more unconventionally toward the future, beyond conflict and unending loss. Obviously, the establishments in both communities are too tied to present "pragmatic" currents of thought and political formations to venture anything more risky, but a few others (Palestinian and Israeli) have begun to formulate radical alternatives to the status quo. They refuse to accept the limitations of Oslo, what one Israeli scholar has called "peace without Palestinians," while others tell me that the real struggle is over equal rights for Arabs and Jews, not a separate, necessarily dependent and weak Palestinian entity.

The beginning is to develop something entirely missing from both Israeli and Palestinian realities today: the idea and practice of citizenship, not of ethnic or racial community, as the main vehicle for coexistence. In a modern state, all its members are citizens by virtue of their presence and the sharing of rights and responsibilities. Citizenship therefore entitles an Israeli Jew and a Palestinian Arab to the same privileges and resources. A constitution and a bill of rights thus become necessary for getting beyond Square 1 of the conflict because each group would have the same right to self-determination; that is, the right to practice communal life in its own (Jewish or Palestinian) way, perhaps in federated cantons, with a joint capital in Jerusalem, equal access to land and inalienable secular and juridical rights. Neither side should be held hostage to religious extremists.

Yet feelings of persecution, suffering and victimhood are so ingrained that it is nearly impossible to undertake political initiatives that hold Jews and Arabs to the same general principles of civil equality while avoiding the pitfall of us-versus-them. Palestinian intellectuals need to express their case directly to Israelis, in public forums, universities and the media. The challenge is both to and within civil society, which has long been subordinate to a nationalism that has developed into an obstacle to reconciliation. Moreover, the degradation of discourse—symbolized by Arafat and Netanyahu trading charges while Palestinian rights are compromised by exaggerated "security" concerns—impedes any wider, more generous perspective from emerging.

The alternatives are unpleasantly simple: either the war continues (along with the onerous cost of the current peace process) or a way out, based on peace and equality (as in South Africa after apartheid) is actively sought, despite the many obstacles. Once we grant that Palestinians and Israelis are there to stay, then the decent conclusion has to be the need for peaceful coexistence and genuine reconciliation. Real self-determination. Unfortunately, injustice and belligerence don't diminish by themselves: they have to be attacked by all concerned.

Source: Edward Said, "The One-State Solution," *The New York Times Magazine*, January 10, 1999.

from *The Los Angeles Times*

4

Timidity Risks More Bloodshed

By Shibley Telhami
May 27, 2001

COLLEGE PARK—Now that the Bush administration has finally reentered Middle East mediation, many analysts' fear of diplomatic failure is pushing them to urge caution. Raising expectations of a breakthrough in the Middle East is certainly foolish, but too much caution will surely not stop the spiraling violence between Israelis and Palestinians. Any delay in restarting peace talks is far more dangerous than any risk of diplomatic failure, for two reasons. The absence of mutual trust virtually guarantees that violence will beget violence without outside diplomatic intervention. More important, the escalation in fighting threatens to transform a nationalist conflict, which is resolvable, into an ethnic-religious one, which is nearly impossible to resolve. Solutions that may be possible today will not be possible after a few more months of bloodshed.

Start with the events of the last two weeks. Israeli soldiers killed five policemen of the Palestinian Authority who had been in charge, among other things, of containing demonstrations and keeping Hamas activists in check. Although they were Yasser Arafat's men, the Palestinian leader's name was not chanted at their funerals. Instead, the name "Brigades," the military wing of Hamas, filled the air. A few days later, Hamas carried out a horrifying suicide bombing in Netanya that killed four Israeli civilians and wounded scores of others. The militant organization claimed the attack was carried out in part to avenge the killings of the policemen. Within hours, Israeli F-16s attacked Palestinian security buildings in Gaza City and the West Bank, the first such use of the jet fighters on the West Bank since the 1967 war.

As a result, a seemingly helpless Arafat was further weakened at home, support for Hamas increased even more and the ability of both Israel and the Palestinian Authority to deter future attacks was significantly eroded. Whatever the goals of Prime Minister Ariel Sharon's government—punitive revenge, stopping the intifada or simply bringing Arafat down—the outcome is the same: an erosion of Israel's ability to deter violence. No person or entity can quickly replace the Palestinian Authority. If the authority collapsed, political fragmentation would likely follow, and Israeli military force is simply not as effective against moving targets.

The current cycle of carnage is tragic enough, but it would worsen significantly without the Palestinian Authority because the character of the conflict would shift from nationalist to ethnic-religious. The breakdown of the Israeli-Palestinian negotiations signaled the beginning of this transformation. Because the conflict was framed as one between two national movements content to find expression in limited states, Palestinians and Israelis could drop their maximum and irreconcilable demands. This approach was reinforced in the Oslo accords in 1993 and helped defuse rising religious forces among Israelis and Palestinians.

But faith in the possibility of "two states for two nations" is eroding. The subjects of the final-status negotiations—the status of Jerusalem and Palestinian refugees—are partly to blame. These issues not only impassioned other Arabs and Muslims but also Israeli Arabs who, during confrontations with Israeli security forces last October, suffered many casualties. The dream of coexistence in a democratic state of equal citizens was shattered for Jews and Arabs alike. Within Israel, mainstream Jewish analysts speak of an "Arab problem" and Arab citizens challenge the Jewishness of Israel.

In the Arab world, the conflict is also being partly reframed, not in relation to Israel or Zionism, but to Jews. This is still the exception, but more common than it was just months ago. It's not a return to the old ways. Pan-Arabism, when it prevailed, opposed the "Zionist entity," not Jews. The dominant political culture differentiated Zionism, as an ideology, from Jews and Jewishness. Pan-Arabists erred in not recognizing Jewish nationalism, but they were not, for the most part, anti-Jewish.

Pan-Arabism gave way to a new vision of a world of states. The rise of the Palestine Liberation Organization in 1974, when it was recognized as "the sole legitimate representative of the Palestinian people," signified the end of pan-Arab aspirations and the birth of a norm of independent states. The Palestinian issue became the responsibility of the Palestinians themselves. States like Egypt could pursue their own policy toward Israel. The "Zionist entity" of pan-Arabist rhetoric increasingly became the "state of Israel."

The nationalist framework was further bolstered when the Palestinian-Israeli conflict was placed in a regional context. During the last 10 years, hope for resolving the conflict was tied to a vision of a U.S.-backed regional order based on negotiations, stability and prosperity. This order was strong enough to help repel an assault by a powerful Islamist movement following the end of the Cold War. But the collapse of hope for Palestinian-Israeli peace exposed disillusionment with the order across the region, which has yet to see prosperity or an end to the Iraq dilemma. Hostility toward the U.S. is now at its highest level in a decade. The escalation of bloodshed in the Palestinian territories further fuels criticism of the order and solidifies ethnic and religious identities even as it highlights the impotence of Arab states to stop the violence.

The transformation of the character of the Israeli-Palestinian conflict is rapidly taking place, but it is still at its beginning. Stopping it requires bolstering, not eroding, the nationalist framework that enables compromise. In concrete terms, the Palestinian Authority must be preserved. For this to happen, Palestinians must come to grips with the reality of Jewish nationalism, and Israelis must not make territorial claims on the West Bank based on religious narrative. Compromise has to be based on U.N. resolutions and the security concerns of both sides. Above all, the temptation to use overwhelming force must be resisted. Otherwise, the ugliness of religious and ethnic conflict will eliminate any hope for peace.

Source: Shibley Telhami, "Timidity Risks More Bloodshed," *The Los Angeles Times*, May 27, 2001.

5

from the Turkish Ministry of Foreign Affairs

Is There a "Kurdish Question" in Turkey?

As the first melting pot and encounter point of many different civilizations and cultures, present-day Turkey contains a multitude of ethnic, religious and cultural elements. Turkey is proud of its great heritage. This centuries-long shared way of life is perfectly second-nature for the people of Turkey.

Yet, different ethnic identities, including the Kurdish, are acknowledged and accepted in Turkey. The state does not categorize its citizens along ethnic lines nor does it impose an ethnic identity on them. Population censuses in Turkey never count people on the basis of their ethnic origins. But, this does not prevent an individual citizen to identify himself or herself in terms of a specific ethnic category. That is a private affair and ultimately a matter of personal preference. Public expressions and manifestations of ethnic identity are prohibited neither by law nor by social custom. Folklore is rich and colorful and local variations, customs and traditions are protected and supported.

Turkey is a constitutional state governed by the rule of law. Democracy rests on a parliamentary system of government, respect for human rights and on the supremacy of law. Multi-party politics, free elections, a growing tradition of local government mark the democratic way of life in Turkey.

Constitutional citizenship is one of the principles upon which the Turkish state was founded. The Turkish Constitution stipulates that the State and the Nation are indivisible, and that all citizens irrespective of their ethnic, racial or religious origin, are equal before the law.

For historical and cultural reasons, and under stipulations of binding international treaties, the concept of "minority" applies specifically to certain groups of non-Moslem citizens. In fact, the social fabric of Turkey is a unique real life case of the OSCE principle that "not all ethnic, cultural, linguistic or religious differences necessarily lead to the creation of national minorities." Our citizens of Kurdish ethnic origin are not discriminated against and they feel themselves to be equal members of the society. Many have risen to the highest positions in the Republic. They share the same opportunities and the same destiny as the rest of the population.

Ethnicity is not a factor in the political geography of Turkey. That is, the predominant majority of the Turkish citizens of Kurdish descent live in western Turkey, with the greatest concentration being in Istanbul. Even in eastern and southeastern Turkey, the Turkish citizens of Kurdish ethnic origin do not constitute a majority. The unitary structure of the State reflects the equality and togetherness of different geographic regions of Turkey.

Therefore, it is simply neither understandable nor acceptable for Turkey to discuss "the respect for social, economic and legitimate political aspirations of Kurds" as if the Turkish citizens of Kurdish ethnic descent constitute a different and separate community. They are citizens of a nation that has been sharing for centuries the same values with respect to language, religion, culture and patriotic identity, common history and the will for a mutual future.

It is of cardinal importance to differentiate between a militant organization, which resorts systematically to terrorism as well as all kinds of organized crime, and the phenomenon of Kurdish ethnicity. It is evident that our citizens of Kurdish ethnic origin are law-abiding people. Most of them live in western Turkey, drawn by economic attraction. They are of their own choice integrated into the society and its economic, social and cultural aspects. In Turkey, citizens of all ethnic origins can rise to the highest political positions and ranks such as cabinet ministers and members of parliament. Throughout the centuries, much mixing has taken place through intermarriages. Progress in industrial, cultural and social fields, as well as urbanization, has also contributed to the voluntary and natural process of integration.

The population in southeast Anatolia, like our citizens in other regions of the country, participate fully in the political life of Turkey; they freely make their voices heard in local administrations, in the municipalities, the Parliament, and the central government through elected representatives. It is nothing out of the ordinary for the individuals of different ethnic origins to participate in the political life of the country. Even the most militant circles concede the fact that there are no obstacles to social mobility of individuals from different ethnic origins to any profession or career, whether public or private.

The fundamental rights and freedoms of all Turkish citizens are secured by the relevant provisions of the Constitution. However, those rights have been threatened by the PKK, creating terror among the populace.

None of our citizens of Kurdish ethnic origin, notwithstanding allegations to the contrary, who publicly or politically asserts his/her Kurdish ethnic identity risks harassment or persecution. However, acts or statements made against the "territorial integrity" of Turkey are subject to legal prosecution under the law. If these allegations were true, none of the publications in Kurdish whose contents are full of assertions of Kurdish ethnic identity would have been tolerated by the authorities.

In the same vein, Turkey is often accused of refusing to negotiate with the terrorist organization PKK. These accusations contradict the fundamental rules of international law. Negotiating with a terrorist organization, responsible for thousands of murders, would be tantamount to justifying and encouraging terrorism.

Source: "Is There a 'Kurdish Question' in Turkey?" posted at www.mfa.gov.tr/grupe/eh/eh01/pkk11.htm.

6

from the Kurdistan Workers Party (PKK)

Party Program

The Tasks of the Revolution In Kurdistan

Our revolution is necessary to arrive, through socialism, at our ultimate goal, namely a classless society. The revolution will realize the following tasks:

A. An end to Turkish colonialism and all forms of imperialist domination over Kurdistan. In order to do this, the following must be achieved.

1. A national, united front of workers, peasants, intellectuals, and other classes and sectors must be developed and expanded.

2. In order to secure the complete organization of the population, associations for workers, peasants, youths, women, and others must be further expanded.

3. The people's war, the fundamental form of struggle against colonialism, must continue on the path to victory, and its base organizations and people's army must be further developed.

4. Conflicts between the various population groups, which are continually fueled by the colonialists and their local agents, must be stopped. Those tendencies which aim at purely local or nationalist organizing must be defeated.

5. All attempts at "special regional status" or "autonomy" which do not aim to break the colonialism of the Turkish Republic, and which in fact are collaborations with colonialism, must be exposed and a decisive struggle must be waged against them.

6. The property of those people who collaborated with the colonialists during the war and who acted against the people must be confiscated and redistributed to poor and needy members of the society.

7. We must create our own institutions of economics, culture, health, and education in order to stop diseases and to fight against the other forms of destruction of nature and people which have been brought about by colonialism.

B. A national, independent, democratic society, ruled by the people, must be established. In order to do this, the following must be achieved:

8. The nationalization of all institutions, including factories, farms, and other establishments, belonging to the colonialists.

9. The abolishment of the colonialist financial and credit system, to be replaced with an independent financial and credit system.

10. There will not be any foreign military bases or other such privileges anywhere in the country.

11. Land reform must be carried out in the interest of the working class.

12. All debts by farmers to creditors and banks will be annulled.

13. As part of the democratization of the society, all hindrances to the organization of working people in the economic, political, and cultural spheres must be eliminated, and such organizations must be granted legal status.

14. New employment possibilities must be created for the workers, bearing in mind that the physical and mental development of workers is of great importance. Furthermore, the work day shall be 8 hours long.

15. The colonialist legal system will be disbanded and replaced by a democratic judicial system.

16. All forms of oppression against women will be stopped, and the equal status of women and men in the society will be realized in all areas of social and political life. Women, who possess an enormous social revolutionary dynamic, will be mobilized towards this aim.

17. All forms of oppression against minority groups, be they national or religious, will be stopped; without lapsing into nationalism, all cultures will be guaranteed cultural freedom; the various cultures in the country will be viewed as a valuable asset.

C. An independent economic structure must be built up. In order to do this, the following must be achieved:

18. The economy will be centrally planned.
19. Public ownership must be differentiated from state capitalism, and state capitalism must be resisted. The guiding principle which is to be realized is that in science, politics, and the production of the society, you can only take as much as you give.
20. In the realm of public property, an emphasis will be placed upon the development of heavy industries.
21. There will be public control over resources, transportation, trade, finance, and means of mass communication.
22. Farmers will be encouraged to form collectives and they will be supported in this effort.

D. In place of colonialist education and culture, a national educational and cultural system must be established. All dialects of the Kurdish language will be allowed to develop, although one will be made into the national language. With respect to the Kurdish language, literature, and history, intensive research will be carried out, and to aid in this, research institutes will be created.

23. The entire population will be able to learn to read and write.

E. For revolution and unity in Kurdistan, the following are needed:

24. The revolution in the various parts of Kurdistan is primarily the task of the people living in those regions.
25. Efforts at pushing through certain reforms under the guise of "autonomy," using the means of the colonialist state, are to be resisted in all parts of Kurdistan.
26. Efforts must be made to increase the support and solidarity between the revolutionary forces in the different parts of Kurdistan.
27. Efforts must be made to bring success to the revolutionary line in all parts of Kurdistan.
28. The basis of unity will be the right to self-determination for the people of all the various parts of Kurdistan.
29. The democratic rights of Kurdish people spread out across the countries of the world must be guaranteed. Kurds living abroad should join with progressive humanity and with the struggle in Kurdistan. Preparations should be made to facilitate these people's return home to Kurdistan.

F. With respect to relations with neighboring peoples and international questions, we must apply the principles of proletarian internationalism. In order to do this, the following must be achieved:

30. Due to the division of Kurdistan between different countries, all relations with revolutionary forces among neighboring peoples will be based on the assumption that all revolutionary movements are themselves responsible for the revolution in their own country; on this basis, different forms of joint struggle at various levels must be developed.
31. Unity with neighboring peoples must be based on the notion that all peoples are independent and free. All forced unities which are not based on this notion must be resisted. Relations with neighboring peoples, in particular with the people of Turkey, will be developed within our vision of a "Federation of the Middle East."
32. Relations with independent countries and their national liberation movements, cooperation with movements of the working class and revolutionary forces across the world, and solidarity with democratic, anti-fascist, environmental, and humanitarian circles must be built up.

January 24, 1995
5th Congress, Kurdistan Workers Party (PKK)

7

from the Kurdish Democratic Party—Iraq

General Information

Revised October 10, 2002

- The KDP combines Democratic ideology and wants to form a society where everyone in Kurdistan can gain freedom, independence and confidence.
- The KDP adopts new policies and strategy which corresponds to the changes in Kurdish and international politics.
- We work hard to strengthen the basis of the Federal state and Kurdistan National Assembly which was made by the sacrifices of thousands of our people and we work hard to mobilize international support for it.
- We have to defend our people's democratic achievements. It is the hope for millions of Kurds who still live under oppression.
- Today, KDP is becoming the strongest political party in Kurdistan.
- KDP has shown to the Kurdish people at home and abroad its commitment to democracy. It illustrated its self-confidence.
- People are having high expectation from the KDP leaders who they believe can lead them safely in this worried times.
- The KDP will continue to support and enhance the democratic process taking place in the country and embodied in the Parliament and Regional Government. It also endeavors to consolidate the unity of the Kurdish people and strengthening the coalition with the other parties in the Parliament and the government.
- The KDP supports the struggle of Kurdish people in Turkey, Iran, Syria and Russia for their just national rights and reiterates its support for the way in which they determine their own future within the state they live in and agreements they reached within their central governments.
- The KDP calls for a national congress of all Kurdish political parties to be convened for setting up future policies and strategies in consultation.
- The KDP believes that the Kurds are one nation. We put forth effort for the strengthening of brotherly relations between our party and the democratic parties and organizations in all parts of Kurdistan. Kurdistan was divided after World War I with no regard to the demands of the Kurdish people for self-determination. The feeling of solidarity and sympathy among people in different parts of Kurdistan is natural and spontaneous, and every uprising that has ever taken place in any part of Kurdistan has always been supported by Kurds in the other parts. The refusal of the ruling governments to recognize this fact in the different regions of Kurdistan still does not change the realities of the situation. The Kurdish people represent the largest single ethnic group in the world to lack their national rights and they are third most populous nation in the middle East; they should therefore play an important role in the whole movement of the oppressed people who struggle for their democratic and national rights and for self-determination.
- Kurds know that the KDP is a party of long standing, founded at the end of World War II. KDP builds upon a continuous struggle and on the personal sacrifice of thousands of Kurdish martyrs, [and] has a wider perspective with the adoption of a completely new strategy for the future.
- The KDP thanks international efforts by the UN, the Allied Forces, and other international organizations and NGO's for their humanitarian aid. The KDP calls upon these bodies to continue their support and urges them to provide aid for reconstruction and development so that the Kurds can become self-sufficient. It calls for the ending of economic [blocks] on Iraqi Kurdistan and the unfreezing of Iraqi assets abroad of which 30% should be used for the reconstruction of Kurdistan or any future financial or trades agreement Iraq might conclude under the supervision of the UN.

Source: Kurdish Democratic Party, 2002. www.kdp.pp.se/generalx.html.

from the American Kurdish Information Network

About AKIN

The American Kurdish Information Network (AKIN) is a non-profit, tax-exempt organization established in 1993 to serve the information needs of this country relative to the Kurds. It does this to fulfill two needs: ours, to hold a mirror to the events that are taking place in the Kurdish regions of Turkey, Iran, Iraq and Syria; and those of the American public, to enable them to have access to collected materials on the subject.

AKIN hopes to function as a bridge over which friendships and knowledge are exchanged to the benefit of both peoples. Its aim is to increase awareness about the Kurds. AKIN is comprised of Kurdish Americans and recent refugees from all walks of life: entrepreneurs, blue- and white-collar workers, students, and American supporters. AKIN collects, translates, and disseminates information about the Kurds. It hopes to become a valuable resource center for policy makers, scholars, and students of the region. At the same time, it seeks to promote understanding between the Kurds and the Americans. In Kurdistan, AKIN wants the killings to stop, peace to prevail, and the will of the people to be respected and accepted.

> *One of the greatest tragedies in Turkey's history is happening now. Our Kurdish brothers are being slaughtered, and apart from a couple of hesitant voices, no one is standing up and demanding to know what the Government is doing.... What will come of all this?*
> —Yasar Kemal, perennial candidate for the Nobel Prize in Literature

AKIN:

- provides commentary on the situation of the Kurds in Kurdistan;
- exposes every form of human rights violation and repression against the Kurdish people;
- informs the public about the political and cultural developments in Kurdistan.

AKIN Aims:

- to disseminate information to the press and media;
- to solicit the support of human rights organizations;
- to secure the interest of Congress, political leaders, and other democratic and progressive forces;
- to publish documents, reports, and books related to the politics and culture of Kurdistan;
- to provide information and give advice to journalists, representatives, and human rights activists intending to visit the region.

AKIN Publications Include:

- press releases and newsletters;
- reports from government officials, lawyers, and human rights organizations concerning human rights abuses in Kurdistan.

> *We must never again leave millions of men, women, and children at the mercy of the Turkish army.*
> —Danielle Mitterand, wife of the former President of France

AKIN Calls For:

- an independent international human rights investigation into widespread human rights abuses committed against Kurds;
- a UN special investigation into torture, arbitrary detention, and extra-judicial executions, and for the perpetrators of these grave violations to be brought to trial;
- the Kurdish people to be allowed to exercise their fundamental universally-acknowledged right to self-determination;
- a negotiated settlement to the Kurdish question, with the governments with jurisdiction over the Kurds agreeing to an open and unconditional dialogue with representatives of the Kurdish people;
- the U.S., Canada, and all European countries to immediately stop supplying weapons to Turkey;
- all nations to end the deportation of Kurdish refugees;
- tourists to boycott Turkey as a holiday destination.

> *... Kurdish people throughout southern Turkey, very ordinary, humble people, have decided they would rather die standing up than spend their lives on their knees ... it was the most extraordinary and inspiring sight of my life.... I hope one day they have the kind of freedoms we take for granted.*
> —Michael Ignatieff, writer and broadcaster

Source: "About AKIN," posted at http://kurdistan.org. www.kurdistan.org

9

from **Mustafa Aydin**

New Geopolitics of Central Asia and the Caucasus: Causes of Instability and Predicament

As the disintegration of the USSR became imminent, fifteen diverse republics simultaneously embarked on the task of building the institutions necessary for independent statehood. As they did so, national minorities rediscovered long-suppressed identities and sought new rights. While the process of nation and state building in the western republics of the former Soviet Union was a quite straight forward matter and went smoothly, it has been a slow and agonising experience in Central Asia and the Caucasus, involving both domestic and regional rivalries as well as international influences and pressures.

The main question is how the newly independent states of Central Asia and the Caucasus are responding to the strains of this transition in their domestic politics and external relations. In general terms, the two regions have dealt with the post-Soviet transition in different ways, and their divergent paths have resulted in different levels of conflict.

To a large extent, Central Asia has thus far avoided major violent upheavals, with the exception of Tajikistan. This relative lack of tension could be attributed to the fact that all of the current heads of state in the region, again excluding Tajikistan, have maintained a degree of continuity with the Soviet era, monopolising power and preserving many of the major institutions. However, their "success" so far in addressing the traumas of post-Soviet transition and ensuring short-term stability has often been dependent upon their well being and individual strength, which is not an adequate basis to ensure long-term stability. In fact, some of Central Asia's authoritarian regimes, seen as helpful for regional stability, may actually be concealing fundamental problems, allowing the seeds of future conflicts to grow.

In contrast, the newly emerged Caucasian leaders discarded the Soviet political tradition and the legacy of the old regime, and instead tried to create their own power bases and institutions. However, the new leaders who earlier nurtured the independence process or came to power immediately after independence, like Zviad Gamsakhurdia in Georgia and Abulfaz Elchibey in Azerbaijan, with their extreme nationalist rhetoric, were lacking both government experience and underlying connections to the local élite and power brokers. Consequently, their challenges to the existing political order resulted in a number of violent clashes, upheavals and, in some cases, civil war, which has over the past decade overwhelmed the Caucasus.

[A] number of dynamic domestic and cultural factors, interacting with outside forces and external influences, has caused confrontation and violence in both Central Asia and the Caucasus. "Certain variables are prominent at different stages, but there are always many underlying factors causing a particular situation to erupt." Therefore, we outline below the factors that may yet come to prominence as the particular situation demands, but all exist in the region even if some of them have remained dormant hitherto as a destabilising factor.

In addition to the challenges of economic and political transition faced by the other newly independent states of the former Soviet Union, the Central Asian and Caucasian states have had to contend with populations searching for and developing a sense of national identity. Thus, from the first day of their independence, they faced the all-imposing necessity of replacing the now "discredited" socialist ideology and its social and economic model with a new thinking that could also help them to define their separate "identities."

Although Central Asia in general and the Caucasus in particular have a long and rich history, and various levels of identification are discernable among the people, the individual states as they arose from communist domination, especially in Central Asia, had no sense of their separate identities in the modern sense. Before the Russian conquest, people mainly identified themselves with their family, clan, tribe, locality and sometimes religion. The creation of five union republics in Central Asia and three in the Caucasus by the Soviet rule, on the other hand, further complicated the issue of national identities. The borders of the union republics, especially in Central Asia, did not seek to create homogeneous republics or confirm with historic quasi-identities. Rather, they divided people and shattered whatever identity and "sense of belonging" existed hitherto, and attempted to replace them with identities flowing from officially recognised republic borders.

The product of this "nationality engineering" was a poisonous mixture of various local, tribal and ethnic groups. Even a casual look today at "the ethnic overlap from one state to another as well as artificial nature of the boundaries between them" clearly indicates to potential crises based on nationality questions for nearly all the Central Asian and Caucasian states, which could easily "destroy whatever political equilibrium exists both within and between them." During the Soviet era, strict totalitarian rule and suppression kept the destabilising character of ethnic and religious diversity under control. However, the root causes of instability were never dealt with, which eventually contributed to the region's turmoil as the forces of destruction were unleashed following the collapse of the Soviet Union without providing adequate mechanisms to cope with them.

As already stressed, the continuity of leadership in Central Asia after independence has to a certain degree provided means to contain or suppress potential ethnic disputes. But, this does not mean that Central Asian states have an easy ride as far as the ethnicity issues were concerned. "As was the case elsewhere in the USSR," the Moslem peoples of Central Asia "had a difficult time preserving their traditional cultures" under Soviet rule. In the post-Soviet era, with their newly earned independence, titular nationalities began to demand certain improvements in their status and culture. In Kazakhstan and Kyrgyzstan, for example, this has already led to tensions over language policy with titular groups demanding to have their language declared the "state language." This gives a favoured status in areas such as education and civil service hiring to the titular groups and threatens ethnic Russians, who usually do not speak these languages. Consequently, "disputes over language policies" and the "selection of state symbols has increased tensions in the multiethnic states of Central Asia, putting strains on the regional stability."

When, in the early 1920s, the central authorities in Moscow drew the political boundaries of the then union republics of the USSR in Central Asia, they paid no attention to local ethnic identities, which were either weak or "in many cases simply did not exist in any modern sense." A number of territories that had existed as single social, political and economic units for centuries and should each have been considered as a whole were divided among different republics, "diminishing the cohesion of these areas." In contrast to this, many areas that had no previous unity of purpose were allocated to a single republic, causing problems of identity and integration. These policies naturally "exacerbated differences among peoples and regions" within these areas and have "contributed to intra-state regional competition" within, and inter-state tension between, the newly independent states of Central Asia.

Today, because of Soviet policies, each of the Central Asian states has significant minority populations. Ethnic Russians make up approximately 35 percent of Kazakhstan's population, and 20 percent of Kyrgyzstan's. There are over 500,000 Uzbeks living in Kyrgyzstan and over one million in Tajikistan. In return, approximately one million Tajiks and just under a million Kazakhs live in Uzbekistan. Although all the post-Soviet states have agreed to honour the existing borders, with such overlapping populations, there is a considerable potential for future claims and for the spread of conflicts from one country to another.

Though cross-border ethnic issues have been avoided thus far, it is still a concern for the future. The Uzbek populations in Tajikistan and Kyrgyzstan, for example, have already started to call for union with Uzbekistan, motivated not only "by ethnic ties, but also by Uzbekistan's growing economic and strategic importance." Coupled with Uzbekistan's self-image as the centre of "Greater Turkestan," these calls have already caused considerable unease among neighbouring countries. In a similar fashion, extreme Russian nationalists in Kazakhstan argue that northern territories should be simply ceded to Russia. Although, most of the Russians in Kazakhstan seem to favour preservation of the status quo, this could change in the longer term if it become economically more convenient to join Russia or ethnic Kazakh nationalism becomes a burden to them.

Outside the borders of the Commonwealth of Independent States, there are over one million Uzbeks in Afghanistan, some 500,000 Turkmen in each of Afghanistan, Iran, Iraq and Turkey, and about two million Tajiks in Afghanistan. Obviously, the situation creates a number of possibilities for the involvement of neighbouring countries in the Afghan civil war. Moreover, there are about two million Kazakhs living in the Xinjiang region of China, which is populated overwhelmingly by approximately eight million Uighurs, whose 250,000 kin are divided between Kazakhstan, Kyrgyzstan and Uzbekistan. Uighurs are known for their long-standing call for independence from China and the creation of "Eastern Turkestan," the west of which falls within the territories of contemporary Uzbekistan and Kyrgyzstan. There has been periodic ethnic unrest in Xinjiang, which the Chinese used force to deal with. The Chinese are extremely agitated about the prospects of further instability spreading from, or being supported by, the newly independent states of Central Asia.

The same kind of ethnic mixture is present in both Transcaucasia and the Northern Caucasus, and these have already caused open conflicts. Although each of the independent Transcaucasian states has its own dominant titular nation, each also has a significant number of minorities. The situation in the region, in contrast to Central Asia where the overwhelming majority is Sunni Moslem, is further complicated by the diversification of religious faiths that are closely related to the separate national-ethnic identities. The Azeris belong to the Turkic race and the majority of them are Shi'ite Moslems, while a majority of the Armenians and Georgians are believers of two branches of the Eastern Orthodox Church. There are Armenians living in Azerbaijan and Georgia as well as over six million Armenian Diaspora all over the world, mainly concentrated in Russia, the US and France.

Azerbaijan contains within its borders the Nagorno-Karabakh Autonomous Region, with a population of approximately 150,000 that are mostly Armenians, while the country's Nakhichevan Autonomous Republic, which consists mostly of Azeris, is a detached enclave sandwiched between Armenia and Iran. Incidentally, President Heidar Aliyev of Azerbaijan is from the Nakhichevan Autonomous Republic. Moreover, because of historical circumstances, the Azeris have a large number of compatriots living in Iran's northwestern province of Azerbaijan, constituting a sizeable ethnic group in Iran (between 15 million and 25 million people), so that the Azeris have a sense of being artificially divided into two states. This complicates Azerbaijan's relations with Iran, which is deeply concerned about the possibility of rising ethnic separatism among its own Azeris who may favour the establishment of "greater Azerbaijan." Georgia for its part includes the Abkhazia Autonomous Republic, the Adzharia Autonomous Republic and the South Ossetian Autonomous Region, with sizeable minorities in each and secessionist movements in Abkhazia and South Ossetia.

More complicated than this is the existing situation in the North Caucasus, astride the southern boundary of the Russian Federation and the Transcaucasus. With its nineteen native national groups (as the last Soviet census recognised in 1989) and a significant ethnic Russian Diaspora as well as non-titular populations of Cossacks, Nogai and a number of others, the North Caucasus is one of the most ethnically and linguistically diverse regions of the world. Embracing as it does three main linguistic groups and almost all religious nuances, the North Caucasus presents a complicated situation where a number of minorities and more than one titular nationality share the same territory. This already intricate condition is further aggravated by the fact that most of the minorities and the titular nationalities are demanding separation from the states, autonomous republics or the regions they are administratively attached. In fact, the Chechens and Ingushs of Checheno-Ingushetia, created in 1934, separated in December 1994 because of the Ingushs' desire to remain within the Russian republican structure and Chechens' to have independence. In contrast, Dagestan is distinguished by its lack of titular national groups and incorporates ten non-titular national groups that are recognised officially as "Peoples of Dagestan." Obviously, all of the North Caucasian "nationalities" are prone to instability and conflict in future, which makes it very difficult for both Russia and external states to come to an understanding of the regional realities.

Source: Mustafa Aydin, "New Geopolitics of Central Asia and the Caucasus: Causes of Instability and Predicament" from mfa.gov.tr/grupa/sam/20.htm

from the Institute for War & Peace Reporting

10

Uzbekistan: Luli Hit the Road

Gypsies return to nomadic way of life, after failure of attempts to integrate them into society.

by Artur Samari
22 September 2003

Tourists visiting the Uzbek region of Samarkand are currently seeing a lot more Central Asian heritage than they bargained for. Every time a group stops to admire one of the many spectacular ancient monuments or medieval buildings that are scattered across the region, they are accosted by groups of colourfully-clad gypsies pleading for money or food.

Known as the Luli in Uzbekistan, they are an ancient nomadic people who have traveled Central Asia for centuries, only acquiring a more sedentary life in Soviet times. But with the post-independence economic downturn in Uzbekistan, the community has been forced to uproot once more, with many turning to begging in order to make ends meet. This is causing a major headache for the Samarkand police, who have had to deploy greater numbers of guards at historic sites to prevent the gypsies pestering foreign visitors for money.

Many other Luli travel to Kazakstan and Russia in search of a better life. The arrival of a large group in the Russian republic of Komi this month sparked a minor diplomatic incident after a local television channel reported that "Uzbeks" were causing a nuisance by begging in the city of Syktyvkar. The Uzbek embassy in Moscow reacted with fury, and several diplomats traveled to Syktyvkar to persuade the local authorities that Lulis—not Uzbeks—were responsible.

It's a harsh reversal of fortune for the minority, which had prospered under the Soviet regime, working in factories and on farms, and even awarded state honours. But with the onset of the country's economic problems, public sector enterprises collapsed and the gypsies along with millions of other citizens found themselves out of work.

Sayd, a Samarkand gypsy who has recently returned from Moscow, told IWPR sadly, "I used to work in a leather tannery and had a good life ... how I wish I could bring those days back. There is no work here for anybody. Begging was how our ancestors lived, and it will help us to survive. I feel embarrassed, asking strangers for food—but I have to eat."

Professor Khol Nazarov, himself a Luli, who has studied the history of gypsies in Central Asia for nearly four decades, told IWPR that the Soviet authorities had made a concerted effort to integrate the gypsy population in everyday life, employing them in factories and giving them plots of land to cultivate.

Luli children were expected to go to school—and any more than two days of unexplained absence would result in a teacher visiting the errant pupil's home. This strict education policy resulted in many gypsy children going on to study at local universities.

"We were glad that our people escaped the ignorance of their forebears," said Nazarov. "Before long, the Luli had its own intelligentsia, teachers, lawyers and doctors, and even the achievements of the working class were recognised with state awards."

But with the economic downturn, all the progress made during Soviet times is beginning to unravel—many gypsies having little option but to resort to begging. "All we achieved in previous years has been ruined," Nazarov told IWPR.

Many Uzbeks look down on the Luli, seeing vagrancy and begging as a conscious choice. According to Nazarov, his people want to live as ordinary citizens, but struggle to do so because they lack the social support they received during the Soviet era.

The hardships experienced now by the Uzbek gypsies has attracted the attention of Samarkand rights activists, who say the authorities should do more for the community. "At the moment, they don't even have a national cultural centre," said Komil Ashurov of the Samarkand Human Rights Centre.

However, even the most sympathetic activists note that the current economic crisis is also hitting the vast majority of the Uzbek population hard.

Source: Artar Samari, "Uzbekistan: Luli Hit the Road: Gypsies Return to Nomadic Way of Life After Failure of Attempts to Integrate them into Society" Sept. 22, 2003 from iwpr.net/index.pl?archive/rea/rea_200309_236_2_en.txt

11

from the Toda Institute for Global Peace and Policy Research

The Civil War in Tajikistan

By Kamoloudin Abdoullaev

One need see only the surface of events to know that Tajikistan is a country in turmoil. Rather than look at the outlines of the civil war in Tajikistan, this article will analyze the background to the conflict and will try to explain the rise of localism and ethnonationalism and the nature of their impact on the prospects for resolution of the conflict.

Background

One hundred years ago, as a result of the Anglo-Russian rivalry known as "The Great Game," a gigantic strategic barrier was established from the Pacific Ocean to the Caspian Sea, that is, from Manchuria to Mongolia to Xinjiang to Afghanistan to Russian Central Asia. This region includes both dependent and independent countries, multiethnic nation-states, and empires, and mixture of beliefs (Islam, Buddhism, Confucianism, Christianity, shamanism) and peoples (Iranians, Turks, Hans, Mongols, and others). The diversity is such that the region came to be known as a new center of gravity in the world, a whirlpool in which met political currents flowing from China, Russia, India, and the Middle East (Lattimore 1950:3). In the western reaches of this center of gravity, in the southeastern part of Central Asia, lies Tajikistan.

When in the second half of the 19th century Central Asia was drawn into political and economic dependence on Russia, the victor in the competition for the region began elaborating a Eurocentric approach to international affairs in that part of the world. The southern frontier of Central Asia—the current Tajik-Afghan border—was treated as the boundary between the capitalist empire and its semifeudal counterpart. Eurocentrism thus emerged as the justification for European (Russian) sovereignty over the region.

From the other side of the frontier, however, some Muslim circles viewed Tajikistan as the front line in the battle against the "infidels." It was not surprising, therefore, when, after the fall of the U.S.S.R. and the removal of communist control, Tajikistan, as a state on the confluence of the Muslim and post-Soviet worlds, once more became a region of chronic political instability challenging human security in that part of the world.

In 1992 conditions deteriorated into a brutal civil war. Up to 50,000 people have lost their lives, and more than 650,000 people—one-tenth of population—have fled in terror. In addition, more than 35,000 homes were destroyed (Richter 1994). It has been one of the bloodiest and costliest wars in the former U.S.S.R.

Ethnonationalism

Today Tajikistan is a country with an ailing political regime, a weak national identity, and a fragile economy. The fundamental challenge facing Tajikistan is localism (mahalgara'i in Tajik). All attempts to develop the economy and culture in my country are undermined by parochial war between rival gangs, clans, and warlords for political and military dominance. From 1992 to 1995 factions from Kulob, Khujand, and Hisar were on one side (which is currently in power), while the groups from Qarategin and Badakhshan opposed them. Officially both the current Tajik establishment and the opposition condemn mahagara'i and have pledged to eliminate this phenomenon. Regrettably, this particular struggle has yet to succeed. This raises the following questions: How do the building of a nation-state and the rise of ethnic consciousness correlate with the growth of localism in the context of the current situation in Tajikistan? What forms of governance will help Tajiks integrate and avoid violent local cleavage?

Ethnicity vis-à-vis Regionalism

The rejection of totalitarian communism engendered a return to the roots of national identity everywhere in the former Soviet world. The paradigm of class consciousness was replaced by one of ethnicity, and ethnic nationalism emerged as the dominant ideology in all of the post-Soviet countries. Replacing the proletariat and his ally the peasant, the "ethnic person" is gradually becoming a central and dynamic figure in events throughout the former U.S.S.R.

Ethnicity may be said to represent an inheritable group solidarity based on common origin, culture, and historic destiny. In other words, ethnicity emerges as a social instinct of a collective way of life. A positive feature of ethnicity is that it is an important means of group adaptation to the surrounding world, helping it to survive under difficult political and economic conditions. Hence, the instinct of group identity appears as an instinct for national self-preservation. As the Central Asians throughout their long history survived wars, occupations, conquests, forced migrations, revolutions and other violent events, many microethnic solidarity groups, originally defined by geographical origin, incorporated themselves into larger groups.

Today there is a sharp crisis in Tajik identity. Growing localism and the lack of a national consolidation have brought the society and the state to the verge of collapse.

In their search for the cause of Tajikistan's underdevelopment and misfortunes, some Tajik writers have looked to the "northern Tajiks," calling them "the fifth column" of an unnamed neighbor (Uzbekistan). These accusations are meant to incite ill will among the "ethnically pure southern Tajiks" toward the "marginal northern Tajiks" (Masov 1995). Thus, discussions about ethnicity and the challenges to harmonious national development are used to achieve political goals. But harmonious national development cannot proceed in an atmosphere of confrontation, during a struggle against "marginals" for "purity of nation," with the search for enemies within the ethnicity. Creating the conditions for human security and development is an urgent aim of the Tajik state and society. Instead of cultivating ill will among ethnicities, we should be cultivating the collective instinct for self-protection, which could lead the people away from destruction and toward the creation of a common will to build a unified Tajik state. Promoting a national orientation in social development will temper localism and prevent the increasing fragmentation of the country and the society.

Risky Choices

Because ethnicism rejects the value of human individuality, it interferes with basic human rights. The "ethnic" person replaces the "social" one.

So, what outcome can we expect from the rise of ethnic consciousness in Tajikistan? The saying, "Let us be friends. Against whom?" has widespread currency in the present climate. Attempts to unite a Tajiki nation could incite a Tajiki-Uzbeki war, provoked by any of the partisan groups under their mendacious slogans of "national reconciliation." Most likely this war would begin with the persecution of northern Tajiks as Uzbek collaborators. The first dangerous signs of that conflict are already evident in the work by Masov cited above (which has warm supporters both in opposition and government) and in the killings of several intellectuals (Professors Iskhaki, Asimi, and Gulomov) from northern Tajikistan in 1996. This path is unlikely, because historically there is not enough discord between the Tajiks and Uzbeks to support the claims of a few Tajik nationalists.

It is clear that ethnic determinism and the rise of ethnic consciousness contain within them the seeds of destruction: chauvinism, nationalism, and localism as well as internal and international war. Because ethnicism rejects the value of human individuality, it interferes with basic human rights. The "ethnic" person replaces the "social" one.

Paths Toward Peace

Regrettably, both sides in the conflict and their external supporters contribute to the current "clash of civilizations" in Tajikistan. The muscovite generals and Uzbekistani establishment raise the dreaded specter of Muslim unrest to justify their deep involvement in the Tajik conflict. Similarly, coreligionists from the Muslim countries support the Tajik fighters (mujahedin) in their struggle against the present pro-Moscow government in Tajikistan.

What then in today's world are the optimum principles for ethnic peace and development in Tajikistan?

1. I believe that unity with diversity is the only path to peace in Tajikistan. Tajiks must be united; therefore the integrating tendencies of national development should be supported and strengthened. It is also clear that to construct a democratic civil society it is necessary to underscore not group rights but individual civil rights, the subject of which is the "social person." The advocates of ethnonationalism should remember that appeals for national purity and the search for enemies will lead only to confrontation, to the final degradation and disappearance of the ethnicity.
2. A human being must be free to determine his ethnic identity, language, culture, and behavioral model. Neither the state, nor any ethnic or ethnoregional group, nor any political party has the right to force a person to follow a particular ethnocultural tradition. Nor do they have the right to prosecute a person or discriminate against him or her for refusing to follow this tradition. All people have this right of choice, so-called marginals as well as those of "pure blood."
3. The most powerful tool to temper ethnonationalist and secessionist movements is an active and broad-based peace dialogue aimed at bringing about national reconciliation. Participants from both sides in the current official inter-Tajik peace talks were drawn only from among the elite rather than from the peoples of the regions of Tajikistan; hence, they have made little progress toward reaching a real accord.
4. The dichotomy Muslim-non-Muslim and Tajik-non-Tajik should be excluded from Tajik politics. Fortunately, cultural pluralism is not new to Tajiks, who historically have faced a variety of cultural influences and have avoided these dichotomies.

I have no doubt that the key to resolving the Tajik conflict lies in improving relations among Tajiks, Uzbeks, and other ethnic groups living in Tajikistan—in moving from hostility and distrust to tolerance and then to cooperation. Tajikistan will survive by developing its identity as a nation based on the principle of free individual ethnocultural self-determination—this is how I picture the social-political development of Tajikistan. It is to be hoped that Tajiks have not exhausted their potential for survival! This is the only way ethnonationalism can lead to cooperative security in that part of the world.

Source: Kamoloudin Abdoullaevg, "The Civil War in Tajikistan," posted toda.org/publications/peace_policy/p_p_s98/abdoulaev.html

Review Questions

1. What does it mean to have self-determination without nationalism?
2. Why are many Southwest and Central Asian nation-states often considered artificial creations?
3. Why did Nasser's pan-Arabism fail?
4. What was the policy towards nationalities in the former Soviet Union?

Discussion/Essay Questions

1. How does the history of Turkish policy toward the celebration of Newroz (described in the profile of Nesibe Yilmaz that began this module) reflect overall Turkish policy toward the Kurdish movement for self-determination? Given this history, why do you think that HADEP chose not to seek permission from the Turkish government for their Newroz party? Do you think that HADEP made the right choice?
2. Consider the example of the Québecois as a nation, which appears to have achieved a high degree of self-determination without achieving its own nation-state. Could the Québec example serve as a model in Israel-Palestine? Could it serve as a model for the Kurds? How would each of the four authors writing on the Israel-Palestine conflict, and the four writing on the Kurdish conflict, respond to this idea?

3. The ninth reading mentions a "Greater Turkestan," but it does not provide any details about what an organized Greater Turkestan would entail. Given what you know about the states of Central Asia both from the readings as well as from your textbook, do you see any group gaining from pursuit of the "Greater Turkestan" ideal? How could this organization be structured so as to minimize ethnic conflict among and within the current states of Central Asia? Be sure to address self-determination on the part of the region's various nationalities.

List of Readings

1. Theodor Herzl, *Der Judenstaat (The Jewish State)*, originally published in 1896, excerpted and reprinted on the website of the American-Israeli Cooperative Enterprise's Jewish Virtual Library, *http:// www.us-israel.org/jsource/Zionism/herzlex.html.*
2. "Declaration of Independence," November 15, 1998, posted on the website of the Palestinian National Authority, *http://www.pna.gov.ps/ subject_details2.asp?DocId=98.*
3. Edward Said, "The One-State Solution," in *The New York Times Magazine,* January 10, 1999, reprinted on the website of High-Or Partnership Enterprises, *http://members.tripod.com/~theHOPE/said.htm.*
4. Shibley Telhami, "Timidity Risks More Bloodshed," in *The Los Angeles Times,* May 27, 2001, reprinted on the author's website, *http://www.bsos.umd.edu/sadat/pub/oped/Timidity%20risks%20more%20bloodshed.htm.*
5. "Is There a 'Kurdish Question' in Turkey?," posted on the website of the Turkish Ministry of Foreign Affairs, *http://www.mfa.gov.tr/grupe/eh/eh01/pkk11.htm.*
6. "Party Program of the Kurdistan Workers Party (PKK): The Tasks of the Revolution in Kurdistan," January 24, 1995, posted on the website of the World History Archives, *http://www.hartford-hwp.com/archives/51/165.html.*
7. "General Information," revised October 10, 2002, posted on the website of the Kurdish Democratic Party—Iraq, *http://www.kdp.pp.se/generalx.html.*
8. "About AKIN," posted on the website of the American Kurdish Information Network, *http://kurdistan.org/aboutakin.html.*
9. Mustafa Aydin, "New Geopolitics of Central Asia and the Caucasus: Causes of Instability and Predicament, posted on the website of the Turkish Ministry of Foreign Affairs, *http://www.mfa.gov.tr/grupa/sam/20.htm.*
10. Artar Samari, "Uzbekistan: Luli Hit the Road: Gypsies Return to Nomadic Way of Life, After Failure of Attempts to Integrate Them Into Society," September 22, 2003, posted on the website of the Institute for War and Peace Reporting, *http://www.iwpr.net/index.pl?archive/rca/rca_200309_236_2_eng.txt.*
11. Kamoloudin Abdoullaevg, "The Civil War in Tajikistan," posted on the website of the Toda Institute for Global Peace and Policy Research, *http://www.toda.org/publications/peace_policy/p_p_s98/abdoulaev.html.*

Websites for Additional Research

1. Two excellent sources on the topics of nationalism and self-determination are The Nationalism Project (*http://www.nationalismproject.org*) and Self-Determination in Focus (*http://www.selfdetermine.org*). The Nationalism Project is a clearinghouse maintained by university-based academics and contains introductory material on nationalism, bibliographies of academic articles, link pages for information on specific nationalist conflicts, and other material. Self-Determination in Focus is a project of the Washington, D.C.-based Foreign Policy in Focus; it seeks to promote nonnationalist models of self-determination as a route to world peace. Along with material on self-determination as a concept, the website contains introductory material on specific regions.

2. For information on the Kurds, a good starting point is the links page at The Nationalism Project's website, *http://www.nationalismproject.org/links/kurdish.html.* The English-language website of one of the main Iraqi Kurdish political parties (The KDP-Iraq, not linked from the Nationalism Project site) can be found at *http://www.kdp.pp.se.* The website of the Turkish Ministry of Foreign Affairs (*http://www.mfa.gov.tr*) contains many English-language statements on the Kurdish situation from the Turkish government's perspective.

3. The following two websites offer a great deal of information on the Caucasus and Central Asia. The first website, *http://www.cacianalyst.org/index.php*, is the homepage of the Central Asia-Caucasus Institute. This site is updated regularly with news briefs and articles about current events in this region. It includes a searchable database of its articles. For more general information and daily news about Central Asia, visit *http://www.times.kg/*. Follow the "Country Guide" link for basic information on each of the region's countries.

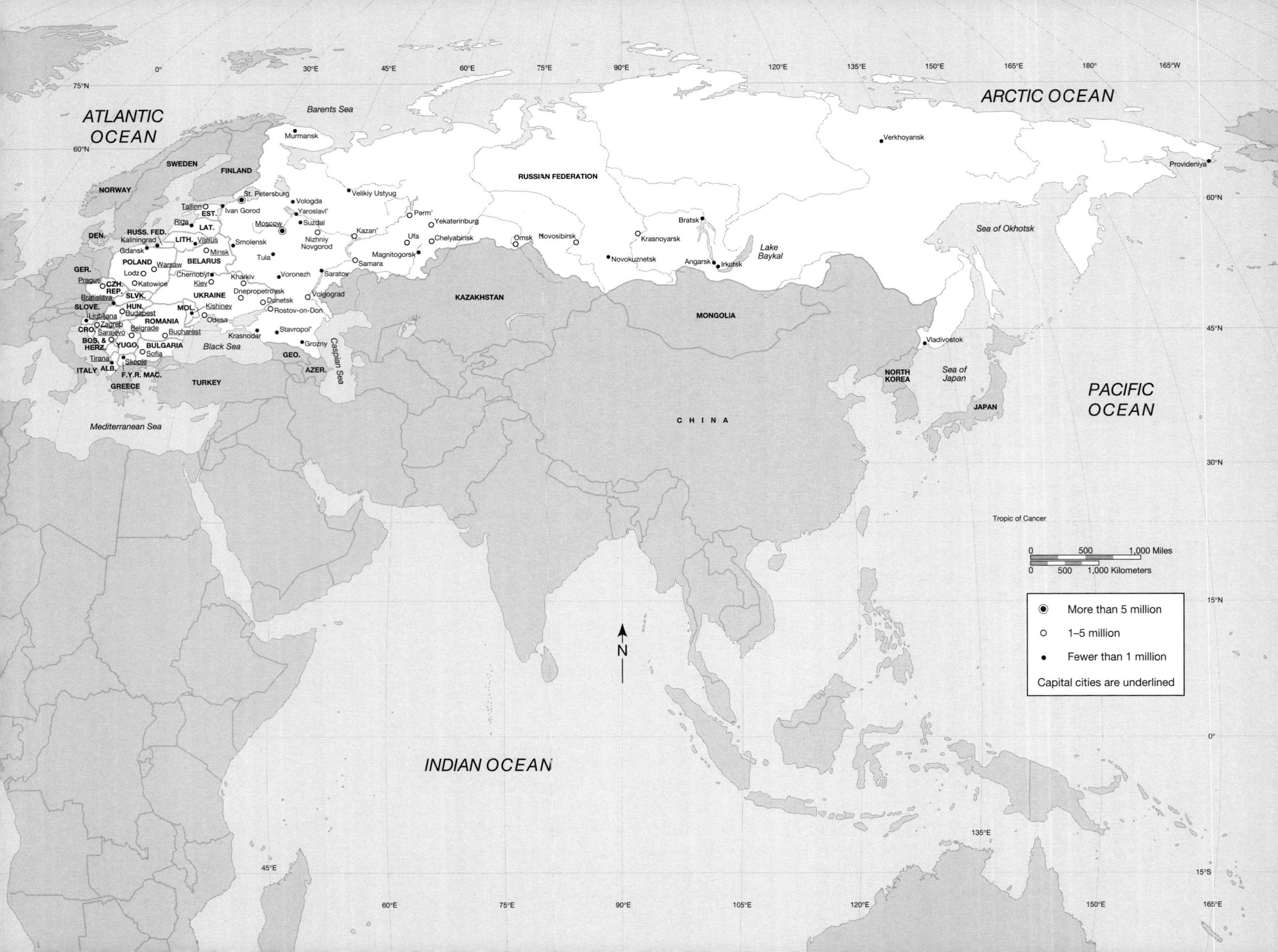

ARCTIC OCEAN
ATLANTIC OCEAN
PACIFIC OCEAN
INDIAN OCEAN
Barents Sea
Sea of Okhotsk
Lake Baykal
Sea of Japan
Caspian Sea
Black Sea
Mediterranean Sea
Tropic of Cancer
RUSSIAN FEDERATION
KAZAKHSTAN
MONGOLIA
CHINA
NORTH KOREA
JAPAN
FINLAND
SWEDEN
NORWAY
DEN.
RUSS. FED.
EST.
LAT.
LITH.
BELARUS
POLAND
GER.
CZH. REP.
SLVK.
UKRAINE
MOL.
HUN.
ROMANIA
SLOVE.
CRO.
BOS. & HERZ.
YUGO.
BULGARIA
ITALY
ALB.
F.Y.R. MAC.
GREECE
TURKEY
GEO.
AZER.
Murmansk
St. Petersburg
Vologda
Yaroslavl'
Suzdal
Velikiy Ustyug
Tallinn
Ivan Gorod
Riga
Moscow
Nizhniy Novgorod
Kazan'
Perm'
Yekaterinburg
Ufa
Chelyabinsk
Magnitogorsk
Samara
Saratov
Kaliningrad
Gdansk
Vilnius
Minsk
Smolensk
Tula
Voronezh
Warsaw
Lodz
Katowice
Prague
Bratislava
Chernobyl
Kiev
Kharkiv
Dnepropetrovsk
Donetsk
Rostov-on-Don
Volgograd
Budapest
Ljubljana
Zagreb
Sarajevo
Belgrade
Bucharest
Kishinev
Odesa
Krasnodar
Stavropol'
Grozny
Tirana
Skopje
Sofia
Omsk
Novosibirsk
Novokuznetsk
Krasnoyarsk
Bratsk
Angarsk
Irkutsk
Verkhoyansk
Provideniya
Vladivostok
N
0 500 1,000 Miles
0 500 1,000 Kilometers
More than 5 million
1–5 million
Fewer than 1 million
Capital cities are underlined
0°
30°E
45°E
60°E
75°E
90°E
105°E
120°E
135°E
150°E
165°E
180°
165°W
75°N
60°N
45°N
30°N
15°N
15°S

European Union Expansion and the Struggle to Define the Eastern Limits of Europe

Companion to Chapter 8
Political Geography

Czechs are the world leaders in beer consumption per head—and many vow that there's no beer like Czech beer. But the fact that many Czech teenagers are regular beer drinkers and some start drinking it even before they are in their teens is allegedly not something that Czech brewers are proud of.

Now the country's five leading brewers, who have approximately 75% of the market, are setting a new trend in the business. They have drafted and adopted a strict new code of ethics that goes beyond the regulations set by the law on advertising. They have committed themselves to an above board advertising strategy that includes not using sex to sell their product, not suggesting that drinking beer makes one "more of a man," and not targeting children and teenagers in their ads.

Czech brewers are trying to change the image of beer. Presently known as a drink consumed in excess by rowdy young men in crowded beer halls, they are trying to market beer as a healthy drink for a wider clientele. *Source:* Jochen Tack/Das Fotoarchiv/Peter Arnold, Inc.

Ivan Sima, secretary of the association of Prague breweries, commented on this new advertising policy in an interview with Radio Prague:

"In terms of sales, [people who drink 10 or 15 half-liter glasses of beer in one evening] make us happy, but it is definitely not something that can be tolerated by society .... In this respect we uphold the motto of the Czech beer association, which says that we want MORE people to drink LESS beer (i.e., for more people to drink it, but in moderation) so in the end we could sell the same volume but consumption would be spread out more evenly ....

"[Regarding the use of sex to sell beer], we use actors in advertising and the new code of ethics does not prohibit the use of actresses, but of course abuse of sexual themes is not allowed ....

"We want to present beer as [something more than a pub drink] .... We want to present beer as a beverage for everyone, to be consumed on a daily basis .... Such a healthy beverage should be drunk by all."

Sima also was asked what impact he thought the Czech Republic's admission into the European Union would have on the Czech beer industry:

"I personally don't expect that too much will change .... Czech beer is a really good product and all the Czech beer exporters sell their brews as super premium lagers—really high priced. They don't sell them in volumes comparable to the cheaper local brands, but I think that the fact that people are willing to pay a high price for Czech beer speaks for itself."

Adapted from: "Czech Brewers Want 'More People to Drink Less Beer'," a radio interview by Daniela Lazarova that aired February 15, 2003 on Radio Prague's "Magazine" program, http://www.radio.cz/en/issue/37569.

Map: Map of Eastern Europe adapted from pp. 98–99 and 152–153 in *World Regions in Global Context* by Marston et al., © 2002. Reprinted by permission of Pearson Education, Inc., Upper Saddle River, NJ. ■

The Cold War Division of Europe

During the Cold War—when the Soviet Union and the United States faced each other as the world's two superpowers—Europe was a divided continent. Western Europe allied itself with the United States under the military umbrella of the North Atlantic Treaty Organization, or NATO. Eastern Europe allied itself with the Soviet Union under the military umbrella of the Warsaw Pact. A few European countries were officially non-aligned, but most of these expressed a strong tilt in one direction or the other. The superpowers, for their part, treated Europe as the likely battleground for any confrontation that was to come, and both sides stationed tens of thousands of troops in the countries bordering the "Iron Curtain" that ran through the middle of Europe, dividing East from West.

During this era, the unofficial geopolitical borders of Europe differed somewhat from those classically found in geography books. Europe was bordered on the west by the Atlantic Ocean, on the north by the Arctic, and on the south by the Mediterranean. The eastern border effectively included all of the Soviet Union, since the Soviet Union was viewed as an aspiring European power that had its sights set on controlling Eastern—and perhaps potentially Western—Europe. The southeastern border effectively included all of Turkey, since this was the southeastern flank of the NATO belt that was designed to contain Soviet expansion. Thus, during the Cold War, Turkey was treated by U.S. military planners as part of the European theatre, even though, by the standard division of continents, the bulk of Turkey lies within Asia. In other words, the military contest between the superpowers defined both the external boundaries of Europe and its most important internal boundary: the Iron Curtain.

Amidst this division of Europe, the idea of Central Europe as a distinct region (or sub-region) faded from popular terminology. Historically, Central Europe had been divided and fought over by the Austro-Hungarians and Germans to the west, the Russians to the east, and the Ottomans (Turks) to the south. These empires represented different religions (Roman Catholic Christianity, Eastern Orthodox Christianity, and Islam, respectively), and the historical extent of each empire is reflected in the religious faiths that are dominant in the countries of Central Europe today.

By the time of the Cold War, it appeared that an end finally had come to the tug-of-war that beset this perpetually "in-between" region. Turkish expansion into Europe was stopped with the defeat of the Ottoman Empire in World War I. The peace that ended World War II divided the remainder into a West (the American sphere) and an East (the Soviet sphere). "West" and "East" rapidly became political shorthand for pro-American and pro-Soviet, even when these terms contradicted geographical reality. For instance, pro-American Greece was typically defined as a part of Western Europe, even though most of the country is located to the east of Czechoslovakia, Poland, and Hungary, all of which were considered part of Eastern Europe because of their pro-Soviet governments. With the division of Europe into "West" and "East," "Central Europe" became obsolete as a concept, although the term "Southern Europe" did gain some usage for describing the non-aligned countries of Albania and Yugoslavia.

After the Cold War: Toward a Reunited Europe?

Between 1989 and 1991, the Cold War came to an end. The Iron Curtain that had divided Europe began to unravel in 1989 when residents of Eastern countries took advantage of temporarily relaxed border controls to stream across to the West in large numbers. This rapidly led to the fall of Eastern Europe's Soviet-aligned governments (in Bulgaria, Czechoslovakia, East Germany, Hungary, Poland, and Romania). Around the same time, the two non-aligned communist governments of Europe—Albania and Yugoslavia—also fell, with the six republics of Yugoslavia dividing into five countries: Bosnia and Herzegovina, Croatia, Macedonia, and Slovenia each became independent, while the remaining two republics—Montenegro and Serbia—remained united in a much smaller Yugoslavia. In 1991, the Soviet Union itself disintegrated. Although the Soviet Union (or USSR) was often simply known as "Russia," in fact Russia was only one of the Soviet Union's fifteen republics, with about half of the Soviet Union's population. In 1991, the other fourteen republics split from Russia, leading to the dissolution of the Soviet Union. Eight of these republics are to the south of Russia, in Asia—the five Central Asian republics and the three Caucasian republics. The other six republics, however, are to Russia's west and are usually considered part of Europe: the three Baltic republics of Estonia, Latvia, and Lithuania, and, farther south, Belarus, Ukraine, and Moldova.

With the fall of the Soviet Union, Europe's external boundaries and internal divisions were again up for grabs. The initial reaction of most people in the parts of the region that had been aligned with Moscow (either as part of the Soviet Union or as a member of the Soviet Bloc) was to look westward, with the hope that Western Europeans would welcome their long-lost cousins. There were many reasons why these countries chose to look westward once Soviet control waned. Those who lived near the Iron Curtain had seen how much better off their neighbors on the other side were and many of them hoped that Western Europe would bring its political and economic systems to the countries that lay to the east. Also, since Western Europe had been the enemy of the Soviets during the Cold War, it seemed natural for residents to turn to the West, as the opposite of the Soviet

Union, once they were free to do so. By the early 1990s, newly emerging post-Soviet governments in the region were also looking to the West for salvation, inviting Western European companies to reinvest in factories and employ workers whose previous workplaces were being shut down.

For their part, most of Western Europe was equally excited about integrating the East into a new, united Europe. Western European firms saw an opportunity for establishing factories in countries where labor was much cheaper than in the West, but from which transportation costs to wealthy Western European markets would be minimal. Additionally, while Eastern Europe was notorious for its inefficient industries, it was famous for its well-trained scientists and engineers, and the prospect of having access to this pool of skilled labor presented another opportunity for Western European companies. Individuals from Western Europe saw a chance for inexpensive vacation sites, and everyone, in the West and the East, saw an opportunity to build a diverse, but integrated, economic region that could rival that of the United States. Thus, the dream of a single, united Europe was born: Just as Central Europe had faded as a concept, now the divisions of West and East would fade too. Western Europe would expand eastward, welcomed by East Europeans who were seeking a better life like that enjoyed in the West. A new, prosperous Europe would be created for all.

The European Union: A Vehicle for Uniting Europe?

Practically all advocates of bringing Eastern Europe into the Western fold acknowledged that, if this were to happen, a key means for bringing it about would be an expansion of the European Union (the EU). The EU was formally created in 1992, but its earliest predecessor was born in 1951, when six countries on the western side of the Iron Curtain—Belgium, France, West Germany, Italy, Luxemburg, and The Netherlands—united their energy industries to form the European Coal and Steel Community (the ECSC). In the decades that followed, the organization's power and scope increased steadily. The ECSC soon merged with a number of other organizations that also had been built around cooperative economic agreements, to form a new comprehensive economic organization, the European Economic Community (the EEC). The EEC set technical standards and facilitated free trade among its member states. Soon, border controls between most of its member states were dropped, so that a citizen of one member state could travel to or even work in another member state without showing a passport or obtaining any special visa.

In 1992, the EEC expanded its scope (and changed its name) one more time, becoming the European Union (the EU). With the emergence of the EU, Western Europe became an integrated and (partially) unified *political* entity as well as an economic region. Some have speculated that the next step is for the EU is to become a single super-state, a United States of Europe. The EU may or may not be going in this direction, but for now (at least) the power exercised by the EU executive in Brussels over the EU member states is much less than that exercised by the president in Washington over the governments of the fifty states that constitute the United States of America. In 2002, the EU made one more major step toward economic integration (and an important symbolic step toward political integration) by introducing a common currency—the Euro—for its member countries to adopt in place of their old national currencies. At the same time as the EU was expanding in scope, it also was picking up more members. From the six countries that originally joined the ECSC in 1951, it had grown to fifteen by 1995.

On May 1, 2004, the EU admitted 10 new countries: Estonia, Latvia, and Lithuania (the three Baltic republics of the former Soviet Union); the Czech Republic, Hungary, Poland, and the Slovak Republic (countries that had formerly been aligned with the Soviet Union); Slovenia (one of the republics that had been part of Yugoslavia); and Cyprus and Malta (two small island-states in the Mediterranean). It was also decided that the EU would further expand in 2007 to include Romania and Bulgaria, pending political and economic reforms in those two former-Soviet Bloc countries, and that the EU would meet again in December 2004 to consider including Turkey.

Prior the 2004 expansion into the former Soviet Bloc, the membership of the EU, Western Europe's most significant economic alliance, was largely, but not entirely, identical to that of NATO, Western Europe's most significant military alliance. Eleven of the EU's fifteen members were also members of NATO, but four EU member states—Austria, Finland, Ireland, and Sweden—belonged only to the EU. Conversely, several NATO members were located outside the EU region, whether to the west (Canada and the United States), north (Iceland and Norway), or east (the Czech Republic, Hungary, Poland, and Turkey). Additionally, some EU members have opted out of specific EU initiatives. For instance, three of the fifteen EU member states—Denmark, Sweden, and the United Kingdom—have chosen to keep using their national currency instead of adopting the Euro. The ten new EU members also have not as of yet adopted the Euro.

Western Views on EU Expansion

In general, Western European governments (and the EU executive, which represents them) were enthusiastic about EU expansion to the 10 new countries and remain enthusiastic about further expansion. Most business interests look forward to new markets where they can sell their goods without tariffs, new undervalued investment sites where they can locate branch plants and offices, and new laborers who can immigrate to the West and work there for lower wages. The first reading is an introductory statement by the EU celebrating and promoting the enlargement process, issued between the 2002 meeting when extension was announced and the 2004 meeting when it actually occurred. The reading stresses the economic potential of the new, expanded EU: "The 25-nation European Union will be the world's largest trading bloc, a single market of 500 million citizens." This document also notes that EU expansion will help foster the rise of democratic political systems and market-oriented economic systems in the newly westernizing countries of the East, as reforms in these areas are being required as a condition for EU membership.

Despite this enthusiasm among business interests and most governments, many ordinary citizens in the West have been less enthusiastic about forming a community with the formerly Communist nations to the east. "Euro-sceptics" in the relatively poor countries within the EU, which historically had been the main beneficiaries of aid from the wealthier EU countries, foresaw that with the admission of new, even poorer countries, their portion of the EU aid budget would shrink significantly. The wealthier EU member states have had their "Euro-sceptics" as well. Although these countries might in the long run gain most from EU enlargement because it would create new export markets for their world-class industries, many of these countries' residents are worried about the cost of bringing Eastern European living standards up to Western levels. They often point to the high price of German unification as an example. In the 1990s, when West Germany incorporated the formerly Soviet-aligned East Germany (forming the present-day state of Germany), West German citizens were faced with a huge reunification bill. Even after spending vast sums on development aid, it still is questionable whether the West German authorities have been successful at creating a unified nation; the eastern parts of Germany remain much poorer than the west, and important cultural tensions between the two areas remain as well. Given this record, some assert that any effort to eliminate the economic and cultural divisions that resulted from fifty years of living under vastly different political and economic systems would probably not succeed, and, even if this effort were to succeed, the cost of eliminating regional economic inequality across the continent would far outweigh any benefit to be gained from a Europe-wide common market and common labor pool.

Citizens of Western countries also have expressed concerns that EU expansion into the poorer countries of Eastern Europe would lead to job loss in the West, as Western companies take their good-paying manufacturing jobs to low-wage sites in Eastern Europe. Here, they point to the similar process that is occurring on the global scale with the flight of manufacturing to East Asian and Latin American export processing zones (see Chapter 11 in the textbook and the accompanying module in this workbook). Conversely, even if firms did not leave for new, lower-cost locations in Eastern Europe, the freedom of movement between EU member states would allow low-wage workers from the East to come to the West, where they could take jobs from higher-paid West European workers. Both of these objections, and responses to them, are brought up in the second reading, an article from the Prague-based magazine, *Transitions*. As this reading notes, Western European trade unions initially opposed EU expansion, citing concerns about both job flight and low-wage immigration. Over time, however, it became clear to the trade unions that the export of jobs and the import of cheap labor was going to happen anyway, so they realized that a better strategy might be to encourage rapid EU expansion so that Eastern European living standards and labor laws could quickly be brought up to Western norms.

Paralleling Western Europeans' economic arguments against EU expansion has been a series of cultural arguments. At one level, some have feared that free migration across an ever-greater zone will lead to the destruction of distinct national cultures (for more on this view regarding migration, see the workbook module accompanying Chapter 3). Another cultural fear relates not to the *expansion* of the EU into more countries but rather to its *intensification* into more aspects of political, economic, and social life. The fear here is that local cultures and laws will be forced to conform to the dictates of EU bureaucrats in Brussels. This sentiment is expressed in the third reading, a public posting by the founder of the Eerst Nederland (Netherlands First) website, which opposes Dutch participation in the EU.

Eastern Views on EU Expansion

As in Western Europe, Eastern Europe's views toward the Western-led EU are mixed. Eastern business interests, like those in the West, were generally strongly in favor of joining the EU. In part, this was because, by being a member of the EU, an East European country would be able to offer an attractive location to foreign investors. For instance, a Polish government official seeking investment by an American firm could remind that firm that, now that Poland is a member of the EU, the American firm can use Poland's cheap labor to obtain tariff-free access to the large and wealthy West European market.

Investment-hungry governments of East European countries also hoped that EU membership would help them with their post-Soviet image problem. Under the Soviet system, and even under the governments that immediately followed the fall of Communism in 1989,

Eastern Europe generally was seen as a difficult place to do business. Factories had received little investment after they were built, workers and factory managers had adapted to a system built around doing just enough work to make distant state bureaucrats happy, and government bureaucracies and corruption created an atmosphere of perpetual crisis. Indeed, immediately after the fall of the Iron Curtain, a number of West European and American companies invested in or pursued joint partnerships with East European manufacturers, only to leave the region a few years later disillusioned with the experience. Once admitted to the EU, however, an East European country's representative would be able to project an image of the country as a "real" European country (and therefore, a safe site for investment). This sentiment is expressed in the fourth reading, a press release from a Baltimore-based international business networking group. In the press release, Jiri Kulis, Economic and Commercial Counselor at the Czech embassy in Washington, is quoted as saying, "The Czech Republic is no longer thought of as part of Eastern Europe, we are part of Europe now."

The next two readings are stories from British newspapers reporting on Eastern views toward European expansion. It should be clear from these two readings that the optimism about joining the EU expressed by the representatives of the Czech Embassy and Czech business leaders is not universally felt by the people of Eastern Europe. One of the key Eastern European objections to the EU parallels a major objection in the West: Fear of a loss of sovereignty. Eastern Europe has had a long history of being on the periphery of one empire or another. Does it make sense to join yet another budding empire as a junior member? As the author of the fifth reading notes, "There are growing reservations about surrendering sovereignty to Brussels so soon after escaping from Moscow." These broad concerns about domination by Brussels are echoed in more regional concerns as individual countries have particular histories of domination by their Western neighbors, and they are wary of setting themselves up for a repeat of this domination. For instance, Czechs fear not only loss of sovereignty to Brussels, but also domination by neighboring Germany, which, until the establishment of the Iron Curtain between West Germany and Czechoslovakia, had often played a commanding role in the Czech economy.

Coupled with these fears is resentment about the economic policy reforms that have been mandated by the EU in return for membership, and dissatisfaction with the level of aid being given by the EU to offset temporary (or, perhaps, long-term) hardships that are being caused by these reforms. In this module's first reading, the EU lists the "Copenhagen Criteria," five conditions for EU membership. The last two of these are that prospective EU members must have the "capacity to cope with market forces and competitive pressures within the Union" and the "ability to take on the obligations of membership, including Economic and Monetary Union." In other words, new member-states must not be so poor that they create an economic drag on the EU as a whole. In order to insure that the new member countries meet these fourth and fifth criteria, the EU also imposes criterion number three: "[The] existence of a functioning market economy."

Prior to 1989, Eastern Europe had anything but a "functioning market economy." The Soviet economic system (applied in the Soviet-aligned states of Eastern Europe as well as the Soviet Union) was based on state planning. A factory was kept open because it produced something that state planners felt that the country needed or because it provided necessary employment in an area, not because the market had determined that this factory necessarily was the most efficient producer. National economies in the West are not completely guided by market forces either, but the market is given a considerable role in determining who produces what, and how.

The transition to a market economy has introduced a number of unpleasant shocks for Eastern European economies. To satisfy EU requirements, Eastern European governments have been forced to implement economic reforms similar to the structural adjustment programs (SAPs) being implemented in the world's less developed countries (see the module accompanying Chapter 9 for more on SAPs), and these reforms are having similar effects of increasing poverty and income inequality. As the sixth reading notes, Eastern European governments are being forced to eliminate subsidies to some heavy industries, previously the engine of these economies, so as to comply with EU "fair competition" rules. Job losses are being coupled with cuts in public services. Under the Soviet system, many essential services such as housing, education, and healthcare had been provided to all for free or at a minimal charge. To build fiscally strong economies, East European countries are increasingly privatizing these services or requiring that they recover their costs through service fees, which means that the poor often are priced out.

As both of these readings note, farmers are also worried by the economic reforms being mandated by Brussels. The EU has required that new member states immediately open themselves up to Western agricultural exports, but, for their first ten years as EU members, these new member states will not be receiving the same level of EU agricultural subsidies that farmers in the Western states are receiving. During these ten years, Eastern European farmers charge, Western farmers will have an unfair price advantage over their Eastern counterparts, and by the time the ten year phase-in of the EU agricultural subsidy has been completed, many small farmers in Eastern Europe likely will have been forced out of business.

Where and What Is Europe?

Paralleling these debates about precisely which states should be admitted into the European Union and what the relationship should be between Brussels and individual, sovereign governments is a series of questions about just what sort of region Europe is. Is Europe a region sharing a common culture? A common heritage? A common outlook? A common political stance toward "non-European" countries? Or is Europe simply a common geographic location? How one answers this question is intimately tied with one's opinion on where the outer borders of Europe should be drawn.

As the seventh reading notes, the Vatican seeks to define Europe as a region with a uniquely Christian heritage, even if it is presently tolerant of all religions. This definition of *what* Europe is implicitly also sets standards for defining *where* Europe is. According to this definition, each of the majority-Christian countries of Europe displays an essential element of "Europeanness" and is therefore deserving of membership in the EU. In this reading, this point is suggested by Stiepan Mesic, the president of Croatia, when he suggests that Croatia should be admitted to the EU in part because of its Christian heritage.

While this definition of Europe as Christendom implies that some new countries (such as Croatia) should be considered for membership in the EU, it also implies that Europe has some predefined limits. Turkey is a predominantly Muslim country that has long sought membership in the EU. It lacks a Christian heritage, but, as its prime minister asserts in the eighth reading, if "Europeanness" implies having attained a level of economic development and having achieved a level of political democracy, then Turkey is very much a "European" country.

Also contributing to this debate, again from the outside, are the two countries that did so much to shape Europe (or the two Europes) during the Cold War: Russia and the United States. On the one hand, Russia sees the expansion of the EU to the Baltic states and Poland as ripe with opportunity. Now the markets of Western Europe abut the Russian border. On the other hand, Russia has a long history of trade relationships with the countries of Eastern Europe and it would like to maintain these relationships even as the East joins the EU. The Russian government fears that EU-mandated changes in its new members' trade and visa policies could result in the creation of a "fortress Europe" from which Russia would be excluded. Thus, Russia's concern, as reported in the ninth reading, is to create a Europe that extends far to the east but that, even more importantly, retains an open relationship with those outside its borders.

The American attitude toward both the "what" and "where" of Europe has always been complex, and it always has been conditioned by American military considerations. During the Cold War, the United States viewed the EU as an important adjunct to NATO. U.S. strategists figured that an economically strong Western Europe would be the best defense both against westward Soviet military expansion and the rise of pro-Soviet communist parties in West European countries. At the same time, though, even during the Cold War, the EU presented a bit of a problem for the United States. If the EU were to succeed too well in fostering economic development in Western Europe, then Europe might develop as a formidable competitor to the United States, economically and, potentially, politically as well. During the Cold War, however, these concerns about EU success were usually given fairly little consideration in light of the EU's role in bolstering Western Europe's position as the United States' front-line ally facing off against a threatening Soviet Union.

In the post-Cold War era, however, these calculations have changed. The United States has few worries about westward aggression from Russia, and thus Western Europe's importance as a strategic zone between Russia and the United States is greatly diminished. Western Europe likewise feels less of a need for U.S. military protection, and it increasingly is pursuing geopolitical goals that differ from those of the United States (expressed, for instance, in the opposition by France, Germany, and several other Western European countries to the 2003 invasion of Iraq).

Parts of Europe, however, remain strategically important for the United States. Although one opponent that bordered Europe (the Soviet Union) is no longer a major strategic concern, another area bordering Europe (the Middle East) increasingly is a region in which the United States wishes to project military power. Thus, while the countries of Western Europe are no longer crucial as a buffer zone for stopping Soviet aggression, the countries of Eastern and Southern Europe are gaining importance for the United States as places from which to launch aggression into the Middle East. The result is that the U.S. government's focus, in the words of U.S. Secretary of Defense Donald Rumsfeld, has shifted from "Old Europe" (the core countries of the EU that abutted the Soviet bloc) to "New Europe" (the formerly Soviet-aligned countries that, especially in the case of Romania and Bulgaria, provide ready access to Middle Eastern battlefields) This shift in the U.S. government's attention is reported in the tenth reading.

For now, this reorientation is military and not economic; the U.S. economy remains closely tied to that of Western Europe, while Bulgaria is relatively unimportant as an economic partner for either the United States or the EU. The formation of the original EU, however, shows that economic alliances often follow military ones. Indeed, history shows that economic institutions often are inspired by perceived military requirements. A further realignment of relations between the United States and various parts of Europe could occur if Bulgaria's and Romania's increasing military cooperation with the United States were to alienate Western European countries that have been trying to develop an independent military policy. If that were to happen, the 2004 expansion of the EU might well be the last. The eastern and southeastern borders of Europe (as expressed by the outer boundaries of the EU) that would emerge from this scenario would be defined not by the limits of Christendom, nor by cultural factors, nor by levels of economic development. Instead the eastern border of Europe would simply be the point at which territory that is "strategically unimportant" to the U.S. military ends and territory that is "strategically important" begins. While this definition is far from any ideal of "Europeanness" proclaimed by EU officials in Brussels, it would continue the Cold War practice of defining and dividing Europe(s) according to outside countries' strategic needs.

Readings

from **the European Union**

1

Activities of the European Union: Enlargement

January 2003

Enlargement of the European Union will be a historic achievement, ending centuries of division. Europe reunited means a stronger, democratic and more stable continent able to gain full advantage from an Internal Market of 500 million people.

Open Doors

The European Union has favoured a steady expansion since the original six Member States joined forces in 1951 to create its forerunner, the European Coal and Steel Community, and then in 1957, the European Economic Community. The founding members called upon the peoples of Europe "who share their ideas to join their efforts."

The Community expanded from six Member States to nine in 1973, to twelve by 1986, and to fifteen in 1995. By then Europe had begun to grapple with the unexpected and unprecedented opportunity to extend European integration into the former Communist central and eastern Europe presented by the collapse of the Berlin Wall in 1989.

May 1, 2004 has been fixed as the date for the next historic enlargement—in time for the new Member States to take part in the European Parliament elections in June. The decision by the European Council at its meeting in Copenhagen in December 2002 that Hungary, Poland, the Czech Republic, the Slovak Republic, Slovenia, plus the Baltic states of Estonia, Latvia and Lithuania, and the Mediterranean islands of Malta and Cyprus should accede to the Union in 2004 was the culmination of a long process of preparation and negotiation.

The EU's enlargement policy that has evolved since 1989 is built on four pillars:

- clear political and economic criteria requiring candidates to respect democratic principles and to operate market economies;
- pre-accession aid programmes to help close the wealth gap between the candidates (13 by 1999 whose per capita gross domestic product is 33% of the EU average);
- encouraging institutional changes in the candidate countries so they can apply and enforce the full range of EU laws;
- Treaty changes to ensure that the functioning of the EU's institutions is not handicapped by a large number of new Member States.

Membership Conditions

In the Treaty on European Union, which came into force in 1993, Article 49 says that any European State which respects the principles of liberty, democracy, respect for human rights and fundamental freedoms and the rule of law may apply to become a member of the Union.

Further clarification was given by the European Council meeting in Copenhagen in 1993 which laid down the basic conditions for membership—the so-called "Copenhagen criteria:"

- stable institutions guaranteeing democracy;
- rule of law, respect for and protection of human rights and minorities;
- existence of a functioning market economy;
- capacity to cope with market forces and competitive pressures within the Union;
- ability to take on the obligations of membership, including Economic and Monetary Union.

A Remarkable Transformation

By the end of 2002—less than 13 years after the break-up of the Soviet Union and the end of the Cold War—eight central and eastern European countries were judged ready for EU membership, together with two small Mediterranean states. Those years witnessed a remarkable transformation in the former Soviet satellites from centrally planned to market economies, and the creation of extensive new trading relationships with the EU.

A "pre-accession" strategy designed by the Union was put in place at the end of 1994. This aimed to provide assistance and promote investment in the candidate countries especially in the environment, transport infrastructure and agricultural modernisation. It also included bilateral trade agreements, political dialogue and mechanisms for bringing their laws and legal systems closer into line with those in the EU.

Preparing for Membership

The primary obligation of EU membership is the adoption of the so-called *acquis communautaire*, which means applying 80,000 pages of EU law, raising the efficiencies of the bureaucratic and administrative systems, strengthening judicial systems and tightening security at the candidates' external borders.

For their part, the existing 15 Member States took steps in the 1999 Treaty of Nice to streamline their decision-making. In the same year, they adopted a six year financial framework allocating more than €3 bn as direct financial support for the candidates.

Benefits Gained and More to Come

The 25-nation European Union will be the world's largest trading bloc, a single market of 500 million citizens.

Prospects for economic growth in the new Member States will be boosted while basic freedoms and fundamental rights will be guaranteed. Existing Member States will benefit from this new prosperity, but the "big picture" benefit for all is seen as the wholesale expansion across the European continent of long-term peace and security.

In central and eastern Europe, benefits are already visible in the shape of stable democracies, recognition of the rights of minorities and high rates of economic growth. The Union has shared in this prosperity and its growing trade surplus with these countries has generated employment and growth in the Member States.

The Next Enlargement

Two other applicant countries, Bulgaria and Romania, are pencilled in for membership in 2007 providing they meet the required standards of preparation in time.

A thirteenth hopeful, Turkey, was given no firm date for accession negotiations, but it has the possibility to begin accession negotiations in December 2004 if the

European Council concludes that it fulfils the Copenhagen political criteria which means making progress in safeguarding human rights, the rule of law, and the protection of minority groups.

Ratification

In many policy areas, there are phasing-in agreements to smooth the transition to EU membership, and all the candidate countries except Cyprus will have to hold a referendum to ratify the Accession Treaty after its signature at a summit in Athens in early 2003.

Source: "Activities of the European Union: Enlargement," posted Jan. 2003 on europa.eu.int/pol/enlarg/overview_en.htm.

from *Transitions* 2

A Western Perspective

Trade Unions Fear the Export of Jobs, and the Import of Cheap Labor

by Anne Dastakian
April 1998

Eight years ago, when France's Renault was vying for a stake in the Czechoslovak car manufacturer Skoda, it ran up against strong opposition from the communist Confederation generale du travail (CGT), France's biggest trade union. "The CGT is trying to convince [its Czech counterpart] to jeopardize our plans," said Raymond Levy, then Renault's chief executive officer. The French unions feared that the deal would endanger jobs, as production would ultimately relocate to the East, where labor is cheaper. A few months later, Germany's Volkswagen won the tender for a stake in Skoda, which is now the Czech Republic's leading exporter.

Now, on the eve of the opening of EU negotiations with the five former communist countries chosen for first-round accession, Western trade unions have no choice but to face the fact that cheaper labor markets have opened up. These markets will have a major impact on the rest of Europe: not only for new member states, but farther east, for non-EU candidate countries such as Romania, Ukraine, or Russia.

Last year, trade unions reacted furiously when Renault closed its Vilvorde factory in Belgium. As justification, the company claimed that Belgian workers, who are paid slightly more than their French counterparts, cost too much. A few weeks later, Renault announced its intention to build a factory in Russia, where average salaries of $200 a month are barely a tenth of Belgium's. Cheap labor in Romania also came under fire in Scandinavia, when in February a British newspaper reported that the Swedish furniture company IKEA contracted with a Romanian manufacturer that paid its workers three Swedish crowns an hour (about 45 cents). Scandinavian wood-industry trade unions immediately threatened a boycott and demanded that "decent salaries" be paid to workers in the East. An IKEA spokesman responded by saying "it is not in IKEA's power to change the standard of living of an entire population."

Romania's average monthly salary of about $60 is far lower than those of the Central and Eastern European countries slated to join the European Union. But their economies still fall significantly short of EU countries': the Czech Republic and Slovenia's gross domestic product is 60 percent that of the EU average (slightly under that of Greece); Hungary's is 37 percent of the EU average; Poland's 31 percent, and Estonia's 22 percent.

Part of the challenge rests in developing economic and social legislation that, according to Jean-Luc Delpeuch, an aide to French Prime Minister Lionel Jospin in charge of relations with the EU, would prevent the new members from being artificially competitive within the Union. "Their salaries already are regularly increasing, and the introduction of the euro should also make massive currency devaluation strategies more difficult," he said.

Cheap-Labor Pains

Eight years of preparation by the EU candidates have helped soothe some of the fears. But countries such as Germany and Austria remain concerned that cheap migrant workers will flood the West. Germany sees this possibility as particularly troubling because unemployment there has soared since reunification and is currently about 12 percent. In late February, Dietrich von Kyaw, the country's ambassador to the EU, said free job movement within the union for Poles would be "unthinkable." Poland's Deputy Labor Minister Irena Boruta reacted strongly, saying her country "will demand the principle of the free movement of labor" at entry talks. Such disputes promise to become increasingly contentious as negotiations for EU membership proceed.

In Austria, the issue of free labor movement has become a rallying cry for the extreme right-wing Freedom Party (FPOe), the country's third-strongest party. Its leader, Joerg Haider, actively campaigns against EU expansion, saying that 8 percent to 9 percent of Austria's labor pool, which includes the unemployed, is made up of non-EU members, the highest rate within the EU. Austria's Labor Ministry is opposed to immediate free movement of labor for new members until their economic and social development is on par with that of the EU's. But it is more than a question of new members potentially dumping cheap labor on the West. Hungary, for example, requires no visas for citizens of the seven countries with which it shares borders, six of which are non-EU members. The concern is that the unemployed as well as criminals from non-EU member states will have new gateways to the West.

Trade unionists in Brussels, however, look beyond such fears and are quick to note that the European Union has already greatly benefited from the opening of new, Eastern markets. The union's exports to former communist countries in 1996 totaled 63.29 billion euro (about $79.4 billion), while those countries' exports to EU countries were 46.97 billion euro (about $58.9 billion). Still, Jean Lapeyre, deputy general secretary of the European Confederation of Trade Unions, admits some "fears" might exist within West European unions. "The EU's enlargement does present some risks, so it must be well prepared, especially from a social point of view," Lapeyre said. He criticizes the European Commission for not adequately preparing for the social consequences of enlargement, which he says will be important to prevent unrest in the region.

Boosting Bargaining Power

Csaba Oery, a member of Hungary's opposition Young Democrats party and the former president of the country's first free trade union, Liga, shares this view. Oery says the development of strong unions in the East has been hampered by the lack of large-scale enterprises in post-communist countries: of the close to 800,000 firms registered in Hungary, fewer than 100,000 have more than ten employees, a size below which union representation is virtually impossible. Only 6,000 enterprises have 50 to 300 employees and 1,100 more than 300. Unions lack members and money; as a result, most of their experts have left, leaving them ill-prepared to negotiate collective bargaining.

Aware of this situation, West European unions have launched intensive cooperation programs since the fall of the Iron Curtain with their counterparts in the East. The programs are aimed at helping Eastern trade unions move away from the discredited state-led organizations they once were.

Western unions see a commonality of interests in working with their eastern counterparts. Syndicates themselves are becoming more international in scope as their employers move with increasing ease across borders, with the ultimate hope being internationally coordinated labor strategies and actions. They also have a clear self-interest in promoting higher wages and benefits among their eastern counterparts, not only out of solidarity, but to ensure that the region does not become a low-wage magnet for manufacturing and service industries, thus luring job-producing investments eastward.

They have, however, encountered difficulties along the way. The rapid privatizations and overnight creation of businesses in the post-communist countries have often resulted in employers who have little preparation, or inclination, to sit at a collective bargaining table. Western unions attempting to promote three-way discussion among the state, employers, and employees in the East have faced numerous obstacles.

"We abandoned a program in the Czech Republic because nobody there was interested in restructuring the steel industry," recalls a French union official. "The Czech unemployment rate was under 3 percent, and they just didn't care. We tried to explain that they should think about it before it's too late, but it was hopeless. Now they have a 6 percent rate and figured out that lots of their reforms were fake. In Poland, we suffered a lack of interest, and a lack of financing to achieve our projects. Differences in mentalities are still very acute."

Regardless of the inherent difficulties, East-West cooperation between unions will undoubtedly continue, because their interests will continue to converge. They hope that the enlargement of the European Union will provide all member states with equal social protection and rights, and that it will foster a level economic playing field not only for investors and potential employers, but for employees as well.

Source: Anne Dastakian, "A Western Perspective: Trade Unions Fear the Export of Jobs and the Import of Cheap Labor," *Transitions*, April 1998.

from Eerst Nederland (Netherlands First)

3

A Message from Mister Opstelten

Hendrik Opstelten

I have been asked on many occasions why I have created this website. As a result, I will will now state for the record my reasoning.

The purpose of my effort is one and only one: To maintain my beloved country. In order to do this, I feel it is imperative that the Dutch nation maintain its sovereignty and identity against the tide of "Multi-nationalism." I am not "Multi-national," I am Dutch. First Dutch, always Dutch. In order to maintain this all important sovereignty, it is also a tenant of my program that the European Union be significantly decreased in its power. I see the EU as a behemoth super-state whose only purpose is to create the largest and easiest to reach free market possible, I feel it sees its citizenry as nothing more than customers and it sees their national uniqueness as a barrier to selling to them, barriers Brussels is determined to tear down. They have guided us into the palm of their hands with promises of peace and prosperity, promises they are obviously incapable of keeping, the truth of which they hide from us. The value of the European Currency has reached a new low, with no upside in sight. Safety on the Continent is no more secure today than ever, although inter-nation conflict is not as foreseeable, international conflict, as would result from a nation's being oppressed and its people rebelling. Whenever I am engaged into discussion by a "Europhile" I am always given read the same sentence, "But Europe must be united." I question why? Why must we unite? The disparity of the European people is so great that the effort to unite us will invariably lead to our homogonization. This is the worst of outcomes imaginable. When I visit France I want to hear French, when I visit Spain I want to hear Spanish, to give up this uniqueness is the path of the lemming, when we loose our identity we have no further purpose. I grew up hearing stories from my father about "The War," that most terrible of events in the scope of Human history which occurred as a result of one man's determination to have a United Europe, a Greater Europe under the command of the German "Super-Man." I heard stories of the Occupation, my family's involvement in the Resistance and the suffering of the Dutch people. I saw tears come to my father's eyes as he spoke of his younger brother David, who died at ten years of age from a simple lung infection because he wasn't able to get medicine. I heard of how my father left training for the priesthood in order to join the fight to save his country. All my family, and all patriotic Dutchmen were forced to become "Soldaaten Van Oranje," "Soldiers of Orange." They fought, not for territorial gain or other selfish reasons, they fought to maintain their nation, to remain Dutch even as they were being forced to learn German and give Hitler Salutes as they walked along the canals as they passed men starving on the streets.

I see that struggle for national preservation repeating itself. Luckily the threat to our bodies isn't as great. There are no bombs being rained upon us, no soldiers charging into our homes. Our bodies aren't at risk, it is our minds that are in danger now, as much as they were in my father's time, our ability to think differently, it is our ability to be ourselves that is under attack. We must be allowed to be Dutch, and have pride in our differences and our unique national identity. The European Union doesn't like to hear those words, they want us to all be the same people, to share the same goals, morals, beliefs and, according to some, language. This must not be allowed to occur. I have been told by some that I am a risk to peace . . . . To this I argue, because I love my own country does not mean I hate all others, what I love is our uniqueness, I understand the value of maintaining peoples' identities, and I love my identity. I would not want it imposed upon any other, when I visit Spain I want to see the Spanish, but I want to be allowed to be Dutch, to do with my in my own country as I see fit, other nations must respect this reciprocal understanding of the difference in people and allow each man and nation to live as he or it sees fit, without hindrance from outside forces lacking in an understanding of each other. That is all I fight for, so that I, my children, and my grandchildren can see that Holland that was so dear to so many that they risked their lives to defend it. In closing . . . my slogan: We are not Germany, We are not only a piece of Europe, We are first Holland, Always Holland, above all Holland. Eerst Nederland!

Source: Hendrik Opstelten, "A Message from Mister Opstelten," geocities.com/eerstnl/speech.html

4

from the World Trade Center Institute

Czech Republic Minister of Trade Comes to Baltimore

Baltimore, July 18, 2003—Czech Republic Minister of Trade, Milan Urban led a 16-person business delegation to Baltimore for a business luncheon on Wednesday. The delegation accompanied Czech Prime Minister Vladimir Spidla who was in Washington meeting with President Bush on Tuesday to discuss Transatlantic ties. "We came to Baltimore to get outside of the Washington beltway and to provide our business delegation an opportunity to meet with major businesses," said Jiri Kulis, Economic and Commercial Counselor at the Embassy of the Czech Republic in Washington, D.C.

Maryland Secretary of State Karl Aumann welcomed the Czech delegation and reminded the attendees that Maryland and the Czech Republic have long standing ties going back to the late 1800s. "We want to develop new partnerships for the State of Maryland and the Czech Republic," said Mr. Aumann. The Secretary and over 50 Maryland businesses were treated to a lunch and an opportunity to hear from senior government and business executives about business opportunities in the Czech Republic.

"This was a great forum for networking," said Clarence Shaw, Principal with The American International Trade Group, an export management company. "I work with small to medium-sized companies who are looking for new partnerships overseas. This meeting provided me with a list of contacts from the Czech Republic who may make the perfect partners for our clients," he said.

"The Czech Republic has received over $36 billion in foreign direct investment of the past ten years," said Martin Jahn, Chief Executive Officer of CzechInvest, a government sponsored investment promotion office in the U.S. Many firms have chosen the Czech Republic as their Central Europe hub due to its location, liberal investment laws and incentives. "Multinationals come to the Czech Republic for many reasons, but perhaps most importantly it is due to our highly technical workforce that ranks among the best in the world," said Mr. Jahn.

One business delegation member, Petr Pokorny, Managing Director of Sigmainvest, Ltd, was looking for partnerships in the environmental industry for market entry into the U.S. The luncheon paved the way for some new contacts for Mr. Pokorny and potential matchmaking with similar Maryland firms looking to do business in the Czech Republic. "We can definitely be helpful to Sigmainvest, Ltd," said Peter Gourlay, Vice President of the World Trade Center Institute (WTCI). "We know the environmental arena well and will work with Sigmainvest and will hopefully find a suitable partner," said Mr. Gourlay. WTCI was the host for the luncheon and worked with the Embassy and the Maryland business community to arrange the forum.

The seeds of the Trade Minister's visit were sown at the annual WTCI *Embassy Day* event held in Baltimore on May 21. "We were very impressed with the caliber of executives at *Embassy Day*, where over 300 business executives and diplomats from 75 embassies were in attendance," said Mr. Kulis. "We sought out WTCI to host this luncheon so that we could connect with real Maryland firms and get the word out to the U.S. business community and through the World Trade Center Institute network," he said. "The Czech Republic is no longer thought of as part of Eastern Europe, we are part of Europe now," said Mr. Kulis.

Headquartered in Baltimore, Maryland, the World Trade Center Institute (WTCI) is the premier international business hub for the Mid-Atlantic region with over 3,000 clients benefiting from its global connections, international business training and customized consulting services. WTCI has an extensive global reach through its 300 world trade centers around the world. As a non-profit, non-political organization, WTCI's overall goal is to enhance international trade, promote local economic development, and ultimately foster economic stability through trade.

Source: "Czech Republic Minister of Trade Comes to Baltimore," July 18, 2003 wtci.org/ress/czechtrade.htm

from *The Daily Telegraph* 5

Angry Poles Lead Protest Against EU Farm Curbs

By Ambrose Evans-Prichard and Julian Isherwood
31 January 2002

East European countries were outraged yesterday over plans unveiled in Brussels to limit farm aid for a further decade after they join the European Union.

They claimed that it relegated them to second-class status, and Poland gave warning that its membership plans could be overturned as a result.

Gunther Verhuegen, the enlargement commissioner, said farmers from the "big bang" group of 10 states scheduled to join the EU by 2004 will receive only a quarter of the vast subsidies given to existing members. The figure will rise to a mere 35 percent in 2006.

"This represents the best possible deal, and not an invitation for haggling," he warned the former Communist states begging for entry to the EU. Full payments would break budget limits.

The move caused widespread anger in Poland, where 20 percent of the 40 million people still have ties to the land. Most have small plots that are unable to compete with the mechanised mass production of EU agro-industrial conglomerates.

Leszek Miller, the Polish prime minister, said the proposal would fire already rising anti-EU sentiment and could result in a resounding rejection of membership in next year's referendum.

"What we have heard from Brussels is extremely worrying," Mr Miller said on a visit to Stockholm. "Polish farmers must be treated like any other farmers in the EU. You cannot start off by dividing members into different categories."

He told Goran Persson, the Swedish prime minister, that Polish farmers—led by the radical leader of the Polish Self-Defence Party, Andrzej Lepper—"would not accept a delay in subsidies." Mr Miller said: "There is a danger that the referendum result will be affected."

Mr Lepper said: "If we are not treated as equals, if the EU tries to exploit us and use us as a dumping ground for its goods, we will start a propaganda war, and make sure that Poles vote No."

Most Poles still favour joining the EU, but there are growing reservations about surrendering sovereignty to Brussels so soon after escaping from Moscow.

Officials said the budget plans risked tipping the balance against the EU in the countryside, where feelings are 50:50. Estonia's EU negotiator, Alar Streimann, said the proposals must be modified. "They are not justifiable," he added.

In a break with the past, the farm aid will be paid according to the area of land rather than as a production subsidy. This is intended to promote rural development instead of creating mountains and lakes of unwanted produce.

Poland, the Czech Republic, Hungary, Slovakia, Lithuania, Latvia, Estonia, Slovenia, Cyprus and Malta are expected to join the EU by 2004.

Romania and Bulgaria will have to wait until later in the decade, once they have established market economies and a proper rule of law.

The European Commission walked a fine line in drafting its proposals. It feared that France, Spain, Italy, Portugal and Greece would sabotage the enlargement process if they thought their farmers and poorer regions were going to see their EU subsidies switched quickly to eastern Europe.

Spain, the current EU president, says it intends to reach a final negotiating position on agriculture by June.

Source: Ambrose Evans-Pritchard and Julian Isherwood, "Angry Poles Lead Protest Against EU Farm Curbs," *Daily Telegraph,* Jan. 31, 2002.

6

from *The Guardian*

Eastern Eurosceptics Threaten EU Expansion

There is plenty of scope for upsets ahead of the opening of EU doors to the former Soviet satellites, writes Ian Traynor.

Friday October 11, 2002

The post-communist countries of central Europe are reacting with quiet satisfaction to the green light from Brussels, confident that years of pain and haggling have generated an unstoppable momentum for their admission to the European Union.

The breakthrough statement in Brussels that eight countries in central and eastern Europe are now "politically" ripe for EU membership comes a month ahead of Nato's summit in Prague at which the three Baltic states, Slovakia, and Slovenia will also be invited to join Poland, Hungary, and the Czech Republic in the military alliance.

The opening of the EU and Nato doors to the former Soviet satellites is the climax to more than 10 years of hope and reform aimed at escaping the Kremlin's orbit and integrating their countries with the west.

"This is a big day for this region," said Laszlo Csaba, an expert in European integration at the central European university in Budapest. "We've been waiting for this since the collapse of communism."

The newly elected Slovak prime minister, Mikulas Dzurinda, said: "Slovakia is now standing in front of the doors leading to the historic enlargement and unification of Europe."

But there is plenty of scope for nitpicking and upsets before the "big bang" expansion is finalised in 2004, adding 75 million citizens to the union.

Freedom of movement, immigration, land ownership, corruption, state subsidies, and farming are all bones of contention waiting to be exploited by political mavericks seeking to ride a growing tide of euro-scepticism in the east.

Latvia and Slovakia have just voted for politicians committed to European integration. But in the Czech Republic and Hungary, for example, the Thatcherite ex-prime minister Vaclav Klaus and the increasingly nationalist opposition leader, Viktor Orban, are morphing into potent leaders of the anti-EU camp.

Both countries are to hold EU membership referendums next year and the results may be closer than expected. Similarly in Poland where Roman Catholic nationalists are imploring the Virgin Mary to protect them against Brussels' godless regime.

"The polls are constantly going up and down. There is no stable rise in the yes vote. It's a really hot issue," said Hanna Ignatiewska, a Polish pollster conducting surveys for the Brussels Eurobarometer monitoring agency.

Across the region with the exception of Hungary, the anti-EU camp is regularly notching up around 40% in the opinion polls benefiting from the bitterness bred by the perceived high-handedness of the Brussels negotiators. The sceptics are being helped by the rightwing parties in neighbouring Germany and Austria, which are deploying scare tactics on immigration and exploiting old second world war grievances against the Czechs, in particular, to build popular sentiment against EU expansion.

But analysts say that the pro-EU elites in the candidate countries, rather than losing next year's referendums, will face an anti-EU backlash after the central European countries are inside the union.

"There's a naive belief that Brussels will inject lots of money into rural Poland or Hungary and put an end to the widespread rural poverty," said Mr Csaba. "This will produce disenchantment."

In the enlargement stakes, Poland is the biggest prize as well as the biggest problem. The expansion slated for 2004 adds 10 members to the union's 15. But in population terms Poland's 38 million represents half the enlargement.

Poland's membership adds hugely to the EU single market, but its size gives Warsaw enhanced clout in the difficult negotiations with Brussels.

Danuta Huebner, Poland's minister for European integration, said she hoped the positive signal from Brussels would have "a mobilising effect" at home.

Corruption, state subsidies to industry, poor public administration and tax collection, and the condition of the farming sector are the biggest hurdles before Poland can become a fully-fledged EU member, the EU commission declared.

Earlier this week under strong pressure from Brussels, the Polish government agreed to close down two steel mills and to cut jobs and subsidies in an attempt to comply with the EU's "fair competition" rules.

Slovakia is embroiled in a similar conflict with Brussels over state aid to steel and car plants.

The biggest threat in Poland stems from the xenophobic Self-defence party of Andrzej Lepper, which is running a virulent anti-EU campaign and regularly coming second or third in the opinion polls.

Mr. Lepper is strong in the countryside. The issue of farm subsidies and the common agricultural policy (CAP) is crucial here, since one in five Poles still works on the land.

Extension of the CAP to eastern Europe will break the EU budget, but reform of the CAP demanded by Germany, the main paymaster, is being stymied by the French.

Chancellor Gerhard Schröder and President Jacques Chirac are to meet in Paris on Monday to try to break the CAP deadlock. But whatever the future of EU aid to farmers, it is certain to be inequitable for the candidate countries, breeding more resentment of Brussels.

For years the central Europeans have been labouring through 80,000 pages of EU laws and regulations to make their systems conform with Brussels. The easy part is passing the laws.

"It's one thing promulgating the legislation," said Mr. Csaba. "But when it comes to implementation, things are not yet what they should be."

Source: Ian Traynor, "Eastern Eurosceptics Threaten EU Expansion," *The Guardian*, 10/11/2002, found guardian.co.uk/elsewhere/journalist/story/0.7792.810368.00.html

7

from the *National Catholic Reporter*

Pope Uses Croatia Trip to Press Europe to Acknowledge its Christian Roots

By John L. Allen Jr.
June 6, 2003

Just days after losing the first round of a high-stakes battle over the role of Christianity in the European Union, Pope John Paul II has begun a five-day visit to Croatia, using the trip in part as a platform to insist anew that Europe must not forget its Christian roots.

"The rich tradition of Croatia will surely contribute to strengthening the union as an administrative and territorial unit, and also as a cultural and spiritual reality," John Paul said during a June 5 welcoming ceremony.

In a 2001 census, 87.83 percent of Croatia's 4.4 million people declared themselves to be Catholic. That's an 11 percent increase from 1991, reflecting in part a strong tie between Catholicism and Croatian nationalism.

The Croatia swing, John Paul's 100th foreign trip, comes one week after a drafting commission working on a new constitution for the European Union released the proposed text of the preamble. Despite months of strong Vatican diplomatic pressure, the preamble makes no specific reference to Christianity.

The proposed text notes that Europe was nourished by "Hellenic and Roman civilizations," as well as "marked by the spiritual impulse that runs through it and whose traces are present in its patrimony, then by the philosophical currents of the Enlightenment."

On the catamaran that carried John Paul II and his entourage from the airport in Krk to the island of Rijeka, the Vatican's Secretary of State, Cardinal Angelo Sodano, challenged this formula.

"It seems to me that in between Greece and Rome and the Enlightenment, there were more than 1500 years of Christian history," Sodano told reporters.

Croatia's president, Stiepan Mesic, picked up on the Christian identity theme in his welcoming remarks.

"We are on the path towards association with the European Union," Mesic said. Croatia aspires to join the EU by 2007. He said that Croatia wants to join Europe "in all its diversity and richness ... departing from its common sources and roots, one of which is Christianity."

France and the northern European nations have opposed references to Christianity on the grounds that Europe should be a pluralistic society without a privileged confession. The Vatican, however, insists that key European values, such as the value of the human person, rest on its Christian roots.

The draft preamble was prepared by a subcommittee of the commission working on the new constitution, and the full body will debate the preamble in coming days. The parliament of the European Union will take up the issue in the fall, so Vatican diplomats still have time to make their case.

In his public comments, the Pope is expected to call simply for further healing of the wounds caused by the war.

Source: John L. Allen Jr., "Pope Uses Croatia Trip to Press Europe to Acknowledge Its Christian Roots," *National Catholic Reporter*, June 6, 2003 found anationalcatholicreporter.org/update/bn0606031a.htm.

from *The Daily Telegraph* 8

West Wants EU to be 'Christian Club' Complains Ecevit

By Boris Johnson
12 July 1999

European Union countries are conspiring to do down Turkey and turn the EU into a Christians-only club, the Turkish prime minister believes.

In an extraordinary attack on Western motives, Bulent Ecevit accused Italy and "other countries" of supporting Kurdish terrorists with the secret intention of delaying modernisation and crippling Turkey's economy.

He also charged some European capitals with Islamophobia—or at least using Turkey's adherence to Islam as an excuse to freeze her out of the EU. As the Turkish parliament prepares to vote on whether to execute the Kurdish PKK leader, Abdullah Ocalan, Mr. Ecevit denied that his country's human rights record was a bar to EU membership.

He said: "We have been faced with very serious separatist terrorism in the last 15 years, but if some western European governments had stopped giving encouragement to the PKK the process of democratisation would have been much wider."

The veteran prime minister said there was a secret agenda in "several countries" to support the PKK and thereby to drain the resources of the Turkish state. The brutal war against Kurdish separatists in south-east Turkey is estimated to cost £5 billion a year.

"Some European countries want to keep Turkey under control and prevent its further strengthening economically and strategically," Mr. Ecevit said in an interview with *The Telegraph*.

Athens provided training camps for the Kurdish terrorists, he said, adding that he was particularly dismayed by the pro-Kurdish sentiments voiced by the Italian government of Massimo d'Alema. Ten years after the European Commission turned down Turkey's application for EU membership, and more than 30 years after Turkey first tried to join, Mr. Ecevit said the Turkish economy had been transformed.

"We are part of Europe historically, geographically, and to a certain extent culturally, and we are far more developed than certain countries for whom the door to membership has been opened," he said, referring to Greece.

"It would serve the interests of western Europe if Turkey became a member, because of our geopolitical position and because the merging of Europe and Asia is a reality of the coming decade." Turkey would act as the bridgehead between Europe and Russia, the new central Asian republics and the Middle East.

Mr. Ecevit said Turkey still faced the prejudice of those who believed in the Little Europe model of Jacques Delors and Helmut Kohl: that the EU should not expand beyond the bounds of Christendom.

"It is not a secret that some in western Europe regard the EU as a Christian institution. This is an anachronistic attitude, and it is an unrealistic attitude," he said. Mr. Ecevit argued that Turkey was the only country with a large Islamic population that had adopted a secular regime, and that this success should be supported.

After decades of disappointment, some Turkish officials are now hopeful that their cause will at last see some progress at the EU summit in December.

Source: Boris Johnson, "West Wants EU to be 'Christian Club' Complains Ecevit," *Daily Telegraph*, July 12, 1999.

9

from **Pravda.RU**

Kremlin Perceives EU's Expansion as Key Aspect of Russia-EU Summit

May 31, 2003

The Kremlin perceives the European Union's expansion and the consequences of such expansion for the entire European system, for Russia's interests and its diverse ties with prospective EU members and the entire EU as a key issue. The Russia-EU summit must attach priority to this issue and some other key issues, as well. This was disclosed to RIA-NOVOSTI at the Kremlin here today.

St. Petersburg will host the eleventh Russia-EU summit May 31, with an unprecedentedly large number of state leaders taking part in it. The summit is to involve 15 heads of state and government from EU countries, the European Commission's President Romano Prodi, as well as Secretary-General Javier Solana of the European Council.

The leaders of 10 countries, which will join the EU in the near future, and which would also be expected to sign the Russia-EU partnership-and-cooperation agreement, were also invited to attend the summit.

This summit will take place during the celebration of St. Petersburg's 300th anniversary. This in itself serves to confirm Russia's historic choice in favor of profound integration into Europe without any demarcation lines, Kremlin officials believe.

This will become the first Russia-EU summit after the signing of an EU-membership agreement by 10 candidate nations, Kremlin people added.

The EU will expand less than a year from now. We must work jointly for the sake of utilizing new opportunities (that will emerge as a result of the EU's expansion) as soon as possible and in full volume, the Kremlin spokesman added.

The present-day contractual base of bilateral relations between Russia and new EU members must be brought in conformity with EU standards. Jurists and sectoral experts will have to work real hard and hand in hand for the sake of accomplishing this objective. All in all, some 100 documents, which were signed by Russia and the afore-said ten countries, will either have to be renounced or drastically modified, a Kremlin spokesman noted.

In Moscow's opinion, this will create all essential pre-requisites for involving future EU members in the afore-mentioned partnership-and-cooperation agreement.

This won't be an automatic process, Kremlin people stressed. Each of the 10 new member-countries will join the partnership-and-cooperation agreement in line with a special protocol, which will be ratified accordingly. Among other things, Russia's Federal Assembly (Parliament) is also going to ratify these special protocols.

Moscow emphasizes the fact that, unlike Russia's interaction with other international organizations and structures, Russia-EU relations directly affect the interests of ordinary people. This implies real-life conditions for business operations, tourist trips, as well as student-and-professor exchanges.

The visa problem is a tell-tale example, the Kremlin believes.

New EU members will simultaneously join the Schengen visa agreement, also abolishing visa-free trips or mutual-trip restrictions between themselves and Russia.

The situation is compounded by the fact that, of the 10 prospective EU members, Cyprus, Malta and the Baltics are quite popular among ordinary Russians. In the long run, some of our citizens would be deprived of these convenient resorts, eventually choosing other resorts. Consequently, prospective EU members would lose a lot of money channelled into their economies by Russian tourists, officials in Moscow stressed.

Moscow is ready to examine Vladimir Putin's initiative about the switch-over to visa-free mutual trips between Russian and EU citizens in a serious and comprehensive manner. Naturally enough, substantial efforts will have to be exerted for the sake of attaining this goal; many current mechanisms will also have to be overhauled. We've got to move in this direction; quite possibly, the sides should accomplish this objective stage by stage, Kremlin officials pointed out.

Russia wants to steadily expand its mutually-advantageous interaction with the EU. Among other things, this can be done by establishing common European economic, energy, law-enforcement and security infrastructures, the Kremlin spokesman pointed out.

At the same time, the Russian side consistently defends its natural political, trade-and-economic and humanitarian interests, Kremlin people stressed.

Source: "Kremlin Perceives EU's Expansion as Key Aspect of Russia-EU Summit," May 31, 2003 newsfromrussia.com/2003/05/31/47696.html

from *The New York Times* 10

U.S. Eyes a Willing Romania as a New Comrade in Arms

By Ian Fisher
July 16, 2003

CONSTANTA, Romania—Kurt Sanger is only a captain and so he will leave to higher-ups the question of whether Romania would make a good ally as the United States sets about a historic reordering of its military, relying less on its old friends in Western Europe and more on new ones in the east. He does, however, have some thoughts about Romania as a place where American soldiers like him may find a new home, perhaps soon.

"Paradise isn't too strong a word," said Captain Sanger, 31, a Marine reservist on leave from his job as a lawyer in Manhattan.

It is cheap, he said, and the food is excellent. So are the beaches near this Romanian port on the Black Sea. But more important, he said, Romanians really like Americans, and his worries, as an officer in charge of protecting United States soldiers on an exercise here, are not the ones he might have elsewhere.

"The biggest threat we face is the stray dogs," he said. "To be able to say that at this point in history and being the military, you've got to consider yourself lucky."

This is the first flush of a new American love affair with Romania, one of the poorest nations in Europe and, until the terror strikes on Sept. 11, 2001, not high on Washington's list as a close military partner. Now it is one of the crucial countries in Europe, along with Poland and Bulgaria, considered for new American bases, probably small installations with rotating troops capable of quickly striking targets in the Middle East or Central Asia, which are closer to here than is Western Europe.

The plans make much strategic sense, military experts say, and were envisioned before the war in Iraq estranged the United States from old allies in Western Europe like Germany and France. But with the bitterness over Iraq lingering, the plans have accelerated.

This shift is giving more solid form to the stinging distinction drawn earlier this year by Donald H. Rumsfeld, the United States defense secretary, between the "old" Europe and the "new," the former Communist bloc states. Planning is under way for a reduction in the more than 60,000 American soldiers in Germany, the outdated cornerstone for Western security during the cold war to contain Russia. Poland, rather than France or Germany, will be heading a sector in Iraq, and Romania, Bulgaria, the Czech Republic, Macedonia and Albania are likely to send troops there.

But the eagerness for a stronger military alliance with the United States also carries risks for the countries involved. For example, their soldiers may die on behalf of an American military engagement.

"The support for intervention in Iraq in Poland is not, and was not, very high," said Jacek Cichocki, acting director of the Center for Eastern Studies, a government-financed research institution in Poland. "And Poles do not see Polish engagement as dying for the Americans. It is rather the engagement of country as the duty of being a member of the coalition."

These nations also hope to join the European Union. The fear, which their leaders rarely state in public, is that their deepening military friendship with the United States could jeopardize their economic integration in Europe. Already the war in Iraq and the International Criminal Court, which President Bush opposes joining, have been areas where the United States has pressed its new allies for support against a strong Western European position.

Charles A. Kupchan, a security adviser on Europe in the Clinton administration, is among those arguing that the Eastern European nations are making a "shortsighted investment."

The North Atlantic Treaty Organization, he contends, is crumbling, as the United States views Europe not as a strategic theater but as a springboard to new threats in the Middle East or Central Asia. Ultimately, he said, the real gains for nations like Poland, Romania and Bulgaria will come from their union with a Europe he sees as more and more willing to check American power.

"They are hoping to keep NATO strong and vital and witness the long-awaited arrival of American troops to Central Europe," said Mr. Kupchan, a professor at Georgetown University and a senior fellow with the Council on Foreign Relations. "But instead they will witness the demise of NATO."

Mircea Geoana, the foreign minister of Romania, is among those arguing that the projected role of his nation helps recreate NATO into a more global organization, prepared to send peacekeepers, for example, to Africa. But Europe cannot take part fully, he argues, without the much stronger American military.

"We know that without some military muscle, and without keeping trans-Atlantic ties strong, Europe will not be able to keep up with its ambitions," he said.

Poland, with 40 million people, a record of reform and a long relationship with the United States, was no surprise as a new American ally.

Not so Romania. Poor and rural, the nation of 23 million people was long regarded in the West as, at best, a hard case and long shot for NATO or the European Union.

Then came Sept. 11, and Romania quickly declared its support—it allowed United States planes to fly over and sent 400 infantry soldiers to Afghanistan. With Iraq as the next target, the United States military eyed Romanian air bases. In exchange, the United States supported Romania for entry into NATO, and the country is now expected to join formally next May.

But the relationship seemed cemented this spring when Turkey, a longtime NATO ally, decided against involvement in the Iraqi war. Almost overnight, the Mihail Kogalniceanu Airport just north of here became a major route for refueling and supply of United States troops in Iraq and neighboring countries. Officials say that up to 3,000 American soldiers could operate there at a time.

Next door in Bulgaria, American troops used a closed airport near Burgas, also on the Black Sea, for communications and logistics, as they also did during the Afghan operation.

"Iraq did provide an opportunity for Romania to demonstrate its capabilities, more so its willingness, to cooperate," said Maj. Gen. Arnold Fields, deputy commander of the United States Marines in Europe. "And Romania stepped up to the plate."

The next step may be more permanent bases here. Nothing is official and military officials here and at the Pentagon stress that planning is in its earliest stages. But, in short, the plans envision a reduction in the forces in Germany in favor of smaller bases in eastern Europe, Central Asia, the Far East and Africa—all filled with rotating troops poised to strike quickly.

The model is quite different from the permanent communities in Germany; soldiers would likely arrive alone, living in spartan quarters. No matter, say Romanians, eager for any economic infusion the new bases would bring.

"The major concern is to speed up the Americans' coming here," said Gheorghe Martin, the prefect of the Constanta area. "We've been waiting for the Americans for 50 years. Now that they are here, can we really ask ourselves if they are welcome?"

In the last few weeks, joint training exercises in Romania have brought together southeast European countries allied with the United States. Technically, it is an exercise in cooperation; practically, it mended Romania's ailing infrastructure, with engineers building a new parking area at the airport capable of handling heavy refueling aircraft and widening runway aprons.

Combat engineers also practiced urban warfare by setting off explosions in abandoned buildings at a military base here—in Germany, American officials noted, many more permits would have been required. They also built a new school and refurbished an old one.

At the new school, in the village of Corbu, north of here, the mayor gave the Americans a warm welcome. Children, some waving American flags, sang, in nice English, "If You're Happy and You Know It," and swarmed around General Fields, the Marine commander, for his autograph.

Lt. Col. David Budak, the Marine reservist in charge of building the school, allowed himself a little high-flown reflection on how much the world had changed since the cold war. "They will know democracy," he said. "They will know opportunity. They will be part of the free world."

Source: Ian Fisher, "U.S. Eyes a Willing Romania as a New Comrade in Arms," *New York Times*, 7/16/03.

Review Questions

1. Why do business interests in both Western and Eastern Europe generally favor EU expansion?
2. Why have Western European labor unions come to favor EU expansion?
3. Why are countries wishing to join the European Union forced to undergo structural adjustment programs similar to those required of non-European LDCs?
4. Why has the United States had an ambiguous attitude toward strengthening the EU?
5. Why is the United States redirecting some of its attention from the "Old Europe" of Germany and France to the "New Europe" of Romania and Bulgaria?

Discussion/Essay Questions

1. Ivan Sima, the Czech beer-breweing industry representative portrayed at the beginning of this module, is proud of his industry's stance against advertisements that exploit sexual images or that encourage abusive drinking or drinking by minors. He also feels that joining the European Union will not affect the export market for Czech beers because these beers typically are higher-priced products that will not be undercut by lower-priced local brands. In fact (although he doesn't state this explicitly), high-end,

low-volume exports to other EU countries might be helped by EU expansion, since Czech beers will not face import tariffs.

However, Sima does not address the effect that EU membership will have on Czech brewers' sales and marketing activities within the Czech Republic. Will the Czech brewers be able to maintain their voluntary advertising standards once the Czech Republic is opened to products by foreign brewers who may not share the same connections with, or concerns for Czech society? Will the Czech breweries, which specialize in higher-priced beers, be able to survive once EU membership forbids restrictions against allowing the import of cheaper, mass-produced brews?

2. Much of the controversy surrounding EU expansion and the possible evolution of the EU into a United States of Europe centers on debates about the scale at which one claims one's political and cultural identity (one's country vs. one's continent-region), the characteristics that make the region distinctive (for instance, its distinct culture, its physical geography, its economic integration, its level of development, its shared history, its unified geopolitical position), and the precise borders that one uses to define the geographical extent of that region. Noting these three debates, define the *one* political region that you feel most represents your identity, discuss its borders, and justify why you feel that your political identity rests at the scale that it does and not at a smaller or larger scale.
3. Many of the readings discuss the economic pros and cons of EU expansion (from the perspective of the 15 pre-2004 EU member states, the perspective of the 10 states that joined in 2004, and the perspective of future EU members). A few readings (especially the third reading) also raise political arguments against expansion. Few of the readings, however, directly make the *political* case *for* expanding both the breadth of the EU (the number of member states) and its depth (its reach in shaping the policies of its member states). Put yourself in the position of an EU executive addressing an audience of Euro-sceptics, either from a Western European member state or a newly joined Eastern European state, in which you try to convince them that EU expansion (in breadth and depth) is a good thing. Make sure that you address your audience's political as well as economic concerns.

List of Resources

1. "Activities of the European Union: Enlargement," posted January 2003 on the website of the European Union, *http://www.europa.eu.int/pol/enlarg/overview_en.htm.*
2. Anne Dastakian, "A Western Perspective: Trade Unions Fear the Export of Jobs, and the Import of Cheap Labor," in *Transitions*, April 1998, posted on the website of *Transitions Online*, *http://archive.tol.cz/transitions/awestper.html.*
3. Hendrik Opstelten, "A Message from Mister Opstelten," posted on the Eerst Nederlands website, *http://www.geocities.com/eerstnl/speech.html.*
4. "Czech Republic Minister of Trade Comes to Baltimore," July 18, 2003, news release posted on the website of the World Trade Center Institute, *http://www.wtci.org/press/czechtrade.htm.*
5. Ambrose Evans-Pritchard and Julian Isherwood, "Angry Poles Lead Protest Against EU Farm Curbs," in *The Daily Telegraph,* January 31, 2002, posted on the website of *The Telegraph, http://www.telegraph.co.uk/news/main.jhtml?xml=/news/2002/01/31/wpole31.xml.*
6. Ian Traynor, "Eastern Eurosceptics Threaten EU Expansion," in *The Guardian,* October 11, 2002, posted on the website of *The Guardian, http://www.guardian.co.uk/elsewhere/journalist/story/0,7792,810368,00.html.*
7. John L. Allen, Jr., "Pope Uses Croatia Trip to Press Europe to Acknowledge Its Christian Roots," in the *National Catholic Reporter*, June 6, 2003, posted on the website of the *National Catholic Reporter*, *http://nationalcatholicreporter.org/update/bn060603a.htm.*
8. Boris Johnson, "West Wants EU to be 'Christian Club' Complains Ecevit," in *The Daily Telegraph*, July 12, 1999, posted on the website of *The Telegraph*, *http://www.telegraph.co.uk/htmlContent.jhtml?html=%2Farchive%2F1999%2F07%2F12%2Fwtur12.html.*
9. "Kremlin Perceives EU's Expansion as Key Aspect of Russia-EU Summit," May 31, 2003, posted on the website of Pravda.RU, *http://newsfromrussia.com/2003/05/31/47696.html.*
10. Ian Fisher, "U.S. Looking to the East for New European Allies," in *The New York Times*, July 16, 2003, posted on the website of *The New York Times*, *http://query.nytimes.com/gst/abstract.html?res=F00712FD3E580C758DDDAE0894DB404482*, reprinted at *http://www.pipeline.com/~rgibson/romania.html.*

Websites for Additional Research

1. The European Union's English-language website, *http://www.europa.eu.int/index_en.htm*, is loaded with information ranging from detailed policy documents to press releases to public relations material on the EU's history and how it works. The section of the website devoted specifically to EU enlargement is at *http://www.europa.eu.int/pol/enlarg/index_en.htm*.
2. The Europ-sceptic Web Resource, *http://www.eurosceptic.org*, is maintained by individuals in the Uni-ted Kingdom who oppose EU enlargement and what they see as the ongoing surrender of British sovereignty to EU bureaucrats. Despite this orientation, the website is "dedicated to critical debate on Europe and helping others continue the debate" and that dedication includes a commitment to providing balanced information. Thus, alongside the links to anti-EU organizations and articles, there is a comprehensive page of links to pro-EU organizations, based in Britain and elsewhere.
3. The Open Society Institute's EU Monitoring and Advocacy Program monitors the development of human rights and rule of law standards and policies both in the European Union and in its candidate and potential candidate countries. The Program's website, *http://www.eumap.org*, contains numerous reports on how individual EU states are subscribing to the EU's stated norms in such areas as anti-corruption policy, the status of women, and minority protection.

More than 5 million
1–5 million
Fewer than 1 million
Capital cities are underlined
Departments are marked with protectorate in parentheses
Example: French Guiana (Fr.)
Caribbean Sea
PACIFIC OCEAN
ATLANTIC OCEAN
Tropic of Capricorn
COLOMBIA
ECUADOR
PERU
BOLIVIA
BRAZIL
GUYANA
SURINAME
French Guiana (Fr.)
PARAGUAY
ARGENTINA
URUGUAY
CHILE
ANDES
Galápagos Is. (Ec.)
Juan Fernández Is. (Chile)
Falkland Is. (U.K.)
South Georgia (U.K.)
TIERRA DEL FUEGO
Cartagena
Lake Maracaibo
Medellín
Bogotá
Cali
Quito
Guayaquil
Iquitos
Ciudad Bolivar
Orinoco R.
Georgetown
Paramaribo
Cayenne
Amazon R.
Manaus
Belém
Fortaleza
Recife
Salvador
São Francisco R.
Brasília
Belo Horizonte
Rio de Janeiro
São Paulo
Curitiba
Porto Alegre
Paraná R.
Uruguay R.
Lima
Arequipa
La Paz
Lake Titicaca
Santa Cruz
Sucre
Potosí
Asunción
Córdoba
Rosario
Buenos Aires
Montevideo
La Plata
Río de la Plata
Valparaiso
Santiago
Concepción
Bahía Blanca
Punta Arenas
N
0 200 400 Miles
0 200 400 Kilometers

Debt in South America

Companion to Chapter 9 Development

Raul is a welder at an automobile transmission factory in São Paulo, Brazil. When the factory was established in the 1960s, it was part of a new Brazilian automobile industry, producing a Brazilian-made car for Brazilian drivers. Now the factory, although still independently owned, performs mostly contract work for Volkswagen. Through the 1990s, these transmissions were installed in cars manufactured in one of the German company's many assembly plants in Brazil. Although the Volkswagens produced in these plants were shipped around the world, most were purchased by consumers in South America's two largest countries: Brazil and Argentina.

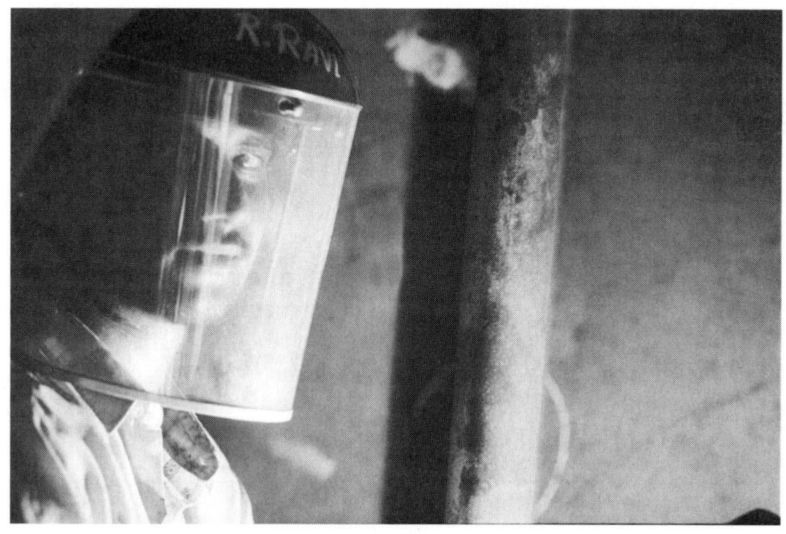

Brazil's industries are threatened by regional structural adjustment programs.
Source: Salvo Photography

In recent years, however, Argentina's economy has entered a period of crisis as its level of indebtedness to North American and European banks has skyrocketed. In order to appease foreign creditors, the Argentine government has instituted a number of "belt-tightening" measures to save the Argentine economy, and one effect of these measures has been to greatly reduce Argentine consumer spending, especially on imports. Faced with a sharp reduction in its Argentine market, Volkswagen has closed down some of its plants in Brazil. Raul's factory is still producing transmissions for Volkswagen, but these transmissions are now shipped to Mexico, where Volkswagen installs them in cars that are then exported to the United States.

Although Raul is no longer producing a Latin American product for Latin Americans, this doesn't trouble him. In fact, Raul is pretty happy. The United States is a rich country, he figures, and Americans will keep buying cars, which means that his job should be secure. The Brazilian government is happy too, because factories like Raul's are bringing U.S. dollars into the country, which are badly needed to pay back Brazil's foreign debt.

Now, though, Raul is beginning to worry. Volkswagen is a global corporation and what is to stop it from finding a new transmission supplier closer to its factories in Mexico? For that matter, Argentina's wounded economy now is ripe with undervalued investment opportunities, and Volkswagen might choose to make its next investments in assembly plants and component manufacturers there. Brazil's leftist government has promised to protect jobs and social welfare provisions, but what can it offer to Volkswagen to keep the car company involved with Brazilian suppliers? ■

Debt and Development

As is noted in Chapter 9, there are many theories about why uneven development occurs between less developed countries (LDCs) and more developed countries (MDCs), and there are several possible routes that LDCs can take in an effort to close the development gap. One major obstacle to development, no matter what development strategy is pursued, is lack of financial resources. Whether countries follow the international trade path or the path through self sufficiency, they must borrow money from MDCs to finance development.

For current LDCs to reach a more advanced level of development, their progress must be built on the foundation of what exists today, and one key element of that foundation is the enormous financial debt owed by LDC governments to MDC-controlled multilateral institutions, governments, and banks. The textbook mentions international debt and structural adjustment programs briefly. This module takes a closer look at how debt burdens one region in particular—Latin America—and how the past may play a role in the future of this region's development.

International Debt: The Past Impacts the Future

Although many countries are in debt, debt burdens, like other burdens, are spread unevenly around the world. The World Bank publishes several different debt statistics and, depending on how one measures debt and its impacts, one could make the case that either sub-Saharan Africa or South America is the most impacted region. In 2000, several sub-Saharan African countries had total debts greater than their gross national income or GNI (a new economic statistic calculated by the World Bank, similar to gross national product) that year. For sub-Saharan Africa as a whole, total indebtedness in 2000 equaled almost two-thirds of that year's GNI. By contrast, only six South American countries are classified by the World Bank as "severely indebted." Before reaching conclusions from indebtedness statistics, however, one should recall that being in debt is not necessarily a bad thing. If a country is in debt it means that money has come in to stimulate the economy, and it may be succeeding. In fact, countries (and individuals) often go to great lengths to be indebted if they can find financing at a very low interest rate; for instance, homeowners sometimes remortgage their houses just so that they can get low-interest loans.

The bigger problem for indebted countries (or indebted individuals) is debt *service*, the payments that must be made periodically to prevent the debtor from defaulting. By this measure, Latin America is in the worst shape. For the Latin America/Caribbean region as a whole in 2000, debt service equaled 35.7 percent of export earnings. Because of the way in which loans are made and paid back, the ratio of annual debt service to annual export earnings is crucial. International loans are almost always made in "hard currency" (universally accepted currency such as dollars, yen, or euros) and they typically must be repaid in hard currency as well. Because the only way that a less developed country can get hard currency is through exports (or through more loans), this means that the average Latin American country took one-third of the hard currency that it had earned through exports and immediately sent it back to the core instead of using it to import goods that could meet consumer needs or spur development.

Furthermore, heavily indebted countries with high debt service burdens can be in the ironic situation that even as they pay their debt service, their total level of indebtedness may actually be increasing. Typically, most of the funds in a debt service payment are used to pay back interest while little progress is made in reducing the principal of the loan. This is similar to making the minimum payment on your credit card debt. Interest continues to build on the remaining principal and, if the interest rate is high enough, the amount of debt may actually increase. Obtaining new loans to service old debts also can lead to a net increase in total debt burden. Thus, there is a net transfer of wealth from LDCs to MDCs (in 2000, less developed countries remitted about $377 billion in debt servicing while receiving only $272 billion in new loans), but at the same time their debt burden continues to grow.

In this module, we look at how South America got into this condition of extreme indebtedness and how it is affecting the region's prospects for development. The module concludes with readings from the ongoing debate over whether some (or all) of the region's debt should be forgiven.

South America's History of Dependence

Although South America's experience with debt is in many ways similar to that of other LDC regions (see Figure 9-1 for a map that indicates the world's MDC and LDC regions), its experience is also somewhat unique. South America's unusually high level of indebtedness is a result of a range of factors including its resource endowment, its history of relations with the core, and its particular cultural geography. This section provides an overview of the geography of South America so as to introduce the special role of debt in the region.

In the late 1400s, as European states' competition for world trade intensified, Europe turned its sights westward. Explorers began looking for a shortcut to China and India (with whom Europe was already trading), and some believed that they could get to Asia quickly by heading west across the Atlantic, rather than south around Africa. When Christopher Columbus "discovered" the Americas in 1492, he thought that he had landed on islands off the coast of India; that's why the Caribbean islands are still sometimes called the West Indies and why natives of the Americas are sometimes called Indians.

A number of European countries—Denmark, England, France, The Netherlands, Portugal, and Spain—soon became active on the Caribbean islands and some of the mainland areas bordering the Caribbean. They established plantations where they forced the native populations to work. After much of the native population was wiped out, Europe began bringing in African slaves to work on the plantations. These plantations focused on tropical cash crops such as sugar, tobacco, and bananas (see the module accompanying Chapter 12 for more on Central America and the Caribbean).

In contrast, in the rest of Latin America (more or less the region that we are here calling South America), only two European states successfully established colonies: Portugal, which colonized what is now Brazil, and Spain, which colonized the remainder of the region. Much of the land in South America was not suitable for the kind of plantation-based tropical agriculture practiced in the Caribbean, so these colonies were governed with a number of different goals in mind:

- *Mining*. Early colonists especially sought silver, which was useful for trading with China. At the time of U.S. independence (1776), the largest city in the Americas was Potosí, a silver mining town in what is now Bolivia.
- *Missionary activities*. Spain and Portugal were both Catholic countries and their governments were in the Americas in part to carry out the holy duty of converting the indigenous peoples.
- *Nonplantation agriculture*. As time went by, some areas of South America began to be developed by the colonists for products—such as coffee, cattle, and wheat—that either could not be grown in Europe or could not be grown in sufficient quantities there.
- *Markets*. Spain and Portugal encouraged settlers and indigenous peoples in the Americas to purchase manufactured goods from their "parent" countries.

Whatever the specific goals of Spanish and Portuguese colonialism, both countries governed their colonies with the general idea that the purpose of a colony was to support its parent country. Thus, a Spaniard sent to Colombia to grow coffee was sent there not to develop Colombia but to generate revenue for the Spanish treasury. To ensure that all profits from the coffee farm found their way back to Spain and did not "leak" to other European countries or stay in the Americas, the farmer was required to trade only with Spanish ships (which bought the coffee at an artificially low price set by the Spanish, since others were not allowed to outbid the Spanish buyers). Additionally, the coffee farmer was required to buy any manufactured products that he wanted from Spanish ships (again, at very high prices set by the Spanish to benefit Spain). And if the farmer wanted to establish his own industrial enterprise so that he would not be so dependent on Spain, the Spanish would close it down. (Portugal's relationship to Brazil was a bit different, but it largely paralleled relations between Spain and its colonies.)

To ensure that this economic system, known as mercantilism, ran as smoothly as possible, the Spanish mandated that the colonies be ruled by Peninsulares—Spanish-born individuals who had no roots in the colonies and whose loyalty would lie solely with Spain. This especially irked the Creoles, or Criollos—individuals who were of pure Spanish descent but whose loyalty to Spain was questioned because they had been born in the Americas. Below the Peninsulares and Creoles in the Spanish colonial social system were the Mestizos, people of mixed European (white) and Native ("Indian") descent; Africans (especially in the areas bordering the Caribbean); and, at the very bottom, the Native peoples, who continue to occupy the bottom of the socioeconomic ladder in Latin American society.

As Creole society advanced through the sixteenth, seventeenth, and eighteenth centuries, wealthy Creoles grew to resent being ruled by a Spanish government that was intent on retaining the colonies as poorly developed areas whose sole purpose was to serve Spain. Inspired by the recent independence of Britain's colonies in North America in 1776 (what became the United States) as well as the French Revolution in 1789, Central and South American Creoles staged a series of revolutions, and, outside the Caribbean, the various countries of Latin America achieved independence between 1810 and 1825. The Creoles were aided by Britain, which was the strongest European country at this time and had an interest in establishing trade and investment links with the region.

Notwithstanding their newfound political independence, the countries of early nineteenth-century South America found that their economies had been badly distorted by three hundred years of colonialism and that they could not easily recover. Although Spain and Portugal had invested heavily in specific areas of the economy that had served their needs (such as building roads from key production regions to ports), they had neglected making other investments that would have been needed for overall national development (such as building roads linking the various parts of the country, or providing health care and education for its citizens). The newly independent countries of South America were in no position to come up with badly needed investment money on their own, so they turned to a new set of outsiders who could "help" them: corporations and individual investors that, at first, were mostly British and then, later on, were increasingly from the United States. This newer system of domination from outside, without formal colonization, is often described as a new round of colonialism, or neocolonialism.

Thus, at the time of independence, South America exhibited a unique combination of factors and people that included the following:

- a productive infrastructure oriented toward extracting resources and wealth from the region
- a local elite (many of whose families dated back to the 1500s) who sought to develop the region
- outsiders from core regions (first Britain and later the United States) who were looking forward to helping the local elite in their efforts at developing the region, so long as they could get a piece of the action

As it turned out, this confluence of factors was a recipe for indebtedness.

From Underdevelopment to Debt

The neocolonialists from Britain and the United States, like the old colonialists from Spain and Portugal, were most interested in specific sectors of South American economies. These economic activities tended to be in the primary sector of the economy: agriculture, mining, fishing, and forestry—industries based on the harvesting of raw nature rather than the processing of manufactured goods. Most investment by outsiders in South America, especially prior to World War II, was in this sector, whether in the form of loans or as foreign direct investment. By the mid-twentieth century, however, many South American academics, businessmen, and politicians were questioning whether this pattern of investment would ever bring about the kind of development that their ancestors had sought when declaring independence more than a century earlier. Many charged that the prevailing pattern of foreign investment by Britain and the United States was just continuing the extraction of resources from the region that had been practiced by the Spanish and Portuguese under mercantilism. Furthermore, the purchasing power of individuals and enterprises in these countries was declining, because the price of imported manufactured goods kept rising while the price of raw-material exports stagnated. South American intellectuals coined the phrase "the development of underdevelopment" to describe what had happened in their region since independence.

In response, during the 1950s and 1960s, a number of South American countries began promoting import substitution industrialization (ISI), a variant of the "self-sufficiency" development strategy described in the textbook. Under ISI, national economies attempted to redirect themselves from producing raw materials for export to producing manufactured goods for internal consumption. The hope was that this strategy would lead to continual growth of the manufacturing sector, so that in the long-term South American countries could begin exporting the high-value manufactured goods that they now were importing.

ISI met a number of obstacles. Markets for locally produced manufactures were often limited and the quality of these goods was often poor, so that the small number of individuals with the money to buy, say, a locally produced automobile might instead pay extra to import a more prestigious and more reliable car produced in an MDC. ISI industries tended to concentrate in major cities, leading large numbers of unemployed workers and landless peasants to flock there in the hope of securing a job in this relatively well-paid sector of the economy. This internal migration has done much to contribute to urban squalor around metropolitan areas of LDCs, described in Chapter 13 of the textbook. The strategy also created, in many countries, politically strong governments that represented very narrow economic interests, a situation that likely has contributed to the region's propensity for frequent and violent changes of government, whether civilian or military. Most significantly, ISI industries required significant amounts of hard currency for importing components, machinery, and technology, but national economies were now being directed away from producing the exports that could earn that hard currency. As a result, South American countries sought to borrow money on the world financial market.

Fortunately, just as South American governments and firms realized that they needed fresh infusions of hard currency in order to support their ISI strategies, American, European, and Japanese banks were flush with money to lend them. In the early 1970s, a spike in petroleum prices led oil-rich countries in Southwest Asia and North Africa to make huge hard-currency deposits with MDC banks. The banks faced a dilemma, however, as there were few lending opportunities at home (largely because the world economy was undergoing a recession thanks to that same spike in oil prices that made so much money available). Since banks try to avoid sitting on money (they obtain revenue only by lending money out), they sought out new customers in the emerging LDC markets. Cash-hungry, relatively highly developed South America seemed like the perfect customer. There was just one problem: No one stopped to consider how these countries would be able to pay the money back. During the 1980s, the situation became worse as commodity prices fell, energy prices rose, interests rates rose (many of the loans were made with floating interest rates), and a sustained global recession led MDC consumers to reduce their purchases of LDC-produced goods.

To Forgive, to Forget, or to Foreclose?

It should come as no surprise that debtor nations want parts of their debt written off; after all, every debtor would like to have her or his loan forgiven. But, more surprisingly, many lender nations and institutions have also joined the call for debt forgiveness.

Among the individuals and institutions with a particular interest in reducing LDC debt are those with a stake in the stability of the world financial system. If, say, Brazil were to default on its debt, then the major commercial banks, such as Citigroup, would have huge losses and might even fail. Your personal bank account would be safe (the U.S. government's Federal Deposit Insurance Corporation insures personal savings accounts up to $100,000), but companies and individuals with savings invested in other financial instruments might suffer major losses. The overall lack of confidence in the financial system that would result from major bank failures would have huge impacts, and, of course, governments in the MDCs don't want this to happen. Additionally, some of the money borrowed by South American governments is owed directly to MDC governments, who stand to lose if repayment is not made. In short, the global financial system is vulnerable to debt defaults by developing countries.

Because of these concerns, the International Monetary Fund (IMF), an organization to which most of the world's governments belong but which is controlled by the wealthiest countries, has agreed to guarantee many Third World loans. This means that if the government of Brazil goes to Citigroup seeking a loan so that it can keep making interest payments on past loans, and Citigroup says that it does not want to assume this risk, the IMF would step in and tell Citigroup that it will cover the new loan in case Brazil defaults. At this point Citigroup can go ahead and lend the money to Brazil, risk-free. The IMF is providing a crucial service to Brazil by agreeing to guarantee the loan, and the IMF has an interest in Brazil being able to make its payments to Citigroup so that the IMF won't have to make the payments on Brazil's behalf. This situation gives the IMF the power and the interest to act as an international debt counselor. (In addition, the IMF sometimes makes its own low-interest loans, especially to the poorest and least developed countries.)

When the IMF gets involved in international debt counseling, it places conditions, known as structural adjustment policies (SAPs), on debtor countries' spending habits. The goal of structural adjustment is to restructure a debtor country's internal budgeting and its economic policy so that it will have the greatest chance of being able to pay back its loans (see the module accompanying Chapter 8 for a discussion of the implementation of SAPs as a precondition for Eastern European countries seeking to join the European Union). Specific policies that debtor countries are required to follow typically include the following:

- Cutting back on social service expenses. Social service expenditures are discouraged by the IMF because they are seen as diverting cash that otherwise could be used to make payments to foreign banks or that could be used to finance industrial enterprises that could generate cash for loan repayments.
- Privatizing state-managed industries and opening these industries, and the economy in general, to foreign investors. The IMF favors privatization because the foreign investments that privatization would attract could be used to develop industries that could generate cash for loan repayments.
- Reorienting the national economy so that it focuses on exports. The IMF favors economies based on exports rather than import substitution, because exports generate the cash that can be used to pay back loans, while production for domestic consumption fails to generate cash that can be sent to foreign banks.
- Devaluing the country's currency. The IMF's support for devaluation is closely linked with its support of production of goods for export. Devaluation has the effect of raising the price of imports and lowering the price of exports. The hope is that this change will make the country's exports more competitive, which should bring in more cash that can be used for loan repayments.

After several decades of SAPs, many economists have concluded that structural adjustment by itself has not achieved its goals. In fact, opponents of structural adjustment note that not only have these policies failed to spur development but they have made life worse for the poor by reducing social services and increasing the price of necessary imported goods.

Many economists, from across the political spectrum, argue that a better solution might be to forgive at least part of less developed countries' debts, so that precious hard-currency reserves can be used to spur economic development rather than just being wired back to the core as interest payments. Even the IMF has instituted a program of partial debt forgiveness for some of the world's most indebted poor countries, although countries are allowed to enroll in this debt-forgiveness program only if they also undertake an IMF-directed structural adjustment program. The G-8, the group of more developed countries that makes most bilateral loans to the less developed countries and that also provides most of the funding for the IMF, similarly has taken a position in support of partial loan forgiveness.

It seems likely that when the bankers at the IMF and the finance ministers of the G-8 countries promote debt forgiveness they are driven not simply by a sense of charity but also by the belief that the principal on the loans will never be paid back anyway and that it is in the interest of global capitalism to have a financially solvent periphery. As Alex Singleton of Britain's conservative Adam Smith Institute wrote in a letter to *The Independent* newspaper in 1999:

> These were often ill-considered loans lent by ill-advised banks to illegitimate governments. The capital is gone, in most cases wasted, and repayment comes from what little income these countries generate. By cancelling that debt on a one-off basis, we not only raise the living standards of the desperately poor, but we give them a chance, and the investment, to embark on that upward path which generates growth, wealth and jobs. Cancellation is [in] our interest as well as theirs.

This point is developed further in the first reading, an article by John Serieux from *Review*, the newsletter of Canada's North-South Institute.

Others have argued for debt relief for what appear to be more purely humanitarian reasons. A global coalition of celebrities, most notably U2 singer Bono, and religious leaders, most notably Pope John Paul II, were leading members of the Jubilee 2000 coalition, a group that argued for debt forgiveness so that poor countries could get a fresh start with the new millennium and whose members continue to argue for debt forgiveness. These humanitarians advocate debt forgiveness as part of an overall strategy that stresses poverty reduction, instead of development, as the primary goal of economic policy. In the second reading, Bono, in a CNN interview, discusses his continuing role in the debt-forgiveness campaign as part of an effort to improve the well-being of people in LDCs.

Other arguments for debt reduction are made in the third and fourth readings. Issues of global finance are rarely paired either with feminism or philosophy, but the third reading reports on how ethicist Alison Jaggar uses principles of feminist philosophy to argue that the core has a moral duty to forgive the debt of countries in the periphery. In the fourth reading, Costa Rican environmentalist Gabriel Rivas-Ducca argues that the need for hard currency for debt servicing has led the countries of Latin America to extract and sell off their natural resources at an irresponsible and unsustainable rate.

Not everyone supports debt relief, however. In the fifth reading, correspondent Geri Smith writes in *Business Week* that Latin America needs more, not less of the kind of changes brought about by structural adjustment programs. If countries were to pursue these market-oriented strategies, she argues, their economies should be strong enough to permit making their required debt servicing payments. In the next reading, from the academic journal *Foreign Policy*, economist William Easterly specifically takes on the IMF's program that combines the "stick" of structural adjustment with the "carrot" of debt forgiveness, arguing that if debt is forgiven the money saved would just be squandered by LDCs' corrupt governments. Instead, he argues, institutions like the IMF should work with political leaders committed to growth-oriented economic policies. In the final reading, Jubilee Research, one of the successor organizations to Jubilee 2000, publicizes a report that, at least for Africa, disputes some of Easterly's assertions.

Readings

1

from the North-South Institute

Debt Forgiveness as Good Economics and Enlightened Self-Interest

By John Serieux , 1999

After the June Cologne summit, G-8 leaders can justly claim to have heeded the call for broader, faster, and deeper debt relief.

At the annual summit, the leaders agreed to new criteria that would expand a 1996 debt relief program, reducing the debt of 33 developing countries (more than the 29 originally chosen) after three years of adjustment (instead of six), and involving more debt forgiveness than before. At first glance this commitment seems like a rather magnanimous gesture on the part of rich nations—moved, no doubt, by the moral fervor of nongovernmental organizations' campaigns for debt forgiveness. Misguided cynics, however, might see this as a gesture of *noblesse oblige.*

But is it fair or prudent to ask Canadian taxpayers, who must continue to pay their private debts, to bankroll the debt of others? The answer is yes, it is both, because it makes good economic sense and, in the long run, it is in the self-interest of the industrialized governments to do so.

Debt forgiveness has a well-established basis both in economic theory and business practice. Once it has been established that the payment of outstanding debt is beyond the ability of a debtor country, it is usually in the creditor's interest to forgive a portion of the debt in order to improve the debtor's ability (and willingness) to service the remaining debt. For private creditors, such as commercial banks, forgiveness is often *de facto*, essentially mandated by the fact that the debt is worth only a fraction of its original value on the open market. This is how a large portion of the debt of several, middle-income, Latin American countries was forgiven in the 1980s. However, when the creditor is a government or multilateral institution such as the International Monetary Fund (IMF) or World Bank, there is no market for the debt. Debt forgiveness, in this case, may simply be explicit recognition of the same reality that the market would have dictated.

In business, bankruptcy procedures apply the same principle of "partial recovery." It is the right of all firms and individuals to declare bankruptcy if they cannot possibly repay their debts. Bankruptcy procedures compel all creditors to accept only partial repayment of the debt owed to them. Market discipline and bankruptcy procedures have nothing to do with charity and everything to do with good economic practice.

A Cost-Saving Measure

Beyond its inevitability, however, forgiving developing-country debt may be a cost-saving measure for Canadians over the long run. For example, in indebted countries, the tax base is simply too narrow to allow debt repayment to be funded fully from new taxes. To raise money to repay debt, cuts in spending will be made in areas such as health, education, and public investment, and more money will be borrowed from the central bank. Therefore, part of the debt repayment burden will be borne by the less fortunate, because they will be denied basic services such as access to healthcare, clean water, education, and because their incomes will be eroded by inflation.

This has potentially disastrous implications for the futures of those countries and ultimately, high costs for Canada and other developed countries. A lack of access to healthcare means less treatment of communicable diseases that cross borders and [will] ultimately result in higher healthcare costs in countries such as Canada. Education and public investment levels are both important determinants of the rate at which these countries will grow. Economic growth means industrialized nations like Canada will benefit from the development of markets for their products and savings in foreign aid dollars. In short, in this increasingly intertwined world, the debt repayment burden is eventually borne by everyone.

For wealthy countries the choice may well be between debt forgiveness that will give these countries a chance to participate more equally in the global economy (with benefits for all) or more money spent in the future building "fortress north" or maintaining high levels of aid. In fact, the amount of debt relief being offered may still be insufficient, whether considered from the perspective of market value or the long-term cost to developed countries.

Even under the revised G-8 plan, few countries can expect to have more than 50 percent of their debt reduced, despite the fact that, in 1994, these 33 indebted countries were only able to meet one-third of required debt service payments. In contrast, in 1987–89 Bolivia (one of the countries currently eligible for debt relief) was able to buy back most of its privately held debt at 11 cents on the dollar. This implies debt forgiveness of 89 cents on the dollar—considerably more generous than the G-8's current proposal.

Both of these facts suggest the G-8 proposals overestimate the proportion of outstanding debt that is serviceable without duress. It is, therefore, highly likely that we will look back one day and conclude that the Cologne debt initiative did not go far enough!

Source: John Serieux, "Debt Forgiveness as Good Economics and Enlightened Self-Interest" *Review: The North-South Institute Newsletter*, 1999.

2

from the Cable News Network

U2 Star Bono: "Drop the Debt"

January 2, 2002

LONDON, England (CNN)—A recent report to the World Health Organization suggests that eight million lives a year could be saved and billions of extra dollars generated if the world's poorest nations spent more on health care.

But according to the "Drop the Debt" organisation, these countries are trapped by the sums they have to spend on servicing debt.

U2 front man Bono has been explaining why he's lent his voice to the "Drop the Debt" campaign, and he recently spoke to CNN's Tom Bogdanowicz.

CNN: Why have you taken up the cause of health in the developing world?

Bono: I'm here because I was part of the Jubilee 2000 Drop the Debt campaign, and the issue of poverty in the poorest countries in the world is completely bound up in health and education. In fact, there's still more of the poorest countries spending more on their debts, their old debts, than on health and education.

Q: You're a rock star—shouldn't you leave issues like health to health and government officials?

A: I absolutely agree with you, and I really would rather not be here in the sense that I know how absurd it is having to listen to a rock star talk about the World Health Organization. But the truth of it is that politics and pop are very similar . . . so I guess I'm the person to try and, you know, sort of try to bring in the wider public.

Q: You're a rock star who has the ear of some politicians, including British Prime Minister Tony Blair. The World Health Organization is looking for some $27 billion extra in aid. What signals are you getting? Are people prepared, are governments prepared to put out this sort of money?

A: This is again where I suppose I have to stress the importance of making this a popular issue. And how do we get the melody line of the suggestions contained in this report out to people in the real world? That's a hard thing to do. But when we do, I think it will become, I think people will be very motivated to do something about it, and if people are motivated about it, then politicians will be.

I have to say, here in London and in the United Kingdom, you have a triumvirate of people who are very committed to these issues. You've Tony Blair at the top, you have (Chancellor of the Exchequer) Gordon Brown who made a historic speech on this subject just recently at the Federal Reserve, and then (overseas aide minister) Claire Short, who I have to say is far more rock 'n' roll than I'll ever be, and between the three of them I think there's a real passion about this.

I don't think they're playing politics, this is too important to play politics about, and I think we'll get that commitment in the UK, and I think we'll get it elsewhere too.

The Americans are a little slow at the moment in committing, but I think in America there's a sort of suspicion about aid, you know, because a lot of it's been wasted over the years. It's like the welfare debate, you know, in America people want to give the fishing rod, not the fish. That's precisely what this report is about, giving people the means with which they can take of themselves.

Q: You're very popular in the United States, but this sort of issue is not. Do you think the United States government is prepared to come on board?

A: I'm going to meet the secretary of the treasury of the United States in Africa, and he told me that the United States is prepared to gather around with continued support where they see progress, and I hope to be able to show them some progress. Places like Uganda, since they had their debts cancelled, have double enrollment in schools. There's been significant achievements in the last few years, I want to show him that.

I've been in Congress and in the Senate, and ... every American politician I could meet, I've met. And I have a feeling that September 11 has changed America forever. I think before, it was a continent acting like an island. I think now it is clear that you cannot be an island of prosperity in a sea of despair. And if September 11 has taught us anything, it's taught us that the world is a much more interdependent and interconnected place.

So I think the Americans will get this. And is there a better or more fitting memorial for the people who lost their lives on September 11 than to see the world significantly changed for the better? You know, not just the pursuit of justice, but the pursuit of a fairer, more inclusive world, because that actually is the route to peace.

Take the country Botswana, for instance. Forty percent HIV infection. You're in a crowded street, in a marketplace, and you look around, and nearly half the people you're standing with have a death sentence. That kind of despair, that kind of lack of hope when you can't get access to medicines to treat you, is exactly the kind of powder keg that set off Afghanistan.

The abject poverty makes people very susceptible to the likes of al Qaeda and ... so I think, if you look at Africa, there's potentially another 10 Afghanistans in Africa. Are we going to leave them the way we left Afghanistan? And is it cheaper to actually prevent the fires from happening than putting them out? I can tell you by a factor of a hundred it is.

Q: You've been involved with the campaign for debt forgiveness. Do you now think that the issue of health in the developing world is perhaps more urgent?

A: I think they're the same thing, as I was saying earlier, you know, a lot of the poorest countries are still paying more (on) servicing old debts than they are on health care and education. That's unacceptable. Health, and particularly HIV/AIDS and malaria, set back development 20 years. So there's no point in us cancelling their debts if there's no one there.

Q: What do you prefer doing, good works or making music?

A: I'd much rather be in a rehearsal room or studio making rock 'n' roll than dressing up in a suit and tie and representing the World Health Organization. I think I'm better at it, I think I'm better suited to it. But you know what—they asked me here and I'm proud of the work I've done over the last few years on the debt campaign. And if they need a voice, they've got mine.

Source: "U2 Star Bono: Drop the Debt" transcript posted January 2, 2002.

3

from the University of Alberta

Debt "Forgiveness" Insults Poor Nations—Philosopher

By Geoff McMaster

January 18, 2002—Canceling the debts of developing nations is a moral obligation rather than an act of generosity by the industrialized north, says a leading feminist philosopher.

Because northern economies are built on the resources of the global "south" (shorthand for the world's impoverished countries), southern countries have already paid far more than they owe, argued Dr. Alison Jaggar, professor of women's studies and philosophy at the University of Colorado at Boulder. Canceling their debts would merely be an initial gesture in a long-term strategy of reparations.

" 'Forgiveness' of debt seems to me an inappropriate and misleading term," Jaggar said at the U of A Thursday during her lecture, *Who Owes What to Whom: a Feminist Perspective on the Southern Debt.*

"It is the north rather than the south that should be given moral, if not economic, forgiveness."

Widely considered to have defined the field of feminist philosophy during 1970s, Jaggar is a scholar of international renown, specializing most recently in moral and philosophical issues raised by the integration of the global economy. She blames free trade and "neo-liberal" globalization for the growing gap between northern wealth and southern poverty, recently described by the United Nations as one of "grotesque proportions."

The world's current debtor nations began borrowing during the 1970s when interest rates were low, but with the debt crisis of the mid-1980s they were surrendering three times as much in debt repayment as they were receiving in foreign aid—a total of some $717 million US per day. By 1997 the total debt load of the debtor countries was $217 trillion.

The only way these countries can make payments on the debt is by exchanging resources for foreign currency, a relationship that renders them dependent suppliers of raw materials to northern industry. Therefore, contrary to the reasoning of the World Trade Organization, integrating into a global economy does debtor nations more harm than good, ensnaring them in a cycle of exploitation passed down through "centuries of colonialism."

"The debt functions as a drain through which the resources of the southern countries are siphoned abroad," said Jaggar. "It functions as a shackle on independent development, keeping highly indebted countries trapped in a global trading system they're not able to abandon if they're going to earn the foreign currency necessary to service their debt."

But her analysis has what she calls a "feminist twist," because roughly 70 percent of the world's poor are women. Missing from any accounting of "who owes what to whom" are the concealed contributions made by legions of women to the world economy, she said, and the circumstances that "keep them trapped in a permanent condition of indentured servitude."

The ideology of neo-liberalism, and the growing push to privatization in the global economy, encourages governments to abandon their social responsibility to provide adequate housing, health care, disability and employment for its people. The lack of such support forces women to adopt "survival strategies" to compensate, such as caring for ailing family members at home.

"In the global south, more domestic work for women has resulted in higher drop-out rates for girls. Less education and longer hours of domestic work obviously contribute to women's impoverishment by making it harder for them to get paying jobs."

Because of a "fiction of accounting" in the global economy, none of the unpaid labour taken by women registers anywhere as wealth or productivity. And yet many of the children given birth to, raised and cared for by women, end up emigrating to northern countries and contributing in significant ways to those economies.

In the United States, for instance, foreign workers make up about 12–15 percent of the workforce, and many economists believe the American economic boom of the 1990s was "fueled by undocumented immigrant labour." In that light, immigration is therefore a form of "reverse foreign aid" to the north.

"Southern women are often portrayed as uncontrolled breeders, producing a burgeoning population that overwhelms the resources of developing countries, but their children are a mainstay of many northern industries ... [women] bear the burden of producing these workers."

Damage to the environment caused by northern industry and military, such as radioactive contamination and global warming, also fails to make any balance sheet of global economics. And the heaviest burden of this damage falls on the poor, especially women.

"When the ongoing environmental destruction by the global north is taken into account, [debt cancellation] starts to look less like charity and more like highly inadequate compensation for damages."

Source: Geoff McMaster, "Debt 'Forgiveness' Insults Poor Nations," *Express News*, posted Jan. 18, 2002. Copyright © 2002 Express News. Used with permission.

from Friends of the Earth International

4

Debt, Energy and Materials Flows

The Looting of Latin America

Gabriel Rivas-Ducca
July/September 2000

As ecologists, we have to go to the roots of problems; in other words, we must be radical. Until now, we have clearly not gone deep enough when addressing the structural economic changes that must be made in order to achieve sustainable societies.

Take, for example, analyses about the impacts of energy and materials flows. Given that one of the conditions for a sustainable society is a minimal use of resources, those of us living in mineral-rich countries are particularly affected by mining's ecological and social rucksacks. We must question such materials flows, both in terms of quantity (the need to reduce the total volume produced) and the direction of movement (from South to North).

As we are also bearing the consequences of the burden of external debt (one that has already been paid many times over), the relation between this debt and materials flows is of primary importance.

As Jacobo Schatan states: "When a country sends a given amount of dollars, yens or marks to pay the interest and amortization of its debt, it is in fact sending abroad a certain quantity of material resources and incorporated human labour." In other words, indebted countries are obliged to exploit their natural resources in order to obtain the hard currency needed to pay their debts.

They may also be obliged to sell state companies in the process of privatization, and this cycle increases as terms of exchange deteriorate, with raw materials becoming cheaper and final products more expensive.

According to the Economic Commission for Latin America, the region's external debt jumped from US$459 billion in 1991 to $750 billion in 1999. Alan Garcia has calculated that between 1979 and 1999 the Latin American countries paid the astronomical amount of $1,165 billion in external debt to the World Bank, the International Monetary Fund and Inter-American Development Bank.

The Real Looting

Researchers like Schatan and Eduardo Gudynas have made very important contributions to understanding the relation between the flows of energy and materials, the ecological and external debt and the impacts on environment and society. Schatan, integrating data from between 1980 and 1995 on external debt figures and energy and materials flows by looking at the 17 most important exported products, has calculated the material balance (the "real looting") of Latin America's external debt.

His conclusions are impressive: the external debt has increased more than three times in physical terms, from 1015 million tons of products that theoretically "had to be" exported in 1980, to 3694 tons in 1995. In contrast, the monetary value of these products has increased more slowly: from US$220 billion in 1980 to $598 billion in 1995.

Despite the fact that an average of 235 million tons of products per year were exported from the region between 1985 and 1996, this enormous amount was not enough to reduce the external debt, which in fact increased from $379 billion in 1985 to $607 billion in 1996. This can be explained by the drop in the average price of an exported ton of basic products from $217 in 1980 to $162 in 1995. But what is even more shocking is that the value of the 235 million tons of goods exported in 1996 is not even enough to pay the accumulated interest on the debt!

What are the environmental and social impacts of this "destruction of resources in exchange for nothing," as Schatan puts it? We can guess the answer by looking at the composition of Latin American exports (73 percent primary products and 27 percent manufactured products) and the impacts associated with the production of raw materials. A typical ton of primary products exported from Latin America consists of the following: 88 percent non-renewable mining products (including oil), 3 percent fish products including shrimp (renewable but highly fragile to overexploitation and produced at the cost of ecosystem destruction), and 9 percent agricultural products (including bananas, coffee, and genetically-modified soybeans, maize and sugar).

Heaven or Hell?

What does the future hold for Latin America? There are two possibilities. The grim version will come about if current rates of energy and materials flows continue, with no external debt reduction and all of the accompanying social and environmental impacts. This scenario leaves Latin American countries more vulnerable to climate change and the other hell-like attributes of a non-sustainable society.

Or, the ideal future: Friends of the Earth's vision for sustainable societies becomes reality due to our struggle, joined by millions. The external debt is dropped, ecological debt recognized, economies become ecologically-driven, external trade has been reduced, many mines have been closed, oil exploration and exploitation have ceased, ecosystems and cultures are respected, environments are healthy, and societies are socially-just. It sounds more or less like heaven. Which future are you working for?

Source: Gabriel Rivas-Ducca, "Debt, Energy and Materials Flows: The Looting of Latin America," *Link*, July/September 2000.

from *Business Week*

5

Down in the Dumps in Latin America

With Growth Slumping, Is the Region Headed for Another "Lost Decade"?

By Geri Smith
July 29, 2002

When night falls in Buenos Aires, poor Argentines start pawing through garbage bags outside restaurants, in search of their family's next meal. In Arequipa, Peru, violent protests recently forced the government to shelve plans to privatize power plants. Many Venezuelans are agitating for the ouster of their President as the currency plummets. Meanwhile, Brazil's Central Bank chief is scrambling to reassure jittery investors that the country's massive public debt is payable even if a leftist triumphs in October's presidential election.

Concerned by the turbulence, U.S. Treasury Secretary Paul H. O'Neill will travel to Brazil and Argentina at the end of July to see how Washington, which has adopted a hands-off approach for most of the past year, can help those beleaguered nations. What he will see isn't pretty: disillusioned voters, panicky investors, rising unemployment, swooning currencies, and leaders desperate for solutions.

It might be tempting for the Bush Administration, its hands full dealing with the domestic U.S. financial crisis, to view Latin America's problems as a regrettable but temporary consequence of unprecedented global economic turmoil. Yet the forces at work in Latin America may be longer-lasting and more pernicious. After a decade of free-market reforms that dramatically improved the region's physical infrastructure but failed to generate sustained growth, jobs, and poverty reduction, Latin Americans are losing faith in the virtues of open markets. Indeed, Latin America appears to be slipping into another "lost decade."

Investors and policymakers remember all too well the previous lost decade. That was in the 1980s, when a string of Latin countries defaulted on their foreign debt, plunging much of the region into a deep recession and starving economies of credit and investment. The human cost was huge. Between 1980 and 1989, per capita gross domestic product fell by nearly 1% per year on average.

Then, Latin America seemed to begin a new chapter. Growth revved up in the mid-1990s, as governments stamped out inflation and courted foreign investors. Peru clocked 8.6% growth in 1996 and Argentina expanded 8% in 1997. But the party did not last long. Today, with a few exceptions, much of Latin America is either stagnant or in crisis. "The region is in very serious trouble," says Riordan Roett, director of the Western Hemisphere program at Johns Hopkins University. "We're in for a long period of stop-start reforms, muddling-through, and populism that will put it even further behind the Asian tigers."

Latin America still has not managed to shake free of its historic boom-and-bust pattern. The double whammy of the Asian and Russian crisis laid much of the region low, and growth has since faltered. Since 1998, it has averaged just 1.5% a year, only slightly above the anemic 1.1% of the 1980s. It's expected to fall to zero this year. "We already have racked up a half-decade of stagnation in per-capita GDP," says José Antonio Ocampo, executive secretary of the U.N.'s Economic Commission for Latin America & the Caribbean (ECLAC). "Some of us thought the expectations raised about the effects of the reforms were unrealistic. Unfortunately, we were right."

Even if growth rebounds to 3% or so—respectable by U.S. standards—it would still fall far short of what Latin America needs to make a lasting dent in entrenched poverty. "We're creating a huge social problem. We need at least 4% to 5% growth a year," says Ernesto Heinzelmann, president and CEO of Brazil's Embraco, a maker of refrigeration systems with annual sales of $646 million. Heinzelmann is speaking of Brazil, but he might as well be referring to the entire continent.

Because most Latin American countries haven't done enough to modernize their economies, build dynamic export sectors, or boost internal savings rates, they remain dependent on foreign capital. "The problem is these countries don't realize that going halfway in reforms is as dangerous as doing nothing—it's like building half a dam," says Drausio Giacomelli, an analyst at J.P. Morgan Chase & Co. in New York. Argentina is the classic example of an incomplete Latin reform. It smothered inflation by pegging its currency to the dollar—but failed to rein in spending and got hooked on foreign borrowing.

The Bush Administration has adopted a tough-love approach, arguing that Latin countries need to do more to help themselves. "Latin America ... made some good reforms in the '90s," says U.S. Treasury Under Secretary John B. Taylor. "[But] there needs to be more in the way of things that will raise growth."

Taylor is right. But it's the rare Latin politician nowadays who dares wave the free-market banner. "It's going to be very hard to win an election if you say you're going to globalize, privatize, and do structural reforms," says Moisés Naim, editor and publisher of *Foreign Policy* magazine and a former Venezuelan Finance Minister.

Populism is on the rise, helping spur a flight of credit from the region. Yield spreads on Latin bonds over U.S. Treasuries, a traditional measure of country risk, have soared above 1,000 basis points. It isn't just Wall Street that's spooked. Even the more steel-nerved multinationals and other foreign direct investors are retrenching. Inflows of foreign direct investment into Latin America were down 30% in 2001 from their 1999 peak of $105 billion. Many expect another sharp drop this year.

The worst-case scenario would be a Brazilian debt default. Some ratings agencies put the risk of that in the neighborhood of 30%. Luiz Inácio "Lula" da Silva, the candidate of the leftist Workers Party, is the current front-runner in the presidential election. Fears that Lula will steer the country onto a populist path have sent the real down 20% since January. That has pushed up the cost of servicing the country's $290 billion public debt, much of which is indexed to the exchange rate. "If the level continues to rise, we'll have a debt crisis," says Alberto Borges Matias, an economics professor at the University of Sao Paulo. Coming hard on the heels of Argentina's $141 billion debt default in December, a Brazilian default could trigger a regional chain reaction.

A dark prospect. Yet this does not have to be Latin America's fate. The region would do well to follow the lead of Chile and Mexico, which have prospered by making exports the pillar of their development. Even they are vulnerable these days, though they are better off than their neighbors. Unless the region's reform movement regains its momentum, these two countries will remain the exception rather than the rule.

Source: Geri Smith, "Commentary: Down in the Dumps in Latin America," *Business Week*, July 29, 2002.

6

from *Foreign Policy*

Think Again: Debt Relief

By William Easterly
November/December 2001

Debt relief has become the feel-good economic policy of the new millennium, trumpeted by Irish rock star Bono, Pope John Paul II, and virtually everyone in between. But despite its overwhelming popularity among policymakers and the public, debt relief is a bad deal for the world's poor. By transferring scarce resources to corrupt governments with proven track records of misusing aid, debt forgiveness might only aggravate poverty among the world's most vulnerable populations.

"Jubilee 2000 Sparked the Debt Relief Movement"

No. Sorry, Bono, but debt relief is not new. As long ago as 1967, the U.N. Conference on Trade and Development argued that debt service payments in many poor nations had reached "critical situations." A decade later, official bilateral creditors wrote off $6 billion in debt to 45 poor countries. In 1984, a World Bank report on Africa suggested that financial support packages for countries in the region should include "multiyear debt relief and longer grace periods." Since 1987, successive G-7 summits have offered increasingly lenient terms, such as postponement of repayment deadlines, on debts owed by poor countries. (Ironically, each new batch of terms and conditions was named after the opulent site of the G-7 meeting, such as the "Venice terms," the "Toronto terms," and the "London terms.") In the late 1980s and 1990s, the World Bank and International Monetary Fund (IMF) began offering special loan programs to African nations, essentially allowing governments to pay back high-interest loans with low-interest loans—just as real a form of debt relief as partial forgiveness of the loans. The World Bank and IMF's more recent and well-publicized Highly Indebted Poor Countries (HIPC) debt relief program therefore represents but a deepening of earlier efforts to reduce the debt burdens of the world's poorest nations. Remarkably, the HIPC nations kept borrowing enough new funds in the 1980s and 1990s to more than offset the past debt relief: From 1989 to 1997, debt forgiveness for the 41 nations now designated as HIPCs reached $33 billion, while new borrowing for the same countries totaled $41 billion.

So by the time the Jubilee 2000 movement began spreading its debt relief gospel in the late 1990s, a wide constituency for alleviating poor nations' debt already existed. However, Jubilee 2000 and other pro-debt relief groups succeeded in raising the visibility and popularity of the issue to unprecedented heights. High-profile endorsements range from Irish rock star Bono to Pope John Paul II and the Dalai Lama to Harvard economist Jeffrey Sachs; even retiring U.S. Sen. Jesse Helms has climbed onto the debt relief bandwagon. In that respect, Jubilee 2000 (rechristened "Drop the Debt" before the organization's campaign officially ended on July 31, 2001) should be commended for putting the world's poor on the agenda—at a time when most people in rich nations simply don't care—even if the organization's proselytizing efforts inevitably oversimplify the problems of foreign debt.

"Third World Debts Are Illegitimate"

Unhelpful idea. Supporters of debt relief programs have often argued that new democratic governments in poor nations should not be forced to honor the debts that were incurred and mismanaged long ago by their corrupt and dictatorial predecessors. Certainly, some justice would be served if a legitimate and reformist new government refused to repay creditors foolish enough to have lent to a rotten old autocracy. But, in reality, there are few clear-cut political breaks with a corrupt past. The political factors that make governments corrupt tend to persist over time. How "clean" must the new government be to represent a complete departure from the misdeeds of an earlier regime? Consider President Yoweri Museveni of Uganda, about the strongest possible example of a change from the past—in his case, the notorious past of Ugandan strongman Idi Amin. Yet even Museveni's government continues to spend money on questionable military adventures in the Democratic Republic of the Congo. Would Museveni qualify for debt relief under the "good new government" principle? And suppose a long-time corrupt politician remains in power, such as Kenyan President

Daniel Arap Moi. True justice would instead call for such leaders to pay back some of their loot to development agencies, who could then lend the money to a government with cleaner hands—a highly unlikely scenario.

Making debt forgiveness contingent on the supposed "illegitimacy" of the original borrower simply creates perverse incentives by directing scarce aid resources to countries that have best proved their capacity to mismanage such funds. For example, Ivory Coast built not just one but two new national capitals in the hometowns of the country's previous rulers as it was piling up debt. Then it had a military coup and a tainted election. Is that the environment in which aid will be well used? Meanwhile, poor nations that did not mismanage their aid loans so badly—such as India and Bangladesh—now do not qualify for debt relief, even though their governments would likely put fresh aid resources to much better use.

Finally, the legitimacy rationale raises serious reputation concerns in the world's financial markets. Few private lenders will wish to provide fresh financing to a country if they know that a successor government has the right to repudiate the earlier debt as illegitimate. For the legitimacy argument to be at all convincing, the countries in question must show a huge and permanent change from the corruption of past regimes. Indeed, strict application of such a standard introduces the dread specter of "conditionality," i.e., the imposition of burdensome policy requirements on developing nations in exchange for assistance from international financial institutions. Only rather than focusing solely on economic policy conditions, the international lending agencies granting debt relief would now be compelled to make increasingly subjective judgments regarding a country's politics, governance structures, and adherence to the rule of law.

"Crushing Debts Worsen Third World Poverty"

Wrong in more ways than one. Yes, the total long-term debt of the 41 HIPC nations grew from $47 billion in 1980 to $159 billion in 1990 to $169 billion in 1999, but in reality the foreign debt of poor countries has always been partly fictional. Whenever debt service became too onerous, the poor nations simply received new loans to repay old ones. Recent studies have found that new World Bank adjustment loans to poor countries in the 1980s and 1990s increased in lock step with mounting debt service. Likewise, another study found that official lenders tend to match increases in the payment obligations of highly indebted African countries with an increase in new loans. Indeed, over the past two decades, new lending to African countries more than covered debt service payments on old loans.

Second, debt relief advocates should remember that poor people don't owe foreign debt—their governments do. Poor nations suffer poverty not because of high debt burdens but because spendthrift governments constantly seek to redistribute the existing economic pie to privileged political élites rather than try to make the pie grow larger through sound economic policies. The debt-burdened government of Kenya managed to find enough money to reward President Moi's home region with the Eldoret International Airport in 1996, a facility that almost nobody uses.

Left to themselves, bad governments are likely to engage in new borrowing to replace the forgiven loans, so the debt burden wouldn't fall in the end anyway. And even if irresponsible governments do not run up new debts, they could always finance their redistributive ways by running down government assets (like oil and minerals), leaving future generations condemned to the same overall debt burden. Ultimately, debt relief will only help reduce debt burdens if government policies make a true shift away from redistributive politics and toward a focus on economic development.

"Debt Relief Allows Poor Nations to Spend More on Health and Education"

No. In 1999, Jubilee 2000 enthused that with debt relief "the year 2000 could signal the beginning of dramatic improvements in healthcare, education, employment and development for countries crippled by debt." Unfortunately, such statements fail to recognize some harsh realities about government spending.

First, the iron law of public finance states that money is fungible: Debt relief goes into the same government account that rains money on good and bad uses alike. Debt relief enables governments to spend more on weapons, for example. Debt relief clients such as Angola, Ethiopia, and Rwanda all have heavy military spending (although some are promising to make cuts). To assess whether debt relief increases health and education spending, one must ask what such spending would have been in the absence of debt relief—a difficult question. However, if governments didn't spend the original loans on helping the poor, it's a stretch to expect them to devote new fiscal resources toward helping the poor.

Second, such claims assume that the central government knows where its money is going. A recent IMF and World Bank study found that only two out of 25 debt relief recipients will have satisfactory capacity to track where government spending goes within a year. At the national level, an additional study found that only 13 percent of central government grants for nonsalary education spending in Uganda (another recipient of debt relief) actually made it to the local schools that were the intended beneficiaries.

Finally, the very idea that the proceeds of debt relief should be spent on health and education contains a logical flaw. If debt relief proceeds are spent on social programs rather than used to pay down the debt, then the debt burden will remain just as crushing as it was before. A government can't use the same money twice—first to pay down foreign debt and second to expand health and education services for the poor. This magic could only work if health and education spending boosted economic growth and thus generated future tax revenues to service the debt. Unfortunately, there is little evidence that higher health and education spending is associated with faster economic growth.

"Debt Relief Will Empower Poor Countries to Make Their Own Choices"

Not really. Pro debt relief advocacy groups face a paradox: On one hand, they want debt relief to reach the poor; on the other, they don't want rich nations telling poor countries what to do. "For debt relief to work, let the conditions be set by civil society in our countries, not by big world institutions using it as a political tool," argued Kennedy Tumutegyereize of the Uganda Debt Network. Unfortunately, debt relief advocates can't have it both ways. Civil society remains weak in most highly indebted poor countries, so it would be hard to ensure that debt relief will truly benefit the poor unless there are conditions on the debt relief package.

Attempting to square this circle, the World Bank and IMF have made a lot of noise about consulting civil society while at the same time dictating incredibly detailed conditions on debt relief. The result is unlikely to please anyone. Debt relief under the World Bank and IMF's current HIPC initiative, for example, requires that countries prepare Poverty Reduction Strategy Papers. The World Bank's online handbook advising countries on how to prepare such documents runs well over 1,000 pages and covers such varied topics as macroeconomics, gender, the environment, water management, mining, and information technology. It would be hard for even the most skilled policymakers in the advanced economies to follow such complex (no matter how salutary) advice, much less a government in a poor country suffering from scarcity of qualified managers. In reality, this morass of requirements emerged as the multilateral financial institutions sought to hit on all the politically correct themes while at the same time trying hard to make the money reach the poor. If the conditions don't work—and of course they won't—the World Bank and IMF can simply fault the countries for not following their advice.

"Debt Relief Hurts Big Banks"

Wrong. During the 1970s and early 1980s, large commercial banks and official creditors based in rich nations provided substantial loans at market interest rates to countries such as Ivory Coast and Kenya. However, they pulled out of these markets in the second half of the 1980s and throughout the 1990s. In fact, from 1988 to 1997, such lenders received more in payments on old loans than they disbursed in new lending to high-debt poor countries. The multilateral development banks and bilateral lenders took their place, offering low-interest credit to poor nations. It's easy to understand why the commercial and official creditors pulled out. Not only did domestic economic mismanagement make high-debt poor countries less attractive candidates for potential loans, but with debt relief proposals in the air as early as 1979, few creditors wished to risk new lending under the threat that multilateral agencies would later decree loan forgiveness.

The IMF and World Bank announced the HIPC initiative of partial and conditional forgiveness of multilateral loans for 41 poor countries in September 1996. By the time the debt relief actually reached the HIPCs in the late 1990s, the commercial banks and high-interest official creditors were long gone and what was being forgiven were mainly "concessional" loans—i.e., loans with subsidized interest rates and long repayment periods. So really, debt relief takes money away from the international lending community that makes concessional loans to the poorest nations, potentially hurting other equally poor but not highly indebted nations if foreign aid resources are finite (as, of course, they are). Indeed, a large share of the world's poor live in India and China. Neither nation, however, is eligible for debt relief.

"Debt Relief Boosts Foreign Investment in Poor Nations"

A leap of faith. It is true that forgiving old debt makes the borrowers more able to service new debt, which in theory could make them attractive to lenders. Nevertheless, the commercial and official lenders who offer financing at market interest rates will not want to come back to most HIPCs any time soon. These lenders understand all too well the principle of moral hazard: Debt relief encourages borrowers to take on an excessive amount of new loans expecting that they too will be forgiven. Commercial banks obviously don't want to get caught with forgiven loans. And even the most charitable official lenders don't want to sign their own death warrants by getting stuck with forgiven debt. Both commercial and official lenders may want to redirect their resources to safer countries where debt relief is not on the table. Indeed, in 1991, the 47 least developed countries took in 5 percent of the total foreign direct investment (FDI) that flowed to the developing world; by 2000 their portion had dropped to only 2.5 percent. (Over the same period, the portion of global FDI captured by all developing nations dropped as well, from 22.3 to 15.9 percent.) Even capital flows to now lightly indebted "safe" countries might suffer from the perception that their debts also may be forgiven at some point. Ultimately, only the arms of multilateral development banks that provide soft loans—with little or no interest and very long repayment periods—are going to keep lending to HIPCs, and only then under very stringent conditions.

"Debt Relief Will Promote Economic Reform"

Don't hold your breath. During the last two decades, the multilateral financial institutions granted "structural adjustment" loans to developing nations, with the understanding that governments in poor countries would cut their fiscal deficits and enact reforms—including privatization of state-owned enterprises and trade liberalization—that would promote economic growth. The World Bank and IMF made 1,055 separate adjustment loans to 119 poor countries from 1980 to 1999. Had such lending succeeded, poor countries would have experienced more rapid growth, which in turn would have permitted them to service their foreign debts more easily. Thirty-six poor countries received 10 or more adjustment loans in the 1980s and 1990s, and their average percentage growth of per capita income during those two decades was a grand total of zero. Moreover, such loans failed to produce meaningful reforms, and developing countries now cite this failure as justification for debt relief. Yet why should anyone expect that conditions on debt forgiveness would be any more effective in changing government policies and behavior than conditions on the original loans?

Partial and conditional debt forgiveness is a *fait accompli*. Expanding it to full and unconditional debt forgiveness—as some groups now advocate—would simply transfer more resources from poor countries that have used aid effectively to those that have wasted it in the past. The challenge for civil society, the World Bank, IMF, and other agencies is to ensure that conditional debt forgiveness really does lead to government reforms that enhance the prospects of poor countries.

How can we promote economic reform in the poorest nations without repeating past failures? The lesson of structural adjustment programs is that reforms imposed from the outside don't change behavior. Indeed, they only succeed in creating an easy scapegoat: Insincere governments can simply blame their woes on the World Bank and IMF's "harsh" adjustment programs while not doing anything to fundamentally change economic incentives and ignite economic growth. It would be better for the international financial institutions to simply offer advice to governments that ask for it and wait for individual countries to come forward with homegrown reform programs, financing only the most promising ones and disengaging from the rest. This approach has worked in promoting economic reform in countries such as China, India, and Uganda. Rushing through debt forgiveness and imposing complex reforms from the outside is as doomed to failure as earlier rounds of debt relief and adjustment loans.

Source: William Easterly, "Think Again: Debt Relief" *Foreign Policy*, Nov/Dec 2001. www.foreignpolicy.com.

from Jubilee Research

7

Debt Relief Works But Success Under Threat, Finds New Report

September 2, 2002

In a major new analysis of the impact of debt relief on African countries, Jubilee Research at the New Economics Foundation today launches a new report: Relief Works.

The report shows that:

- In countries granted debt service relief, education spending has risen from $929 million (or less than the amount spent on debt service) to $1306 million—more than twice the amount spent on debt service.
- Spending on health has also risen from $466 million (or less than half of debt service payments) to $796 million—30% more than debt service.
- Despite concern from critics of debt relief, spending on the military has not increased beyond the average of 2% of GDP.

Jubilee Research director, Ann Pettifor, one of the co-founders of Jubilee 2000, says: "This report reflects the great achievements of a world-wide peoples' campaign to achieve justice for debtor nations. Critics of debt relief tend to argue that it simply supports corrupt regimes or gives governments more money to spend on defence. But our research shows that relief has been used well; and is leading to greater investment in health and education—not the military."

However Relief Works—African proposals for debt cancellation and why debt relief works also demonstrates that much more debt cancellation is necessary, if the proposals made by Presidents Thabo Mbeki and Obasanjo are to be implemented. In their NEPAD document, these presidents call for debt payments to be limited to a "proportion of budget revenues." Research undertaken by the New Economics Foundation shows that if this measure is used, then at least another $100bn debt must be cancelled.

The report underlines that:

- Social spending is still desperately below what is needed to meet the Millennium Development Goals—especially on health
- New debt is already being piled up—even in those countries that have had full debt cancellation, like Uganda. Projections show their debt levels will increase over the coming years

Romilly Greenhill, author of the report, says: "Unless western creditors provide the debt cancellation called for by African leaders, the NEPAD proposal is bound to fail. If this happens, what has been described as the 'best chance in a generation to do development differently' will be just one more squandered opportunity.

"It would be devastating for those countries if the international community fails to follow the leadership offered by African presidents in NEPAD, and act to end this inhumane system of punishing the poor for their poverty."

Source: "Debt Relief Works but Success Under Threat, Finds New Report," News release Sept. 2, 2002.

Review Questions

1. How did South America get so deeply into debt?
2. Why do some banks and governments in the MDCs support forgiveness (or at least partial forgiveness) instead of simply foreclosing, as they would do to an individual or company unable to make debt service payments?
3. What are structural adjustment programs, and why do lending countries and institutions frequently require borrowing countries to undertake these programs?
4. How might South America's history of colonization and independence explain why its countries are even more heavily in debt than those in other LDC regions?

Discussion/Essay Questions

1. Consider the case of Raul, the Brazilian welder whose biography was presented at the beginning of this module. How does his work history reflect the succession of development strategies in South America? Do you think that his life will be better or worse once the Brazilian economy undergoes "structural adjustment"?
2. Using the information in the textbook and in the readings, formulate a strategy for growth in one South American country. Would you seek debt forgiveness? What other obstacles to further development need to be overcome? Why are some of these issues more or less important in South America than they are in other parts of the world also being considered for debt relief?
3. The IMF has asked you, the president of the United States, for advice regarding forgiving South American countries' debt. Because much of the IMF's funding comes from the U.S. government, you can be sure that the IMF will listen to what you have to say. As president of the United States, should you organize a debt-forgiveness program, and, if so, what conditions should be required of countries that participate so as to ensure a better future both for inhabitants of the United States and for inhabitants of South America whose countries' debts are being forgiven? Remember that your goal is not only to assist in the development of South America but also to support U.S. banks that have invested in South America and the people of the United States who demand economic security. You know that if you sacrifice too much money, attention, or sovereignty to global causes, your opponent in next year's election will accuse you of forgetting that your first responsibility is to the United States, but if you concentrate solely on affairs at home your opponent will accuse you of being blind to the new era of globalization.

List of Readings

1. John Serieux, "Debt Forgiveness as Good Economics and Enlightened Self-Interest," in *Review: The North-South Institute Newsletter*, 1999, posted on the website of the *North-South Institute*, *http://www.nsi-ins.ca/ensi/publications/review/v3n2/02.html.*
2. "U2 Star Bono: Drop the Debt," January 2, 2002, transcript of an interview posted on the website of the Cable News Network, *http://www.cnn.com/2002/WORLD/europe/01/01/bono.debt/.*
3. Geoff McMaster, "Debt 'Forgiveness' Insults Poor Nations—Philosopher," in *Express News*, January 18, 2002, posted on the website of the University of Alberta, *http://www.expressnews.ualberta.ca/expressnews/articles/news.cfm?p_ID=1778&s=a.*
4. Gabriel Rivas-Ducca, "Debt, Energy and Materials Flows: The Looting of Latin America," in *Link*, July/September, 2000, posted on the website of Friends of the Earth International, *http://www.foei.org/publications/link/94/e94gabrielflows. html.*
5. Geri Smith, "Commentary: Down in the Dumps in Latin America," in *Business Week*, July 29, 2002, posted on the website of *Business Week*, *http://www.businessweek.com/magazine/content/02_30/b3793094.htm.*
6. William Easterly, "Think Again: Debt Relief," in *Foreign Policy*, November/December 2001, posted on the website of *Foreign Policy*, *http://www.foreignpolicy.com/issue_novdec_2001/easterly.html.*
7. "Debt Relief Works But Success under Threat, Finds New Report," news release from the New Economics Foundation, September 2, 2002, posted on the website of Jubilee Research, *http://www.jubileeplus.org/media/reliefworks300802.htm.*

Websites for Additional Research

1. Jubilee Research, a successor organization to Jubilee 2000, maintains a website of clippings from newspapers from around the world at *http://www.jubilee2000uk.org/media/daily/dailylist.htm.* This clipping service, which is updated daily, is broadly about international economics issues, but many articles focus specifically on debt and development. The newspapers range across the political spectrum (and cover the range of opinions on debt forgiveness). The rest of the Jubilee Research site provides a huge quantity of information on debt and its impact on developing countries, from country statistics to comment pieces. Unlike the clipping service, most of the site is explicitly pro-forgiveness. For the Jubilee Research homepage, go to *http://www.jubilee2000uk.org* and follow any of the numerous links.
2. A fact sheet on the IMF web page promotes the IMF's partial debt relief program for countries that agree to adopt structural adjustment programs; see *http://www.imf.org/external/np/exr/facts/hipc.htm.* Follow links from this page for more on the IMF's position on debt, and also check out the IMF's homepage at *http://www.imf.org.*
3. The U.S. Department of the Treasury maintains a fact sheet responding to frequently asked questions about its position on debt and developing countries at *http://www.treas.gov/education/faq/international/hipc.html.*

MAURITANIA
Nouakchott
MALI
Tombouctou (Timbuktu)
Gao
NIGER
CHAD
CAPE VERDE
Praia
Dakar
SENEGAL
GAMBIA
Banjul
Bamako
Niamey
Ouagadougou
BURKINA FASO
Bissau
GUINEA-BISSAU
GUINEA
Conakry
Freetown
SIERRA LEONE
CÔTE D'IVOIRE
GHANA
TOGO
BENIN
Porto-Novo
NIGERIA
Abuja
Kano
N'Djamena
Lake Chad
Yamoussoukro
Kumasi
Ibadan
Lagos
Monrovia
Buchanan
LIBERIA
Abidjan
Accra
Lomé
Cotonou
Port Harcourt
CAMEROON
Douala
Yaoundé
Malabo
Sekondi-Takoradi
Gulf of Guinea
EQUATORIAL GUINEA
SÃO TOMÉ AND PRÍNCIPE
São Tomé
Libreville
GABON
REPUBLIC OF CONGO
Equator
CENTRAL AFRICAN REPUBLIC
Bangui
Ubangi River
Congo River
Kisangani
DEMOCRATIC REPUBLIC OF CONGO
Brazzaville
Kinshasa
Pointe Noire
ANGOLA (CABINDA)
Luanda
ANGOLA
Cassinga
ATLANTIC OCEAN
Red Sea
Nile River
Blue Nile
White Nile
ERITREA
Massawa
Asmara
DJIBOUTI
Djibouti
Gulf of Aden
Berbera
Addis Ababa
ETHIOPIA
SOMALIA
Mogadishu
UGANDA
Kampala
Entebbe
Lake Rudolf (Turkana)
Lake Victoria
KENYA
Nairobi
RWANDA
Kigali
BURUNDI
Bujumbura
Kigoma
Arusha
Mombasa
Mpanda
Zanzibar
Dar es Salaam
TANZANIA
Lake Tanganyika
Lake Nyasa
SEYCHELLES
Victoria
Moroni
COMOROS
MALAWI
Lilongwe
Lubumbashi
ZAMBIA
Lusaka
Zambezi River
Blantyre
Nacala
MAYOTTE (FRANCE)
Lake Kariba
Harare
Caprivi Strip
ZIMBABWE
Maun
Bulawayo
MOZAMBIQUE
Mozambique Channel
Antananarivo
Toamasina
MADAGASCAR
MAURITIUS
Port Louis
RÉUNION (FRANCE)
NAMIBIA
Windhoek
Walvis Bay
BOTSWANA
Limpopo River
Gaborone
Tropic of Capricorn
Pretoria
Maputo
Johannesburg
Mbabane
SWAZILAND
Kimberly
Bloemfontein
Maseru
Orange River
Durban
LESOTHO
SOUTH AFRICA
Cape Town
Port Elizabeth
INDIAN OCEAN
More than 1 million
500,000–1 million
Fewer than 500,000
Capital cities are underlined
N
0 250 500 Miles
0 250 500 Kilometers
50°W
40°W
30°W
20°W
10°W
0°
50°N
40°N
30°N
20°N
10°N
0°
10°S
20°S
30°S
10°E
20°E
30°E
40°E
50°E
60°E

Genetically Modified Foods in Sub-Saharan Africa

Companion to Chapter 10 Agriculture

Mina had come from West Africa to study plant genetics at a large state university in the United States. For the people of her country, as for those in many poorer nations, genetic engineering held out the promise of greater crop yields and the possibility of feeding millions of underfed and starving people.

Now, Mina is on the floor of the greenhouse where she had been conducting her research. The greenhouse had been broken into overnight. Outside the vandals had spray-painted "Stop Genetic Mutilation!" on the walls. Inside it was chaos.

African agronomists seek to feed the hungry while maintaining economic independence. *Source:* © FK Photo/CORBIS

Mina watched Professor Bob Milikin slowly enter the greenhouse, shaking his head as he stepped over the debris and surveyed the damage. There had been a number of attacks around the country, but he had never imagined it might happen here. The irony was that in their case only 15 percent of the uprooted plants were genetically engineered. The rest had been developed using traditional breeding techniques. The plants that had been genetically modified were part of an experiment testing potential genetic engineering techniques for reducing the use of pesticides. He couldn't understand it. "This was research to *benefit* the environment," he said aloud to no one in particular. "To find a way to develop a plentiful, safe, healthy crop without using so many chemicals."

Mina thought of other benefits of genetically modified foods. They could be engineered to deliver more nutrients, reduce spoilage, curtail chemical contamination, even provide immunization against disease. But Mina knew that there was growing opposition in the United States and Europe to biotechnology. She had a friend, Erik, who was vehemently opposed to corporate biotechnology. Erik had e-mailed Mina on the dangers of corporate mergers that concentrated plant breeding and genetics in the hands of a few large transnational corporations. Erik was outraged that these companies plundered genes from countries in Africa, Latin America, and Asia. Erik charged that the companies were patenting the genes they found and holding them hostage, making indigenous farmers buy back the rights to grow their own seeds.

Erik had appealed to Mina's sense of history, telling her that she of all people, whose country had been under the yoke of a European imperialist power for generations, should refuse to aid and abet this new form of corporate imperialism.

Now, Mina is looking at the devastation around her. She is wondering if Erik is right. She also is wondering if Professor Milikin's research grant will be renewed. Her funding is provided through that grant, and now she may have to return to her country empty-handed.

Source: Adapted from "Frankenfoods? The Debate over Genetically Modified Crops," by Bill Rhodes et al. ■

Global Concerns about Genetically Modified Foods

The protest Mina experienced was not that unusual. There have been frequent protests in North America and Europe against the production and sale of food products using genetically modified organisms (GMOs, or GM crops). Occasionally the protests turn destructive, as in the preceding story. In recent years, protesters from Scotland dumped sacks of GM crops outside government buildings in London, activists damaged 2,400 square yards of corn in France, and genetically altered beet fields were trampled in California.

While these protests have been most prevalent and passionate in Europe, some of their characteristics are global. They tend to draw attention from the media, and they often include illustrative words such as "frankenfood" or "farmageddon." They also bring attention to questions that probably *should* be the focus of debate: Are genetically modified foods poised to feed the world's starving populations and save its endangered environments, or is their introduction into agricultural systems and food chains an ecological catastrophe in the making? Is their production by MDC-based companies a means of helping those in poorer countries or a means of dominating them? Do genetically modified foods address the root causes of starvation in much of the world? Are these foods really that new, or are they simply the latest phase in a continual process of agricultural innovation? In this module, we will attempt to shed some light on the genetically modified foods debate, with a focus on its implications for sub-Saharan Africa.

Basics of Genetic Modification

A genetically modified organism is an organism that has had its DNA modified in a laboratory instead of through more traditional cross-breeding practices. Whereas cross-breeding of plants takes many attempts to acquire the plants with the "right" characteristics, technologies developed in the 1970s and 1980s allow scientists in a laboratory to fine-tune this process, replacing certain aspects of a plants DNA so as to produce a desired change in the plant's characteristics. Genetic modification is a very specific form of scientific agriculture, and to understand the debate about genetically modified organisms one must understand the distinction between GMOs and other agroscience terms such as *biotechnology* and the *Green Revolution.*

Agriculture, by definition, occurs when humans manipulate nature so as to produce crops (or animals) that suit their desires. This is not by any means a new practice. For instance, cattle herders long ago realized that if they desired large cows it would be best to allow only the largest bulls and cows to breed with each other. Although engineering the selective reproduction of plants is more complicated, agriculturalists have been guiding plant reproduction according to similar principles for thousands of years.

Biotechnology goes one step beyond simple agricultural engineering, as it occurs when inputs and influences from one species are used to affect changes in another species. Although biotechnology in food-crop production is most often associated with the post-1950 Green Revolution, it too has been practiced in at least a rudimentary form for thousands of years. The Institute of Food Technologies notes that humans have been employing biotechnology since around 1800 B.C., when they began using yeast to leaven bread and ferment wine.

Biotechnology (and selective breeding as well) received a major boost in the nineteenth century when the Austrian botanist Gregor Mendel developed the field of genetics, whereby one can assess the statistical probability of certain traits emerging in future generations of an organism based on their prevalence in past and present generations. Genetic science is sometimes applied to human reproduction as well as animals and crops. A doctor, for instance, may use genetic probability calculations to advise a couple that, even though they are perfectly healthy, the presence of a serious hereditary disease such as Down's Syndrome in their siblings (or their parents' or grandparents' siblings) means that there is a relatively high likelihood that their child will have the disease.

The Green Revolution was a result of experiments designed to develop a higher yield form of wheat and later corn and rice. In the 1960s, population appeared to be growing at a faster rate than food could be produced, leading many to predict the onset of a population crisis predicted by Thomas Malthus in the late eighteenth century (see Chapter 2 in the textbook and the accompanying module in this workbook). Malthus argued that while population increases geometrically, food production increases only arithmetically. In an effort to forestall the crisis of famine that was likely to ensue, scientists rushed to develop new methods of agricultural production that would enable them to grow more food on the same amount of land. This effort was directed in particular toward LDCs which had even higher rates of population growth, but not the resources to import food. New varieties of seeds were developed to increase crop yield. This higher yield came with a price tag however; Green Revolution crops had to be more heavily watered, fertilized and treated for pests. This, in turn, required the consolidation of small subsistence farms into large landholdings, which disrupted both the environment and the social fabric of agricultural communities. Ultimately, the Green Revolution had mixed results. Production of grains increased, but some have argued that the Green

Revolution's benefits are outweighed by its negative social and environmental effects.

While farmers practicing traditional selective breeding of plants and animals favor certain genetic traits over others, neither they nor practitioners of traditional biotechnology (nor those using Green Revolution seeds) actually *modify* those genes through direct human intervention. The production of genetically *modified* (GM) crops became possible only in 1973, with the development of technologies for transferring recombinant deoxyribonucleic acid (rDNA) from one organism to another. Over the next two decades, this technology improved, so that by the late 1990s genetically modified organisms, also called "genetically engineered" or "transgenic" organisms, had become fairly common, at least in the United States.

The revolutionary nature of GM crops becomes apparent as soon as you consider examples of some of the crops that have been created using GM technology. If a school of flounder were to swim in the vicinity of a field of tomato plants, baby flounder and baby tomato plants would result, but there would be no possibility of any genetic mixing between the flounder and the tomato. However, scientists using rDNA transfer technology can insert flounder genes into tomatoes (and have done so) in an attempt to manufacture a frost-resistant tomato.

Since their introduction in the 1990s, the prevalence of GM crops has increased dramatically. According to *The New York Times*, as much as 40 percent of soybean, cotton, and corn acreage planted in the United States in 1998 was grown from bioengineered seeds. The anti-GMO activists of GeneticEngineering.net assert that GM crops can be found in more than 60 percent of all processed foods marketed in the United States. The Oak Ridge National Laboratory reports that more than 100 million acres in the United States were planted with transgenic crops in 2000.

Opponents of GM crops point to many of the same objections that also have been raised with regard to Green Revolution hybrids. One of the major criticisms of Green Revolution crops is that they can produce an outstanding yield, but only if considerable money is spent up front engineering the environment, typically on construction of irrigation works and application of herbicides, pesticides, and fertilizers. Farmers can go the budget route and hope that nature left to its own devices will produce the perfect environment for their Green Revolution crops, but if the year is exceptionally wet or dry, or hot or cold, the farmer might be left without anything to market. While farmers always have engaged in this kind of risk taking, as seeds become more refined the farmer's risk increases, both because these seeds (and their supporting infrastructure) typically are more expensive and because they often are so particular about their environmental conditions. Thus, the increased expenses associated with Green Revolution seeds have contributed to the decline of the family farm and the rise of farms controlled by large corporations that can make the necessary investments for engineered seeds to produce their potentially high yields. Ideally, GM seeds, in contrast with Green Revolution seeds, should not need such extensive inputs of fertilizers, irrigation, herbicides, and pesticides. Many critics, however, fear that GM crops will only reproduce the problems caused by the Green Revolution, but on a greater scale.

Highly engineered seeds place an additional burden on the small farmer in that, in some cases, they do not reproduce by themselves. This means that the farmer must continually buy new seeds from the manufacturer, resulting in long-term dependence. Indeed, there have been charges that GMO manufacturers intentionally insert "terminator genes" into their crops so that they will not be able to reproduce. In other cases, such as Monsanto's "Roundup Ready" seeds, GM seeds are designed to work with a specific pesticide or herbicide that is manufactured by the same company that manufactures the seeds. This creates a degree of dependency among farmers who previously were dependent only on the vagaries of nature and who perhaps had buffered that dependency by planting crops whose genetic codes (and environmental needs) varied greatly.

The following table of potential benefits from GM crops and some of the controversies that surround growing and consuming them originally appeared on the Oak Ridge National Laboratory's website *(http://www.ornl.gov/TechResources/Human_Genome/elsi/gmfood.html)*:

Benefits
Crops
Enhanced taste and quality
Reduced maturation time
Increased nutrients, yields, and stress tolerance
Improved resistance to disease, pests, and herbicides
New products and growing techniques
Animals
Increased resistance, productivity, hardiness, and feed efficiency
Better yields of meat, eggs, and milk
Improved animal health and diagnostic methods

(Continued)

Environment

"Friendly" bioherbicides and bioinsecticides

Conservation of soil, water, and energy

Bioprocessing for forestry products

Better natural waste management

More efficient processing

Society

Increased food security for growing populations

Controversies

Safety

Potential human health impact: allergens, transfer of antibiotic resistance markers, unknown effects

Potential environmental impact: unintended transfer of transgenes through cross-pollination, unknown effects on other organisms (e.g., soil microbes), and loss of flora and fauna biodiversity

Access and Intellectual Property

Domination of world food production by a few companies

Increasing dependence on industralized nations by developing countries

Biopiracy—foreign exploitation of natural resources

Ethics

Violation of natural organisms' intrinsic values

Tampering with nature by mixing genes among species

Objections to consuming animal genes in plants and vice versa

Stress for animal

Labeling

Not mandatory in some countries (e.g., United States)

Mixing GM crops with non-GM confounds labeling attempts

Society

New advances may be skewed to interests of rich countries

Source: Oak Ridge National Laboratory

Many of these potential benefits of GM crops, and the controversies surrounding them, are discussed in the remainder of this module, as our focus turns to sub-Saharan Africa.

Famine and Hunger in Sub-Saharan Africa

Famine is an event with a rapid onset during which many people die of starvation. In recent years, sub-Saharan Africa has been the world region that most consistently has experienced famine conditions. While there is no question that, in any given year, a significant number of people in one country or another of this region suffer from malnutrition (and, in some instances, experience famine), there is considerable debate about why these food shortages occur. Two frequent explanations are overpopulation and drought. However, neither of these explanations is as straightforward as it appears. Sub-Saharan Africa, for the most part, is a relatively sparsely populated region. While there is considerable variation within the region, many areas could be at least as agriculturally productive as areas in other regions that support much larger populations. As for drought, much of the area has long had extensive year-to-year variation in rainfall, and agricultural systems historically have adapted to these conditions.

While the region's recurrent famines may have been intensified in recent decades by an increase in population and perhaps by an overall decline in rainfall due to global climate change, it seems likely that social factors also are responsible for the recurrent food shortages in sub-Saharan Africa. Many countries in the region rely on agriculture for the bulk of their exports. Since one of the strategies to pay back international debt leads national governments to focus their economies on maximizing exports (see Chapter 9 in the textbook and the accompanying module in this workbook), agricultural production in staple grains such as sorghum and millet frequently is shifted to export-oriented specialty crops such as coffee, flowers, or cotton. In some cases, countries continue to grow large quantities of staple grains, but this grain is no longer consumed by the local population. Instead, it is fed to cattle, which are then exported as beef. Even in instances where a country produces enough food for local consumption, if the transportation infrastructure is geared toward exporting (as has been the case throughout sub-Saharan African countries since the era of colonialism) there may not be a means for distributing locally produced food to the hungry before it spoils.

The drive to produce crops so as to maximize export earnings (rather than to feed villages) also has led many farmers (and government policymakers) to favor high-risk strategies in the hope of obtaining high yields. Historically, farmers have understood that while it would be nice to have an agricultural surplus to sell, one must at least produce enough to feed one's family, and so farmers typically planted several different crops (and several strains of each crop). That way, even if some crops fell victim to pests or bad weather, others would survive. As has already been noted, Green Revolution seeds hold out the promise of greatly expanded yields, but only if the conditions are right. With Green Revolution seeds, farmers often "put all their eggs in one basket" (and they also often must make large investments in irrigation systems, fertilizers, and pesticides so as to "line" that basket), a dangerous strategy if one is living on the edge of famine. This agricultural strategy also is likely to concentrate power in the hands of those who have the wealth (or political connections) with which to provide the "lining."

Political factors also may be contributing to a decline in food availability for local populations. When wars displace large numbers of people, crops go unharvested, and people go without food. In some cases, governments have stockpiled large surpluses of crops, which can then be redistributed in return for political loyalty or hoarded as a means of raising prices so that companies controlled by governments (or by politically connected merchants) can make huge profits.

In the drier parts of sub-Saharan Africa (particularly the Sahel, the area just south of the Sahara), these declines in agricultural production are often associated with desertification. Desertification can be both a result and a cause of famine. The Sahel region has always had highly variable rainfall, and people in the region traditionally have mostly been pastoralists, seminomadic people who move southward with their animals in years when the desert encroaches and return northward when the desert recedes. Now, however, many of these pastoralists have been transformed into settled agriculturalists, encouraged by governments to intensively farm a plot of land in order to produce a crop that can be exported for foreign exchange, even though the land that they are farming is not suitable for this kind of intensive farming. The result is that, during a dry year, what had been an inconvenience requiring moving south is now a crisis of desertification. The usual response of farmers is to work the land that much harder, in an attempt to eke whatever crops they can out of it. This leads to erosion of the temporarily arid, but potentially fertile topsoil that remains, thereby breaking the cycle of recovery that normally would have occurred within a few years when the rain came back. Thus, disruption of the traditional pastoral system has led to a situation in which the local people are now committed to obtaining their livelihood from a plot of land that, for the foreseeable future, is doomed to low productivity. In short, policies encouraged by global political and economic pressures plus "normal" drought lead to food shortages, which lead to agricultural practices that result in long-term desertification, which leads to lower agricultural productivity among an agriculturally dependent population, which leads to famine.

GMOs as a Solution to Famine?

The reality of famine intersected directly with the debate over genetically modified foods in the fall of 2002, when southern Africa was suffering from a famine and the United States proposed donating genetically modified crops as food aid. Africans were torn on whether to accept these donations. Would modified DNA from GM seeds mix with—and transform—genetically "natural" local crops? Even if that were to happen, would that necessarily be a bad thing? Unlike Green Revolution crops, ideally GM crops would require few pesticide or fertilizer inputs and could be engineered specifically for local weather conditions. Thus, they could provide a means for African self-sufficiency in food production. On the other hand, the entire process would be controlled by transnational corporations (TNCs) that hold genetic codes as their corporate property, and many Africans feared that the introduction of GM crops would be just the latest event in a long history of foreigners seizing local resources and selling them back at a profit.

The first reading is from the *UNESCO Courier*, a magazine published by the U.N.'s Educational, Scientific, and Cultural Organization. Although the article does not specifically address the issue of GMOs as food aid, it asks the question, "Can genetically modified organisms feed the world?" In this reading, you meet Kanayo Nwanzé, who heads the Association for the Development of Rice Growing in Côte d'Ivoire, West Africa. Nwanzé sees a role for GMOs in Africa, but he questions who will see most of the benefits from these engineered foods. Nwanzé notes that one can make considerable progress in increasing agricultural productivity without entering the environmentally and politically risky area of genetic manipulation; he himself is working on a genetically unmodified variety of rice that grows quickly without irrigation.

The second reading reports on the dilemma faced by southern African countries in 2002, as they grappled with the U.S. offer of GM food aid. In the end, some countries chose to accept any food offered; some agreed to accept genetically modified food for consumption, but only if it was milled so as to prevent local farmers from planting seeds from the donated GM food; and one country—Zambia—banned GM food outright. This reading also brings up a concern of African countries quite apart from their worries about GMOs fostering dependence on foreign TNCs. Most African countries send the bulk of their agricultural exports to Europe, and most European countries have banned the importing (or growing) of any GM crops. Thus, even if African farmers (or governments) saw nothing wrong with GM crops, they were hesitant to accept food aid that could lead to genetic manipulations in their own crops for fear of losing their primary export market.

The third reading, a 2000 article from *The Ecologist* by Canadian economist Michel Chossudovsky, anticipates the crisis of 2002 and urges African countries not to accept GMO food aid. Chossudovsky views the introduction of GMOs into Africa as one component of the structural adjustment programs that have been introduced by Western financial institutions in order to maintain the dependency of LDCs (see the modules accompanying Chapters 8 and 9 for more on structural adjustment programs). Specifically, Chossudovsky argues that GMOs will be used to maintain a situation wherein Africa continues to provide export crops at low cost while remaining dependent on the world's wealthy countries and their TNCs for meeting local food needs.

The next reading is from the Council for Biotechnology Information (CBI), a U.S.-based public relations group funded by the major American and Canadian biotechnology companies and their trade association, the Biotechnology Industry Association. Like the second reading, this reading addresses the debate that raged in 2002 when southern African countries were deciding whether to accept GM food aid for famine relief. Quoting extensively from Tanzanian pediatrician Michael Mbwille, the CBI argues that while Europe can afford to ban crops that pose unknown risks, Africa is facing starvation and does not have that luxury. For Mbwille, unlike Chossudovsky, this is not an issue of long-term domination of one's economy by foreign corporations but rather an issue of short-term risk calculation in the midst of starvation.

The final reading is also generally in favor of GMOs. The author, Jennifer Thomson, a professor of microbiology at South Africa's University of Cape Town, defends her country's policy of allowing production of GM crops. She refutes a number of the claims made by opponents about GMOs' dangers, argues that South Africa's system of regulating GMOs can work, and, like Mbwille and the authors of the CBI article, she asserts that Africa cannot afford to be as cautious as Europe in adopting a technology that, as she sees it, has relatively few risks.

Readings

1

from *UNESCO Courier*

Can Genetically Modified Organisms Feed the World?

Philippe Demenet, UNESCO Courier *journalist*
September 2001

The controversy over biotechnologies is raging. Advocates claim they're the only answer to malnutrition, while opponents warn that drought-resistant millet and vaccinated yams will only increase poverty.

Near Africa's mighty Niger River, farmers are anxiously waiting for rain to fall before they sow millet or sorghum, then hoe, harvest, feed their families and replenish their granaries. Meanwhile, researchers in Japanese, Chinese, Philippine, European and U.S. laboratories are making strides in sequencing the 12 chromosomes and 50,000 genes composing rice, the matrix of all grains and a staple for three billion human beings. In five to ten years, they hope to know enough to genetically modify not only rice, but millet, sorghum, manioc and sugar cane as well. The aim is to make them "naturally" resistant to drought, soil salinity, viruses, blights and other scourges.

Will these genetically modified organisms (GMOs) really guarantee "food security" in the short term for the world's 826 million undernourished individuals? Will they help the small-scale farmers cultivating the Niger's barren, powdery soil to feed their families? The controversy is raging. In its 2001 report, the United Nations Development Programme (UNDP) says they will, emphasizing GMOs' "unique potential" to feed the world. In 50 years, the Earth's population will have soared to nine billion—three billion more than today. And most of the newcomers will increase the already overwhelming pressure on the southern countries' much-depleted soil. The alarm has already been sounded for sub-Saharan Africa where, unlike India and China, the population growth rate is still sky-high and the number of undernourished people is barely declining. GMO supporters say only a major, revolutionary "technological leap" will enable the planet to feed all its children.

Experimenting with Miracle Seeds

But others strongly disagree, arguing that low food production is not what causes malnutrition. There is enough to eat in the southern countries, they say. But the world's poorest people, those with neither money nor land, living in disintegrating, war-torn countries, simply have no access to food. They argue that land-use conditions must change, poor people must have access to credit and local markets and small landowners must be freed from money-lenders. Better use could be made of traditional seeds instead of importing high-risk technology with unpredictable consequences, most of whose patents belong to giant multinationals.

Advocates of the GMO revolution work in biogenetics laboratories, multinational seed and agrochemical companies, genome research, American foundations and some UN agencies, while most skeptics are out in the field. A case in point is Kanayo Nwanzé, a Ph.D. in agronomy who heads Adrao (Association for the Development of Rice Growing in West Africa) in Bouaké, Côte d'Ivoire. "Are GMOs being developed for the needs of small-scale farmers or multinational corporations?" he asks. "If we manage to negotiate with these patent-holding multinationals a technology that meets the needs of small-scale farmers and that's not under license, then 'Yes!,' GMOs will have a role to play in Africa. But their impact will have to come under careful scrutiny and the region's countries must have safety rules and the means to enforce them."

Adrao researchers have experience with miracle seeds. With international funding, they have developed a revolutionary variety of rice called Nerica. Genetically unmodified, it is the result of a conventional cross between a high-yield but fragile variety of Asian rice and a local variety that has had 35 centuries to adapt to Africa's stressful environment. Nerica offers tremendous possibilities. It reaches maturity in 90 days instead of the usual 120 to 150, resists insects, yields three tonnes per hectare with neither fertilizers nor irrigation—compared with 1.5 tonnes for traditional varieties—and grows like a weed. Ideally, it should improve life for hundreds of thousands of small-scale farmers who practice pluvial rice farming on plots of land ranging in size from 20 to 200 square metres and help West Africa's countries drastically cut their rice imports, perhaps even to export the grain.

Fear and Red Tape

However, this breakthrough is having difficulty leaving the laboratory. There are approximately 3,000 variants of Nerica, and for four years Nwanzé has been trying to involve small-scale farmers in selecting which ones to market. But in the summer of 2001, only a thousand Côte d'Ivoire farmers grew this "miracle rice" on a total of just one hectare.... Red tape, administrative delays and lack of communication between ministries and farmers, seed-certification organizations and rural credit institutions are to blame. A "technological leap" is not likely to help matters much. On the contrary. "If we offer a farmer genetically modified seeds, he'll say 'no thanks, I don't want to kill myself!"' says the head of Adrao.

Several African, Asian and South American countries have already passed laws regulating GMO production. But can they enforce them? Which laboratories will monitor the changes in biodiversity that might result from unforeseen crossbreeding between GMOs and related wild species? And with what funding? Who will see to it that pollen from GMOs capable of transmitting their defense mechanisms against insects and viruses do not spread? Researchers retort that it would be a mistake to focus on first-generation GMOs, which are necessarily flawed. "Soon we'll see the appearance of the second, third and fourth generations, which will meet developing countries' needs better," says Jean Claude Prot, who is currently dissecting rice's 12th chromosome at the Development Research Institute, a French public outfit affiliated with the International Rice Genome Sequence Project.

Biogenetics makes it possible, for example, to insert an insect's gene into a plant or vaccines into bananas or potatoes. To create a variety of rice that needs no more water than a camel (instead of the 4,000 to 5,000 litres necessary to produce one kilo), or again, to enrich plants with vitamins and minerals and develop others that revitalize acidic soil devastated by over-farming. So why not carry out the wildest projects?

It is easy to understand the enthusiasm of certain researchers and philanthropic institutions. For example, the Rockefeller Foundation sees what it calls the biotech-driven "second Green Revolution" as a way to make up for the first one's mistakes and tragedies. True, in the 1960s, the Green Revolution helped double food production by creating high-yield varieties of wheat and rice, keeping pace with the world's population growth. But these seeds, which need plentiful inputs (irrigation, fertilizers, herbicides and pesticides), have primarily benefited farmers who could afford to invest. Africa and the poorest areas in Asia and Latin America were left out. Moreover, the results for beneficiaries, such as China and Vietnam, are mixed: traditional varieties have vanished, irrigation has increased soil salinity, and farmers have over-used herbicides and pesticides to the detriment of their health and the environment.

Great Expectations

Supporters of the "second Green Revolution" say GMOs should spark an explosive increase in yields without inputs and in extreme farming conditions. But will they benefit the poorest farmers? Until now, multinational agrochemical companies converted to the "life sciences" have focused all their investments on intensive crops with close connections to industry and built a wall of prohibitively-expensive patents around their discoveries. Only publicly-funded research has taken an interest in indigent farmers living in tropical areas. Strapped for cash, the public sector has been forced to sign cooperation agreements with the private sector at the risk of losing its independence.

Major biotech companies are accused of producing "Frankenstein Food," especially in Europe. They quickly caught onto the image-enhancing advantages of helping to develop GMOs for the Third World. In 2000, amid a blaze of publicity, biotech heavyweights granted the free use of 70 patents to help develop a genetically-modified variety of rice enriched with beta carotene. The grain was heralded as a "miracle rice" capable of conquering Vitamin A deficiency, which kills one to two million children each year. But so far, the "golden rice" has fallen far short of expectations. The publicly-funded, Philippines-based International Rice Research Institute (IRRI) estimates that it will take five to ten years before seeds can be distributed free to farmers with annual incomes of under $10,000, in line with the agreements signed with industry. Furthermore, it is still unclear how much of the rice has to be consumed to make up for a Vitamin A deficiency.

For non-governmental organizations campaigning to protect the environment and preserve biodiversity, such as the Rural Advancement Foundation International (RAFI) network, this deal "could kneecap other, low-tech and more cost-effective solutions, such as re-introducing the many vitamin-rich food plants that were once cheap and available."

Lessons from the Past

Will GMOs help wipe out malnutrition? The golden rice episode has set the tone for the debate. Advocates say it would be utopian to wait for a better world while existing technology can help solve the problem here and now. Some argue that lack of food is simply a problem of unequal distribution.

"If poor people were not poor, they could buy the food they need," says Rockefeller Foundation president Gordon Conway. "This is true, but oversimplistic. There are no signs the world is about to engage in a massive redistribution of wealth."

Adversaries of the "GMO revolution" have diametrically opposed priorities: equity first, technology later. Otherwise, they say, the same mistakes that were made during the 1960s will be repeated. The Green Revolution "increased production and the number of poor people at the same time," says Jean-Pierre Roca, director of France's Institute for Training and Support to Development Initiatives. "Where there isn't a credit system, middlemen and the powerful are the ones who appropriate improved seeds and the use of pesticides. The poorest farmers must go into debt and sell their land to wealthier ones. GMOs are hazardous if measures aren't taken to go along with them."

A pragmatic Nwanzé says that "GMOs aren't a priority. First it is necessary to improve the conditions of agricultural production and soil management, keep the ground from getting hard after clearing and decrease rice imports by impoverished West African countries. All that can be achieved without GMOs, which run the risk of impoverishing biodiversity."

Source: Philippe Demenet, "Can Genetically Modified Organisms Feed the World?" *UNESCO Courier,* September 2001. Copyright © 1998 UNESCO.

from *The Christian Science Monitor*

2

Even Hungry Africa Wary of Gene-Modified Food

A Shipment of Grain Sits in South Africa as 12 Million People in the Region Face Shortages.

By Nicole Itano

LUSAKA, ZAMBIA—Under a grass-thatched shelter just off the main road that heads east from Lusaka, Josephine Musopelo waits with neighbors for the return of her husband. Two days ago, he left for Zambia's capital city, some 35 miles away, in search of food for his hungry village.

People here remember the yellow corn distributed seven years ago during the area's last major drought, and Mrs. Musopelo's husband has gone to find its source. In Southern Africa, where most people eat white corn, the yellow variety is considered animal feed, not fit for human consumption. But Musopelo is desperate.

"If we eat this pitiful stuff," she says, gesturing at a bag of small fruits the family has been pounding into mash for the children, "we will eat anything."

But will they eat genetically modified (GM) corn? That's the question her country's leaders are hotly debating as a shipment, partially filled with GM corn, sits 1,000 miles away in a South African port.

Until now, the scientific debate over the risks, and benefits, of GM foods was something fought over largely in the streets of Paris and the dinner tables of Iowa. Suddenly, it is a life and death decision for ordinary Zambians, and it threatens to derail international efforts to avert a famine in Southern Africa. The governments of countries like Zambia find themselves in the difficult position of either accepting a technology into their country before they have determined if it's safe, or turning away grain that could save lives now.

The US has offered to provide half of what Zambia needs to feed those who are hungry, and one-third of what is needed regionally. At least some of that is corn that has been genetically modified.

Although none of the seven African countries targeted for emergency food aid has officially said they will reject American aid, at least one shipment of food has already been diverted from Zimbabwe, in part due to GM concerns. Several other countries are seriously considering turning away the food. Zambia's president, Levy Manawasa, said last week that despite its need, his country would not accept American food aid if it cannot be proven safe.

People have been selectively breeding crops for thousands of years to improve yields or adapt plants to new environments. New technology now allows the genes from one organism to be inserted into another.

But the technology is controversial. Some scientists, particularly in Europe, worry that GM food could cause allergic reactions in humans or that it could pollute the environment by cross pollinating with natural varieties. Others say it could provide the solution to feeding the world's hungry by making plants that are drought and pest resistant.

On the streets of Lusaka, where much of the debate is fueled by misinformation and public hysteria, people are primarily concerned about whether the food will make them sick.

In a poor rural settlement on the outskirts of Lusaka, George Chilumbo is skeptical of the help pouring in for the estimated 2.3 million people in Zambia, and more than 12 million regionally, tottering on the edge of starvation. Mr. Chilumbo doesn't understand the details of genetic modification. All he knows is that some people think the food could be dangerous.

"We think that if we eat that food, bad things could happen to us, like cancers could grow in our stomachs," he says, to the nods of his friends. "I would rather starve than eat that food."

Since there is thus far no evidence that GM foods have a harmful effect on people, and Americans have been eating it for some time, those fears are likely to be superceded by the immediate need for food.

"What I personally find so disturbing about this whole GM debate is the fact that we've been eating it for the past five years," says Allan Reed, director of the United States Agency for International Development in Zambia. "For someone to say that this food is being dumped, or that it has been rejected by the United States, is simply not true."

Even before the aid debate put genetic modification in the headlines, farmers here have been bitterly divided over the issue. Some, like those in the cotton industry, say its introduction is necessary to keep them competitive in the international market. Others in the country's growing organic export market worry that allowing GM crops into the country could close the European Union (EU) market to them. They fear that GM crops will either mix with their own or, at the very least, hurt their image among anti-GM European consumers.

"You can understand the Zambian officials' concern because on one hand donors and others are telling them, 'Look, you've got to improve your agriculture by going for niche markets like the organic market,' " says Rob Tripp, a researcher at the Overseas Development Institute. "On the other hand there's a real fear that they could get cut out of these markets in places like the EU, which are creating increasingly strict GM regulations."

Lovemore Simwanda, who is leading the investigations into GM food for the Zambian National Farmers Union, worries the food shortage is forcing the country to make a decision on GM too quickly and without the proper information.

"The American government wants to push us into accepting this," he says. "They think we've got hunger and that we're going to be forced into accepting their food and ultimately GM."

One potential short-term solution is to mill the corn before distribution, since one of the main concerns in Zambia is that seed intended as food aid will be hoarded and planted next year, potentially affecting local maize crops with GM strains.

Namibia has milled grain for several years due to GM concerns. But such a solution is expensive—as much as $25 per metric ton—and could cause delays in getting the food to people already living on the edge. Additionally, some United Nations officials privately say such a move is opposed by the United States which fears such a compromise implicitly acknowledges a problem with the food.

Even if it is not milled, the American aid poses little threat to Zambia's crops, say experts. American corn is specifically bred for the conditions in the Midwest and is unlikely to thrive in Zambia. Many experts say GM corn is likely already present in Zambia since South Africa is one of the largest producers of GM food in the world and a major trade partner with Zambia.

The GM debate hasn't yet filtered down the road towards Malawi where Mrs. Musopelo and her neighbors are waiting for help. They don't understand what a gene is or how parts of one plant can be inserted into another. There are no words in their language for such technologies. All they knows is that their families are hungry.

"Our survival depends on help from outside," says Mrs. Musopelo's neighbor, Michael Simwinga. "If you don't assist us, our children and ourselves will die."

from *The Ecologist* 3

Sowing the Seeds of Famine in Ethiopia

By Michel Chossudovsky
Professor of Economics, University of Ottawa
September 2000, revised September 2001

The "economic therapy" imposed under IMF–World Bank jurisdiction is in large part responsible for triggering famine and social devastation in Ethiopia and the rest of sub-Saharan Africa, wreaking the peasant economy and impoverishing millions of people.

With the complicity of branches of the US government, it has also opened the door for the appropriation of traditional seeds and landraces by US biotech corporations, which behind the scenes have been peddling the adoption of their own genetically modified seeds under the disguise of emergency aid and famine relief.

Moreover, under WTO rules, the agri-biotech conglomerates can manipulate market forces to their advantage as well as exact royalties from farmers. The WTO provides legitimacy to the food giants to dismantle State programmes including emergency grain stocks, seed banks, extension services and agricultural credit, etc., plunder peasant economies and trigger the outbreak of periodic famines.

Crisis in the Horn

More than 8 million people in Ethiopia—representing 15% of the country's population—had been locked into "famine zones." Urban wages have collapsed and unemployed seasonal farm workers and landless peasants have been driven into abysmal poverty. The international relief agencies concur without further examination that climatic factors are the sole and inevitable cause of crop failure and the ensuing humanitarian disaster. What the media tabloids fails to disclose is that—despite the drought and the border war with Eritrea—several million people in the most prosperous agricultural regions have also been driven into starvation. Their predicament is not the consequence of grain shortages but of "free markets" and "bitter economic medicine" imposed under the IMF—World Bank sponsored Structural Adjustment Programme (SAP).

Ethiopia produces more than 90% of its consumption needs. Yet at the height of the crisis, the nationwide food deficit for 2000 was estimated by the Food and Agriculture Organization (FAO) at 764,000 metric tons of grain representing a shortfall of 13 kilos per person per annum.[1] In Amhara, grain production (1999–2000) was twenty percent in excess of consumption needs. Yet 2.8 million people in Amhara (representing 17% of the region's population) became locked into famine zones and are "at risk" according to the FAO.[2] Whereas Amhara's grain surpluses were in excess of 500,000 tons (1999–2000), its "relief food needs" had been tagged by the international community at close to 300,000 tons.[3] A similar pattern prevailed in Oromiya, the country's most populated state where 1.6 million people were classified "at risk," despite the availability of more than 600,000 metric tons of surplus grain.[4] In both these regions, which include more than 25% of the country's population, scarcity of food was clearly not the cause of hunger, poverty and social destitution. Yet no explanations are given by the panoply of international relief agencies and agricultural research institutes.

The Promise of the "Free Market"

In Ethiopia, a transitional government came into power in 1991 in the wake of a protracted and destructive civil war. After the pro-Soviet Dergue regime of Colonel Mengistu Haile Mariam was unseated, a multi-donor financed Emergency Recovery and Reconstruction Project (ERRP) was hastily put in place to deal with an external debt of close to 9 billion dollars that had accumulated during the Mengistu government. Ethiopia's outstanding debts with the Paris Club of official creditors were rescheduled in exchange for far-reaching macro-economic reforms. Upheld by US foreign policy, the usual doses of bitter IMF economic medicine were prescribed. Caught in the straightjacket of debt and structural adjustment, the new Transitional Government of Ethiopia (TGE), led by the Ethiopian People's Revolutionary Democratic Front (EPRDF)—largely formed from the Tigrean People's Liberation Front (PLF)—had committed itself to far-reaching "free market reforms," despite its leaders' Marxist leanings. Washington soon tagged Ethiopia alongside Uganda as Africa's post—Cold War free market showpiece.

While social budgets were slashed under the structural adjustment programme (SAP), military expenditure—in part financed by the gush of fresh development loans—quadrupled since 1989.[5] With Washington supporting both sides in the Eritrea-Ethiopia border war, US arms sales spiralled. The bounty was being shared between the arms manufacturers and the agribusiness conglomerates. In the post—Cold War era, the latter positioned themselves in the lucrative procurement of emergency aid to war-torn countries. With mounting military spending financed on borrowed money, almost half of Ethiopia's export revenues was earmarked to meet debt-servicing obligations.

A Policy Framework Paper (PFP) stipulating the precise changes to be carried out in Ethiopia had been carefully drafted in Washington by IMF and World Bank officials on behalf of the transitional government, and was forwarded to Addis Ababa for the signature of the Minister of Finance. The enforcement of severe austerity measures virtually foreclosed the possibility of a meaningful post-war reconstruction and the rebuilding of the country's shattered infrastructure. The creditors demanded trade liberalization and the full-scale privatization of public utilities, financial institutions, State farms and factories. Civil servants including teachers and health workers were fired, wages were frozen and the labor laws were rescinded to enable State enterprises "to shed their surplus workers." Meanwhile, corruption became rampant. State assets were auctioned off to foreign capital at bargain prices and Price Waterhouse Coopers was entrusted with the task of coordinating the sale of State property.

In turn, the reforms had led to the fracture of the federal fiscal system. Budget transfers to the State governments were slashed leaving the regions to their own devices. Supported by several donors, "regionalization" was heralded as a "devolution of powers from the federal to the regional governments." The Bretton Woods institutions knew exactly what they were doing. In the words of the IMF, "[the regions'] capacity to deliver effective and efficient development interventions varies widely, as does their capacity for revenue collection."[6]

Wrecking the Peasant Economy

Patterned on the reforms adopted in Kenya in 1991, agricultural markets were wilfully manipulated on behalf of the agribusiness conglomerates. The World Bank demanded the rapid removal of price controls and all subsidies to farmers. Transportation and freight prices were deregulated serving to boost food prices in remote areas affected by drought. In turn, the markets for farm inputs including fertiliser and seeds were handed over to private traders including Pioneer Hi-Bred International which entered into a lucrative partnership with Ethiopia Seed Enterprise (ESE), the government's seed monopoly.[7]

At the outset of the reforms in 1992, USAID under its Title III program "donated" large quantities of US fertilizer "in exchange for free market reforms:"

> [V]arious agricultural commodities [will be provided] in exchange for reforms of grain marketing ... and [the] elimination of food subsidies ... The reform agenda focuses on liberalization and privatization in the fertilizer and transport sectors in return for financing fertilizer and truck imports.... These program initiatives have given us [an] "entrée" ... in defining major [policy] issues....[8]

While the stocks of donated US fertiliser were rapidly exhausted, the imported chemicals contributed to displacing local fertiliser producers. The same companies involved in the fertiliser import business were also in control of the domestic wholesale distribution of fertiliser using local level merchants as intermediaries.

Increased output was recorded in commercial farms and in irrigated areas (where fertilizer and high yielding seeds had been applied). The overall tendency, however, was towards greater economic and social polarisation in the countryside, marked by significantly lower yields in less productive marginal lands occupied by the poor peasantry. Even in areas where output had increased, farmers were caught in the clutch of the seed and fertilizer merchants.

In 1997, the Atlanta based Carter Center—which was actively promoting the use of biotechnology tools in maize breeding—proudly announced that "Ethiopia [had] become a food exporter for the first time."[9] Yet in a cruel irony, the donors ordered the dismantling of the emergency grain reserves (set up in the wake of the 1984–85 famine) and the authorities acquiesced.

Instead of replenishing the country's emergency food stocks, grain was exported to meet Ethiopia's debt servicing obligations. Close to one million tons of the 1996 harvest was exported, an amount which would have been amply sufficient (according to FAO figures) to meet the 1999–2000 emergency. In fact the same food staple which had been exported (namely maize) was re-imported barely a few months later. The world market had confiscated Ethiopia's grain reserves.

In return, US surpluses of genetically engineered maize (banned by the European Union) were being dumped on the horn of Africa in the form of emergency aid. The US had found a convenient mechanism for "laundering its stocks of dirty grain." The agribusiness conglomerates not only cornered Ethiopia's commodity exports, they were also involved in the procurement of emergency shipments of grain back into Ethiopia. During the 1998–2000 famine, lucrative maize contracts were awarded to giant grain merchants such as Archer Daniels Midland (ADM) and Cargill Inc.[10]

Laundering America's GM Grain Surpluses

US grain surpluses peddled in war-torn countries also served to weaken the agricultural system. Some 500,000 tons of maize and maize products were "donated" in 1999–2000 by USAID to relief agencies including the World Food Programme (WFP), which in turn collaborates closely with the US Department of Agriculture. At least 30% of these shipments (procured under contract with US agribusiness firms) were surplus genetically modified grain stocks.[11]

Boosted by the border war with Eritrea and the plight of thousands of refugees, the influx of contaminated food aid had contributed to the pollution of Ethiopia's genetic pool of indigenous seeds and landraces. In a cruel irony, the food giants were at the same time gaining control—through the procurement of contaminated food aid—over Ethiopia's seed banks. According to South Africa's Biowatch: "Africa is treated as the dustbin of the world.... To donate untested food and seed to Africa is not an act of kindness but an attempt to lure Africa into further dependence on foreign aid."

Moreover, part of the "food aid" had been channeled under the "food for work" program which served to further discourage domestic production in favour of grain imports. Under this scheme, impoverished and landless farmers were contracted to work on rural infrastructural programmes in exchange for "donated" US corn.

Meanwhile, the cash earnings of coffee smallholders plummeted. Whereas Pioneer Hi-Bred positioned itself in seed distribution and marketing, Cargill Inc. established itself in the markets for grain and coffee through its subsidiary Ethiopian Commodities.[12] For the more than 700,000 smallholders with less than 2 hectares that produce between 90 and 95% of the country's coffee output, the deregulation of agricultural credit combined with low farmgate prices of coffee had triggered increased indebtedness and landlessness, particularly in East Gojam (Ethiopia's breadbasket).

Biodiversity Up for Sale

The country's extensive reserves of traditional seed varieties (barley, teff, chick peas, sorghum, etc.) were being appropriated, genetically manipulated and patented by the agribusiness conglomerates: "Instead of compensation and respect, Ethiopians today are ... getting bills from foreign companies that have 'patented' native species and now demand payment for their use."[13] The foundations of a "competitive seed industry" were laid under IMF and World Bank auspices.[14] The Ethiopian Seed Enterprise (ESE), the government's seed monopoly, joined hands with Pioneer Hi-Bred in the distribution of hi-bred and genetically modified (GM) seeds (together with hybrid resistant herbicide) to smallholders. In turn, the marketing of seeds had been transferred to a network of private contractors and "seed enterprises" with financial support and technical assistance from the World Bank. The "informal" farmer-to-farmer seed exchange was slated to be converted under the World Bank programme into a "formal" market oriented system of "private seed producer-sellers."[15]

In turn, the Ethiopian Agricultural Research Institute (EARI) was collaborating with the International Maize and Wheat Improvement Center (CIMMYT) in the development of new hybrids between Mexican and Ethiopian maize varieties.[16] Initially established in the 1940s by Pioneer Hi-Bred International with support from the Ford and Rockefeller foundations, CIMMYT developed a cosy relationship with US agribusiness. Together with the UK-based Norman Borlaug Institute, CIMMYT constitutes a research arm as well as a mouthpiece of the seed conglomerates. According to the Rural Advancement Foundation (RAFI), "US farmers already earn $150 million annually by growing varieties of barley developed from Ethiopian strains. Yet nobody in Ethiopia is sending them a bill."[17]

Impacts of Famine

The 1984–85 famine had seriously threatened Ethiopia's reserves of landraces of traditional seeds. In response to the famine, the Dergue government through its Plant

Genetic Resource Centre—in collaboration with Seeds of Survival (SoS)—had implemented a programme to preserve Ethiopia's biodiversity.[18] This programme—which was continued under the transitional government—skillfully "linked on-farm conservation and crop improvement by rural communities with government support services."[19] An extensive network of in-farm sites and conservation plots was established involving some 30,000 farmers. In 1998, coinciding chronologically with the onslaught of the 1998–2000 famine, the government clamped down on seeds of Survival (SoS) and ordered the programme to be closed down.[20]

The hidden agenda was to eventually displace the traditional varieties and landraces reproduced in village-level nurseries. The latter were supplying more than 90 percent of the peasantry through a system of farmer-to-farmer exchange. Without fail, the 1998–2000 famine led to a further depletion of local level seed banks: "The reserves of grains [the farmer] normally stores to see him through difficult times are empty. Like 30,000 other households in the [Galga] area, his family have also eaten their stocks of seeds for the next harvest."[21] And a similar process was unfolding in the production of coffee where the genetic base of the arabica beans was threatened as a result of the collapse of farmgate prices and the impoverishment of small-holders.

In other words, the famine—itself in large part a product of the economic reforms imposed to the advantage of large corporations by the IMF, World Bank and the US Government—served to undermine Ethiopia's genetic diversity to the benefit of the biotech companies. With the weakening of the system of traditional exchange, village level seed banks were being replenished with commercial hi-bred and genetically modified seeds. In turn, the distribution of seeds to impoverished farmers had been integrated with the "food aid" programmes. WPF and USAID relief packages often include "donations" of seeds and fertiliser, thereby favouring the inroad of the agribusiness-biotech companies into Ethiopia's agricultural heartland. The emergency programs are not the "solution" but the "cause" of famine. By deliberately creating a dependency on GM seeds, they had set the stage for the outbreak of future famines.

This destructive pattern—invariably resulting in famine—is replicated throughout Sub-Saharan Africa. From the onslaught of the debt crisis of the early 1980s, the IMF–World Bank had set the stage for the demise of the peasant economy across the region with devastating results. Now, in Ethiopia, fifteen years after the last famine left nearly one million dead, hunger is once again stalking the land. This time, as eight million people face the risk of starvation, we know that it isn't just the weather that is to blame.

Endnotes

1. Food and Agriculture Organization (FAO), Special Report: FAO/WFP Crop Assessment Mission to Ethiopia, Rome, January 2000.
2. Ibid.
3. Ibid.
4. Ibid.
5. Philip Sherwell and Paul Harris, "Guns before Grain as Ethiopia Starves," Sunday Telegraph, London, April 16, 2000.
6. IMF, Ethiopia, Recent Economic Developments, Washington, 1999.
7. Pioneer Hi-Bred International, General GMO Facts, *http://www.pioneer.com/usa/biotech/value_of_products/product_value.htm#*.
8. United States Agency for International Development (USAID), "Mission to Ethiopia, Concept Paper: Back to the Future," Washington, June 1993.
9. Carter Center, press release, Atlanta, Georgia, January 31, 1997.
10. Declan Walsh, "America Finds Ready Market for GM Food," *The Independent*, London, March 30, 2000, p. 18.
11. Ibid.

12. Maja Wallegreen, "The World's Oldest Coffee Industry in Transition," *Tea & Coffee Trade Journal*, November 1, 1999.

13. Laeke Mariam Demissie, "A Vast Historical Contribution Counts for Little; West Reaps Ethiopia's Genetic Harvest," *World Times*, October, 1998.

14. World Bank, Ethiopia-Seed Systems Development Project, Project ID ETPA752, 6 June 1995.

15. Ibid.

16. See CIMMYT Research Plan and Budget 2000–2002, *http://www.cimmyt.mx/about/People-mtp2002.htm#.*

17. Laeke Mariam Demissie, op. cit.

18. "When Local Farmers Know Best," *The Economist*, 16 May 1998.

19. Ibid.

20. Laeke Mariam Demissie, op. cit.

21. Rageh Omaar, "Hunger Stalks Ethiopia's Dry Land," BBC, London, 6 January, 2000.

22. An earlier version of this article was published in *The Ecologist*, September 2000.

Source: Michel Chossudovsy, "Sowing the Seeds of Famine in Ethiopia," *The Ecologist,* Sept. 2000. This chapter was first published in *The Globalization of Poverty and the New World Order,* 2nd Edition, Global Outlook, Shanty Bay, Ont., 2003.

4

from the Council for Biotechnology Information

African Doctor Says Biotechnology Could Help Feed the Hungry

Pediatrician says "there is something insane about food aid rotting while people starve."

In September, starving people in the village of Monze, Zambia, looted storage sheds filled with thousands of tons of U.S. corn donated to help ease southern Africa's worst food crisis in years.

Why were people who are literally starving denied access to food that could save their lives?

Because Zambia's president considers the corn—some of which has been grown from biotech seeds—unsafe. He and other African leaders say anti-biotech activists told them the corn contains "poisons" that could harm people and "contaminate" native crops.

With more than 14 million people starving in Zambia and other African countries, the controversy over whether to accept food that is eaten everyday in the United States, Canada and other parts of the world, has raised the ire of people like Dr. Michael Mbwille, a pediatrician from Tanzania and African editor of the Food Safety Network. The network is a nonprofit coalition based in Washington, D.C. that seeks ways to improve global food security.

"There is something insane about food aid rotting while people starve due to disinformation campaigns," says Mbwille. "Death by starvation and the long-term costs of malnutrition have been deemed less offensive than hypothetical, unsubstantiated food safety ills."

Zambia, which made its decision not to allow its citizens to eat U.S. corn final on October 29, is the only African country to reject food aid outright. So despite the fact that more than 2 million Zambians are starving after two years of drought, tons of donated grain are locked away in sheds like those in Monze.

Many biotech opponents who initially lobbied for Zambia's decision now say they've changed their minds and would allow the U.S. grain into these countries to help alleviate the famine. But many still oppose giving African farmers access to biotechnology and argue that years of additional testing is needed to ensure the safety of biotech crops and foods.

A growing number of African farmers and African scientists disagree. They say the continent needs more than food aid to ease a famine; it needs tools like biotechnology to improve agricultural productivity so that fast-growing populations can see the same benefits realized by farmers and consumers elsewhere.

"Biotechnology is an essential tool, along with other methods, if Africa is to achieve food security," says Mbwille.

In Zambia, the need is even more immediate. "Zambians are not asking for fancy things like wheat flour or apples, but mere maize to fill their stomachs," says Petronella Chisanga, a Zambian civic leader and chair of the country's Non-Governmental Organizations Coordinating Committee. "Can somebody tell us what is wrong with eating the modified food?"

Famine in Southern Africa

Six countries—Lesotho, Malawi, Mozambique, Swaziland, Zimbabwe and Zambia—are experiencing famine, primarily caused by a drought that's spanned the last two growing seasons. About 14.5 million people in those countries are at risk of starvation, according to the United Nations.

None of the affected countries allow their own farmers to grow biotech crops (South Africa is the only place on the continent that does). All except Zambia, however, have accepted biotech food aid to feed their people—Mozambique and Zimbabwe on the condition that the corn is milled before it's distributed, which prevents farmers from saving and replanting the seeds.

Scientific groups, scholars, even American farmers have urged Zambia to reverse its policy, noting that nonbiased observers from the United Nations to the National Academy of Sciences are on record affirming the safety of biotech crops. Zambia's President Levy Mwanawasa has maintained the ban, however, despite the fact that the very same *Bt* corn donated to his country makes up one-third of the corn grown in the United States, and has been in commercial use since 1996. "I'm not prepared to accept that we should use our people as guinea pigs," he's said.

His controversial position has sparked debate not only about the needs of his country, but also the future of biotechnology in Africa—a debate picked up by the World Summit on Sustainable Agriculture which convened in August in Johannesburg, a short distance from the epicenter of the famine.

At the summit, antibiotech groups such as Greenpeace and the Friends of the Earth articulated why, in their view, biotechnology is wrong for Africa and the developing world.

Chief among those reasons is that the technology has not been proven safe both for people and the environment—either by producing unintended allergic reactions in people or by creating "superweeds" through the flow of pollen to nearby weeds.

Those views were challenged by African scientists, farmers and other biotech supporters as unsubstantiated by facts and inhumane in the face of famine and chronic hunger problems. For African farmers and those who depend on their crops, the challenges of daily life—getting the tools they need to grow successful crops and get enough calories to lead healthy, productive lives—are the primary concern.

Hypothetical Risks vs. Empty Stomachs

The famine in southern Africa is not an anomaly, but a severe symptom of a continent-wide hunger problem.

Simply put, African farmers aren't growing enough food. So many African people don't have enough to eat. Over the last two decades, while Africa's population has increased, agricultural productivity per farm worker has declined. Corn yields in places like Zimbabwe are less than half what they are in places like Iowa. One in three Africans is malnourished.

"As a medical doctor with a specialty in pediatrics who has worked throughout Africa and Asia, I can personally attest that the ravages of malnutrition and hunger in childhood are carried throughout the life of the victim," says Mbwille. "Children will suffer throughout their lives from illnesses due to the associated developmental impact of malnutrition."

To feed those who are malnourished, African farmers need to boost yields. Biotechnology has helped farmers in other parts of the world do just that. A June 2002 study by the National Center for Food and Agricultural Policy (NCFAP), for example, found that six biotech crops planted in the United States produced an additional 4 billion pounds of food and fiber on the same acreage, and improved farm income.

In South Africa, farmers like T.J. Buthelezi also have dramatically increased yields by planting biotech crops. Buthelezi was one of a contingent of African and Indian farmers who rallied at the World Summit in Johannesburg to demand access to biotechnology for the developing world.

Buthelezi, Mbwille and other African proponents of biotechnology argue that given the everyday suffering on their continent from lack of food, it's imperative to take advantage of a technology that boosts productivity with little or no demonstrated risk.

"Those who profess that there is a 'poetry' to subsistence agriculture and promote a return to the past doom those in the present as well as future generations to suffering," says Mbwille. "I would ask that the members of Greenpeace, Friends of the Earth and those that write them checks move to places like Zambia and watch their own children die from starvation facilitated by their ill-conceived views of the world."

Mbwille points out that biotechnology foods have undergone more testing than any food product in history, and that no deaths or illnesses—"not even a single headache"—has been associated with the technology. Meanwhile, the death toll from famine and the debilitating effects of malnutrition mount daily in Africa and other underfed places.

Developed World Laws Stifle Developing World Needs

Supporters of biotechnology say African farmers are being victimized by countries in the developed world that have, in the face of all scientific evidence, implemented tight restrictions on importing biotech foods—restrictions that keep Africa tied to what Mbwille calls "medieval farming practices."

Much of the resistance comes from the European Union (EU), which has a history of food safety scares unrelated to biotechnology. While acknowledging there's no scientific evidence of risk, the EU has imposed a *de facto* moratorium on approvals of new biotech crops (several varieties of biotech soybeans and corn were approved for import before the moratorium). The EU also has pressured African growers to avoid planting biotech crops if they want access to European markets.

The need to protect future exports to the EU is one reason cited by Zambian President Mwanawasa for refusing biotech food aid, which he fears will be planted by farmers eager to use the pest-resistant seeds. It's also why Mozambique and Zimbabwe have taken the costly step of milling donated biotech corn so it can't be replanted, a step some experts say is keeping food from hungry people.

Europeans do not have to worry about having enough food to eat and therefore feel less pressure to embrace a new agricultural technology, Mbwille says. But their African neighbors need to improve farm productivity as a partial solution both to chronic hunger and to poverty: A more productive agricultural sector is a critical building block for economic growth.

Europe's laws, in Mbwille's view, help keep Africa tied to conventional crops and archaic methods that result in huge crop losses from pests, disease and poor growing conditions when what the continent really needs is the best technology available to grow more and better food for its own people.

"It's abhorrent that in opposing biotechnology, activist groups and the organic lobby in Europe and also in North America argue that Africa risks losing a currently nonexistent 'export' market," he argues. "In promoting Africa as a source for their current food fad, they're promoting starvation and suffering."

A recent editorial in *Nature Biotechnology* concurs with Mbwille's view, noting that while developed countries can afford to approach new technologies with extreme caution if they wish, "applying the same Alice-in-Wonderland principle of zero risk to life in the countries of sub-Saharan Africa is not only inappropriate—it is unconscionable. Wrangling over barriers to trade at the World Trade Organization is one thing. Increasing the likelihood that millions will starve is an entirely different matter."

Overstressed Resources, Underfed People

Opponents of biotechnology say it threatens Africa's environment and may pose health risks. However, recent studies on the effects of biotech crops after six years of use in the United States show that they in fact help the environment. Biotech crops have reduced pesticide use, for example, by 46 million pounds according to the NCFAP study.

Farmers who plant biotech crops say that by reducing the need for some chemical inputs, encouraging the use of conservation tillage and making marginal land more productive, biotechnology enables them to grow more while leaving a smaller imprint on the land. Meanwhile, scientists have so far found little evidence for some of the specific claims made against biotechnology, such as superweeds and loss of genetic diversity.

Africa's low-yield farming practices, on the other hand, can exact a stiff environmental toll. Lacking access to new technologies, African farmers have to clear more land and farm it more intensively with little hope of moving beyond subsistence to profitability. Roughly 12 million acres of African forest is lost every year. The Food and Agricultural Organization (FAO) estimates that an average of 23,000 square miles of damaged land goes out of production every year.

"The rural environment is being devastated in Africa because farmers there have not yet found any way to increase production other than through an extension of their low-yield farming practices into the forest (which means cutting more trees and destroying more habitat) or onto fragile, drought-prone or sloping lands (which invites more soil erosion and watershed destruction)," writes Dr. Robert Paalberg, a professor of political science at Wellesley College.

Mbwille says that biotechnology offers a better, sustainable alternative. "African scientists support biotechnology as a way we can address our food security locally with virus- and disease-resistant crops, grow more food with fewer chemicals and grow food in harsh conditions without relying on Western aid."

"GM seeds would be a great solution for arid areas," says Kenyan scientist Dr. James Ochanda. "We need to harness the technology for staples like cassava, millet and sorghum."

Indeed, a global partnership made up of 30 of the world's leading experts on cassava was just announced Nov. 5, 2002, to promote and coordinate a global investment in the genetic improvement of cassava.

While Zambia's President Mwanawasa worries about whether biotech food is safe, researchers are also exploring ways biotechnology can proactively address food-related health problems such as allergies. One team of public- and private-sector researchers, for example, has just succeeded in disarming the P34 gene in soybeans, one of nature's most widespread allergens.

With malnourishment affecting more than 30 million children in sub-Saharan Africa alone, the greatest health risk is not doing everything possible to improve their diets, says Mbwille.

Food Security for the Future

When the famine in southern Africa eases, the continent as a whole will still face long-term hunger and food-production problems. By 2050, the world's population will increase by another 3 billion people. To feed its share of that increase, the FAO says Africa needs to boost food production by 300 percent.

The "Green Revolution" of the 1960s and 1970s that introduced high-yield farming to much of the developing world largely bypassed Africa. But Mbwille and other experts think that biotechnology—"technology in a seed" that doesn't require farmers to buy expensive equipment or adopt complex growing procedures—will help Africa's small farmers dramatically expand production in sustainable ways over the coming decades.

"Biotechnology puts the information, resources and tools right into the seed, making it much easier for subsistence farmers with less education than Western farmers to gain the benefits of high-yield, low-input agriculture," says Mbwille.

Some of the most innovative research going on in biotechnology could be especially beneficial to Africans. Scientists are optimistic, for example, that they can develop crops which are resistant to droughts like the one affecting Zambia and its neighbors. So-called "edible vaccines" are also in the works—using staple foods to deliver inexpensive, effective vaccines for hepatitis and other illnesses, many of which disproportionately affect low-income Africans.

Edible vaccines and drought-resistant crops are in the future. Today, Zambians need the biotech food aid that's been donated to help them survive. And a growing number of Africans say that biotechnology is essential for their continent to work its way out of poverty and achieve a plentiful, stable food supply.

"I know of no true African farmer who does not desire technology tools, from biotechnology to computer technology, as a way to help Africa break its cycle of poverty and suffering," Mbwille concludes. "When addressing hunger, poverty and despair, excluding options simply is sinful."

Source: "African Doctor Says Biotechnology could Help Feed the Hungry," posted at www.whybiotech.com/index.asp?id=2217. Reprinted courtesy of The Council for Biotechnology Information.

5

from **Jennifer A. Thomson**

The Genetically Modified Foods Debate in South Africa

February 2000

The debate has been heating up in South Africa during the past year. There have been numerous radio talk shows, TV programmes, newspaper articles, live debates, workshops, lectures, etc. I was invited by the World Economic Forum to participate in a debate on the subject at their January meeting in Davos, Switzerland. At the end of the session 82% of the people present voted in favour of GM foods.

I thought it might be helpful to spell out the South African situation and I will do so under the headings of FICTION and FACT.

	Fiction	**Fact**
1	A large number of genetically modified crops are commercially available in South Africa.	Cotton and yellow maize, resistant to insects, are commercially available. The latter accounts for about 2% of the total crop. A number of GM crops are undergoing field trials. The 1998 trials of insect resistant cotton in the Makatini Flats region of KwaZulu Natal showed a significant decrease in the use of insecticides and an increase in yield of between 18 and 23%. The highest increase in yield was achieved by the small scale farmers involved in the trial and one woman farmer stated that she has 30,000 rand in the bank that she hadn't expected.
2	The sale of herbicide resistant crops will result in a huge increase in the use of the specific herbicide, which will be environmentally damaging.	Herbicide resistant crops will allow farmers to spray either before the crop is planted or when the crop is still young. This will result in the use of less herbicide. Data from the USA show that in addition to decreased use of herbicide there is less soil erosion. When herbicide sensitive crops are planted, fields are tilled, weeds allowed to grow, sprayed and then fields are left for the herbicide to dissipate (during which time rain can wash top soil away), before crops are planted. In the case of herbicide resistance, crops and weeds grow up together, are both sprayed and the dead weeds can act as mulch.
3	South Africa has no legislation restricting the release of GMOs, which will result in multinationals "dumping" GM food here.	The GMO Act was passed by Parliament in May 1997 and the Regulations in November 1999. The Executive Council, Registrar and Advisory Committee have all been appointed. Contravention of the Act can result in a fine or imprisonment of up to four years.
4	Farmers, and in particular, small-scale farmers, will be forced to buy GM seeds.	Market forces will prevail. If the GM seeds provide a better yield, farmers will buy them—if not they will buy seed from other companies.
5	It would be easy to separate engineered from non-engineered foods.	Many food items on supermarket shelves contain soybean, from canned soups to baby foods. America is one of the largest suppliers of soybean and some of their exported soybean may have been genetically modified for insect or herbicide resistance. However, all the soybeans are pooled. It is possible, but very expensive to separate the GMOs from the non GMOs. Obviously, the soybeans will have been cooked during the preparation of the food items, a process which denatures proteins, including the ones produced by the introduced gene(s).
6	GM insect-resistant crops contain lectins and lectins are known to be potentially toxic to animals and man.	This was particularly worrying as it was said with great conviction by a Greenpeace representative from London on Will Bernard's "Talk at Will" show on SAfm. Indeed, lectins may be toxic to animals and man, and this was precisely the reason behind the tests done by the Rowett Research Institute. But the GM insect-resistant crops produce the *Bacillus thuringiensis* (bt) toxin which is NOT a lectin and although toxic to certain specific insect species, is NOT toxic to animals or man.
7	It is easy to detect GM foods in any product on a supermarket shelf.	It is possible to do this but it is a costly exercise. At present this expertise is not widely available in South Africa.

8	It is easy to label GM foods on supermarket shelves.	As explained above, detection of GM foods in a given commodity is expensive and cannot be done, at present, in South Africa. A general label, such as "This product might contain plant material which may have been subject to genetic modification" is not very helpful. Such labeling will also tend to "demonise" GM foods in the mind of the public whereas it is the opinion of the vast majority of scientists, in South Africa and abroad, that GM foods are safe.
9	Genes from GM crops can be passed on to other plants resulting in environmental havoc.	Plants can only be pollinated by closely related crops. Therefore the GMO Advisory Committee looks very closely at the potential of GM crops to cross-pollinate other plants, especially weedy relatives. Fortunately major food crops such as maize do not have weedy relatives.
10	The production of GM crops will result in a decrease in biodiversity.	Hundreds of new varieties of crops are produced every year by conventional breeding. This increases biodiversity. GM crops will simply add to this.
11	Insects will rapidly develop resistance to the Bt toxin in insect resistant crops, so planting crops containing this gene is a waste of time. In addition, organic farmers who spray their crops with Bt will no longer be able to use this form of biological control.	This is certainly a concern and requires Integrated Pest Management. Whenever a Bt crop is planted farmers should plant a certain % of non-Bt plants to reduce the risk of the development of insect resistance. The best way is to require seed companies to sell a correct mixture of Bt and non-Bt seeds. This is what South African regulators are investigating.
12	"Terminator" gene technology—which results in seeds produced from a crop being sterile—will force small farmers to continue buying seed from multinationals, rather than being able to plant some of what they produce.	So-called "terminator" technology has been patented, by the US Department of Agriculture and one commercial company. However, it has not yet been perfected, let alone been used anywhere—and may never be used, thanks to public pressure.
13	Genetically modified foods are inherently allergenic and/or harmful.	There is no evidence whatsoever that GM foods in general are any different [from] "normal" foods in terms of toxicity or allergenic potential. Many of the genes used to modify plants occur naturally in plants, or the viruses or the microorganisms that infect them or are associated with them, meaning humans have already been exposed to them.
14	Non-target, beneficial insects will be killed by eating insect resistant plants.	In fact quite the opposite is happening. Because Bt crops are not sprayed, beneficial insects are returning, together with bird species, some of which have not been seen in the area for many years. We are hoping that field analyses will be done in South Africa to establish hard facts in this regard. Data from the USA show an increase in insectivorous insects and a concomitant decrease in pests such as spider mites.

15	Non-target insects will be killed by eating pollen containing insect toxins.	There was a glasshouse study in the USA in which Monarch butterflies were fed pollen from Bt-containing maize. Not unexpectedly they died as they are sensitive to the particular Bt used in the maize. However, the feeding was totally artificial with doses far exceeding those that would be encountered in the field. Field trials have recently been completed and it is clear that even inside maize fields the build-up of Bbt pollen would not be sufficient to pose a threat to Monarch butterflies.
16	There is enough food to feed South Africa and sub-Saharan Africa—it is just a question of distribution.	This is a naïve attitude considering transportation problems on the sub-continent, wars and corruption, to name but a few impediments.

In conclusion, GM crops and foods are just one part of the overall strategy to ensure sufficient food for South Africa and the rest of the continent. Europe has enough food—they don't need GM foods. But we have different needs. Don't throw out GM foods simply because other countries don't want them.

Source: Jennifer Thomson, "The Genetically Modified Foods Debate in South Africa, Revised February 2000."

Review Questions

1. Many reasons have been given for why sub-Saharan African countries should not accept GMO food aid. What are some of these?
2. What is terminator gene technology and how does it fit in to this debate?
3. What are the differences between Green Revolution seeds and GMO seeds?
4. How can desertification be both a cause and an effect of famine?

Discussion/Essay Questions

1. Mina's education at the American university is being funded by a European overseas aid agency. When a report on the attack on Professor Milikin's greenhouse appeared in the *Chronicle of Higher Education*, Mina was quoted in the article. The aid official responsible for Mina's funding recognized Mina's name and was horrified to find that his country, which has strongly opposed GMOs, was funding the training of someone who would go on to promote their production in Africa. He sent a letter to Mina informing her that her aid would be withdrawn if she continued to engage in GMO-related research.

   Mina can choose two routes in her letter of response: She can appeal the aid administrator's decision, arguing that the research in which she is engaging will help her country achieve food security, or she can agree and ask to be transferred to work with a professor at another university where the research is devoted solely to advancing agricultural production without genetic modification. Write Mina's letter.
2. The argument is made in the final two readings that those of us who live in developed nations where food security is not an issue should not be telling

countries with food shortages whether they should be growing, consuming, or exporting genetically modified crops. What do you think about this argument?

3. Michel Chossudovsky (the anti-GMO economist who wrote the third reading) and some of the people quoted in the initial readings assert that if one is serious about solving the world's food needs one must address the uneven political and economic relations that lead so many people to be hungry. Chossudovsky argues that if the world grows more food in a way that reproduces these uneven power relations it will just exacerbate the problems of powerlessness that make people vulnerable to famine. How would the acceptance of GM crops or seed "reproduce uneven power relations" and what would you advise the governments of countries of sub-Saharan countries to do with respect to GM seed?

List of Readings

1. Philippe Demenet, "Can Genetically Modified Organisms Feed the World?" in the *UNESCO Courier*, September 2001, posted on the website of UNESCO, *http://www.unesco.org/courier/2001_09/uk/planet.htm.*
2. Nicole Itano, "Even Hungry Africa Wary of Gene-Modified Food," August 6, 2002, posted on the Website of *The Christian Science Monitor*, *http://www.csmonitor.com/2002/0806/p01s03-woaf.html*
3. Michel Chossudovsky, "Sowing the Seeds of Famine in Ethiopia," in *The Ecologist*, September 2000, revised version (September 2001) posted on the website of the Centre for Research on Globalisation, *http:// www.globalresearch.ca/articles/CHO109B.html.*
4. "African Doctor Says Biotechnology Could Help Feed the Hungry," posted on the website of the Council for Biotechnology Information, *http://www.whybiotech.com/index.asp?id=2217.*
5. Jennifer A. Thomson, "The Genetically Modified Foods Debate in South Africa—Revised February 2000," posted on the website of the Microbiology Department at the University of Cape Town, *http://web.uct.ac.za/microbiology/gmos.htm* (original version published as "UCT Prof Throws Light on GM Foods Debate," in *Monday Paper: Weekly Newspaper of the University of Cape Town*, March 22–29, 1999, *http://web.uct.ac/za/general/monpaper/99-no6/gmfoods.htm*).

Websites for Additional Research

1. For relatively balanced information on genetically modified organisms, the best websites are generally those sponsored by scientific associations or institutes. In the United Kingdom, the Institute of Food Science & Technology, the organization of professional food science researchers, has issued a statement on genetic modification and food at *http://www.ifst.org/hottop10.htm.* In the United States, although most research on GM crops has been conducted by corporate laboratories (in contrast with Green Revolution research), some has been conducted by the Federal Government's Human Genome Project at the Oak Ridge National Laboratory. These researchers have issued an information page at *http://www.ornl.gov/TechResources/Human_Genome/elsi/gmfood.html*, which contains an excellent set of links to other sources of information on ethical issues and policy debates surrounding GM crops.
2. For the anti-GM position, see *http://www.connectotel.com/gmfood.* The site has links to a year's worth of articles from newspapers around the world dealing with the GM food debate. The site also includes links to stories about companies that work with biotechnology such as Monsanto. Additional links to news stories about GM foods can be found at another anti-GM site, *http://geneticengineering.net.* For

the pro-GM position, visit the website of the Council for Biotechnology Information at *http://www.whybiotech. com*. The Council was established by the Biotechnology Industry Group, the association of North American biotechnology companies.

3. Several aid agencies have devoted portions of their websites to providing information on southern Africa's food crisis. Particularly informative are the sites maintained by the U.S. Agency for International Development (*http://www.usaid.gov/about/africafoodcrisis*) and the International Federation of Red Cross and Red Crescent Societies (*http://www.ifrc.org/WHAT/disasters/response/safrica.asp*). (Note: These sites were being maintained in 2004, but they may be taken offline if and when southern Africa's famine situation subsides.)

MONGOLIA
Ulan Bator
Ürümqi
Kashi
Harbin
Shengyang
N. KOREA
Beijing
Tianjin
Dalian
Pyongyang
Seoul
S. KOREA
Pusan
Jinan
Yellow R. (Huang He)
Lanzhou
Xi'an
CHINA
Chengdu
Yangtze R.
Wuhan
(Chang Jiang)
Shanghai
Chongqing
Lhasa
Irrawaddy R.
Kunming
Taipei
Guangzhou (Canton)
Hong Kong
Taiwan
Red R.
Mandalay
Hanoi
Haiphong
BURMA
LAOS
Luang Prabang
Chiang Mai
Gulf of Tonkin
Hainan
Luzon Strait
Bay of Bengal
Vientiane
Mekong R.
Rangoon (Yangon)
THAILAND
Hue
Da Nang
Luzon
Baguio
Philippine Sea
Quezon City
Manila
Bangkok
Chao Phraya R.
VIETNAM
PHILIPPINES
CAMBODIA
Mindoro
Andaman Islands
Phnom Penh
Ho Chi Minh City (Saigon)
South China Sea
Palawan
Visayas
Iloilo
Cebu
Cebu City
Negros
Andaman Sea
Gulf of Thailand
Spratley Islands
Sulu Sea
Cagayan
Mindanao
Davao City
Phuket
Penang (George Town)
Bandar Seri Begawan
Kota Kinabalu
SABAH
Banda Aceh
Strait of Malacca
Ipoh
MALAYSIA
Kuala Lumpur
Medan
Melaka (Malacca)
SARAWAK
BRUNEI
Celebes Sea
PACIFIC OCEAN
SINGAPORE
Singapore
Kuching
Borneo (Kalimantan)
Manado
Moluccas
Ternate
Halmahera
Sumatra
Riau Islands
Pontianak
Padang
Balikpapan
Makassar Strait
Sorong
Sulawesi (Celebes)
Palembang
Banjarmasin
Buru
Ceram
Ambon
Jayapur
Irian Jay
INDONESIA
N
Telukbetung
Jakarta
Java Sea
Ujungpandang
Banda Sea
Bogor
Bandung
Semarang
Surabaya
Java
Flores Sea
Bali
Lombok
Flores
Dili
Denpasar
Sumbawa
EAST TIMOR
Arafura Sea
Merauke
Sumba
Timor
Timor Sea
INDIAN OCEAN
More than 1 million
500,000–1 million
Fewer than 500,000
Capital cities are underlined
0 200 400 Miles
0 200 400 Kilometers
Tropic of Cancer
Equator
Tropic of Capricorn
140°
40°N
20°N
10°N
0°
10°S
20°S
130°E
120°E
80°E
90°E
100°E
110°E

Export Processing Zones in East and Southeast Asia

Companion to Chapter 11 Industry

Most athletic footwear sold in developed countries is produced in Asia's export processing zones. *Source:* Mike Yamashita/Woodfin Camp & Associates

Xiaoling and **Jingfang** live and work at a garment factory in the Guangdong province of southern China. This garment factory makes athletic shoes for Reebok, Adidas, and Kickers, most of which are shipped to the United States. The two women are in their twenties, as are most of the other workers at the factory. They normally work from 7:45 in the morning until 5:30 in the afternoon, but the workday can be extended by three to five hours during the peak season. They must apply for an exemption if they do not want to or cannot work the extra hours. Normally, they work six days a week, but the workday may be extended to seven days if extra orders need to be met.

Xiaoling and Jingfang each have had one month of their wages withheld by management as a "deposit." If either woman wants to receive her deposit, she must apply for permission to quit one month before leaving. If a worker is fired (or if she leaves before waiting for permission to be granted), her employer keeps the deposit.

During the peak seasons (usually August and December) workers like Xiaoling and Jingfang may make up to $125 for the month. During slow times, though, the pay is not so good. It can be as low as $25 per month, since workers are paid at a piece rate.

When they are through with their shift, Xiaoling and Jingfang return to their dormitory, which is provided for a small fee by their employer. The monthly rent amounts to around US$2.20, but they have to share their sleeping quarters (a 24-square-meter room with bunk beds) with ten other women. They also have only one bathroom on their floor.

Most of the workers in this factory are women, and most are young. In fact, the employment advertisement Xiaoling and Jingfang replied to requested women between ages 17 and 21. When they began working at the factory they each signed a contract, although neither has seen a copy since then.

Source: Adapted from the report on working conditions at the Heng Yu Garment Factory in Panyu, China, produced in 2000 by the Clean Clothes Campaign. ■

Commodity Chains and the Global Assembly Line

The textbook's discussion of site factors in industry location includes an introduction to the global clothing industry (see Chapter 11). This discussion includes the string of events that take place in the production of a piece of clothing that you might buy at the local mall or discount store. This string of processes includes, but is not limited to, the spinning of yarn, the weaving of fabric, and the manufacturing of the finished garment. This string or "chain" of events that includes all the processes involved with production of an item from its extraction to its final destination as a finished product is called a commodity chain. Xiaoling and Jingfang are part of the commodity chain that results in the sneakers that you wear to school or the gym.

Every product that you consume goes through a commodity chain before it gets to you. Although you don't actually "see" financing, design, component production, assembly, or marketing when you purchase a product (or "commodity"), these processes are embedded in every product. As an individual consumer, you are the last link in the commodity chain. Countless unseen workers, in one way or another, contribute to the product that you consume, a concept that was articulated particularly clearly by Karl Marx in his classic study of political economy, *Capital.*

When Marx was writing, in the mid-nineteenth century, most commodity chains were contained within one area and one firm. Back then, you might have known the individuals who produced many of the goods you were consuming. Today, that is rarely the case. Think of those "inspected by no. 21" labels in clothing; you are so far from knowing who "no. 21" is that he or she doesn't even have a name! Manufactured goods are being produced on a global scale as never before. Increasingly, manufactured goods are designed in one country, components are manufactured in a second country, final assembly is completed in a third country, and the product is sold in a fourth country. The bank (or investors) supporting the entire project may be located in a fifth country.

Frequently, each of these processes in the commodity chain is organized by a different firm, as semifinished products are traded between component producers and assemblers. Three main processes are involved: vertical disintegration, economies of scale, and transnationalization.

Vertical Disintegration

Historically, at least since the nineteenth century, most firms have tried to be as large as possible and control all links on the commodity chain. This was known as vertical integration. Not only did General Motors own component manufacturers (such as GM's Delphi and Delco divisions), but at one point GM even owned sheep ranches in Wyoming to provide wool for its car seats! Retail dealers were independently owned franchises, but they were closely controlled by GM. Corporations benefited from this business strategy in several ways: Quality of components could be assured, production of components could be organized to fit the needs of the finished product, and costs could be reduced by cutting out the "middleman."

Now, however, firms increasingly are pursuing an alternate strategy: vertical disintegration. New products are often designed at small independent research labs (sometimes affiliated with universities). The design is then sold to a large well-known company that organizes financing, production, and wholesale marketing. The actual manufacture of the product, however, often is carried out by a number of smaller firms, although the big firm may still engage in final assembly. And retailing is then done by other companies that do nothing but operate stores (think of how automobile superstores are replacing manufacturer-affiliated dealers).

With vertical disintegration, the big firms can make someone else bear the risk for time and money spent in research and design that fails to pay off. They can also play suppliers against each other, canceling a contract with one supplier if its costs get too high and instead buying from another. Intense competition between suppliers to produce cheaper inputs is often referred to as a "race to the bottom" as component suppliers all seek to cut costs as much as possible so as to be able to sell to the big firms. If component suppliers become overenthusiastic in their race to the bottom and abuse workers' rights, the company whose name is on the consumer product can deny responsibility since the abuse did not take place in "its" factory.

Economies of Scale

Even as different links in the commodity chain are controlled by different firms, within each link there tend to be fewer and fewer firms as individual producers take advantage of economies of scale. An economy of scale is simply the condition wherein the average cost of a product goes down when you produce many units of the product. For instance, it may cost $100,000 to produce your first television but only $50 to produce each additional one, because most of the cost goes into the research and development of the first unit. If you produce only 100 TVs, your average cost is $1,050 but if you produce 3,000 TVs your average cost falls to $83 because you are able to spread the fixed costs of research and development over many units.

As technology develops, an increasing percentage of the cost of manufactured goods is in design or in the original purchase or engineering of production machinery. In such an environment only the biggest producers survive, and these producers then have the capital with which to organize further vertically disintegrated commodity chains. The one exception to this trend is at the very low-technology links within the commodity chain, where production of even the first unit is very cheap. At this lowest end, small firms are constantly entering the market and undercutting existing producers, engaging in the "race to the bottom" that results in a continual downward push on wages and working conditions.

Transnationalization

As the world economy becomes characterized by large firms (economies of scale) organizing many component producers (vertical disintegration), these firms are able to draw from a global field of component producers. For instance, when General Motors needs a low-tech component (say, door handles), it seeks bids from a number of component suppliers around the world. These firms—and, in effect, these places—all compete with each other to win GM's favor. It's in General Motors' interest to have these competing component manufacturers spread out over many world regions so that regions can compete with each other to lower wages and thereby present GM with the lowest product cost. Thus, places as well as firms join the "race to the bottom." As this competition intensifies around the globe, the lengths of commodity chains increase, and complex consumer products (such as cars or computers) often contain components manufactured in more than a dozen countries.

Amid the convergence of vertical disintegration, economies of scale, and transnationalization, the world economy is increasingly dominated by transnational corporations (TNCs, sometimes also known as multinational corporations or MNCs). TNCs receive specific advantages due to their internationalized form. TNCs are able to trade within the firm with a minimum of regulation. The free market does not set the price for the goods transferred through intrafirm trade (trade that is within a single company); corporate managers do.

It has been estimated that more than one-third of the world's trade is intrafirm, and intrafirm trade constitutes more than two-thirds of trade that crosses national borders. Through international, intrafirm trade, a TNC can manipulate the global economy to its advantage, in several ways:

- Prices of goods sold from a subsidiary in one country to a subsidiary in another country can be set artificially low or artificially high, thereby shifting profits to the country with the lowest tax rate. Through use of this practice, known as transfer pricing, many of the U.S.'s largest and most profitable TNCs avoid paying any U.S. corporate income tax.
- This shifting of profits from country to country can also be used to make a national subsidiary look unprofitable when unions ask for wage increases. Conversely, it can be used to make a national subsidiary look profitable when seeking loans or investments.
- Shifting of costs and profits between countries is used to enable TNCs to profit from differences in currency rates.

As TNCs spread their manufacturing across national borders and form strategic alliances with their competitors to offset the costs of research and development, it can become difficult to distinguish the nationality of a company or a product. If you go to the website of the Ford Motor Company *(http://www.ford.com)*, you will find, across the top of the page, the nameplates for Ford's Lincoln and Mercury brands, but also the nameplates for the Japanese firm Mazda, the Swedish firm Volvo, and the British firms Jaguar, Aston Martin, and Land Rover. While these firms' production lines are only partially integrated with Ford's, Ford owns a substantial portion of each of these firms and there is considerable trading of technology and components. Similarly, if you go to the General Motors website *(http://www.gm.com)* you will find, interspersed with G.M.'s U.S. brands (Buick, Cadillac, Chevrolet, GMC, Hummer, Oldsmobile, Pontiac, and Saturn), the names of GM's international subsidiaries, including Opel (in Europe), Vauxhall (in the United Kingdom), and Holden (in Australia). More surprising, you will also find the names of some of GM's "competitors": the Italian firm Fiat; the Japanese firms Isuzu, Suzuki, and Subaru; and the Swedish firm Saab, all of which are partially owned by GM.

Even when there is no cross-ownership, "competing" firms often work together. GM and Toyota have a joint factory in California that formerly produced the (almost identical) Toyota Corolla and Geo/Chevrolet Prizm and now produces the Toyota Matrix and its near-twin, the Pontiac Vibe. Which car is more "American": a Japanese-designed Honda Accord manufactured at a Honda factory in Ohio; a Ford Focus designed by Ford engineers in Germany and manufactured at a Ford plant in Mexico; or a Mazda B2300 pickup truck that, although bearing a Japanese nameplate, is simply a rebranded version of a U.S.-designed Ford Ranger built at a Ford factory in Missouri?

TNCs and Foreign Direct Investment

TNCs can be as powerful as countries and their annual sales are often greater than the annual production of the countries within which they operate. Many less developed countries attempt to attract investment by TNCs, called foreign direct investment (FDI), in an attempt to inject life into stagnant or depressed economies. Countries typically hope for more than just jobs from TNC investments. The hope is that FDI also will lead to growth of new supplier companies, as well as a transfer of technology, and that this should then lead to all-around economic growth.

Others, however, suggest that investments by TNCs may do more harm than good in LDCs. TNCs tend to locate their lowest-value, least skilled assembly processes in the periphery, so there is little technological skill development among the local population. Furthermore, since so many LDCs have large numbers of unemployed laborers, TNCs can play these countries against each other, precipitating the "race to the bottom" discussed earlier. The result is that workers in these plants are paid just enough to live on (if that), and they have no extra money to spend in the local economy. Tax revenues from TNCs are often minimal as well, since extensive tax breaks are often given to entice a TNC to locate its factory in that country and not another. Indeed, an LDC may lose money on a TNC factory because often the country has to build expensive infrastructure (buildings, roads, power grids) in order to get the TNC to locate there. Finally, because the products that are assembled in these factories are often made from imported high-technology parts (as in the case of computer and electronics assembly), TNCs frequently do little to support a local economy of supplier firms and workshops.

Export Processing Zones

In order to attract foreign direct investment from TNCs, many countries have established export processing zones (EPZs). EPZs are industrial parks designed specifically for foreign companies that wish to establish export-oriented industries. Usually they are located next to ports or airports to facilitate easy import of components and export of final products. Basic infrastructure is usually provided free to foreign companies, and often other special incentives are given as well, such as tax exemptions, provision of security, and construction of workers' dormitories. The workers in EPZs are overwhelmingly young and female; in LDCs (as in MDCs) young women are the least expensive laborers and they have a reputation among employers for being easy to manage.

EPZs are ideal for industrial activities that use large numbers of unskilled laborers and whose product is relatively lightweight (to lower shipping costs, since by definition EPZs are producing goods for export). For this reason the most common EPZ industries are clothing/footwear manufacture, electronics assembly, and data processing.

EPZs in East and Southeast Asia

Although EPZs have achieved some success in Central America and the Caribbean (especially along the Mexico-U.S. border, where they are known as *maquiladoras*), the development strategy has been particularly popular in East and Southeast Asia. In the 1950s and 1960s corporations from the United States, Japan, and Europe began to seek out cheap production sites overseas. Africa and South Asia were going through decolonization at the time, and their governments generally were considered too weak to provide the kind of security (and control of potentially rebellious workers) that foreign corporations required. Latin America had long been independent, but most governments there were seeking development on their own terms, with national industries producing for national markets (see Chapter 9 and the accompanying module in this workbook). Several countries of East and Southeast Asia, however, had large populations that could be enticed to leave rural areas, and strong governments that could promise to provide a stable legal system and, if necessary, put down worker rebellions. In addition, U.S. Cold War strategy saw East and Southeast Asia as a key geostrategic region, and extra money was pumped in to bolster national governments (which, in turn, could further assist in the management of EPZs) in return for continued integration into the U.S.'s economic and political system. Thus were born Asia's "four tigers" of industrialization: Hong Kong, Singapore, South Korea, and Taiwan. Although development in each of these four areas came at a cost, by almost any measure, it has been successful, to the point where each of the original "tigers" now has its own TNCs that are seeking out low-wage overseas locations for making investments. Seeing the success of the "four tigers," most other countries in the region have attempted to mimic their example, although generally with less success. Today, EPZs can be found in Indonesia, Malaysia, Thailand, China, and Vietnam, among other countries.

Supporters of EPZs stress that they bring income to parts of the world that are at the early stages of development and, pointing in particular to the "four tigers," they note that EPZs can advance nations up the development ladder. This argument is made in the first reading by Richard Bolin, the director of the U.S.-based World Economic Processing Zones Association (WEPZA), an organization of EPZ managers from around the world.

The second reading takes a markedly different view on EPZs, by linking them with the presence of sweatshops. This reading, from UNITE, the union representing garment and textile workers in the United States and Canada, notes that sweatshops are factories in which vulnerable workers are used to produce garments under harsh conditions. While sweatshops are particularly common in EPZs, they exist globally, even in the heart of some of the world's wealthiest cities. This point is important because it suggests that sweatshops will not disappear simply because a nation develops. The link between EPZs and sweatshops in East Asia is given a human face in the third reading, a *New*

York Times report from 2003 on working conditions in one of China's export-oriented industries.

The next two readings suggest that EPZs may not be achieving all that they were supposed to achieve economically. In the fourth reading, the state-run Vietnam News Agency echoes that country's government's generally supportive opinion of EPZs. Nonetheless, a careful reading of the news report reveals that the government is disturbed by the lack of links between EPZs and the rest of the nation's economy. According to the news story, EPZs are neither buying from domestic suppliers nor selling to domestic consumers; instead they are just being used by foreign companies to assemble imported components for re-export. Such a function may bring jobs to Vietnam, but it will not to do much to aid transformation of the overall economy. The fifth reading, a summary of a report presented at a conference sponsored by the International Labour Organisation (a U.N. affiliate), discusses the impact of EPZs on women in Thailand and suggests that, due to the "race to the bottom" among EPZ-sponsoring countries, Thai women are not benefiting as much as they might from the EPZs where so many of them are employed.

As countries and firms at the bottom of the economic ladder attempt to outcompete each other, there has been considerable pressure around the world to halt the worst excesses of globalization that occur within EPZs. Because so many of its products are made by EPZ-based contractors, Nike has become a lightning rod for organizations seeking to limit abuse by EPZ employers. Nike's contractors operate more than eight hundred factories around the world, with more than 600,000 workers in more than fifty countries. Nike itself has relatively few employees and does almost none of its own manufacturing (it mostly engages in research, design, marketing, and financing). Nonetheless, it has obvious leverage over its many EPZ-based contractors, and, because it is such a highly visible brand name, the corporation has been identified by labor reformers as one that could be pressured to place its corporate weight behind changes in EPZ practices.

In response to criticisms of its contractors, Nike has adopted a code of conduct that it claims it requires its contractors to follow. (The Nike Code of Conduct can be found on the web at http://www.nike.com/nikebiz/nikebiz.jhtml?page=25&cat=compliance&subcat=code.) According to the sixth reading, however, an article from the magazine *Multinational Monitor*, these codes of conduct are not adequately enforced. Labor-rights supporters seeking to develop a more vigorous contractor monitoring system have developed the tactic of using universities to pressure garment manufacturers (such as Nike) to enforce their codes of conduct through independent monitoring. Universities have some leverage over the garment corporations because companies like Nike sell products with licensed university logos, and they achieve much of their marketing by placing their corporate symbol (such as the Nike swoosh) on college team uniforms. The seventh reading is from the Worker Rights Consortium (WRC), an organization funded by more than one hundred U.S. universities to independently monitor conditions in factories that produce the goods that ultimately bear university logos. The WRC gathers information about overseas factories so that universities can enforce university codes of conduct, if necessary by withdrawing contracts from garment firms that fail to demand ethical standards of their contractors.

Over the past years, several college campuses have had divisive battles over whether to join the WRC or its industry-supported counterpart, the Fair Labor Association. The anti-sweatshop/pro-WRC activists' strategy of pressuring universities to pressure large manufacturers to pressure their contractors appears successful, but it is quite indirect. A more direct attempt at changing conditions in EPZs is illustrated in the final reading, a 2002 report on a violent strike at an EPZ in China.

Readings

from the World Economic Processing Zones Association

1

Editorials

Editorial 1: The WEPZA Family

The star performer in attracting foreign investment and technology to developing countries (creating jobs, improving skills, earning hard currencies and giving poor people hope for the future) during the past 50 years has been the Economic Processing Zone (EPZ). This child of the 3000-year-old Free Zone with modern technology in the form of factories and many kinds of economic services linked by intermodal transport systems energizes the "supply chains" now in vogue. To use Peter Drucker's terms, it is "the Ugly Duckling that no one loves"—the overlooked "management innovation" already proven. "To create wealth, jobs and incomes in desperately poor countries, it is the only poverty program that works." The EPZ is ready to improve economies everywhere for the 21st Century.

Whole countries have come out of poverty only through using EPZs since 1950—Puerto Rico, Taiwan, Singapore, Ireland, Mexico, Korea, Dubai, UAE and currently Coastal China. Mexico alone created one million export worker jobs with our help in the last decade in their maquiladoras which are their version of EPZs. At least 40 other developing countries are on their way to success through intelligent application of EPZs. The key is their efficiency in their freedom to access the Global Market through Global Production and Added Value inside their territories.

There is only one organization in the world which has consistently and effectively encouraged EPZs since 1978 and that is WEPZA, the independent non-profit private World Economic Processing Zones Association. We are now an organization of 40 of the world's leading EPZs with 150 production sites in 38 countries and with 800,000 production workers. WEPZA has published journals, books, and 18 conference memoirs with a wealth of practical and statistical information about EPZs. See our website at *http://www.wepza.org*. To counter the false "job loss" arguments of trade unions (really "job upgrades") and defend against rising trade-obliterating protectionist moves of rich countries, WEPZA knowledge is essential.

WEPZA is poised to lead both poor and rich countries to their wealthier destinies through research, education and cooperation in the next hundred years. Corporations with global ambitions, countries which yearn to be serious about successful development, and world multilateral organizations which want to raise the social and economic lives of millions will do well to heed our message and aid us when you can. The WEPZA Family welcomes you.

Sincerely yours,

Richard L. Bolin
Director, Secretariat

Editorial 2: Why EPZs Work

The Key Question Today:

How will the advanced nations ever progress if the poor countries can't pay?

Wrong Answer, Today:

The top 31 OECD countries can trade among themselves and sell at discount prices or donate surpluses to poor countries. Witness 1) the present sale of discounted AIDS drugs to Africa and 2) plans to credit poor countries for their trees which produce oxygen from carbon dioxide to "fight" global warming. To count on charity in this way is wrong and contrary to the way history has clearly shown to our generation that economic progress is made through productivity and efficiency in the open market. These two short-term responses to perceived negative circumstances are defensive—not creative. They do little to help poor nations in the long run. What is needed is self sufficiency in poor countries.

Right Answer, For the Last 50 Years:
Boost the economies of poor countries so they can afford to pay their way and stand on their own feet. Economic Processing Zones (EPZs) offer risk-reduction services making it easy for advanced corporations to invest in poor countries. Production Sharing among global factories in many countries teaches workers, technicians, engineers, managers, politicians and the general public how to make money from the Global Market.

Economic improvement starts on Day One with new jobs, and is clearly shown to occur significantly within a decade from a cold start—by spreading rapidly as EPZs are duplicated throughout the poor country. The foreign exchange gained by the poor country during this first period buys advanced products from advanced nations (e.g., aircraft). This enlarges the world market available at full prices to the advanced nation's workers, technicians, engineers, managers, politicians and the general public. Better jobs and value added are created in advanced nations by this process. Also, they benefit from the lowering of prices on consumer goods such as apparel and electronics assembled in the EPZs.

Meanwhile the poor countries enjoy low-priced consumer goods of high quality and attractiveness produced in their own EPZs while their treasury collects tariffs on the imported components for this portion. At the same time the EPZ improves the infrastructure, worker skills, training, competitive position and hope for the future within its population. It prepares itself to enter more rapidly and with higher technology into the global market in the second generation.

Within a single worker's lifetime career, we have seen changes that are literally amazing with respect to EPZ product and service sophistication. In the second generation in Mexico a world market design center for engineering products now employs up to 3000 local engineers (equivalent to the 3-year output of the country's engineering schools when the process started 35 years ago, but less than 1/3 of annual output, today—students see they can count on employment as engineers if they excel).

But it all can start with one small EPZ instead of a huge program, which becomes a target for every advocacy group concerned with policy decisions of the next 50 years, which can kill good ideas with their invective. A single EPZ is politically possible in many countries because, if it fails, it is no big mistake for anyone. Creation of a single modest EPZ is sufficient to begin the process of proving the poor country's case in only a few short years. It will be a competitive successful source of products and services for the global market if it is carefully designed, hires quality personnel, has the full backing of government, operates in a corruption-free business climate, has a competitive set of incentives to meet competition (unique to meet its own needs), and takes all the risks necessary to attract early investors to prove its case. It must be promoted by experts paid as officers of the EPZ.

Economic development led by the EPZ has occurred in 8 important countries: Puerto Rico, Taiwan, Singapore, Korea, Ireland, Mexico, Dubai, UAE and currently, Coastal China. To use Professor Peter F. Drucker's words in a letter dated July 25, 2001: "To create wealth, jobs and incomes in desperately poor countries, it (WEPZA) is the only poverty program that works."

Sincerely yours,

Richard L. Bolin
Director, Secretariat

Source: Richard Bolin, "The WEPZA Family" and "Why EPZ's Work." Reprinted by permission of World Economic Processing Zones Association and The Flagstaff Institute. Used with permission.

2

from the Union of Needletrades, Industrial, and Textile Employees

Stop Sweatshops!

What Is a Sweatshop?

The word "sweatshop" was originally used in the 19th century to describe a subcontracting system in which the middlemen earned their profits from the margin between the amount they received for a contract and the amount they paid workers with whom they subcontracted. The margin was said to be "sweated" from the workers because they received minimal wages for excessive hours worked under unsanitary conditions.

Today's subcontracting system functions on a global basis. Large clothing companies produce apparel in 160 countries, often with shockingly low wages and horrible working conditions. Apparel workers in Bangladesh earn 20 cents an hour, and in the free trade zones in El Salvador they earn 56 cents an hour, just to give two examples. The clothing companies then export that apparel to 30 developed countries, like the United States and Canada. Meanwhile, apparel workers in the developed world are forced to compete against those conditions.

A sweatshop is characterized by the systematic violation of one or more fundamental workers' rights that have been codified in international and U.S. law. These rights include the prohibition of child labor, forced or compulsory labor and discrimination in employment based on any personal characteristic other than the ability to do the job; the right to a safe and healthy work environment that does not expose workers to degrading or dangerous working conditions; freedom of association and the right to organize and bargain collectively. A sweatshop is also characterized by wages that do not permit workers to feed, clothe and shelter themselves and their families, and hours of work so long that education and a decent family life are out of reach.

Sweatshops in the U.S. are often lawless operations in other ways, evading not only wage and hour laws, but also paying no taxes, violating fire and building codes, seeking out and exploiting undocumented immigrants and operating in the underground economy, hidden from public view.

Sweatshops are most prevalent in apparel manufacturing, but sweatshop conditions exist in an increasing number of manufacturing and service industries. Subcontracting is being used for auto parts, building maintenance and many kinds of public sector work with the explicit goal of lowering wages and benefits. Even among professionals, there is a trend towards replacing corporate staff with freelance consultants without job security or benefits. Apparel sweatshops are just the extreme version of the general lowering of living standards and corporate attempts to evade responsibility for workers and working conditions.

Why Have Sweatshops Returned?

Today's apparel sweatshop is a product of the global economy. Large retailers and manufacturers, seeking greater profits in a highly competitive industry, contract production to thousands of contractors and subcontractors located wherever labor costs are low, whether in Malaysia or Honduras, Los Angeles or New York.

An increasingly concentrated retail market has changed the structure of the apparel industry. As early as 1993, the five largest retail companies accounted for almost half of total retail sales—$168 billion. Today, these giant retailers often bypass the traditional apparel manufacturer, ordering their own "private label" or "house brand" goods directly from contracting shops at home and abroad.

Changes in immigration laws in the mid-1980s made it illegal to employ undocumented workers. This did not stop the employment of undocumented immigrants; it just drove that employment underground. Once these contractors were breaking the immigration law, it was a simple step to breaking wage-hour, tax and labor laws. And undocumented workers' fear of discovery helped keep sweatshops a dirty secret until recently.

But immigrants do not cause sweatshops. Rather, it is the low prices set by the retailers and manufacturers at the top of the apparel industry chain that create the conditions for sweatshops to flourish and for immigrants to be exploited. The decade of cuts in funds for government monitoring of industry lawbreakers makes sweatshops a highly profitable and low-risk "solution" to the competitive pressures of the global apparel industry.

How Many Sweatshops Are There in the United States Today?

Sweatshops are part of the underground economy and operate in violation of many laws. Transactions are often in cash, and sweatshops actively work to avoid detection, so it is difficult to count them. It is possible to make an educated guess, however. There are fewer than 1,000 garment manufacturers who parcel out production to about 22,000 contractors, some of them legitimate and some of them sweatshops. Each contracting shop typically employs 25 to 50 workers (but can have as many as 100 workers). There has been a three-fold increase in the number of tiny subcontractors (under 20 workers) from 1977 to 1992 at the same time that larger shops (with better conditions) have closed down. The GAO has reported estimates of sweatshops in major garment producing centers:

- New York City: In 1994, the GAO reported an estimate that 4,500 of 5,000 garment shops were sweatshops.
- Miami: 400 of the total 500 garment shops are sweatshops.
- El Paso: 50 of 180 are sweatshops.
- New Orleans: 25 of 100 apparel firms are sweatshops.
- The GAO also reported that there are apparel sweatshops in parts of New Jersey, Chicago, Philadelphia, San Antonio, and Portland, OR.

Source: "Stop Sweatshops! What Is a Sweatshop?" posted at www.uniteunion.org/sweatshops/whatis/infosheet.html.

from *The New York Times* 3

China's Workers Risk Limbs in Capitalist Export Drive

by Joseph Kahn
April 7, 2003

YONGKANG, China—In his 17 days of molding tool boxes, Wang Chenghua learned to work like a metronome. He slipped strips of metal under a mechanical hammer with his right hand, then swept molded parts into a pile with his left. He did this once a second for a 10-hour shift, minus a half-hour lunch.

Just before lunch on the 18th day, he lost the beat. The hammer, backed by 4,000 pounds of pressure, ripped through the middle and ring fingers of his right hand, reducing them to pulp.

Mr. Wang, 26, now spends his days in the orthopedic ward of the Yongkang First People's Hospital, where wall posters advertise reattachment surgery for molders, millers, pressers and lathe operators who, unlike Mr. Wang, salvaged their digits and limbs.

"The work is so boring it is almost impossible to keep your mind on it," Mr. Wang recalled one recent afternoon while resting on a dirty cot in his crowded hospital room. "But if you let your mind wander for just a second, it's over."

Yongkang, in prosperous Zhejiang Province just south of Shanghai, is the hardware capital of China. Its 7,000 metal-working factories all privately owned make hinges, hubcaps, pots and pans, power drills, security doors, tool boxes, thermoses, electric razors, headphones, plugs, fans and just about anything else with metallic innards.

People all over the world use Yongkang-made parts. They are the nuts and bolts of hundreds of brand name products, like Bosch, Black & Decker, and Hitachi.

Yongkang, which means "eternal health" in Chinese, is also the dismemberment capital of China. At least once a day someone like Mr. Wang is rushed to one of the dozen clinics that specialize in treating hand, arm and finger injuries, according to local government statistics.

Unofficial estimates run as high as 2,500 such accidents here each year. Lawyers, journalists and doctors say hundreds or thousands of injured workers are left out of the official count either because the boss does not have insurance, or because workers were not hired legally, or because city officials are under pressure to show that they have safety under control.

The reality, all over China, is that workplace casualties have become endemic. Nationally, 140,000 people died in work-related accidents last year up from about 109,000 in 2000, according to the State Administration of Work Safety. Hundreds of thousands more were injured.

Safety officials call this alarming. But the steps the government has taken to address the issue occasional propaganda campaigns and isolated crackdowns on irresponsible bosses have done little to solve fundamental problems in China's emerging capitalist culture, where high rates of injury and death are tolerated as the price of economic progress.

In what has become the world's factory for labor-intensive products over the last decade of surging economic growth, an endless supply of transient workers line up for jobs offering about 50 cents an hour, about one-twentieth of the average American manufacturing wage.

Millions of eager Chinese entrepreneurs help fill the shelves of Wal-Mart and Target with sneakers and DVD players that get better and cheaper year after year.

Last year China surpassed Japan as Asia's biggest exporter to the United States, and it replaced the United States as the largest exporter to Japan.

But safety has suffered in this relentless drive, not least in Yongkang. Some factories resemble operations at the dawn of the industrial age, where migrant laborers use rudimentary machines that can sever the limbs of those who succumb to momentary distractions. Workers usually have only elementary school educations and get no training. They do not have the right to organize unions or, in some cases, even discuss workplace hazards.

Xu Xing Metals seems typical. Its 50 workers press metal into strips for thermal cups, working in a dimly lit shed where bars cover broken windows. The floor rumbles with the pulse of six metal presses. Workers have to shout over the roar to be heard.

Li Jiahao, a 38-year-old master craftsman at Xu Xing, said he had long worried about safety, especially since his boss discovered he could save money by using wastewater instead of fresh water for the machines' cooling systems. Wastewater clogged the machines' pipes, requiring constant cleaning, Mr. Li said.

"I told him somebody was going to get his arm caught unclogging the filter," Mr. Li said. "He said I had no right to speak."

Mr. Li was scooping grime out of a cooling pipe one day in mid-March when the machine's rollers clipped his shirt, dragging his arm into the metal press, elbow first. When co-workers wrenched him free, he could see his hand popping out of his shirt but felt nothing, the arm was hanging by a few tendons. It had to be amputated.

Recuperating at the First People's Hospital, where he is teaching himself to write and use chopsticks left-handed, Mr. Li said his boss at Xu Xing, Hu Xu, had paid for his medical care but refused to discuss compensation. "The boss claims this was operator error, not a safety problem," Mr. Li said.

Mr. Hu declined to discuss the case in a telephone interview. Pressed to explain what happened to Mr. Li, he said, "Why don't you ask him?" and hung up.

The riskiest jobs in Yongkang, as in war, go to green recruits, fresh from the farm. Young migrants are hired at the train station to run metal-stampers and molders, high-pressure hammers driven by flywheels. Many new workers do not last a month.

Ling Banghua, 23, a native of rural Jiangxi Province, arrived in Yongkang in early March. He was spending his fourth day making tops for bicycle pumps when a mold the size of a hockey puck took three fingers off his left hand. He has pleaded with his boss to give him enough money for a bus ticket home.

"No one needs a one-handed migrant worker," he says.

Mr. Wang, the tool-box maker, never got paid in his 18 days at Hua Xin Electronics and ran up a bill at the company canteen. Though the law mandates compensation for injured workers, Mr. Wang, a native of Guizhou Province, one of the poorest in China, did not know he had to sign a contract to be legally employed. The young man, who speaks heavily accented Mandarin, cannot afford a lawyer.

Mr. Wang's unwashed hair has melded into a pompadour and his bandaged hand is stained with dirt and yellow disinfectant. Sitting on his hospital bed as fellow patients gathered around, he described his discussions with his boss, Shi Yanxin. The boss, Mr. Wang said, told him that his finger injury did not merit any hardship pay beyond covering medical fees.

"I joked with him that I wish I lost more fingers maybe he'd give me compensation," Mr. Wang said. "I just hope he gives me enough to get home."

Mr. Shi denied that he had decided not to pay compensation.

"It depends on how well he recovers," he said in an interview. But he said the accident was Mr. Wang's fault.

"With these guys, you tell them to pay attention and they don't listen," Mr. Shi said. "They have no culture or education. They are told many times to be safe and they just don't get it."

Reducing injuries may require more than increased diligence by workers, however. It is also a legal, economic and political problem.

By law, the most dangerous machines have to be sold with infrared or temperature-sensitive devices designed to shut the machine down when hands or limbs extend past a safety zone.

But in Yongkang's open-air market, where salesmen haggle with customers over the price of pressurized hammers and metal lathes the size of pickup trucks, the law is negotiable. Few customers pay extra for safety screens.

"The customer is God," said the manager at Yangli Group's showroom. "The law and the way things are done here are two different things."

Yongkang's government also does little to enforce a statutory schedule of worker compensation pegged to the degree of injury. Legally, for example, the loss of all fingers on one hand is a sixth-degree injury, mandating compensation of 200,000 yuan, or about $24,000.

In practice, few receive payments of this size without going through lengthy and costly arbitration or court hearings, which can take months or years. Most owners reach settlements with their employees for nominal amounts and pay their bus fare out of town.

By that standard, Huang Ruirong, a 35-year-old migrant from Anhui Province, made out well. After he lost most of the index, middle and ring fingers on his right hand working at a exercise equipment factory, he studied the law. He remained in Yongkang, unpaid, for four months after his injury last year to go through arbitration.

In the end, he went home with 23,000 yuan, about $2,800 less than he thought he deserved but far more than his boss initially offered.

Back in Anhui, however, the money has vanished. His aging parents needed medical care. He has had to pay fellow villages to take care of his family's land because he could not plant or harvest rice with a crippled right hand. Recently he traveled to Shanghai in search of jobs for handicapped migrants, but found none.

"Some people tell me I have an unlucky fate," Mr. Huang said. "But I don't think it's fate. I think it's an accident. Accidents can be avoided."

Source: "China's Workers Risk Limbs in Capitalist Export Drive," Joseph Kahn, *New York Times*, 4/7/03.

4

from the Vietnam News Agency

Export Processing Zones Proposed to Become Effective Export Market

May 19, 2001

Economists have proposed measures to make export processing zones, EPZs, the most effective market for Viet Nam-made exports, while enterprises outside EPZs should work as material suppliers.

Measures focus on developing closer relations between enterprises inside and outside the export processing zones in an effort to alleviate high import costs for materials that could be supplied with in the country.

A proper legal system to boost their coordination is necessary, economists said. The inflow of goods from outside into the EPZs, for example, should follow the same rules as those circulated outside the EPZs, they argued.

Rich human resources, including high skills in embroidery, electronics assembly, and precise engineering, in areas nearby should be tapped to work as sub-contractors for EPZs so as to gradually establish a system of satellite workshops.

Loose coordination of action between enterprises inside and outside the EPZs in Ho Chi Minh City has caused heavy losses.

Two good examples are the Tan Thuan and Linh Trung EPZs, which are listed as the most successful in Viet Nam. They have drawn in more than 160 foreign investors, who have poured in more than 763 million dollars and provided almost 63,000 people with jobs.

These two giants have gained more than 263 million dollars worth of exports in the first four months of this year, an increase of 24.6 percent over the corresponding period last year. Thus far, their export revenues have added up to the municipal total amount of 2.1 billion dollars.

However, the bilateral trade turnover between enterprises inside and outside the EPZs remains modest. Domestic sales, for example, made up just 1.3 percent of the EPZs' export revenue of 741 million dollars in 2000. In return, they bought just 36.34 million dollars worth of materials and parts from domestic producers, or 5.6 percent of their total imports of 609 million dollars.

The EPZs massively import from other countries raw materials that are produced by Vietnamese companies in need of markets. This situation must be addressed.

The production relationship between the two sectors has been deteriorating. The total value of sub-contracts transferred from the EPZs to outside producers fell to just 4 million dollars in 2000 from 6.6 million dollars in 1999. The sub-contracts the EPZs received from the outside businesses were worth just 158,384 dollars last year, the lowest in five years, from 2 million dollars in 1999.

Economists reportedly await the effect of the Prime Minister's recent decision to regulate sales from the EPZs into domestic markets so that they may be competitive with those from other countries, removing some obstacles.

Source: "Export Processing Zones Proposed to Become Effective Export Market," posted May 19, 2001 at www.saigonnet.vn/english/bs/investment/may01/19.htm.

from the International Labour Organization

5

Thailand Country Report

Based on conference presentations by Jaded Chouwilai

Friend of Women Foundation

Women comprise 90% of Thailand's export-oriented workforce. Electronics is the most common industry in the zones. Thailand is considered to be one of the best, if not the best, investment areas in the region because of its pro-business policies, generous incentive system (including investor holidays), and a relatively skilled and inexpensive labour force.

Women workers are campaigning against the displacement of workers through technology. They feel that they have no job security. When high technology has been introduced into the factories, women workers have been displaced.

Women are also saddled with an unequal burden of household work. Most Thai men have not adjusted to the new roles that women have taken on in the workplace.

Women workers are campaigning for gender-related demands such as maternity leave, health and safety, and increased wages. Thailand's minimum daily wage, which ranges between 110 Baht (US$4.40 at a rate of 24.706 Baht/$US1) and 135 baht (US$5.40), only covers a single person's living costs.

Workers have demanded wage increases every year. Their demands are always met with threats from companies that they will move their operations to China, where the daily wage is US$1.22, or Vietnam, where it is US$0.81.

This year, labour unions asked Thailand's national wage board for a 15% increase in the minimum wage. The Board decided to increase the daily minimum wage by half the amount requested by the unions. Effective 1 July, the official minimum daily wage will range from 110 baht (US$4.48) to 145 baht (US$5.86), depending on the region.

Meanwhile, Thailand's most powerful industrial group, the Federation of Thai Industries, is seeking a 35% to 40% cut in the current daily minimum wage in designated areas along Thailand's border with Burma and Laos. The border zone proposal is intended to target migrant workers.

In its rush towards NIC status, the government has created many industrial zones in rural areas. But economic growth is meaningless for the zone workers as long as they are forced to pay the high costs that come with it: work-related sickness and injuries, pollution of their water and air, and destruction of their traditional economies and way of life.

Source: Jaded Chouwilai, "Thailand Country Report" posted at www.itcilo.it/english/actrav/telearn/global/ilo/frame/epzthai.htm.

6

from *Multinational Monitor*

Nike Does It to Vietnam

By Jeff Ballinger
March 1997

Nike, long in the vanguard of U.S. companies producing in Asia, is now leading Corporate America's charge into Vietnam. Twenty-five thousand young Vietnamese workers currently churn out a million pairs of Nikes every month.

The lure of Vietnam is obvious. The country's minimum wage is $42 a month. At that rate, labor for a pair of basketball shoes which retail for $149.50 costs Nike $1.50, 1 percent of the retail price. Vietnamese newspapers report that Nike contractors even cheat many workers out of the paltry minimum wage.

Nike workers in Vietnam are also subject to other labor rights violations. The Vietnamese press contain frequent allegations of verbal, physical and sexual abuse of workers, charges echoed by Thuyen Nguyen of the New York City-based Vietnam Labor Watch. Nguyen also says that Nike contractors require overtime work far in excess of permissible limits.

Vietnam is not Indonesia or China, however, and Nike has found itself more vulnerable to criticism in Vietnam. Unlike China and Indonesia—which together account for about 70 percent of Nike's shoe production—Vietnam has a vocal immigrant community in North America which stays in touch with developments back home. Vietnamese leaders appear concerned about safeguarding some vestige of the nation's socialist principles in the face of a flood of foreign investment. That lingering socialist legacy distinguishes the country from both China, where exploitation of workers in foreign shoe companies is rife and the regime does not permit critical news reports, and Indonesia, which years ago banned newspapers from reporting on the harsh conditions and minimum wage violations at sports shoe-producing factories.

Tour of Duty

With the spotlight suddenly turned on its Vietnamese operations, and increasingly shining on its Indonesian factories, Nike has been caught flat footed. Last August, the company had to hurriedly dispatch two outside directors for a tour of production facilities in

Indonesia and Vietnam just weeks before the company's annual shareholder meeting. The tour was prompted by a resolution filed by the General Board of Pensions of the United Methodist Church calling on the company to address the abusive practices of its contractors.

Upon her return, Jill Ker Conway, a Nike director, best-selling author and visiting professor at the Massachusetts Institute of Technology, told shareholders at the annual meeting that the young Indonesian and Vietnamese workers were treated fairly. She said Nike contractors provide adequate compensation, safe working conditions and health care that is superior to what the workers received in the villages that most of them have recently left.

Conway countered reports that Nike contractors pay workers less than the monthly minimum wage of $65 a month in Indonesia and a third less in Vietnam. "I was informed of minimum wage regulations by Nike" and contractors' adherence to them, Conway told the shareholders. When told about wage violations uncovered by CBS just days before, Conway expressed confidence in Nike's expatriate staff. Since the shareholders' meeting, further reports have alleged ongoing wage violations by Nike contractors. In January 1997, the Independent Sports Shoes Monitoring Network, a collection of respected Indonesian non-governmental organizations, charged that cheating continues despite the presence of Nike-hired "monitors" from the accounting firm Ernst and Young.

In an interview with *Multinational Monitor*, Conway expressed utmost confidence in her powers of observation—even though Nike organized her tour and provided translators in both countries and she did not contact any of the non-governmental organizations in Indonesia which have been critical of labor practices in the Nike factories for years.

The workers she chatted with in factories, lunch rooms and dormitories "did not appear to be intimidated," Conway says. Workers do want a reduction in overtime, she acknowledges, though she did not mention this criticism to shareholders. Conway stands by her statement to shareholders that the length of annual leave for the Indonesian workers making Nike shoes is more than 30 days, though dozens of workers interviewed in November, two months after her visit, said the actual amount is 10 days.

The second outside director to tour Nike's Asian facilities, Georgetown University basketball coach John Thompson, told shareholders that he was "far more satisfied with my own conscience" after the visit. He expressed relief that he observed no one being "overly abused." He also said, "I did not see the kinds of things that I had heard about."

What Conway and Thompson did not uncover was widely reported in the Vietnamese press, starting in April 1996. Vietnamese news accounts reported that a Nike contractor's Korean supervisor was found guilty of beating 15 Vietnamese about the head with a shoe "upper," and that another Korean supervisor was charged with sexual molestation.

Nike Chief Executive Officer Phil Knight did mention these incidents at the shareholder meeting. Knight told the shareholders that "one Vietnamese worker was hit on the arm." About the sexual molestation case, Knight said, "There was perhaps some misappropriate behavior. And then he touched a part he should not have."

But the Vietnamese news accounts reveal that the incidents Knight belittled were serious. In the sexual molestation case, according to the news reports, factory managers tried to buy the young women's silence, and the Korean manager fled to Seoul after charges were filed against him.

Lacking independent unions and meaningful corporate codes of conduct to discipline management, some workers for Nike contractors are turning to the courts, with mixed results. A Vietnamese court recently found the Korean supervisor guilty of beating workers, and extradition may be sought for the sexual molester who fled. In Indonesia, 24 discharged Nike workers are challenging the legality of their dismissal before the

Dress Code

Nike's code does not address the issue of corporal punishment or verbal abuse by supervisory staff, frequent occurrences in Nike's contractor factories, which are often run by brutal managers from Korea or Taiwan. Donald Katz's book, *Just Do It* describes the Korean supervisors' approach in Indonesia as "management by terror and browbeating." Nike recommends the book to young people doing research on sports footwear.

The Athletic Footwear Association's (AFA's) Guidelines for Contractors, to which Nike has been a signatory since 1993, does contain a corporal punishment clause. But the guidelines do not include monitoring or enforcement provisions, says Greg Hartley of the AFA. "First, we have no budget for monitoring," Hartley says. "Second, these are voluntary guidelines." Asked about intimidation and corporal punishment in Nike plants, Hartley acknowledges that "factories have been run like this for years." Fixing the problem could take years, he says. "That's the culture they are working with," he says of the major athletic shoe companies such as Nike that are hooked up with Korean and Taiwanese contractors. The cultural background of these managers "has to be respected as well" by Western-based multinationals, Hartley concludes.

Nike officials fiercely resist calls for independent monitoring by human rights and religious groups. They insist that "social audits" performed by the accounting firm Ernst and Young are sufficient.—J.B.

country's Supreme Court. The Indonesian workers face a long wait, according to their lawyer, Apong Herlina. The Supreme Court's docket is crowded with hundreds of cases, and "only 26 decisions were issued last year," Herlina says. The case "could be settled if Nike took it up with their contractor," says Herlina, but this seems unlikely. Last summer, Nike refused even to see one of the dismissed workers when she visited the company's headquarters in Beaverton, Oregon.

Stopping Nike from Trampling Labor Rights

That the Vietnamese press has published reports detailing abuses in Nike contractor plants is significant because it demonstrates a government disapproval of these practices. The government tightly controls the Vietnamese media, and the appearance of dozens of stories about these incidents in recent months suggests the government is not as willing to ignore abusive behavior as its counterparts in Indonesia and China. Whether disapproval of abusive labor practices will override the government's feverish desire for foreign investment remains to be seen, but it certainly enhances the leverage of labor rights activists working to eradicate the most brutal practices connected to one of the world's greatest corporate beneficiaries of economic globalization.

Source: Jeff Ballinger, "Nike does It to Vietnam," *Multinational Monitor*, March 1997.

7

from the Worker Rights Consortium

Key Principles and Investigation and Enforcement Mechanisms

Key Principles Underlying the WRC's Work

1. It is the responsibility of the licensee to ensure that workers are not exploited.

Sweatshops are proliferating and the living standards of workers declining because the abuse of cheap labor is a competitive strategy of many companies in the industry. Reversing this trend requires that licensees themselves make dramatic changes in how they produce (or "source") their products.

2. It is the University's role to define expected standards for treatment of workers and to hold licensees accountable. The test of the system is what happens to real workers in real factories.

Universities are responsible for enforcing contract provisions that require decent working conditions in factories producing their licensed products. Companies must develop their own internal process to comply with these provisions. The university's role should be limited to requiring outcomes: the decent working conditions mandated by their Codes of Conduct. As the verifier of compliance, the University should not have a stake in company-controlled monitoring, through its participation or endorsement.

3. Information is critical to uncovering problems, evaluating violations of the Code, and verifying that improvements have been made.

Sweatshops flourish when they are hidden. Greater transparency is a core requirement of an effective verification system. Companies that are required to reveal their contractors will have greater incentive to assure that these contractors abide by the university principles. Moreover, information gathered should be made available to workers and worker-allied groups in the producing regions, serving as an additional check on company-reported data.

4. Enforcement should be based on citing companies for violation—and using the licensing agreement to hold the licensee accountable for such violations. The WRC shall not play the role of certifying "good" companies.

Given that there are tens of thousands of garment factories nestled in dozens and dozens of countries, it is impossible for the University to know with certainty that all factories are complying with conditions set by the Code. Experience has shown that factories are often "cleaned up" for short periods of time, but then return to significantly violating the Code. One-time investigations often just cover up poor working conditions. Hence, certifying "compliance" of an entire corporation or factory is ultimately impossible and only extends the probability that the name of the University will be lent to companies that are still profiting off of abusive working conditions.

5. Limited university funds can improve working conditions if channeled to bring the experiences of workers to the public. The WRC shall engage with licensees to achieve the most effective remediation of abusive working conditions. The resources of the WRC shall be managed independently of the licensees and the economic activity surrounding collegiate licensing, and by people whose clear mission is to be responsive to the concerns of workers.

There is a need for "responsive" investigations of alleged workplace abuses, so that workers and their allies have a way to substantiate claims of abuse. University decision-makers similarly need a way to verify workers' complaints so that they can take action to enforce the licensing agreement. Furthermore, there is a need for "pro-active" investigations, to shine the spotlight on workplaces in countries that suppress worker organization, and to do further fact-finding about the workplaces of licensees that have demonstrated patterns of violations. The combination of maximum public information, and limited but completely independent investigations provides a powerful incentive to licensees to make systemic changes in how they source their products.

6. When abusive conditions at a particular worksite are exposed to public view, the licensee has an obligation to use its leverage to correct conditions—and not to "cut and run" from that site.

Failure to abide by this principle should be seen by Universities as a serious violation. Otherwise, there would be a perverse incentive for workers not to report conditions.

7. Misreporting of factory locations and conditions is a serious offense that undermines the possibility of enforcement.

8. Verification procedures must be continually reevaluated.

Due to the highly varied and constantly developing situations in apparel-producing regions across the world, verification methods must be flexible and responsive. The Worker Rights Consortium will work to develop and revise its programs in consultation with workers, human rights groups, and other interests, including apparel-producing companies.

WRC Investigation Mechanisms

1. To verify worker complaints.

One role for the investigative agency is to verify complaints when they surface. In the debate on campuses, this has been referred to as the "fire alarm" method for triggering inspections: workers are empowered to pull the alarm, signaling to the agency and the public that a violation has occurred. The agency is then responsible for verifying the complaint. Both verification and pro-active inspections must include both on-site inspections and off-site worker interviews.

2. To proactively investigate conditions.
   Additional resources will go into unannounced spot investigations at the places most at risk: countries and regions that suppress workers' rights, and companies with a pattern of violation. Again, the power of these investigations is that the licensees are expected to comprehensively change their labor and sourcing practices to ensure compliance—any violation is evidence of negligence or worse on the part of the licensee.

3. To act as a watchdog once abusive conditions have been exposed.
   Once violations at a site have been confirmed, the pressure on the licensee will be to improve conditions, rather than to shut down factories where violations have been found. Otherwise, there would be an incentive for workers not to report abusive conditions. Licensees will not be allowed to distance themselves from violations by claiming that fault lies with a subcontractor—instead, licensees are fully responsible for all factories producing their products. The Consortium will also play the critical role of keeping the spotlight on the licensee's response to exposure of bad practices at specific sites. The Consortium will require reporting of remediation steps, will conduct further inspections, and will evaluate whether or not a company has adequately remediated identified violations.

4. To catalyze research relevant to improving working conditions.
   We recognize that there are many outstanding questions that require further research. It is the unique role of Universities to be centers of inquiry, and therefore one additional role of the Consortium is to promote research activities into questions such as setting living wages, methods for investigating workplaces, etc.

5. To work in partnership with indigenous worker-allied groups when carrying out investigations and research initiatives, and to help build their capacity to further participate in and direct the process.
   The Worker Rights Consortium does not aim to set up a permanent system of factory policing, contributing to the privatization of national labor law enforcement. Rather, the Consortium aims to build capacity and open up the space for workers and their allies to advocate on their own behalf. The Consortium will engage in efforts to educate workers about their own rights and consistently partner with local worker-allied groups for investigations. The scope of such educational efforts and the mechanisms by which workers will file complaints are still issues being debated by organizations in producing regions. These efforts will need to be responsive to the political dynamics specific to each area. They will be developed, reexamined, and revised based upon further consultation with worker-allied groups and practical experience.

WRC Enforcement Mechanism

Currently, the Worker Rights Consortium does not mandate specific penalties to be applied to licensees that are in violation of Codes of Conduct. Instead, it is left to the individual college or university to determine a plan of recourse. However, the Consortium will establish a system of guidelines for such penalties with which to advise participating universities that wish to impose sanctions in response to violations. Failure to disclose required information, along with Code violations, will instigate the recommendation of sanctions.

It must be noted that the ultimate penalty is termination of the licensing agreement when a pattern of violations is shown. It is likely there will be clear cases of such patterns, and termination of contract with specific licensees will motivate the rest of the industry to improve. Termination is not a first step in response to violations, but must be one that universities are willing to take.

Source: "Key Principles Underlying the WRC's Work," "WRC Investigation Mechanisms," and "WRC Enforcement Mechanism." Copyright © Worker Rights Consortium. Used with permission.

from the *South China Morning Post*

8

Scores Hurt in Three-Day Riot at Hong Kong-Owned Factory

June 29, 2002

Thousands of workers at a Hong Kong-owned textile factory in Guangdong fought running battles with security guards in a three-day riot that left scores injured, media and officials said.

Paramilitary police had to be called to the Huiyang Nanxuan Woolen Textile company in Shuikou on Wednesday to put down one of the most violent cases of labour unrest in recent years, local officials said.

There were reports of deaths but these were denied by police and government officials.

Local officials said they did not know what sparked the violence at the factory. But the Yangcheng Evening News said it began on Monday after security guards armed with sharpened iron piping beat up one worker.

The plant, with a workforce of 15,000, then descended into chaos as guards and workers fought running battles around its huge compound, the Guangzhou-based paper said.

It quoted witnesses as saying several workers had been killed and others forced to jump off dormitory buildings.

A Shuikou police officer said at least 20 workers had been detained. "Two of our officers were badly beaten when they confronted security guards at the factory wielding iron sticks," he said.

Security guards and workers also hurled rocks and smashed some police vehicles, he said.

Pictures carried in the Yangcheng Evening News showed trails of blood on the ground in the compound and armed riot police guarding the entrance, as well as a burned-out truck and bandaged workers.

Kayce Law, chief executive officer of the factory's owner, the Hong Kong-based Nanxuan Industrial Co Ltd, said that on Wednesday factory and office equipment and windows were smashed and a truck belonging to a supplier was torched. Reports that workers had been attacked with pipes were denied.

A plant manager, Wang Tingqiong, said 63 people were injured in the incidents, including about a dozen guards.

The newspaper said police tried to interfere but factory security guards locked the gate of the compound and beat them back until riot police arrived from Huiyang on Wednesday.

A Shuikou government official said production at the factory had resumed on Thursday.

Source: "Scores Hurt in 3-day Riot at Hong Kong-Owned Factory," 6/29/02 posted at http://iso.hrichina.org/iso/news_item.adp?news_id+848.

Review Questions

1. How are export processing zones a product of globalization?
2. What is a sweatshop?
3. Why do so many governments promote EPZs as a route to development?
4. Why do promoters of the WRC feel that U.S. college students can improve the lives of factory workers in China?
5. How does Nike embody the linked processes of economies of scale, vertical disintegration, and transnationalization? How does General Motors embody these same linked processes?

Discussion/Essay Questions

1. Xiaoling and Jingfang are having a whispered conversation as they inspect shoes at their factory (they're speaking in a local dialect from the rural region where they both were born, so the boss can't understand them). The boss has told his workers not to speak with any outside inspectors who might visit the factory. He warns the workers that if they complain to these inspectors, not only will he beat them, but if the inspectors bring back a bad report to Nike, Nike will take away the company's contract and all of the workers will be unemployed.

   Xiaoling nodded dutifully while the boss said all this, but now she's telling Jingfang, "This is our chance. We're in a globalized economy and we need to get the power of Western consumers and big Western companies behind us." Jingfang disagrees: "The boss doesn't care what people in the United States think. If we're going to change things here, we should go out on strike like they did over at the Nanxuan Wool Textile Factory." A third worker, Jialing, overhears the conversation and pitches in: "Quiet, both of you. Conditions here are bad, but having to go back to the countryside because you've been banned from holding a job in the EPZ would be even worse, and you know that's what will happen to you if you start talking to outsiders." Whose argument makes the most sense? Why?
2. One of the benefits of EPZs is lower prices for consumers in developed nations. Would you be willing to pay more for clothing to ensure that the people who made your clothes had more pay, or worked fewer hours? Where would you draw the line?
3. The reading from the Vietnam News Agency stresses the need for more links between an EPZ and the rest of the country's economy. How do you think that Richard Bolin (the author of the editorials from the World Economic Processing Zone Association) would respond? What do you think could be done to improve the relationship between EPZs and their host countries?

List of Readings

1. Richard L. Bolin, "The WEPZA Family" and "Why EPZs Work," posted on the website of the World Economic Processing Zones Association, *http://www.wepza.org/editorials.html.*
2. "Stop Sweatshops! What Is a Sweatshop?" posted on the website of the Union of Needletrades, Industrial, and Textile Employees (UNITE), *http://www.uniteunion.org/sweatshops/whatis/infosheet. html.*
3. Joseph Kahn, "China's Workers Risk Limbs in Capitalist Export Drive," April 7, 2003, posted on the website of *The New York Times, http://www.nytimes.com/2003/04/07/international/asia/07CHIN.html.*
4. "Export Processing Zones Proposed to Become Effective Export Market," May 19, 2001, posted on the website of the Vietnam News Agency, *http://www.saigonnet.vn/english/bs/investment/may01/19.htm.*
5. Jaded Chouwilai, "Thailand Country Report," posted on the website of the International Training Centre of the International Labour Organization, *http://www.itcilo.it/english/actrav/telearn/global/il/frame/epzthai.htm.*
6. Jeff Ballinger, "Nike Does It to Vietnam," in *Multinational Monitor*, March 1997, posted on the website of *Multinational Monitor, http://multinationalmonitor.org/hyper/mm0397.07.html.*
7. "Key Principles Underlying the WRC's Work," "WRC Investigation Mechanisms," and "WRC Enforcement Mechanism," posted on the website of the Worker Rights Consortium, *http://www.workersrights.org/key.asp.*
8. "Scores Hurt in 3-Day Riot at Hong Kong-Owned Factory," in *South China Morning Post*, June 29, 2002, posted on the website of Human Rights in China, *http://iso.hrichina.org:8151/iso/news_item.adp?news_id=848.*

Websites for Additional Research

1. For a classic introduction to commodities and the ways in which they embody the labor that enters a product at each point along the commodity chain, see Karl Marx's "The Fetishism of Commodities and the Secret Thereof" (Chapter 3 of *Capital*, Vol. 1), reprinted on the World Wide Web at *http://www.wsu.edu/~dee/MODERN/FETISH.HTM*.
2. The website of the International Labour Organization's Bureau for Multinational Enterprise Activities has issued a report with detailed information about EPZs at *http://www.transnationale.org/pays/epz.htm*. The report contains a list of other names for EPZs, such as *special economic zones* in China and *industrial estates* in Thailand. It also includes statistics, such as the number of people working in EPZs in each country and prominent industries involved. The report notes both positive and negative aspects of EPZs, although on balance it is pro-EPZ. Other sources supporting EPZs (or the behavior of companies in EPZs) include the website of the World Export Processing Zones Association (WEPZA) at *http://www.wepza.org* and the "manufacturing practices" section of the Nike Inc. website (*http://www.nike.com/nikebiz/nikebiz.jhtml?page=25*, or just go to *http://www.nikebiz.com,* click on the "responsibility" tab, and then click on "workers & factories".
3. For news on various antisweatshop campaigns, see the websites of the Worker Rights Consortium (*http://www.workersrights.org*), United Students Against Sweatshops (*http://www.usasnet.org*), the anti sweatshop campaign of the Union of Needletrades, Industrial, and Textile Employees (UNITE) (*http://www.uniteunion.org/sweatshops/index.htm*), UNITE's *Behind the Label* antisweatshop webzine (*www.behindthelabel.org*), or the AFL-CIO's antisweatshop campaign (*http://www.aflcio.org/corporateamerica/stop/ index.cfm*).

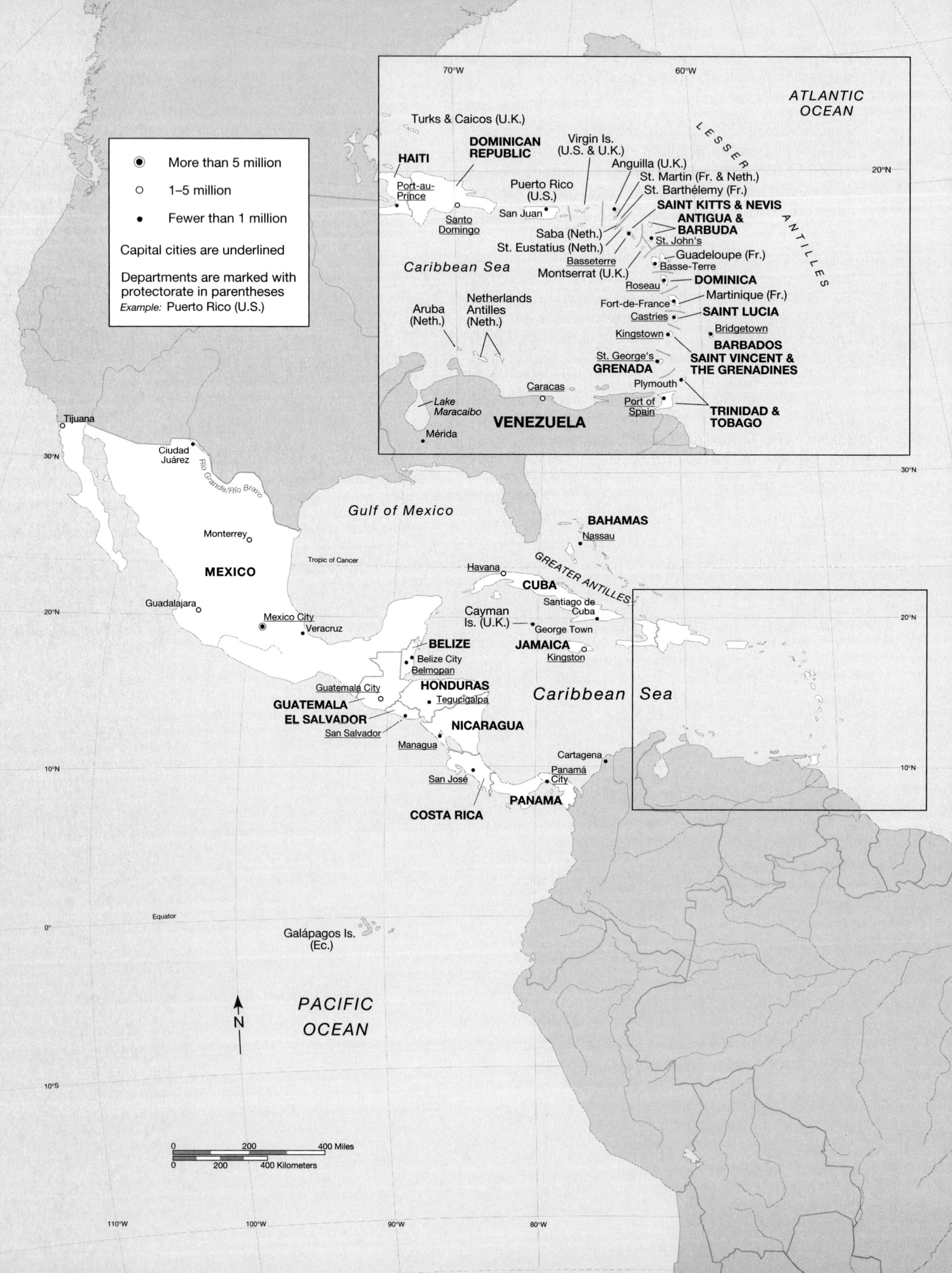

More than 5 million
1–5 million
Fewer than 1 million
Capital cities are underlined
Departments are marked with protectorate in parentheses
Example: Puerto Rico (U.S.)
70°W
60°W
ATLANTIC OCEAN
Turks & Caicos (U.K.)
LESSER ANTILLES
HAITI
DOMINICAN REPUBLIC
Virgin Is. (U.S. & U.K.)
Anguilla (U.K.)
St. Martin (Fr. & Neth.)
St. Barthélemy (Fr.)
20°N
Port-au-Prince
Santo Domingo
Puerto Rico (U.S.)
San Juan
SAINT KITTS & NEVIS
ANTIGUA & BARBUDA
St. John's
Saba (Neth.)
St. Eustatius (Neth.)
Basseterre
Montserrat (U.K.)
Guadeloupe (Fr.)
Basse-Terre
Caribbean Sea
Roseau
DOMINICA
Martinique (Fr.)
Fort-de-France
SAINT LUCIA
Castries
Bridgetown
Kingstown
BARBADOS
Aruba (Neth.)
Netherlands Antilles (Neth.)
St. George's
GRENADA
SAINT VINCENT & THE GRENADINES
Plymouth
Caracas
Lake Maracaibo
Port of Spain
TRINIDAD & TOBAGO
VENEZUELA
Mérida
Tijuana
Ciudad Juárez
Rio Grande/Río Bravo
30°N
Gulf of Mexico
BAHAMAS
Nassau
Monterrey
Tropic of Cancer
MEXICO
Havana
GREATER ANTILLES
CUBA
Guadalajara
Santiago de Cuba
Mexico City
Veracruz
Cayman Is. (U.K.)
George Town
BELIZE
JAMAICA
Belize City
Belmopan
Kingston
Guatemala City
HONDURAS
GUATEMALA
Tegucigalpa
Caribbean Sea
EL SALVADOR
San Salvador
NICARAGUA
Managua
Cartagena
10°N
San José
Panamá City
PANAMA
COSTA RICA
Equator
0°
Galápagos Is. (Ec.)
PACIFIC OCEAN
N
10°S
0
200
400 Miles
400 Kilometers
110°W
100°W
90°W
80°W

Tourism in Central America and the Caribbean

Companion to Chapter 12 Services

Margarita, a young woman from the highlands of Guatemala, appears to be dressed in indigenous clothes, selling traditional local clothing to tourists. To the untrained eye, both the *blusa* (non hand-woven blouse) that she is wearing and *huipils* (hand-woven blouses) that she is selling look like those that her ancestors must have worn. The reality, however, is more complicated.

Vendors selling "indigenous" clothing are a common sight in Central America. *Source:* Jennifer Burtner

During the 1960s and 1970s, indigenous women weavers from San Juan La Laguna, in the highlands of Guatemala, began crossing Lake Atitlán to the tourist community of Panajachel, becoming marketers for the community's surplus production. By the late 1980s, weaving for international consumers was commonplace, with more than half of all indigenous women living in Guatemala's western highlands producing textiles for tourist or export markets.

Tourists and exporters influence the goods produced through what they buy and order. However, many stylistic creations are not the work solely of foreign designers, but the result of intercultural cycles of innovation and adaptation. These feedback systems have produced many hybrid forms, such as the *huipil Juaneras* (women from San Juan) created specifically for foreigners. Unlike the earlier *huipiles* (embroidered blouses) that identified the wearer's home community (as well as his or her age, gender, and economic status), this hybrid "*no pertenece a ningún pueblo*" (doesn't belong to any town/people). An integration of stylistic elements from different towns (hand-woven *lienzos* copied from Zunil; embroidery modeled after the *cofradía sobrehuipiles* from Sololá), these pieces try to appeal to foreigners' cultural beliefs, aesthetic preferences, and wallets. "All natural" 100 percent cotton fibers and dyes replace the synthetic colorfast yarn used by contemporary indigenous peoples. Muted earth tones and pastels replace the region's bright reds. Simpler patterns and looser weaves requiring less time and materials lower costs.

As international market tastes turned toward more "authentic" goods in the early 1990s and tourists made increasing offers to buy the *huipiles* displayed not on the rack but on the marketers' backs, Juaneras decided to begin selling their own *huipiles.* At first they sold the few older pieces they still owned, then those made for special occasions but rarely worn. By 1991, rows of red San Juan *huipiles* lined the racks of the roadside market stalls in Panajachel, and young Juanera like Margarita's daughters donned the cheaper surplus/seconds hybrids their mothers had woven for tourists and exporters, but now were unable to sell.

Source: Adapted from "No Pertenece a Ningún Pueblo" by Jennifer Burtner. Copyright © Jennifer Burtner. Used with permission. ■

In this module, we examine one of the world's most important service industries—tourism. The textbook introduces the chapter on services (as it does every chapter) by discussing how one can think about it geographically, focusing on questions such as why certain services are clustered in some areas and not others. The chapter ends with the notion that even though globalization is visible on the landscape, individual places still offer distinctive services that attract tourists.

Perhaps nowhere on earth is tourism as important as in Central America and the Caribbean. The region offers several characteristics that make it a good location for the tourism industry. The climate is warm most of the year, and the many small islands provide an abundant supply of coastline. Another important factor is the region's proximity to its customer base. According to the World Tourism Organization, nearly half of the money spent on international travel is accounted for by residents of 5 countries—the United States, Germany, the United Kingdom, Japan, and France. Of these, the United States provides most of the tourists to the Caribbean. The region is also close enough to Europe to get a sizable share of tourists from there as well.

Even though the Central America-Caribbean region as a whole has seen a decline in tourism arrivals and receipts since 2001, tourism is still being pursued as a major component of regional economic development strategies, and some countries have been able to improve their economies as a result of tourism. Because the region has many unique natural and cultural places, it is also a place where we can observe the negative aspects of tourism. In the module that follows, we begin by discussing the vulnerabilities of tourism as a service industry. We then turn our attention to the effect that tourism can have in degrading local culture and the environment, and we conclude by examining some of the steps that can and are being taken to make the tourism industry more beneficial and sustainable.

Tourism and the Local Landscape

Tourism is one of the world's largest industries, and it has many direct and indirect impacts on people and the environment in tourism destination regions. Although many of these impacts are positive (economic growth and more jobs), tourism is not without costs.

Defining tourism is not as easy as it sounds. Tourism is more than traditional activities such as visiting beaches, national parks, theme parks, and resorts. Business and conference travel is counted as tourism, increasing the scope of tourism greatly. For our purposes, we can think of tourism as any travel event in which the purpose is to get oneself (rather than a product) to a destination. This means that tourism always involves encounters between different people from different places. *Difference* is at the heart of tourism, as people encounter different environments, cultures, and so on. These encounters provide opportunities for economic development and cross-cultural understanding, but also for social and environmental destruction amid the clash of cultures and social orders.

The effect of tourism on locales is highly debated. Supporters of tourism note that it is a clean way to improve economic livelihoods, in comparison with manufacturing industries. It brings jobs to locales that may need economic development. Others, however, argue that tourism does little to help local economies because the profits are usually exported to the home offices of the companies that build and manage the tourism resorts. Critics also note that tourism is a dangerous development strategy because when local residents become dependent on tourism revenues, they make themselves vulnerable to external forces over which they have little control. For instance, bad weather (such as the hurricanes that are common in the Caribbean and Central America) can ruin a tourism season and greatly reduce expected revenues. A poor economy in the tourism-sending country or political unrest in a tourism-dependent country's capital city or in a neighboring country can also keep tourists away. Other unanticipated events such as terrorist attacks or illness outbreaks can lead potential tourists to reconsider their travel plans. Like clothing or food fashions, a particularly "hot" tourism destination can fall out of popularity abruptly. With places growing closer together through mass media and advancements in transportation, places that may have been considered quite exotic in the 1950s may now be considered commonplace destinations. This is the same phenomenon that is mentioned in Margarita's story at the beginning of the module. Just as people will travel elsewhere to find more "authentic" items and experiences, so will they choose new destinations that are perceived as more exotic or adventurous.

Furthermore, as is often the case in the service sector, the tourism-related jobs that are actually available to local residents tend to be low-paying and low-skilled. In less developed countries, many of the goods used at tourism resorts are imported, to match the level and type of consumption to which tourists from wealthy countries are accustomed. As a result, the desired effect of tourism stimulating local supplier industries is often minimal. And finally, because tourists often represent a lifestyle very different from that of the host country and place large demands on natural resources, tourism can threaten the host country's environmental and cultural integrity.

The Tourist Gaze

Often, travel marketers construct a romantic image of the place that they are trying to sell to the traveler, the consumer. This image typically plays upon what the traveler both expects and wants in a vacation destination. Think about commercials that you have seen on television or in magazines. We are shown images of Caribbean islands with white sand beaches, and homes with varying shades of pastel coats and whitewashed, sun-drenched porches. Each island is depicted as a different sort of paradise.

Life for the residents of those touristed places may not be much of a paradise, however. In many cases, much of the money generated by tourism leaves the local economy—especially when tourists spend most of their money at foreign-owned resorts and hotels. Many hotels located in Central America and the Caribbean are not locally controlled, but are instead part of large multinational conglomerates. For example, Nestlé, which is based in Switzerland, is perhaps best known for its chocolate, but it also operates the Stouffer's line of hotels, which includes several in the Caribbean and Mexico.

People who work for hotels or tour groups may make more money than they would if they were engaging in other activities such as agriculture, but local wages paid still often account for a small percentage of the tour operator's profits. In 1994, Richard Gehrmann, an Australian political scientist, wrote in the journal *Social Alternatives* about the cultural and economic impact of tourism in Indonesia, a frequent destination for tourists from Australia. A tour company there runs overnight excursions up a mountain so that tourists can see the sun rise over a volcano. Each tourist pays 18,000 rupiah for the experience, but the local tour guide receives only 6,600 rupiah for his work (worth about US$3.00 in 1994, or 74 cents in 2004). Because one cannot support a family on 6,600 rupiah a day, many tour guides follow a night spent guiding tourists over mountains with a day of hard labor in the fields.

Gehrmann also discusses the cultural changes that are taking place on the Indonesian island of Bali. He writes of the village of Kuta, once a fishing village but now a popular destination for Australian tourists where "tourist operators speak with Australian accents." In Kuta, one can participate in a pub crawl, enter a vegemite sandwich-eating contest, and view a Balinese dance and cockfight while drinking a Foster's. As this example demonstrates, although tourists seek difference and encounters with "local" landscapes, customs, and people, they seek these encounters on their own terms and, in the process, they often transform those "different" people and places into something rather more familiar.

Part of the tourist experience involves getting away from one's everyday life, but tourists still want some things to be familiar and to feel comfortable while on vacation. Much as when you walk into a shopping mall in a different city and the stores and layout seem very familiar, resorts on different islands are very much the same, and sometimes devoid of local character. Web pages and tourism brochures promoting all-inclusive resorts illustrate this concept. One resort in Jamaica boasts its Thai restaurant. Another, in the Dominican Republic, advertises a buffet with rotating themes including Italian cuisine. Clearly, tourists going to these resorts are seeking something more familiar than a "pure" Jamaican or Dominican experience (if there is such a thing as a "pure" local experience). Familiarity is also achieved through reproducing architectural styles, such as the classical or plantation look. This is the same phenomenon that is discussed in Chapter 4's section on the globalization of popular culture and the diffusion of uniform landscapes.

Often tourism uses part of the cultural landscape as a marketing tool. Think about the islands of Hawaii. Many of us are familiar with the Hawaiian Islands—if only through popular media images. One image that probably comes to mind as you think about Hawaii is a beautiful young woman dancing the hula. The hula is one facet of native Hawaiian culture that has been packaged and sold to tourists as part of the island getaway. Hotels often put on displays of the dance for visitors.

Some islanders feel that the hula has been corrupted by Westerners who have stripped this traditional dance of its essence. The following passage is taken from an opinion piece published in 2002 in Hawaii's daily newspaper, *The Honolulu Advertiser*, in a special section reflecting on Hawaii fifty years after achieving U.S. statehood. The opinion piece was written by Kahu Charles Kauluwehi Maxwell Sr., a Hawaiian priest and cultural specialist:

> Our culture is being used by everyone in the world, and words that are sacred to us such as kahuna and aloha are so badly misused by everyone. It is now the "in" thing to use a Hawaiian word because it is "exotic." What is even more comical but sad is that there are those who move here because they "feel" this is their homeland, while those whose homeland this really is move to the Mainland to provide better for their families.
>
> For shame! If anyone can feel Hawaiian, or be "Hawaiian at heart," where does this leave us as true Hawaiians, as Kanaka Maoli?
>
> Our ancient hula is learned and performed in Japan, with taiko drums, no less. People teach hula today and call themselves kumu hula regardless whether they read it in a book or watched a videotape. Those who contribute to behaviors like these have cheapened the sanctity of hula, a rich cultural practice.
>
> (Excerpt from opinion piece by Kahu Charles K. Maxwell, Sr., *Waikiki Times*, August 11, 2002. Copyright © 2002 Charles K. Maxwell, Sr. Used with permission.)

Father Maxwell likely would identify this text, from the Honolulu section of the website all-hawaiihotels.com, as illustrative of the "cheapening" of sacred native cultural practices for tourist consumption and—in this case—titillation:

> Then you begin observing people in this airport. The most striking aspect of the people is that their racial origins are diverse and are primarily from the Orient. Aside from the Portuguese, the main groups were Polynesians,

> Japanese, Chinese, and Korean. This mixture of races has produced women of legendary beauty, with coal black eyes and olive thighs, who hula dance through the longings of the male visitor. Of course, the men are handsome also, and, as the woman visitor may eventually learn, while observing the rippling muscles of the beach boy paddling the outrigger canoe, the men were the original hula dancers.

In the worst cases, a subculture of prostitution emerges when the free-spending, wealthy tourist, freed from the rules that restrain behavior at home, encounters the objectified, impoverished native.

Beyond these cultural costs, tourism can also lead to environmental degradation. Many islands (in the Caribbean and elsewhere) have limited freshwater supplies and, while local residents have adapted their lives to these water-scarce environments, tourists often overwhelm local capacity with huge demands for water and for disposal of sewage and solid waste. Tourists also often damage historical and environmentally sensitive sites in the process of visiting them. Also, tourism often requires a great deal of infrastructure to support it, which may require clearing forests, or building on beaches or wetlands.

In short, while tourism can bring economic benefits, these benefits frequently come at a social cost. Given that the economic benefits of tourism are themselves often limited, some have questioned whether tourism should ever be pursued as a development strategy.

Ethical Tourism and Ecotourism

Tourism is often promoted because of the large number of jobs that it generates throughout the service sector—jobs in hotels, restaurants, and retail shops and in the various means of transporting people. Tourism may also be promoted as a means of preserving local culture and the natural environment. The challenge for governments is to promote tourism in such a way that its benefits are maximized while its drawbacks are minimized. Two forms of "good" tourism have been promoted—ethical tourism and ecotourism—and many tourism projects attempt to combine the best from both of these tourism programs.

Ethical tourism focuses on the behavior of the tourism provider and the tourist. Tourism providers are encouraged to purchase locally made inputs, pay decent wages to their local employees, and operate with host communities in a nonexploitative manner. Tourists are encouraged to respect local customs and venture outside resorts so as to gain an appreciation for the "real" communities that they're visiting. Ethical tourism packages frequently include educational components and visits with "regular" people.

Ecotourism is nature tourism with an ethical spin. Nature tourism, in and of itself, is not necessarily ethical. For instance, a safari is a nature tourism adventure, but it results in killing animals as trophies, depriving local residents of these animals that may be a crucial food source. Ecotourism, by contrast, involves building an appreciation for a place's nature while leaving that nature undisturbed for continued use by the people who live among that nature. While this is a noble goal, it is not always easy to achieve in practice. It can be difficult to make a place's nature available to tourists (such as by putting it in a national park) while still making it available to local residents who depend on this nature for their sustenance.

In the Caribbean, several organizations are involved in promoting tourism that is beneficial to both local people and the community. The Caribbean Tourism Organization (CTO), the World Tourism Organization (WTO), and the World Trade and Travel Commission (WTTC) all have supported "sustainable" travel, a concept that joins the goals of ethical tourism and ecotourism. As the CTO notes, a well-managed tourism program can promote conservation of environmental, historical, and cultural resources by making them economically valuable to local residents. It also can foster intercultural respect and understanding while bringing economic benefits.

A number of organizations now certify hotels and tour providers that market environmentally and culturally responsible "products." Currently, twenty-four European countries participate in the Blue Flag Campaign, a certification program for sustainable tourism that focuses mostly on the physical environment and the quality of care given to protecting coastal resources. In 2004, 2,312 beaches and 602 marinas were awarded the Blue Flag, a dramatic increase over a decade, indicating heightened interest from tourist facilities in adopting more efficient and environmentally sound practices.

Sustainable tourism standards also are being promoted by Green Globe 21, an organization created to research and develop standards for the tourism industry around the world. Green Globe 21 was conceived following the 1992 U.N. Conference on Environment and Development (the "Rio Summit"), when the World Travel and Tourism Council decided it needed to devote more attention to environmentally sustainable tourism. In the Caribbean, Green Globe 21 has partnered with the International Hotels Environment Initiative (IHEI) and the Caribbean Alliance for Sustainable Tourism (CAST). CAST is a nonprofit organization that works with businesses and communities to "ensure social responsibility and environmental care" for both locals and tourists.

The United Nations Educational, Social, and Cultural Organization (UNESCO) also began the World Heritage Site program in 1972 as a means of furthering its mission to preserve and protect the natural and cultural landscape worldwide. Sites can be either cultural or natural (or a combination of both, although few sites are designated as both cultural and natural world heritage sites). Presently, more than 700 cultural and historical sites are listed, including close to 50 in the Caribbean and Central America. In Mexico, for instance, the pre-Hispanic city of Teotihuacan, "the place where gods were created," is a World Heritage Site. It was chosen because of its archaeological significance and great pyramid monuments. The Viñales Valley in Cuba's Pinar Del Rio province is another example. This site was chosen for its unique karst (limestone) landscape as well as the architecture of the local farms and villages, music, and traditional tobacco farming methods. The World Heritage committee also maintains a list of sites in danger of disappearance or rapid deterioration due to urban or tourism development projects, armed conflict, or environmentally hazardous events, and some sites receive preservation and improvement grants from the committee.

Tourism in the Caribbean and Central America

The travel and tourism industry is the fastest-growing sector of the global economy. The World Resources Institute estimates that it generated US$3.5 trillion and almost 200 million jobs globally in 1999. Coastal tourism accounts for a major portion of the gross domestic product in many small island nations. This important industry and the economic well-being of the people in the Caribbean and Central America depend on the region's diverse and delicate ecosystems: white sandy beaches, azure waters, and tropical flora and fauna.

Although the specific situation varies from country to country, in general the countries of the Caribbean are highly dependent on tourism. Some of the larger islands in the Caribbean have more diversified economies, but for small islands tourism is the dominant economic sector. Most countries in the Caribbean obtained their independence only recently (and several islands are still dependencies of European countries or the United States). Colonized by outsiders, these countries were remade to serve the colonizer. Native populations were decimated (and in some cases completely wiped out) by European diseases; slaves and indentured servants were brought in from Africa and, in some cases, South Asia to work the fields; and plantations and farms were established for growing export crops such as sugar, tobacco, coffee, and bananas. Although some countries have been able to diversify their economies with assembly of garments and electronic products (following the example of many East and Southeast Asian countries; see the module accompanying Chapter 11), and a few have been fortunate enough to have significant mineral or timber resources, most have been left to market their most saleable resources: sun, sand, and sea, all just hours by jet from the wealthy population centers of the United States, Canada, and increasingly Europe.

Countries in Central America are less dependent on tourism than those of the Caribbean. A remnant of their colonial and neo-colonial past—several countries in Central America are still heavily dependent on agricultural exports. According to the CIA World Factbook, 50 percent of the labor force in Guatemala is still involved in agriculture, even though tourism is increasing in importance. In Nicaragua, employment in the services is roughly equal to that in agriculture, although there too tourism is on the rise. Nicaragua is still not a favored destination for tourists from the United States, due to political unrest in the late twentieth century. The political situation in El Salvador also makes tourism there an unlikely prospect for future development. As a whole, tourism grew in Central America throughout the 1990s, and it is a primary source of income for Costa Rica and Belize. Many of the countries are encouraging tourism to rain forests, beaches, and cultural sites. Whether they choose to promote tourism or not, the countries of Central America tend to have more diverse natural resources than the islands of the Caribbean. As a result, their economies have more potential to diversify, and those countries that pursue tourism may be able to accommodate tourists while minimizing the import of goods that often are needed to accommodate tourists.

The readings for this module focus on various issues of tourism in Central America and the Caribbean, with an emphasis on the advantages and disadvantages of tourism. The first two readings deal with the responsibilities of the tourist and the operator of the tourist facility. The first reading, the Ten Commandments of Eco-Tourism, is a document published by the American Society of Travel Agents (ASTA), designed for distribution by travel agents to tourists who are preparing to take environmentally sustainable holidays. The second reading is the Charter of Cultural Tourism from the International Council on Monuments and Sites (ICOMOS), a nongovernmental organization dedicated to the protection of world sites and monuments. The charter begins with the premise that tourism is an economic and cultural fact. It then discusses some of the actions that should be taken to maximize the positive effects of cultural tourism.

The next two readings look at cultural tourism from two different perspectives. The third reading is an academic critique of cultural tourism. It asks the question, "Is cultural tourism on the right track?" The author, Mike Robinson, suggests that due to the inherently unequal power relationship between the tourist and the "native," it is impossible for tourism to promote true cultural understanding. Robinson also mentions one point that was brought up earlier in the module: that tourists get staged "authenticity" rather than rich meaningful experiences. The fourth reading, from *UNESCO Courier*, profiles Gilbert Trigano, co-founder of one of the world's most famous networks of resorts—Club Med. As one might expect, Trigano's view of tourism is considerably more positive than Robinson's, although Trigano also is suspicious of tourism that seeks to deliver a product of "authenticity."

The next three readings turn to ethical tourism and ecotourism. The first of these readings examines a success story offered by the RARE Center for Tropical Conservation. A U.S.-based nongovernmental organization, RARE's mission "is to protect wildlands of globally significant biological diversity by empowering local people to benefit from their preservation." RARE does this by working "with local communities, non-governmental organizations, and other stakeholders to develop and replicate locally managed conservation strategies." This reading presents the story of Project Eco-Quetzal, a "low impact tourism" project in Guatemala.

The next reading—reading number six—is a promotional leaflet for a vacation that combines the goals of cultural appreciation, environmental sustainability, and contribution to a local economy. The leaflet promotes tourism in the small town of Cruz Verde in the Dominican Republic, so that tourists can learn about (and simultaneously support) a community that is guided by explicitly environmental principles.

In the seventh reading, journalist Sue Wheat adopts a more sceptical attitude toward ecotourism and ethical tourism. She charges that even well-intentioned tourism enterprises that have been designed to increase understanding of a place's culture and nature can be horribly disruptive and exploitative. She concludes by suggesting that in some locations ecotourism may be just as detrimental to indigenous communities as the "traditional" tourism that it is replacing.

The final three readings turn our attention to the cruise ship industry, a particularly crucial sector of the tourism economy in the Caribbean. Reading number eight is a paper prepared by Graeme Robertson of the Lighthouse Foundation, an organization dedicated to promoting awareness of the significance of the world's oceans and supporting sustainable development. The paper discusses a number of reasons why the cruise ship industry may not be the best vehicle for development for small Caribbean island nations. The ninth and tenth readings are speeches mady by the tourism ministers of two Caribbean island nations—Jamaica and St. Lucia. Both ministers welcome tourism development on their shores but, while the Jamaican tourism minister has unreserved praise for the tourism industry, the St. Lucian minister subtly requests that the cruise industry make a greater effort to be "seen as giving back more to the island."

Readings

1

from the American Society of Travel Agents

Ten Commandments of Eco-Tourism

Thank you for booking your travel with us. Whether you're traveling on business, pleasure or a bit of both, all the citizens of the world, current and future, would be grateful if you would respect the ten commandments of world travel:

1. Respect the frailty of the earth. Realize that unless all are willing to help in its preservation, unique and beautiful destinations may not be here for future generations to enjoy.
2. Leave only footprints. Take only photographs. No graffiti! No litter! Do not take away souvenirs from historical sites and natural areas.
3. To make your travels more meaningful, educate yourself about the geography, customs, manners and cultures of the region you visit. Take time to listen to the people. Encourage local conservation efforts.
4. Respect the privacy and dignity of others. Inquire before photographing people.
5. Do not buy products made from endangered plants or animals, such as ivory, tortoise shell, animal skins, and feathers. Read Know Before You Go, the U.S. Customs list of products which cannot be imported.

6. Always follow designated trails. Do not disturb animals, plants or their natural habitats.
7. Learn about and support conservation-oriented programs and organizations working to preserve the environment.
8. Whenever possible, walk or use environmentally-sound methods of transportation. Encourage drivers of public vehicles to stop engines when parked.
9. Patronize those (hotels, airlines, resorts, cruise lines, tour operators and suppliers) who advance energy and environmental conservation; water and air quality; recycling; safe management of waste and toxic materials; noise abatement, community involvement; and which provide experienced, well-trained staff dedicated to strong principles of conservation.
10. Encourage organizations to subscribe to environmental guidelines. ASTA urges organizations to adopt their own environmental codes to cover special sties and ecosystems.

Travel is a natural right of all people and is a crucial ingredient of world peace and understanding. With that right comes responsibilities. ASTA encourages the growth of peaceful tourism and environmentally responsible travel.

Source: "Ten Commandments on Eco-Tourism," published by the American Society of Travel Agents.

from the International Council on Monuments and Sites

2

Charter of Cultural Tourism

Revised January 1996

ICOMOS aims to encourage the safeguard and to ensure the conservation and promotion of monuments and sites—that privileged part of the human heritage.

In this capacity, it feels directly concerned by the effects—both positive and negative—on said heritage due to the extremely strong development of tourist activities in the world.

ICOMOS is conscious that today—even less than theretofore—the isolated effort of any body, however powerful be it in its own sphere, cannot validly influence the course of events. This is why it has attempted to participate in joint reflection with the large world and regional organizations which in one capacity or another share in its preoccupations and which are likely to contribute to the implementation of a universal, coherent and efficacious effort.

The Representatives of these bodies, [who] met in Brussels, Belgium, on 8 and 9 November 1976 at the International Seminar on Contemporary Tourism and Humanism, have agreed the following:

Basic Position

1. Tourism is an irreversible social, human, economic and cultural fact. Its influence in the sphere of monuments and sites is particularly important and can but increase because of the known conditions of that activity's development.
2. Looked at in the perspective of the next twenty-five years, in the context of the phenomena of expansion which may have heavy consequences and which confront the human race, tourism appears to be one of the phenomena likely to exert a most significant influence on Man's environment in general and on monuments and sites in particular. In order to remain bearable, this influence must be carefully

studied and at all levels be the object of a concerted and effective policy. Without claiming to meet this need in all its aspects, the present approach, which is limited to cultural tourism constitutes, it is believed, a positive element in the global solution which is required.

3. Cultural tourism is that form of tourism whose object is, among other aims, the discovery of monuments and sites. It exerts on these last a very positive effect insofar as it contributes—to satisfy its own ends—to their maintenance and protection. This form of tourism justifies in fact the efforts which said maintenance and protection demand of the human community because of the socio-cultural and economic benefits which they bestow on all the populations concerned.
4. Whatever, however, may be its motivations and the ensuing benefits, cultural tourism cannot be considered separately from the negative, despoiling or destructive effects which the massive and uncontrolled use of monuments and sites entails. The respect of the latter, just like the elementary wish to maintain them in a state fit to allow them to play their role as elements of touristic attraction and of cultural education, implies the definition and implementation of acceptable standards.

In any case, with the future in mind, it is the respect of the world, cultural and natural heritage which must take precedence over any other considerations however justified these may be from a social, political or economic point of view.

Such respect cannot be ensured solely by policies regarding the siting of equipment and of guidance of the tourist movements based on the limitations of use and of density which may not be disregarded without impunity.

Additionally one must condemn any siting of tourist equipment or services in contradiction with the prime preoccupation due to the respect we owe to the existing cultural heritage.

Basis for Action

Resting on the foregoing,

- the bodies representing tourism, on the one hand, and the protection of the natural and monumental heritage, on the other, [are] deeply convinced that the protection and promotion of the natural and cultural heritage for the benefit of the many cannot be ensured unless it be in an orderly fashion, i.e., by integrating cultural assets into the social and economic objectives which are part of planning of the resources of the states, regions and local communities,
- acknowledge with the greatest interest the measures which each of them states he is prepared to take in his own sphere of influence as expressed in the appendices to the present Declaration,
- appeal to the will of the states to ensure the fast and energetic implementation of the International *Convention for the Protection of the World Cultural and Natural Heritage* adopted on 16 November 1972, and of the *Nairobi Recommendation*,
- trust that the World Tourist Organization, fulfilling its aims, and Unesco in the framework of the Convention mentioned above, shall exert all efforts in cooperation with the signatory bodies and all others who in future may rally to ensure the implementation of the policy which the signatory bodies have defined as the only one able to protect Mankind against the effects of tourism's anarchical growth which would result in the denial of its own objectives.

They express the wish that the states by the means of their administrative structures, of tourist operators' organizations, and users' associations, shall adopt all appropriate measures to facilitate the information and training of persons travelling for tourist purposes inside and out of their country of origin.

Conscious of the acute need which obtains now to change the attitude of the public at large towards the phenomena resulting from the massive development of touristic needs, they express the wish that from school age onwards children and adolescents be educated to understand and respect the monuments, the sites and the cultural heritage and that all written, spoken or visual information media should express to the public the elements of the problem thereby efficaciously contributing to effective universal understanding.

Unanimous in their concern for the protection of the cultural patrimony which is the very basis of international tourism, they undertake to help in the fight initiated on all fronts against the destruction of said heritage by all known sources of pollution; and they appeal to the architects and scientific experts of the whole world so that the most advanced resources of modern technology be used for the protection of monuments.

They recommend that the specialists who shall be called upon to conceive and implement the touristic use of the cultural and the natural heritage should receive training adapted to the multi-faceted nature of the problem, and should be associated from the outset in the programming and performance of the development and tourist equipment plans.

They solemnly declare that their action is to respect and protect the authenticity and diversity of the cultural values in developing regions and countries as in industrialized nations since the fate of mankind's cultural heritage is of the very same nature everywhere in the face of tourism's likely expansion.

Source: "Charter of Cultural Tourism" posted at www.icomos.org/tourism/tourism_charter.html.

from *UNESCO Courier*

3

Is Cultural Tourism on the Right Track?

by Mike Robinson, director of the Centre for Travel and Tourism at the University of Northumbria, UK
August 1998

Tourism has long been assumed to promote cultural understanding and peace, but in fact it often chips away at cultures and leads to conflict.

Among the Toraja people of Sulawesi, Indonesia, not all was going well with tourism. In fact, resentment became so great over the way in which sacred funeral ceremonies were being adapted to meet tourists' needs that in the late 1980s, a number of Toraja communities simply refused to accept tourists.

The Toraja example highlights the dilemma that faces contemporary cultural tourism. On the one hand, tourists increasingly seek exotic and often unique cultural spectacles and experiences, and are willing to pay a premium to do so. But on the other hand, the very presence of tourists can chip away at local culture and essentially re-invent it to fit the exigencies of the tourism industry.

The result is that host communities find culture and traditions under threat from the purchasing power of the tourism industry. Neither are tourists better off from the cultural viewpoint. Instead of getting rich and authentic cultural insights and experiences, tourists get staged authenticity; instead of getting exotic culture, they get kitsch.

With nearly one billion international trips expected in 2000, the impact of tourism on culture has become so palpable that the question arises as to whether or not we can continue along the current path without something having to give. More than ever, we must find a way to achieve sustainable cultural tourism.

Surprisingly, and in contrast to the attention given to the natural environment in the sustainable development debate, very little energy has been devoted to this end. A major reason for this lethargy appears to lie in our basic assumptions about tourism.

The predominant notion is that tourism generates cultural harmony. This idea derives from the romantic (and elitist) traditions of travel in the eighteenth and nineteenth centuries and is today enshrined in the World Tourism Organization's mission statement, which includes the goal of fostering international peace and understanding.

But claims that tourism is a vital force for peace are exaggerated. Indeed there is little evidence that tourism is drawing the world closer together. The truth is that a host of cultural conflicts have developed around tourism.

Little thought is given to the fact that tourism is one globalizing influence which can initiate dramatic and irreversible changes within the cultures of host communities. Unfortunately, while the idea that we should respect cultures and cultural rights may be present, the idea that we should sustain cultures is not fully developed. Nor is there any clear indication of which cultures we are speaking of.

Perhaps the most obvious conflict is between tourist and host. This is in part engendered by the fundamental difference in goals: while the tourist is engaged in leisure, the host is engaged in work. While the tourist arrives with loads of expectations, many of the local stakeholders often have no idea of what to expect.

Another source of conflict is between the often persuasive and economically powerful developers and operators of the international (though mainly first world) tourism industry and the host country. Tourism can turn local cultures into commodities, that is, consumer items much like any others. Religious rituals, ethnic rites and festivals continue to be reduced and sanitized to conform with tourist expectations, resulting in what one scholar has dubbed "reconstructed ethnicity."

An Unequal Relationship

Part of the conflict stems from the fact that packaging culture begins well away from the cultural site. Cultures are reduced to a two-dimensional world carried by glossy brochures presenting idyllic locations and generally reducing distinctive cultures to superficial and readily substitutable narratives.

The flow of tourism receipts is mainly to the developed world—where the majority of tourism businesses are located—creating a permanent backdrop for conflict.

Another level of conflict is found among different sectors of the host community. For example, locals working in the tourism industry might have different goals from those of agricultural workers in the same community. Access to tourism employment, which in developing countries also often means access to relatively high wages, may be skewed to certain social and ethnic groups.

The attraction of generating hard currency relatively quickly and often with minimal investment compared to, say, establishing a manufacturing industry, is a powerful argument for governments of both developed and developing nations seeking to develop tourism.

However, the various levels of tourism-related cultural conflict force us—or should force us—to question the very foundations of cultural tourism.

Active collaboration with local cultures must be at the centre of any efforts to promote sustainable cultural tourism. However, to date the extent of collaboration remains narrow and almost a token afterthought following environmental and economic considerations. One study has shown that in New Zealand Māori economic involvement in the tourism industry amounted to less than one percent.

Though definitions might differ, I would argue that at its heart, sustainable cultural tourism recognizes the value of cultural diversity and needs to provide local cultures with a forum in which they can participate in decisions that affect the future of their culture. In other words, host cultures should be empowered to say no or yes to tourism, and in the latter case, to set guidelines for tourism if they so wish.

There are examples where the redistribution and ownership of resources are being addressed in tourism in such a way that indigenous peoples are beginning to move from being the providers of cultural experiences for tourists, to having an ownership and management role in tourism.

Such examples are encouraging, though still few and far between, and even then, largely shaped by first world value systems. It is the allocation of cultural rights and subsequent respect for and protection of those rights which underpin sustainable development and should underpin the notion of sustainable tourism. Those cultural rights need to be accompanied by rights in other areas. Armed with land, resources and intellectual property rights, communities and cultures can not only influence the direction and pace of tourism developments, but also provide or withhold consent for them.

Except for some rare cases, I am not at all confident that at present we are on the right track at all. The tourism industry and the governments and organizations which empower it cannot, and arguably would not, engage in dramatic structural and intellectual reshuffling which would put the notion of cultural consent at the centre of a collaborative process.

Under the current self-regulatory approach, the industry can encourage local community participation in the management of tourism resources and can aim to include non-traditional decision-makers in the development process. The problem is that it does things in a way that is designed to serve both the economic goals and desires of the tourism industry and the dominant first world value systems which it represents.

One of the implications of a sustainable tourism constructed around the idea of cultural consent is that tourism may be rejected outright. More likely is that via more equitable collaborations, the nature, extent and type of tourism development will be adopted to suit the cultural needs of the host community. Either way the challenge is to establish mechanisms which will involve local cultures and transfer to them the right to decide on the type and extent of tourism which they wish within the economic, environmental and cultural limits which they have set.

Source: Mike Robinson, "Is Cultural Tourism on the Right Track" *Unesco Courier,* July/August 1998. Copyright © 1998 UNESCO. Used with permission.

from *UNESCO Courier* 4

An Empire Built on the Sands

By Amy Otchet, UNESCO Courier journalist
August 1998

Mass tourism is unavoidable and even desirable according to Gilbert Trigano, one of the Club Méditerranée's founders.

He built an empire on the sands of some of the world's most beautiful beaches by wheeling and dealing with kings and presidents. He made a fortune by selling a sun-splashed brand of hedonism with the trademark bikini-clad couple frolicking along a lick of white sand. Meet Gilbert Trigano, one of the two men responsible for starting the world's most famous network of resorts, the Club Méditerranée—better known as the Club Med.

At the age of 78, Trigano's club-days are over but he has found a new role as a consultant to government officials on the tourism industry, which may send shivers up the spines of those devoted to the new school of cultural tourism. Club Med's hallmark lazy get away in which visitors need not exchange money, let alone venture to a local restaurant, is often cast as the dark side of the industry. "It's become a sin just to use the expression 'mass tourism'," says Trigano, "but that's what we are faced with. Should we be nostalgic and try to return to the past when people didn't travel in groups or should we try to better control the flows?"

The Spirit of the Times

Ironically, Trigano cannot help but hark back to Club Med's early days in the 1950s, when the resorts consisted of tented camps on pristine beaches of the Mediterranean. "The real success of the club may not have been in the quality of the services but in the way it represented the spirit of the times. We [Trigano and his business partner Gérard Blitz] were young men, survivors of the war. We made a profession out of offering others what we wanted for ourselves: a chance to discover the sea, to breathe deeply and live healthily. There was an incredible desire to discover new things in life and new people."

Trigano maintains that the same trends are intensifying today. Yet there is a hollow ring to his talk about discovering other cultures. The clubs are and have always been removed from surrounding life. The flash of these pleasure domes often seems indecent in poor countries. "It is very easy to criticize along these lines," says a visibly aggravated Trigano, but the bottom line is that tourism brings much-needed money and jobs.

"Besides do you really think that anyone can understand the culture of a country like Mauritius after visiting for eight or 15 days," says Trigano. "People at the club basked in the sun of Spain. They tasted the tomatoes of Tunisia and the grilled mutton of the Berbers in Morocco. They didn't discover these countries but they did get a taste of them."

According to Trigano, it would be irresponsible for the club to dive any deeper into local culture. "In Indonesia, I used to pay actors to stage traditional marriages because I was furious that people wanted to penetrate Buddhist temples to see a ceremony. This would have been totally inappropriate."

Trigano has just returned from Palestine, enchanted by the cultural heritage. Yet he is also concerned, explaining that the impending recognition of the formal state of Palestine will unleash a wave of tourists. "Now is the time to act," says Trigano. "It would be a bit far-fetched for me to play the role of the moralist" and tell the Palestinians what to do, he says. "It is a lot more plausible for me to offer my experience as to how a 'good idea' can get out of hand."

"My role used to be to look for isolated sites of exceptional beauty," says Trigano, and then develop them by brokering deals with the government to bring electricity, drinking water and even small airports. "It is possible to discover and develop a site without destroying it," he insists. But what of places like Cancún in Mexico, the Moroccan beach resort of Agadir, or Tunisia's Isle of Jerba? In all three cases, Club Med was the first to lay the foundations for the tourism industry. Indeed, with a certain pride, Trigano explains that when he and his team arrived over 30 years ago, "there was nothing." Yet he is also the first to admit that "they are now concrete horrors." Was this inevitable? "Not entirely," he says. For Trigano, the error lies not in developing a site but in overdeveloping it—and to avoid this, "authorities must have the courage to issue laws" to regulate the industry's growth. "I'm not ashamed of anything I have built," says Trigano. As for the others. . . .

Source: Amy Otchet, "An Empire Built on the Sands" *Unesco Courier*, July/August 1998.

from the RARE Center for Tropical Conservation

After Two Years: Project Eco-Quetzal

Summary

Since 1990, Project Eco-Quetzal (PEQ) has been working to protect the Quetzal's cloud forest habitat in the mountains of Guacax, Caquipec, and Yalijux in Alta Verapaz, Guatemala. PEQ works with the Q'eqchi'es in their villages to promote sustainable use of the area's natural resources by identifying alternative sources of income, such as sustainable agriculture and production of handicrafts, as well as raising local awareness of the value of the cloud forest through environmental education and research and monitoring programs.

In 1997, PEQ began to investigate low impact tourism. Today, families in two communities welcome tourists into their homes where they are able to experience the daily life of the Mayan people. The following year, the families chose not to cut the forests on their property, allowing it to stand for the benefit of the foreign visitors. In 1999, the number of participating families increased and now PEQ is considering expanding the program to other communities. Another product of the tourist program is the creation of a tour guide program for the region. By promoting the entire region, the hope is to attract more tourists to the cloud forest.

Introduction

Over the course of the past ten years, PEQ has become very well informed of the ecological, social, and cultural situation of the Q'eqchi'es villages in the Alta Verapaz mountains, one of the poorest regions of Guatemala. Working with these communities, they look for solutions to the local environmental and socioeconomic problems. The poverty, lack of alternative sources of income, and population growth result in the conversion of forest to cornfields. PEQ works in about 72 km^2 with a wider zone of influence reaching 200 km^2 Now there are 80 landholders (indigenous and ladino) who are in the process of declaring these 72 km^2 a private reserve protected under Guatemalan law. This reserve will protect a diverse group of threatened species in danger of extinction, such as howler monkeys, salamanders, giant ferns, orchids and a variety of medicinal plants.

Low Impact Tourism: "Adventures in the Cloud Forest"

The 37 villages surrounding the protected cloud forest find themselves in one of the most remote regions in Guatemala. The area is extremely poor, has a great need for medical services and a high birth rate. At an altitude of 2,500 m, the area has no electricity, potable water, communication or access to motor vehicles. As with other indigenous groups in Guatemala, this population was displaced from lower, more fertile lands, to the very tops of mountains and cloud forest. They are traditionally a farming people, basing their culture on seasonal crops such as corn, beans and malanga. Because of the high population growth rates, these agricultural systems are no longer sustainable and pose a serious threat to the cloud forest. Considering the dire need for additional sources of income and the great demand for special and interesting tourist destinations, PEQ decided to implement a program called, "Adventures in the Cloud Forest." It is the first program in Verapaz and was developed with the local families.

Overview of Actions Taken

Phase I

During the pilot year (1997), 68 tourists visited the community of Chicacnab. After a participatory evaluation of the social, historical, economic, and ecologic situation of the community, the Tourism Committee decided that instead of building facilities exclusively for tourists, they would train families to receive tourists in their homes. The courses included themes such as cooking and hygiene and workshops for nature guides to bring tourists to the cloud forest. The Tourism Committee was involved in all of the decision-making for the program, including the prices and types of services.

The first phase also included publicity. Tour operators, travel agencies, Spanish schools, restaurants, and cafeterias throughout the country distributed pamphlets, calendars, and posters. Various guide books included the program Aventures in the Cloud Forest in the listings on Coban and Verapaz. Coban is 200 km from Guatemala City (3 or 4 hours by car). Hotels and restaurants also supported the program and passed on information to visitors.

Phase II

In 1998, 181 tourists visited the community. This represents a 266% increase over the pilot year. There was a promotional campaign for PEQ in tourist centers throughout the country. Guides continued to be trained by a biologist in birdwatching and description of the forest's flora. Mothers of families attended courses on cooking. Each participating family earned approximately US$156 from tourism in 1988, the equivalent of 52 days of work on a regional coffee plantation.

Phase III

Due to the success of the program, it was extended to a second village, San Lucas. The emphasis there was to improve the standard of living of participating families.

Currently, PEQ is working with families to improve accomodations for tourists. This involves constructing a private room with a comfortable bed, hot water and decoration typical of the region. The average tourist stays two nights in Chicacnab, but with an improvement in service, it is hoped the average stay will increase.

With the ultimate goal of forming a microenterprise to replace the Tourism Committee in Chicacnab, PEQ is working in conjunction with the families. Adults are learning to use a radio to communicate with PEQ offices in Coban, without the help of any project employees. This way they know when tourists are coming and can organize guides and families.

The number of visitors so far has been stable in 1999, but the visitor profile has been changing. Last year the average tourist was a backpacker of 25 years interested in Mayan culture and nature. This year the average tourist is 32 years old and is a professional who wants to see the Resplendant Quetzal. This change indicates that the program is reaching a wider audience year and has an image that also attracts more demanding tourists in terms of service. Quetzal watching is a yearround activity with best viewing during the dry season between February and June. During other months they are developing a weaving and handicraft school with the women.

The guides are actually the parents in the family where the visitors stay. They are continuously trained by a biologist in bird and other animal watching. This year the biologist plans to include the guides in a monitoring program.

Evaluation

The tourism program is valued participatively with the families. Early on there were not only positive opinions towards visitors in Chicacnab, but jealousy as well. Of the 80 families, only 20 could receive tourists because they had virgin forest on their property. This resulted in conflict among the families.

In March a meeting created a fund in PEQ for community projects that benefit all the community. Today each tourist contributes 20 Quetzales to this fund. In May 1999 about 1000 quetzales were collected which will be given to the community if they can present a concrete project such as building a school.

Problems between families and visitors over cultural and social aspects are played out in workshops using popular theatre. In small plays acted out by PEQ personnel and volunteers, conflicts are discussed such as those involving food, different ideas about hygiene, health, and comfort of the tourist. The works motivate the families to express their opinions and and help PEQ understand the social dynamic in the community.

The program in general is being constantly evaluated by an ecotourism student (Master's candidate in natural resource management). This person is noting the reduction in environmental impact and the benefits to the local people. The evaluation report will be delivered in September 1999. As a result of the participatory evaluations and the work of the student, it is becoming clear that competition between the communities is necessary to raise the quality of service. If only one community has PEQ's support, it would be hard to raise the quality. With the inclusion of San Lucas, Chicacnab has to better organize itself in order to develop future tour operations. Another product of the tourist program is the creation of a tour guide program for the region, which would compete in the regional and national market to achieve greater success.

Lessons Learned

PEQ's recommendation to other NGOs that aim to work with communities is to plan the entire program, each action, and each step with the community. Although the process will be much slower this way, in the end the program will be sustainable, continuing long after the NGO leaves. It is only if the program is born of the same community and is developed by its members that it has a hope of success in the long term.

Source: "After Two Years: Low Impact Tourism Program in Q'eqchi'es Communities in Alta Verapaz, Guatemala" posted at www.rarecenter.org/content/e_case_template.cfm?cs=37.

from the Cruz Verde Foundation

6

A Small Town Dominican Wonder

The Story

Cruz Verde is a small town wonder nestled in the Monte Plata region of the Dominican Republic. It is rich in culture, ecological awareness and great people. About five years ago a German Foundation called Friedrich Naumann started a sustainable community project here and it has changed the town's outlook and enriched their hopes.

Friedrich Naumann actually built upon a long tradition of farming methods, knowledge and philosophy. They taught them about the growing worldwide environmental concerns, fine tuned their organic agricultural techniques and assisted in acquiring loans for projects like the ecological center. They also showed them alternative energy sources and recycling. But most importantly, they helped them focus on their goals with a vision of economic sustainability and ecological harmony.

The Tour

This all-inclusive tour will allow you to visit Cruz Verde and be able to experience a bit of Dominican countryside life. You will visit the Bucaro Ecological Center, the Organic Sugar Cane Processing Plant and Farms, a local woman's humble home where you will be able to get a "taste" of real "campo" life. She tends to an organic garden, makes delicious organic wine from a variety of fruits and is part of a micro-enterprise that makes beautiful paper crafts made from recycled paper and natural products. You will also be able to eat a typical Dominican meal with all its countryside flavor! The Cruz Verde tour will allow you to observe its local ecosystem with a diverse mix of medicinal plants, tropical fruits, reptiles (including bats!) and birds. Of course, there are also lots of cows, horses, chickens, pigs and ducks. This tour is impossible to describe in full with all its benefits and joys, but rest assured that it is really educational, enriching and fun!

Our tours are put together in order to help the town's economic sustainability. All proceeds go to each micro-enterprise and the collective, the Cruz Verde Foundation, excluding taxi services and tour operator fees which is mandated for insurance purposes.

We would like to take this opportunity to invite you to help Cruz Verde and its collective and come share some time with us!

One Day Tours—Cost: $59 US

8:00 am—Pick up at Hotel (Santo Domingo Area Only—includes Boca Chica and Juan Dolio)

10:00 am—Welcoming at Cruz Verde

10:30 am–12:30 pm—Bucaro Ecological Center

12:30 pm–2:00 pm—Lunch

2:00 pm–3:00 pm—Organic Sugar Cane Processing Plant and Farms

3:00 pm–5:00 pm—Veba's House

5:00 pm—Departure to Hotel

Two Day Tours—Cost: $89 US

Day One

8:00 am–5:00 pm—See above.

5:00 pm–9:00 pm—Free Time

9:00 pm–?—Dancing in the Town's Hall with Lessons (if you like).

Day Two

8:00 am–9:00 am—Breakfast

10:00 am–12:00 pm—Ceramic Making Workshop

12:00 pm–2:00 pm—Lunch

2:00 pm–5:00 pm—Recreation of Choice (River Bathing, Horseback Riding etc.)

5:00 pm—Departure to Hotel

Discounts

10% Discount applies for groups of 5 or more (maximum 15 for less ecological impact). There are some people who enjoy Cruz Verde so much that longer stays have been arranged. Long Term Stays—Cost: $150 US per month, per person. Includes Lodging only. $215 US per month, per person. Includes Lodging and three meals per day.

Volunteering

If you are interested in volunteering or doing some sort of study here in Cruz Verde there is a 50% discount applied on the "Long Term Stays Rate." There is a minimum of 15 community hours per week and 5 family hours per week. For more information please email: *Mauricio_Fabian@yahoo.com.*

The Commitment

My name is Mauricio Fabian. I came here to Dominican Republic to help my family and am living in Santo Domingo. I am organizing this tour and am committed to this project. I will be happy to answer any questions and will be meeting you when you come. I also would like to ask for a commitment from you. You see, we do not have access to any credit card services or any other way of getting your deposit via e-commerce so I would like a commitment from you when you make your reservations. There are no telephone lines in Cruz Verde so that means I have to take the time to go and let them know you are coming. So please don't let them down. Thank you, and I hope to see you soon!

Source: "Cruz Verde: A Small Town Dominican Wonder" posted at http://interconnection.org/cruzverde/index.htm.

from *The Guardian* 7

Sold Out

A rapid growth in ecotourism has been at the expense of indigenous peoples.

Sue Wheat
Wednesday May 22, 2002

Earlier this year, 250 Filipinos were evicted from their homes. Their lake-shore village of Ambulong, in Batangas province, was attacked by hundreds of police, who demolished 24 houses. Many people were reported wounded, four seriously and one with a bullet wound. Cesar Arellano, of Pamalakaya, a Filipino human rights organisation, said: "The people are not leaving—they have set up camp. They are going to fight for their land."

The intention of the authorities was to clear people to make way for a major business venture—not oil, logging or mining, but ecotourism, which is growing massively around the world and is now backed by governments, world bodies and international banks. This year has been declared by the UN the international year of ecotourism and this week, a world summit is being held in Quebec to consider the problems and potential for the fastest growing sector of the world's largest industry.

According to many conservationists and tour operators, this "benign" version of tourism offers a way to fund environmental protection, stimulate the incomes of the poor and encourage cultural exchange. Ecotourists are thought to spend considerably more than mass tourists and for debt-strapped developing countries, having people visit, look at things that require minimal investment and pay lots of money for the privilege, can seem manna from heaven.

Nature is a money spinner. Ecuador earns over $100m a year from 60,000 visitors to the Galapagos, for instance, and Kenya as much income from its safari holidays. But the stakes are now getting higher and the dispossession of people from their land is increasingly associated with ecotourism.

The cases are widespread. In the Moulvibaza district of Bangladesh, over 1,000 families of the Khasi and Garoare indigenous groups face eviction from their ancestral lands for the development of a 1,500-acre eco-park. "We were born here and grew up here. We have been living here for hundreds of years ... we will not leave this forest," said Khasi headman Anil Yang Yung in a public demonstration during a hunger strike in Dhaka last February. "We cannot survive if we are evicted from the forest."

In Brazil, two fishing villages near the coastal resort town of Fortaleza are fighting for their land. In one, Tatajuba, which was recently voted one of the world's top 10 beach sites by the Washington Post, a village of 150 families has gone to the courts to try to show that a real estate agency illegally took possession of publicly-protected land where they live. A company wants to build a 5,000–hectare "ecological resort catering for 1,500 tourists" in their place.

In Prainha do Canto Verde, a village of 1,100 fishing and farming families, the community is also defending itself against speculators who, they say, bought beach land deceptively from fishing families and then registered the land for clearance. "It wasn't illegal, but the fishing families can't read and didn't know what was happening," says Rene Schaerer, a US public policy adviser working with the community.

Governments in developing countries, keen to modernise, often say that "primitive" subsistence activities are incompatible with conservation. These were the arguments given to evict the Masai in East Africa and the Bushmen of Botswana, but the reality is that many indigenous and other poor communities are living on prime areas of ecotourism real estate and speculators want the land without getting involved in land rights claims.

Much ecotourism development comes as part of "development packages" funded by international banks. The Asian Development Bank is funding a $1.2bn scheme in south-east Asia, which includes an ecotourism development that may affect many hill tribes. The Inter-American Development Bank has been the focus of protest by the Tatajuba and Prainha do Canto communities, and the World Bank is funding an eco-park in Karnataka state in India that involves a long-running land rights protest with indigenous communities.

"There are many cases in Indonesia where whole communities have been evicted—more accurately, driven out, beaten up and their possessions destroyed or looted—to clear land for tourist developments," says development consultant Sean Foley, who worked in Indonesia—including for the World Bank—during the 80s and 90s.

The likelihood of mass evictions in Asia are "absurd," says Warren Evans, director of environment and social safeguards division at the Asian Development Bank. "We are looking at ecotourism as a part of rural development, poverty alleviation and to strengthen conservation. Any activity the ADB is supporting has to follow our safeguard policies on environment, indigenous peoples and involuntary resettlement."

Few poor communities are set against ecotourism, but they almost all want to be able to control it. "We were about to start community ecotourism on our lands, as Bushmen in Namibia have done," said a Khwe bushman. "But then the intimidation, torture and evictions started again. The government did not want to lose tourism business to us."

An international indigenous people's forum on tourism held last month in Mexico found feelings running high, says Deborah McLaren, a native American with the Rethinking Tourism Project, an international network of indigenous peoples campaigning on tourism. "We want to bring these issues to light. Communities are being oppressed. Governments and industry have corrupted the whole idea of ecotourism and it is proving just as destructive as any other industry. But somehow, no one wants to hear that."

Source: "Sold Out: A Rapid Growth in Ecotourism Has Been at the Expense of Indigenous Peoples" by Sue Wheat, May 22, 2002.

from **the Lighthouse Foundation**

Cruise Ship Tourism Industry

by Graeme Robertson

Facts and Figures

There is a paucity of studies and little academic literature on cruise ship tourism despite it being the fastest growing sector in the tourism industry. With an 8% annual growth since 1980, it has increased at almost twice the rate of tourism overall. A record 8.5 million people took cruises worldwide in 1997. The North American market (which includes the Caribbean) is the dominant one, and in 1997 it grew by 8.6% to reach a record 5.05 million cruise passengers. In 1998, 71 cruise ships (which can carry over 93,000 passengers) from 24 lines plied the Caribbean, some year-round and some seasonally.

Although North America now accounts for almost 80% of total cruise passengers, this dominance is expected to decline as other markets mature. International cruise revenue is estimated at US$17 billion a year. The Caribbean is likely to maintain its position as the most popular cruise destination in the world because of increasing preference for shorter cruises and an ever-younger market. The 2–5 day cruise accounts for some 37% of the total product.

The Caribbean, accounting for 50% of capacity in 1999, is the most popular cruise ship destination. Its convenient proximity to North America makes it an easily accessible "pleasure periphery" for that market. Miami has ensured its place as the major hub from which most ships into the region operate, with up to 30 departures a week. Other major destinations include the Mediterranean (15%), Alaska (8%) trans-Panama Canal (6%), west Mexico (5%) and northern Europe (4%). The length of the cruise season in these locations, however, is determined by climatic conditions.

The South Pacific as a destination attracts only 2.2% of the world's biggest and most lucrative cruise market, North America, and Australia's own cruise passenger generating capacity has remained consistently low and very specific in its product requirement.

The "big three" cruise companies are Carnival, Royal Caribbean International and Princess that collectively control over two-thirds of the North American market. Star Cruises, a Malaysian-based company, which caters primarily to Asian tourists, aims to be the fourth largest. On the Asian ships a high percentage of passengers cruise in order to access gambling facilities that are not readily—or legally—available in their home countries, while the ships provide many activities to occupy their families.

Trends and Perspectives

The mid 1970s was a period of international crisis in fuel supplies and costs; this affected itinerary planning for the Pacific considerably, with some shipping lines withdrawing from the market altogether or canceling cruises, both moves having long reaching effects on island economies. It was estimated that cruise passengers spent $40 a day in ports and the cancellation of two cruises in 1974 for instance, was estimated to mean loss of $100,000 to small businesses in various Pacific ports. P&O reduced the speed of its ships, which meant arriving at ports a little later and leaving earlier. The reduced port time affected the income of people like handicraft sellers and transport operators, the two sectors of Pacific communities which managed to benefit directly from tourism.

Numerous Caribbean islands now receive substantially more cruises than they do stopover tourists. Since stayover tourists are far more economically beneficial to Caribbean countries than cruise day visitors, the shift away from the former towards the latter represents an alarming trend for the region.

Competition to add exotic ports of call has been intense, with Cunard adding 55 and Crystal Cruise Lines 24 new ports of call in 1998 alone.

The sharp increases in passenger capacity have been made possible by larger and still larger ships, whose economies of scale have produced record profits for the largest cruise lines. The size of many new ships, however, also dictates their cruising routes: they are simply too big to pass through the Panama Canal into the Pacific and are therefore restricted to the Caribbean and Mediterranean "ponds." They have flatter hulls than their predecessors and this, coupled with increasingly tall superstructures, means they are unsuitable for dealing with oceans and seas in those regions of the world that experience severe winds and currents e.g., the Pacific. (See very recent article from IMO News which reports on measures to ensure that the safety regime keeps pace with this trend toward ever larger vessels)

The cruise sector's ability to increase its passengers has been based on its success in reaching beyond its traditional upper and upper-middle class base into the middle-class mass market. Both the average age and the average income of cruise passengers have fallen steadily.

An important part of the strategy of the mass-market cruise companies has been to define land-based resorts such as Orlando and Las Vegas as their competition and to market their ships themselves as resort destinations. The ship is sold as the primary destination, not the ports it docks at. Indeed, "destinational cruising"—where the ports are central to consumer choice and experience—is now considered within the sector to be a niche market.

The hotel and entertainment giants have been increasing their presence in the cruise business. It has been widely commented that the new mass-market ships seem more like floating theme parks, artificial islands largely replacing real-life destinations.

The rhetoric of globalization is very much evident in the pronouncements of cruise leaders. Cruise ships represent the ultimate in globalization: physically mobile; chunks of multinational capital; capable of being "repositioned" anywhere in the world at any time; crewed with labor migrants from up to 50 countries on a single ship; essentially unfettered by national or international regulations.

Globalization of the cruise sector has also led to increased internationalization of ownership and further concentration in this business, with a massive shakeout reducing the number of players. The pace of mergers, acquisitions, and bankruptcies has been dizzying over the past two decades.

Globalization is also seen in the construction of a new terminal in the Grenadines. This will host mainly US-based cruise ships flying foreign flags and is being jointly financed by the European Investment Bank and the Kuwaiti Fund for Arab Economic Development, and constructed by a Kuwaiti firm.

Mexico and Central America

In 1998, 742 cruise vessels were reported in Cozumel and the maximum at one time was 11, although there are berths for up to 13 ships. With larger ships, many arriving at once, the numbers stress the existing capacity of the small town of San Miguel, as well as the available infrastructure.

Last year Ted Manning visited the San Blas islands of Panama where even their small ship (700) exceeded the population on the visited islands several fold—and at times two or more larger ships visit at the same time—totally transforming the port island into a circus of vendors.

Michael Lueck makes reference to a tv programme The Environmental Tourist—An Ecotourism Revolution, which was produced by the National Audubon Society and TBS Productions in Washington, D.C. It has a section on an American diving cruise ship that goes to Belize and spits out about 300–350 divers at a time. He comments "this is an enormous amount and the worst thing is that the ship doesn't even dock in Belize. There is absolutely no benefit for the local people at all but the negative impacts on the fragile reef are tremendous." He didn't know if the ship is still operating because local people lobbied against it in a big way. (Belize Cruise Tourism Policy)

Caribbean

Fantasy theming and simulation are endemic on most cruiseships. The ultimate in fantasyscapes on Caribbean cruises is not on the ship, however. It is to be found on "fantasy islands," privately owned by the cruise companies, off-limits to all but their passengers and employees, and marketed as the true Caribbean experience — only better. Of the 8 major cruiselines operating regularly in the Caribbean, six own private islands which they include among their ports of call. They are Half Moon Cay, Casaway Cay, Great Stirrup Cay, Princes Cay, Serena Cay, Coco Cay or at Labadee. The last is not actually an island but a piece of Haiti, surrounded by a ten-foot high iron wall, patrolled by armed guards. Disney dredged sand from the Casaway Cay bay and then ground it up further to make the island's beaches conform to a touristic image of Edenic perfection. It goes without saying that the development of private island destinations has been alarming to Caribbean countries, for in essence a local port is being cut out of the cruise itinerary in the process. The company reaps the economic rewards of renting their passengers everything from snorkelling equipment to cabanas to small boats, and selling them drinks and souvenirs at company-owned shops and markets. The already limited contribution of cruise passengers to local Caribbean economies is further eroded.

A further development of enclave-based encapsulation of cruise tourists (and their dollars) is the creation of private clubs for passengers in Caribbean ports of call.

To some extent Caribbean destinations are imitating the cruise ships, introducing theming in the port city landscapes (such as Aruba, whose main street feels very much like a theme park) and creating manmade, artificial attractions, divorced from the geographical environment, as in St. Maarten.

The image of Dominica calling itself "the Nature Island of the Caribbean" and free from mass tourism was severely dented in the late 1980s. It is the cruise-ship tourists (and their highly visible profile) who some critics believe have compromised the government's previous commitment to ecotourism. Cruise-ship arrivals rose spectacularly from 11,500 in 1986 to 124,765 in 1994. With a new cruise-ship berth at the Cabrits National Park, in the north of the island, and an improved deep-water facility near the capital, Roseau, up to 1,000 people per day pour off the cruise ships. Most of them take a whirlwind tour by minibus of a few of the island's best-known and most accessible sites. Ken Dill, a Dominican tour operator who specializes in hikes to the interior and nature tours warned that "if cruise ships develop to four or five a day and 4 or 5 times a week, it will be a turn-off for the ecotourists." Dill has to make sure his own customers do not bump into the cruise tourists when they visit the Emerald Pool, a short nature trail in the Morne Trois National Park leading to small waterfall. The carrying capacity of three sites, including the Emerald Pool, is being studied but there are some forestry experts who consider that places like the Emerald Pool will have to be "sacrificed."

In Grenada, the Tourist Board has called for stiffer penalties for crimes against tourists when one cruise line threatened to stop calling at St. George's because of passenger harassment by vendors.

Patti Pattullo writes that "it is the dumping of cruise ship waste that has been the focus of most concern. Cruises to the region have recorded a phenomenal growth rate and it has been estimated by the IMO that up to two kilograms of waste per person per day is generated. While some cruise ships have their own waste-processing facilities, many more do not." She makes the excellent point that "the attempt to clean up the ocean has also put extra strain on the land-based facilities of islands. In fact, the reason why not all countries have signed MARPOL is that to sign it would increase the pressure on their own land dumps. By not signing it, countries are not obliged to provide waste-disposal facilities and can refuse to accept garbage from cruise ships. Yet according to the IMO, which with the World Bank is organizing a project to deal with ship-generated wastes, this tempts cruise ships to dump at sea, whether legally or illegally."

Conflicts and Benefits

The economic impact cruise lines make to the Caribbean is matter of controversy. Pattullo writes: "Conflicting statistics (from using different methodologies and multiplier effects), major leakages of spending, especially of duty-free goods, and a generally low contribution to the overall income generated by tourism in the Caribbean are themselves indicators of the economic limitations of the cruise industry. But even more fundamentally, who earns the money spent by the cruise industry? Who benefits from the government's expenditure on port and shopping facilities and such expenses as extra police security?" Moreover, "the extent of the interlocking of interests between cruise ships and local big business at the expense of local small business is at the heart of the debate about the cruise industry's economic contribution to the region."

Globalization detaches economic life from the constraints of geography—physical, cultural, political—and nowhere is this more evident than in Caribbean cruise tourism. The companies are entirely non-Caribbean. Their destinations are increasingly under their direct ownership and control; Caribbean cruises are taking on elements of "cruises to nowhere." The ships' laborforce is overwhelmingly non-Caribbean. What these ships do in the Caribbean Sea (including dumping) is outside the jurisdiction of Caribbean states. Meanwhile, the Caribbean Hotel Association complains helplessly about the unequal playing field it shares with the largely unregulated and untaxed cruiseships and worries about the stated ambitions of leaders to "empty out" its hotels. Indeed, the Caribbean Tourism Organization reports that the proportion of North American tourists who spend at least one night on land has declined from 61.8% in 1987 to 48.6% in 1998. Citing declining demand, American Airlines in 1998 forced Antigua, Grenada and St Lucia to pay it a subsidy of $1.5 million each in order to maintain daily nonstop jet service from Miami. With stopover passengers vastly outspending cruise visitors and therefore greatly preferred and sought after, smaller Caribbean countries find themselves having to subsidize their transport in order to get them.

Source: Graeme Robertson, "Cruise Ship Tourism Industry," from lighthouse-foundation.org.

9

from the Jamaican Ministry of Industry and Tourism

Broadcast to the Nation by the Minister of Industry and Tourism Hon. Aloun Ndombet-Assamba

Wednesday, September 24, 2003

It is my pleasure today, to launch National Tourism Awareness Week 2003. This incorporates the World Tourism Organization's annual celebration of World Tourism Day on September 27.

There is a growing recognition all around the world, of tourism as a means of achieving economic growth and self-sufficiency at the national as well as community level.

This year's World Tourism Organization's theme, "Tourism: A Driving Force for Poverty Alleviation, Job Creation and Social Harmony" is therefore apt, and propels people everywhere to capitalize on the tremendous possibilities for development and advancing ourselves through tourism.

Here in Jamaica, our industry plays a vital role in our economy. Tourism contributes 50 percent of total gross foreign exchange earnings and generates gross earnings of 1.3 billion US dollars annually for our country.

A total of 75,000 persons are directly employed in the tourist sector and another 90,000 jobs are generated through indirect employment.

We Jamaicans are an enterprising people and tourism involves entrepreneurial endeavour at every level, from the major investor in hotels and attractions to the small craftsperson, store owner, taxi operator and restaurateur.

The linkages with other productive sectors such as agriculture, manufacturing and transportation are also tremendous. For example, the tourist industry purchases close to 2 billion dollars worth of locally produced food and 4 billion dollars worth of manufactured goods annually.

Tourism is unparalleled in encouraging nations to embrace each other's cultural differences, exchange ideas and trade spaces in the true spirit of "One Love."

This is indeed a "people centered" industry in which all members of our society should see themselves as participants.

This will be more readily achieved as we broaden the reach and benefits of tourism through implementation of our comprehensive Master Plan for Sustainable Tourism Development.

This plan was completed after many months of consultation with the public and was recently approved by Cabinet. It is designed to open up new opportunities in tourism both in the traditional resorts and new areas across the island, which will be increasingly accessible as our highway network is improved.

These are indeed exciting and critical times for tourism. Over the last several months, we have seen remarkable and encouraging improvement in both stopover and cruise ship arrivals to our island. Projections for this year are for 1.35 millions stopover arrivals, a 7.1% increase over 2002, and a record one million cruise passengers visiting our shores!

The industry continues to attract major investment such as that announced by the Pinero group out of Spain which plans to construct three new hotels in the St. Ann area over the next three to four years. We must protect and build on these gains.

As we turn the spotlight on the industry this week, I should like to salute our many partners and tourism workers in both the private and public sectors, through whose will, creativity and commitment, this industry continues to develop.

I urge all directly involved in tourism to continue giving of your best to this industry and Jamaicans as a whole to ensure that our island-home is safe, hospitable and inviting. We will thus achieve improved quality of life for ourselves and ensure that visitors from near and far increasingly flock to our shores to enjoy our outstanding natural beauty and diverse cultural offerings.

I look forward to the continuation and acceleration of our combined efforts to move this industry forward and to the benefits which, undoubtedly, will be reaped by all Jamaica.

Source: Aloun Ndombet-Assamba, "Broadcast to the Nation by the Minister of Industry and Tourism, Hon. Aloun Ndombet-Assmba," 9/24/03 on jis.gov.jm/MinSpeeches/html/20030921T100000-0500_409-JIS_Broadcast_to_the_Nation_by_the_minister_of_industry_and_tourism_hon_Aloun_Ndombet_Assamba.asp

from the St. Lucian Ministry of Tourism 10

Address by Tourism Minister Hon. Menissa Rambally to Meeting of St. Lucia Cruise Industry Stakeholders and Officials from the Florida Caribbean Cruise Association (FCCA)

June 28, 2001

On behalf of all St. Lucia I extend warm greetings to visiting associates from the Florida Caribbean Cruise Association.

I use the term "associates," because it correctly implies the elements of partnership, cooperation, confidence and mutual trust and respect that have characterized St. Lucia's relationship with the FCCA over the past four years.

In many respects, therefore, today's meeting can be likened to that of a meeting of the board of directors of a very successful enterprise.

The enterprise, in this instance, is the cruise industry of St. Lucia, and the board of directors are the stakeholders here assembled, comprising resident and overseas based individuals, all committed to safeguarding the gains that have been made and maximizing the potential that exists for this all important sector.

This, I am persuaded, is not an occasion for hard sell-marketing, because we are all convinced of the importance of this sector and have been beneficiaries of the phenomenal growth that has taken place in cruise tourism in recent years.

Therefore, as directors with a vested interest in the success and continued growth of this sector, I urge that we use this occasion to iron out recurring impediments to the smooth running of the sector and safeguard against unnecessary perils.

You would note, ladies and gentlemen, that I have repeatedly used the terms "we" and "our," because that is precisely how we must perceive our relationship if we are to move this sector to the next level. We are partners in an enterprise with seemingly limitless possibilities for growth. The challenge is for us to work together, recognizing and respecting each other's particular and peculiar interests.

For example, there is the lingering perception of cruise tourism not contributing immensely to the sustainable development of our region in general and our island in particular. The facts, I am persuaded, reveal the complete opposite. Nevertheless, the existence of this commonly held view, highlights the need for a heightened public awareness programme showcasing the net contribution of cruise tourism to the socio-economic development of our country.

In this regard, social partners such as the FCCA, SLASPA and other high profile interests, need, I believe, to be seen to be giving back more to the island.

Consideration should be given at this meeting to identifying and contributing significantly to major national causes that help to create an enabling environment for the sector. The whole question of disaster preparedness and mitigation is one area that comes immediately to mind.

Then there is the perennial perceived jealousies that exist between stakeholders in the cruise and land base tourism sectors. Most of you here assembled have the enviable fortune of being involved in but one sector, which is Cruise. However, there are persons like myself who must, of necessity, pursue the interest of all sectors.

We must work steadfastly at coming up with a formula for maximizing the twin potential of converting cruise passengers into long-stay visitors and vice-versa.

St. Lucia's social and economic development is dependent upon our recording consistent rates of growth in both cruise and land-based tourism.

To this end, the challenge is hereby made for us to use the occasion of this meeting to identify realistic proposals for translating years of vocal support and well-wishes, one sector for the other, into practical initiatives that will bear fruit and bring benefits to all concerned.

Ladies and gentlemen, the current growth pattern in cruise ship and passenger arrivals is heartening, indeed overwhelming, particularly at a period when land-based tourism in the Eastern Caribbean is experiencing noticeable declines.

I recall Ms. Michelle Paige saying to the region not so long ago, that the problem in subsequent years would not be a lack of arrivals, as the FCCA had undertaken to so market this region and expand the capacity of passenger vessels as to ensure virtual year round activity at our ports of entry.

Those of us, who took the FCCA at its word and upgraded facilities, have had no reason for regret.

I do not think I would be found guilty of boasting if I said here that when it comes to readying oneself for the opportunities that cruise tourism offer, St. Lucia, in the context of the Caribbean, has been a distinct leader.

However, we cannot become complacent. We have an opportunity today to review our product and determine what changes and improvements are necessary to not only stay ahead of the competition, but also attend to and meet the needs of a discerning traveling public.

In response to the challenges thrown out by the FCCA over the years, St. Lucia has remained focused, putting in place the physical infrastructure required and the diversity of onshore activity necessary to ensure the safety and comfort of visitors.

At the same time, we have implemented policies to maximise opportunities and benefits to stakeholders such as taxi drivers, ground handlers, tour guides, activities/sites operators, gift/souvenir shops, craft vendors etc.

With specific reference to the provision of shore-side excursions, it is evident that the town of Soufriere is approaching the limits of its carrying capacity, requiring the identification, development, packaging and selling of a wide variety of new tour products in other parts of the island.

Over the past three (3) years, through the introduction of the St. Lucia Heritage Tourism Programme (SLHTP), we have sought to respond to the challenge of an over concentration of visitors in one area.

Today cruise visitors have, not only a wide dispersal of geographical areas, but also a multiplicity of tour activities to choose from.

With respect to harassment, though the situation in Port Castries and the other tourist centres is not perfect, we are moving to great lengths, involving the expenditure of scarce monetary resources, in putting in place measures to ensure a safe, hassle-free experience by cruise and indeed all visitors to our shores.

In this regard, a substantial portion of the National Conservation Authority's budgetary allocation is spent on the employment and operation of the Rangers and Hostess programme.

As a result of these measures, we are pleased to note that incidents of visitor harassment have been on the decline, particularly in the last year.

The Ministry of Tourism is instituting a destination-wide system of minimum standards for the tours and activities sector. This initiative targets all relevant sectors, such as training, capacity building and certification for vendors and taxi drivers.

It also involves imposition of Environmental Management Systems (EMS) for sites and attractions, to ensure on-site compliance with visitor safety and environmental quality, and, National training and certification of tour guides.

The St. Lucia Air & Seaports Authority (SLASPA) has expended an estimated $50 million in expanding port facilities for on-shore visitor activity, such as duty-free shopping, as well as berthing infrastructure to accommodate the new generation mega ships. For example, the extension to La Place Carenage is expected to include a state-of-the-art, walk-in animated interpretation centre on the history of Castries. Thus adding to the diversity of offering and a "must-see" attraction for cruise passengers.

Noting the potential of expanded tour activity to rural communities, and the popularity of West Coast tours, we are in the process of refurbishing the facility at Anse La Raye and constructing three new jetties in the coastal villages of Laborie, Canaries and Choiseul.

In all cases the feasibility study have been completed, and upon construction these facilities should serve to open up even more tour opportunities, ensuring benefits to local entrepreneurs, rural communities and the St. Lucian economy.

Fellow stakeholders, multi-sectoral collaboration is absolutely necessary if we are to add further value, maximise passenger satisfaction and increase shore side expenditures.

To this end, this meeting is most welcomed as it provides an opportunity for the sharing of information and experiences and the adumbration of new ideas, all geared at the continuous improvement of our product.

Ladies and gentlemen, it is against this backdrop and with a feeling of great expectation, that I again welcome all here assembled, especially our visiting associates from the Florida Caribbean Cruise Association.

I look forward to a wholesome exchange of views on how we can accelerate our quest to make St. Lucia a sustainable, world-class cruise destination.

I thank you.

Source: Menissa Rambally, Address by Tourism Minister Hon. Menissa Rambally to Meeting of St. Lucia Cruise Industry

Review Questions

1. Why might it not be ideal for a small, isolated Caribbean island to rely on tourism for bringing about revenue and national development?
2. What are some of the things that you can do (and not do) as an environmentally responsible and culturally sensitive tourist?
3. How does tourism both market and change local landscapes?
4. What is Project Eco-Quetzal? What are some of the things it has done for the local community?

Discussion/Essay Questions

1. Imagine the following scenario involving Margarita, the Guatemalan woman profiled at the beginning of this module: Upon finding out that neither the dull-colored *huipil* that she is wearing nor the bright red *huipiles* that she is selling are truly "authentic," tourists from the United States stop patronizing her roadside stand. Instead, they go deeper into the hills searching for a more "pure" Guatemalan culture. How should Margarita respond to this, using points raised in some of this module's readings?
2. Although they come from very different perspectives, Mike Robinson, the author of the third reading, and Gilbert Trigano, the co-founder of Club Med portrayed in the fourth reading, both are skeptical of the potential for culturally sensitive tourism like that being implemented by Project Eco-Quetzal in Guatemala (reading number 5) and by Mauricio Fabian who is leading tours of Cruz Verde in the Dominican Republic (reading number 6). How do you think that Fabian and the organizers of Project Eco-Quetzal would rebut Robertson and Trigano, and how do you think Robertson and Trigano would each respond?
3. The author of the eighth reading, Graeme Robertson, calls the cruise industry "the ultimate in globalisation." What reasons are given for this? In what ways does this nature of the cruise industry provide support for many of the negative impacts of tourism as an economic development strategy? If you were the minister of tourism for a Caribbean island, would you encourage use of one of your ports?

List of Readings

1. "Ten Commandments on Eco-Tourism," published by the American Society of Travel Agents, posted on the website of the Global Development Research Center, *http://www.gdrc.org/uem/eco-tour/10-command.html.*
2. "Charter of Cultural Tourism," posted on the website of the International Council on Monuments and Sites, 1996, *http://www.icomos.org/tourism/tourism_charter.html.*
3. Mike Robinson, "Is Cultural Tourism on the Right Track?" in *UNESCO Courier*, August 1998, posted on the website of UNESCO, *http://www.unesco.org/courier/1999_08/uk/dossier/txt11.htm.*

4. Amy Otchet, "An Empire Built on the Sands," in *UNESCO Courier*, August 1998, posted on the website of UNESCO, *http://www.unesco.org/courier/1999_08/uk/dossier/txt41.htm.*
5. "After Two Years: Low Impact Tourism Program in Q'eqchi'es Communities in Alta Verapaz, Guatemala," posted on the website of the RARE Center for Tropical Conservation, *http://www.rarecenter.org/content/e_case_template.cfm?cs=37.*
6. "Cruz Verde: A Small Town Dominican Wonder," posted on the website of Interconnection, *http://interconnection. org/cruzverde/index.htm.*
7. Sue Wheat, "Sold Out: A Rapid Growth in Ecotourism Has Been at the Expense of Indigenous Peoples," May 22, 2002, posted on the website of *The Guardian*, *http://travel.guardian.co.uk/ecotourism/story/0,8945,719581,00.html.*
8. Graeme Robertson, "Crusie Ship Tourism Industry," posted on the website of the Lighthouse Foundation, *http://www.lighthouse-foundation.org/lighthouse-foundation.org/eng/forum/artikel00304eng.html#fragment3d.*
9. Aloun Ndombet-Assamba, "Broadcast to the Nation by the Minister of Industry and Tourism Hon. Aloun Ndombet-Assamba," September 24, 2003, posted on the website of the Jamaica Information Service, *http://www.jis.gov.jm/MinSpeeches/html/20030921T100000-0500_409_JIS_BROADCAST_TO_THE_NATION_BY_THE_MINISTER_OF_INDUSTRY_AND_TOURISM_HON__ALOUN_NDOMBET_ASSAMBA.asp*
10. Menissa Rambally, "Address by Tourism Minister Hon. Menissa Rambally to Meeting of St. Lucia Cruise Industry Stakeholders and Officials from the Florida Caribbean Cruise Association (FCCA)," June 28, 2001, posted on the website of the Government of St. Lucia, *http://www.stlucia.gov.lc/addresses_and_speeches/tourism/address_by_tourism_ministerat_meeting_of_st__lucia_cruise_industry_stakeholders_and_the_fcca_-_june_28,_2001.htm*

Websites for Additional Research

1. The Caribbean Tourism Organization is an industry association devoted to promoting tourism in the region. Its website, located at *http://www.onecaribbean.org*, is loaded with economic data on tourism as well as forecasts and other statistics. There also are links to documents devoted specifically to sustainable tourism in the region.
2. The Caribbean Alliance for Sustainable Tourism (CAST) is attempting to ensure socially responsible and environmentally safe tourism in the region. Its efforts are detailed on its website, *http://www.cha-cast.com.*
3. Globally, a number of organizations are attempting to establish standards for environmentally and culturally sensitive tourism, and certification systems for hotels and tour operators that conduct their business in an environmentally sustainable and/or ethical manner. See the website of the International Hotels Environment Initiative (IHEI) at *http://www.ihei.org*, the website of the Foundation for Environmental Education's Blue Flag Campaign at *http://www.blueflag.org*, and the website of Green Globe 21 at *http://www.greenglobe.org*. For UNESCO's World Heritage Committee, see *http://whc.unesco.org.*

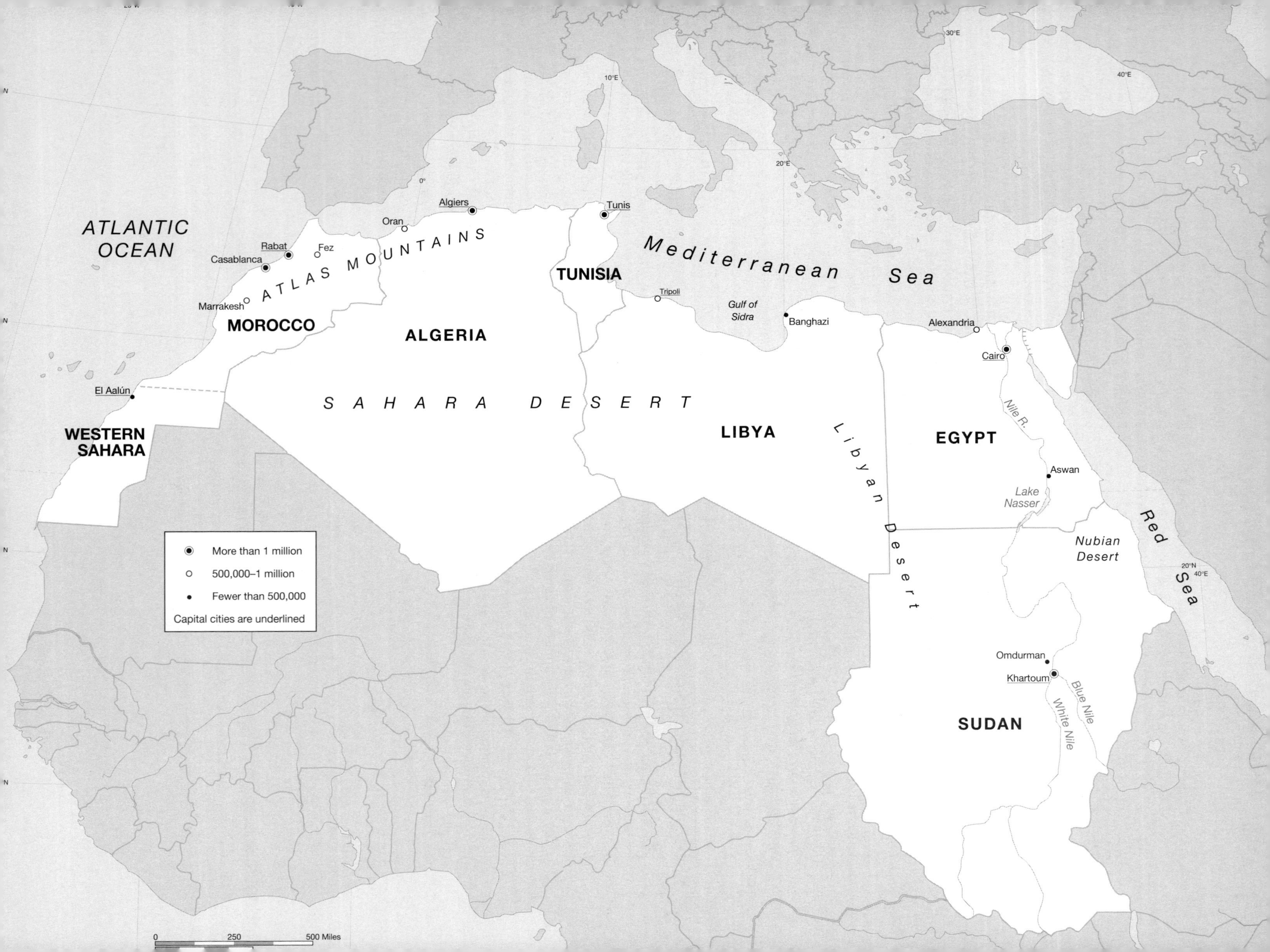
ATLANTIC OCEAN
Mediterranean Sea
Red Sea
Rabat
Casablanca
Fez
Marrakesh
Oran
Algiers
Tunis
Tripoli
Gulf of Sidra
Banghazi
Alexandria
Cairo
Nile R.
Aswan
Lake Nasser
Nubian Desert
Omdurman
Khartoum
Blue Nile
White Nile
El Aalún
ATLAS MOUNTAINS
SAHARA DESERT
Libyan Desert
MOROCCO
WESTERN SAHARA
ALGERIA
TUNISIA
LIBYA
EGYPT
SUDAN
More than 1 million
500,000–1 million
Fewer than 500,000
Capital cities are underlined
0
250
500 Miles
0°
10°E
20°E
30°E
40°E
20°N

Primate Cities in North Africa

Companion to Chapter 13
Urban Patterns

Urban planners seek to construct medium-sized cities in North Africa.
Source: Dynamic Graphics/Dynamic Graphics, Inc.

Waffa is a consultant for an engineering firm in Benghazi, the second-largest city in Libya. Recently, she led a team that had been contracted by the Libyan government to study the possibility of designing a satellite area for the city of Benghazi. The problem confronting Benghazi was a common one for the nations of North Africa. Without middle-size cities for rural Libyans to migrate to, large numbers are moving to the cities of Tripoli and Benghazi, overwhelming these cities' abilities to provide jobs, housing, or essential services such as water and sewage.

The initial plan was to build a new area outside of Benghazi to relieve it of some of the strain from this migration. However, Waffa's study found that the construction of a satellite city would actually *increase* the physical problems that the city would face over the long term. Waffa suggested instead that the Libyan government build an entirely new city to absorb the country's growing urban population. The location and structure of this new city would be determined based on political and economic criteria. It would be pre-planned and it would have an identity independent of any of Libya's pre-existing cities.

Waffa examined the successes (and failures) of other planned cities in Libya as well as other large cities in North Africa and Southwest Asia. Her report had several recommendations for this new city. Besides considering Islamic building requirements, Waffa suggested that it be a medium-sized city with a population of around 150,000. She also recommended that the city host a range of economic activities and that it be governed by regulations and zoning ordinances that encourage investment. The report stressed the importance of establishing an independent authority for the new city. Above all, the report urged that the city be part of a regional and national urban development strategy.

Source: Adapted from "Benghazi City, Urban Problem: New City! How?" by Fatma Giaber of Al-Emara Engineering Consultants. ■

Problems of Urbanization

There is no one precise definition for an urban area. An urban area is an agglomeration of people, a space in which people cluster together to gain advantages from proximity. Although an urban area is often thought of as a political unit with "city limits," a city is rarely contained within its defined, political boundaries. Agglomerations (or urban areas) occur in all different sizes. Individual towns and villages represent the lowest level of urban areas, while huge metropolitan areas such as the Los Angeles basin represent large-scale agglomerations.

Individuals cluster in space (or agglomerate) for a number of reasons—usually economic, but sometimes military or political. Historically, many cities have developed at places that are conducive to transportation and trade. World cities such as New York and London developed on the basis of waterborne commerce (rivers gave access to the interior of the country while seaports allowed for easy import and export of goods). The establishment of railroad lines allowed these cities to further their advantage. Later technological developments such as the automobile, the telegraph, the telephone, air transport, and the Internet have provided cities with even greater advantages, as corporations demand state-of-the-art technologies and connectivity that are cost-effective only in the largest cities. Often investments in transportation systems reinforce an urban area's initial transportation advantage. In some cases, however, changes in transportation modes have led to a city's decline. For instance, the world is littered with sleepy villages (or ghost towns) that were booming metropolises in the nineteenth century when river transport predominated but were then bypassed by emergent railroad and highway systems.

While most cities have grown because they consolidated economic advantages, other agglomerations have grown into cities for reasons that were not primarily economic. Islamabad, the capital of Pakistan, was created for political reasons, as a *forward capital.* A forward capital is created when a capital city is relocated to an unsettled or contested part of a country in order to establish or maintain a governmental presence there. The location of Islamabad represented a move inward to a disputed part of Pakistan. Brasilia is another example of a forward capital (see the discussion of Brasilia in Chapter 3 of the textbook).

While cities need transportation and communications infrastructure to develop, they also need human capital in the form of a trained workforce. Cities attract migrants from rural areas with the promise of opportunities to find jobs, make money, and perhaps build a better life. While the dynamic of in-migration is a component of urbanization everywhere, it is especially profound (and its impacts are especially problematic) in less developed countries. Rapidly urbanizing LDCs typically suffer from a huge (and continually growing) gap between rural and urban lifestyles, and economic growth of the city frequently is inadequate to provide a revenue base for the services necessary to sustain its burgeoning population. This module focuses on an area of the less developed world in which urbanization has been particularly intense: North Africa.

Global Urbanization and the Primate City

The textbook introduced the concept of the primate city in Chapter 12, primacy being the condition when the population of the largest city in a country is more than twice that of the country's next largest city. Primate cities are especially common in LDCs, where they serve as centers of political and economic power and attract the vast majority of migrants from rural areas.

When they occur in LDCs, primate cities typically trace their histories to colonial efforts to establish a base for penetration and extraction of resources. In some cases, colonial powers selected already established towns for their bases, while in others they chose new sites based on the transportation advantages offered. During the colonial period, the dominant mode of transport was water, so the locations that attracted the attention of colonial powers typically were on coastlines or navigable waterways. Roads, and later railroads, were developed to bring goods to the port city, but other towns were not interconnected. The entire transportation network was designed to get goods out of the region, onto boats, and shipped back to the colonizing country. Linking interior towns was not high on the priority list, unless these towns happened to produce goods that the colonialists sought to transport to the port city for export.

Whether or not internal towns were also developed, the transportation system that resulted was focused on the port city, in many cases transforming it into a primate city. With advantages from the start, the primate city continued to grow. When colonial powers abandoned their colonies in the post–World War II era, new leaders inherited countries that, in most cases, contained one major city that dominated economic and political life. As population growth increased, so did migration to the primate city. Contemporary urban migration frequently is motivated by a combination of "push" factors, such as increasing land concentration in rural areas, which pushes rural people off of what had been their family plots, and "pull" factors, such as the distant hope that one may get a job in the tiny but lucrative sector of the economy populated by foreign business people who live and conduct business in the primate city.

History of Urbanization in North Africa

The Middle East (North Africa plus the adjacent region of Southwest Asia) is often called the "cradle of civilization" because many of the world's first cities grew there. It was a region in which empires flourished, and trade was important for centuries. Egypt was the location of one of the earliest city-based empires, built around the Nile River.

Notwithstanding the period of the Egyptian Empire, the region more often has been under the rule of outsiders, typically from Southwest Asia or Europe. Historic rulers of the region included the Assyrians, the Persians, the Romans, the Byzantines, and the Ottomans. In the nineteenth and twentieth centuries, the region fell within the control of various European countries. Italy controlled what is now Libya; France controlled Morocco, Tunisia, and Algeria; Spain controlled Western Sahara; and the United Kingdom controlled Egypt and Sudan. With the exception of Western Sahara, all of the countries in the region became independent between 1951 (Libya) and 1962 (Algeria). Western Sahara finally became independent from Spain in 1976, but almost immediately thereafter it was occupied by neighboring Morocco, which retains control of the country.

Today, North Africa is a region characterized by primate cities and many small rural villages. All of the highly populated primate cites are located on the Mediterranean or on the Nile River. The following table provides population statistics on the two largest metropolitan areas in each of the six North African countries (omitting Western Sahara).

The second to last column of the table shows that, with the exception of Morocco and Libya, all of the countries exhibit extreme primate-city features, with their largest areas considerably more than twice the size of the next largest urban area, and even in Morocco and Libya the largest city is more than twice the size of the second city. In fact, Tunisia and Sudan have no other significant urban areas. The final column reveals that, with the exception of Sudan, the region is highly urbanized (and, with an annual urban growth rate of almost 5 percent, Sudan appears to be well on the road to urbanization as well). In short, the countries of North Africa tend to have large portions of their populations crowded into one central city. This creates a host of problems for both urban and national planning, as urban populations outstrip the capacity of their cities to provide housing, jobs, schools, or physical infrastructure (fresh water, sewage, transportation, and so on), and as urban populations develop cultures and economies that increasingly are distinct from those of the rural hinterland.

		Largest Metro Area >500,000		***Next-Largest Metro Area >500,000***			
Country	**Country Pop. (1000s)**	**City**	**Pop.(1000s) [% of Country]**	**City**	**Pop.(1000s) [% of Country]**	**Ratio of Largest Metro Area to Next-Largest**	**Total % Urbanized [Annual Urban Growth]**
Egypt	69,296.0	Cairo	15,892.4	Alexandria	4,442.6	3.6:1	44%
			[22.9%]		[6.4%]		[2.3%]
Libya	7,250.8	Tripoli	2,357.0	Benghazi	1,124.4	2.1:1	85%
			[32.5%]		[15.5%]		[2.9%]
Tunisia	9,879.6	Tunis	1,660.3	—	—	>3.3:1	62%
			[16.8%]				[2.5%]
Algeria	33,577.5	Algiers	3,917.0	Oran	794.2	4.7:1	57%
			[11.1%]		[2.4%]		[3.6%]
Morocco	30,456.9	Casablanca	3,397.0	Rabat	1,636.6	2.1:1	52%
			[11.2%]		[5.4%]		[3.2%]
Sudan	37,985.9	Khartoum	5,717.3	—	—	>11.4:1	31%
			[15.1%]				[4.9%]

Country and city population data are 2003 projections as calculated at *http://www.world-gazetteer.com*. Urbanization and urban growth data are 1990–1999 figures from the United Nations.

Environmental Problems of Rapid Urbanization in Primate Cities

Although the readings in this module focus on social problems encountered in rapidly growing primate cities in North Africa, large cities (including primate cities in LDCs) also face a number of environmental problems. Many of these environmental problems are also discussed (in a more general context) in Chapter 14 of the textbook.

One environmental problem cities often face is waste disposal. Cities typically create more waste than they can dispose of. A case in point is the ill-fated "Garbage Barge" that floated around the world for sixteen years trying to find a final resting place. The barge, called the *Khian Sea,* left Philadelphia in 1986 with seven tons of ash from incinerated household trash. Over the next sixteen years, the *Khian Sea* tried to unload its cargo in Africa, the Caribbean, the Philippines, and a host of other nations with no luck. In the summer of 2002, the garbage finally was disposed of permanently in a landfill outside of Philadelphia, not far from its place of birth.

While this story is not typical, it does demonstrate the difficulties that urban areas face as they attempt to dispose of their solid waste. Although cities generate huge amounts of solid waste, landfills are extremely unpopular and unwanted in densely populated areas because they generate large amounts of traffic (from garbage trucks), they smell, and they take up space that might otherwise be used for housing or industry. (Additionally, beyond the local scale, landfills create methane gas, which contributes to the greenhouse effect.) As transportation costs have decreased, many cities in more developed countries have taken to transporting their garbage surprisingly long distances, to communities that have the space for burial in landfills and, typically, are so impoverished that they accept the short-term benefits of payments for accepting waste in return for what will almost certainly be a long-term devaluation of their property. Sometimes these waste-accepting communities are in distant LDCs (that's where the "Garbage Barge" was trying to unload its waste) but often they are in poor, rural areas of the same MDC that generates the waste. For instance, much of New York City's waste is trucked to landfills in impoverished areas of Pennsylvania, Ohio, and Virginia.

Some cities have turned to incineration as an alternative to landfills, but incinerators can cause more problems than they solve by producing toxic gases and emitting them into the air. Additionally, the ash from incinerators frequently is a highly concentrated blend of toxins that, if not buried in high-technology landfills, can leach into the ground and poison an area's soil and water systems.

Because cities accumulate many stationary and mobile sources of air pollution, air quality issues also are particularly problematic in urban areas. Some of the places we think of as having the worst problems with air pollution are also among the world's largest cities (such as Los Angeles and Mexico City). While activities within the city generate air pollution, other factors also can lead to a reduction in urban air quality. For instance, cities in valleys are more likely to experience problems, while windy conditions may assist in diluting the pollution.

A third environmental problem in urban areas involves the treatment of wastewater and the management of clean supplies of drinking water. Human settlements and water systems have a dynamic relationship that is exaggerated due to urbanization. People everywhere need fresh drinking water, which is provided by lakes, rivers, and groundwater sources. In rural areas, the pollution of these water sources from human and animal waste and from the runoff of agricultural chemicals typically is offset by rainfall that contributes "clean" water, so that the water remains fit for human consumption with little direct intervention. In urban areas, however, the management of the water system is much more complicated.

Water coming out of the urban system cannot simply be cycled back into the system. Because cities are characterized by impermeable surfaces (such as paved streets), rainwater that falls in cities tends to run along the surface for a while, where it picks up oil, chemicals, and organic materials that have accumulated on roads and other paved surfaces. Cities channel this water into stormwater systems that typically terminate in water treatment plants. Along the way, this water is joined by the output of residential and commercial buildings' sewage systems carrying human waste. At the treatment plant, wastewater is treated to remove harmful microorganisms that can cause illness or even death. The wastewater is treated through a process that allows materials to settle out in stages, and sometimes a disinfectant such as chlorine is added. Finally, the treated water is released into the environment (typically into a river or the ocean), where it gradually combines with the global water system that provides freshwater to cities around the world.

This system of water management generally works, but when it fails the results can be disastrous. Water treatment systems in rapidly growing cities often are strained to capacity, and a sustained flood, which greatly increases the amount of water in the system that must be treated, can push a municipal water treatment system over the limit. To cope with floods, many newer cities separate their systems for processing stormwater runoff (whose flow varies greatly depending on weather conditions) from their systems for processing sewage (whose flow is predictable and can be particularly harmful if allowed to enter the general water system untreated). Many older cities, however, do not have the funds to build new, separate stormwater systems.

Social Problems of Rapid Urbanization in Primate Cities

One should not put too much emphasis on the "magic number" for primate cities, when the largest city in a country is more than twice the size of the next largest. Primate cities emerge for a number of reasons, and the problems associated with rapid urbanization occur whether or not this urbanization is concentrated in a primate city. Nonetheless, the cultural divide between the city and the rest of the country is likely to be more severe if urbanization is concentrated in one, usually coastal, city rather than in a few cities scattered around the country. Likewise, the strains of urbanization, such as housing shortages, are likely to be greatest if urban growth occurs all in one city rather than in a number of cities.

Many of these strains are detailed in the first reading, a report by the U.N. Environment Programme that briefly summarizes some of the reasons why the countries of North Africa have urbanized so quickly as well as detailing some of the problems faced by new urban residents. Besides discussing the problems of unplanned growth, the reading mentions some of the efforts being made by the governments of countries in this region to help the urban poor.

The second article focuses on Cairo, the region's only megacity (a megacity usually is defined as a city whose urban area population is greater than 10 million). Cairo functions as a primate city not just for Egypt but for the entire Arab world, which spans the regions of North Africa and Southwest Asia. Its metropolitan-area population of almost 16 million is more than twice that of the next-largest Arab city (Baghdad, with 6.8 million) or the next-largest city in North Africa (Khartoum, with 5.7 million). Even as Cairo continues to grow as a center of fashion, entertainment, and business, it also grows as a city of slums, pollution, and traffic jams. The second reading introduces the numerous aspects of Cairo, focusing both on the vitality and the weighty problems encountered in such a rapidly growing center. The article is from *Aramco World,* a magazine published by the Saudi national petroleum company that attempts to educate westerners about the realities of life in the Arab world.

One of the key problems faced by rapidly urbanizing cities in LDCs is housing. When people are drawn to cities that cannot accommodate them, they create squatter settlements. The textbook discusses squatter settlements and emphasizes the prevalence of these settlements in the less developed world. As many as 85 percent of people in Addis Ababa, Ethiopia for example live in squatter settlements. These areas have different names around the world, but the conditions are similar. In Tunisia, these locations are called gourbevilles. Cairo has the City of the Dead. In the third reading, Mahmoud Kassem of the *Cairo Times,* an Egyptian weekly newsmagazine, introduces this shockingly different type of settlement.

Recognizing the problems of rapid urbanization is not enough. The challenge is to limit the worst aspects of urbanization without stalling national development programs. One option could be for governments to encourage prospective urban migrants to stay in rural areas where at least they might be able to feed themselves and where they won't strain urban services. However, few national governments in poor countries have the power to keep people out of the cities even if they wanted to. Another option would be for governments to build new cities in the hinterland and encourage growth in these locations instead of in the congested old cities. This was the strategy favored by Waffa in the personal profile that began this module. Alternately, governments could build satellite cities for new immigrants, near, but not actually within, the primate cities. As the *Cairo Times* reading shows, however, a project along these lines in Egypt was not terribly successful. Maybe the best that can be done is to empower city governments to extend basic infrastructure to the outskirts, effectively transforming the temporary squatter settlements that are already forming there into permanent extensions of the city. This seems to be the strategy presented in the fourth reading, a 2002 press release from the World Bank announcing a new loan to Tunisia for extending urban services to low-income areas of Tunis.

The final readings shift focus to the other side of urbanization: the part of the country that is not incorporated into expanding cities, and the people who remain behind in these rural areas. As a country's soul (and its economic orientation and site of political power) moves from the agricultural or pastoral countryside to the crowded cities, what becomes of those who are left behind? What cultures are lost as countries transform themselves from rural to urban?

Of all the cultural types in the world, perhaps those who are least amenable to urbanization are nomads. If urbanization is a process of agglomeration—a clustering in place—then nomads represent the opposite of urbanization as they move about. North Africa, the site of some of the world's most rapid urbanization processes, also happens to be home to many traditionally nomadic or seminomadic societies. The final readings turn to two of these societies—the Tuareg and the Berbers (or Imazighen)—and the problems that they are facing as the worlds around them urbanize.

The first of these readings, from the Natural History Museum of Los Angeles County, describes the traditional lifestyle of the Tuareg, perhaps the most nomadic people of the region. Although much of the area traversed by the Tuareg is in Mali and Niger, south of the region covered here, they also are found in parts of Algeria and Libya.

The next reading examines the plight of the Berbers (or Imazighen), an ethnic group in parts of North Africa, especially Morocco, Algeria, and Tunisia. Although members of this ethnic group prefer to be called Imazighen (the word *Berber* is derived from the Greek word for barbarian), the term *Berber* still is commonly used in the West. Although the Berber/Imazighen people are a minority that is sometimes discriminated against in the region (especially in Algeria, where the government has been attempting to "Arabicize" them), they comprise nearly 40 percent of the population of Morocco, so they are not a small minority. In this reading, Lucy Jones of the *Washington Report on Middle East Affairs* discusses the decline of Berber/Imazighen communities in Tunisia due to lack of resources and employment opportunities amid a rapidly urbanizing national society. It offers some Berber/Imazighen viewpoints on the decline of their culture in Tunisia. Although this reading mentions a few steps that are being taken to avoid such a loss, the tone is not very hopeful. The last reading, however, sounds a more positive note, as the author asserts that urbanization (and nationalization) in Libya has occurred in a way that has preserved village culture.

Readings

1

from the United Nations Environment Programme

Africa Environment Outlook

Past, Present and Future Perspectives

Northern Africa

With 64 percent of its population living in towns and cities, Northern Africa is the most urbanized sub-region in Africa. Within the sub-region, urbanization rates vary between countries—36 percent in Sudan to 95 percent in Western Sahara. Most of this growth has occurred in the last two decades, although urban populations have increased steadily since the colonial phase. In 1990–1998, the average rate of urbanization reached 2 percent per annum and it is predicted that this will continue over the next 15 years, which means that by 2015, 70 percent of Northern Africa's population will live in cities.

Urban growth in Northern Africa is partially the result of rural-urban migration, but natural urban growth and reclassification account for more than 70 percent of urban development. Furthermore, the new structural adjustment programmes adopted by most countries in the sub-region have brought new frontiers in industrial development in the last 10 years. By 1990, Alexandria (Egypt), Algiers (Algeria), Cairo (Egypt), Casablanca and Rabat (Morocco), Tunis (Tunisia), Tripoli (Libya), and Khartoum (Sudan) had populations of more than one million.

The Significance of Urban Areas in Northern Africa

Old and well-established urban centres such as Cairo, Casablanca, Alexandria, and Tripoli continue to thrive and have retained their character despite various economic, social, cultural and political changes. However, urban agglomerations and "mega-cities" are important features of recent urbanization trends, as well as heavy industrialization. Most infrastructure and societal services are centred in these cities and they make the greatest contribution to national products. As a consequence, they play a vital part in economic development by providing opportunities for investment and employment. They have also gained a key political and administrative role.

Urban Areas and the Environment in Northern Africa

Despite the gains outlined above, rapid urbanization in Northern Africa has also given rise to significant environmental and social problems, and is characterized by increasing urban poverty, emergence of informal settlements and slums, shortage in basic urban services, and encroachment on agricultural land. Between 15 and 50 percent of city residents in the sub-region are urban poor living in squatter settlements, illegal subdivisions, sub-standard inner-city housing, custom-built slums, and boarding houses. It has also been indicated that cultural heritage sites are especially at risk from uncontrolled development and environmental degradation in urban areas. While there have been national and international efforts to conserve this heritage, greater efforts are required to ensure its protection.

Water supply and sanitation rates are higher in Northern Africa than in some other sub-regions, but they are nonetheless variable and many of the poorer residents of urban centres are without reliable services. In 2000, most urban residents had access to improved water supply (ranging from 72 percent in Libya to 98 percent in Algeria) and sanitation (87 percent in Sudan and 100 percent in Morocco). Casablanca's wastewater network, however, cannot cope with the volume of wastewater produced, despite recent upgrading by the municipality. As a consequence, large quantities of wastewater are discharged into the sea.

Unplanned Settlements in Northern Africa

Cairo has several large informal settlements on the city periphery and these accounted for 84 percent of urban growth between 1970 and 1981. In 1992, there were 400 shantytowns in Casablanca, housing more than 53,000 families, and in Algiers, 6 percent of the population are reportedly squatters. These settlements suffer from a shortage in basic infrastructure and hence have major health problems. Their illegal status compounds the problem, and residents often have to be moved to more appropriate sites where local authorities or municipalities can provide them with basic urban services including water supply, transport or health care. More investment is required to provide these services to the whole urban population.

There has been land use planning and zoning in most of the cities in the sub-region. However, this has not, in some cases, prevented chaotic expansion and densification of cities. It is now the norm to find residential zones next to industrial sites or industries enveloped by housing estates, with all of the potential risks to the environment and to health which that entails.

In 1979, the Egyptian government implemented a strategy for improving the living conditions of urban slum dwellers in Cairo, by moving them to alternative locations. Residents of Eshash El-Torgman and Arab El-Mohamady were moved to El-Zawia El-Hamra, Ain Shams, or Madinat El-Salam. Their original residences were close to the city centre and thus to the job market, whereas the resettled areas were far out of town. Reports of the negative social impact of this move prompted the government to change its approach to informal settlement upgrading, and to adopt a new twofold strategy: clearance for state land (squatter settlements), and upgrading for private land (informal settlements). As a result of social pressure, however, no areas have been cleared, but provision of infrastructure to informal settlements has made significant strides. In Morocco, settlements have been upgraded in Agadir (see below) and, in Casablanca, the state issued a plan, in 1992, to build 200,000 houses for low-income people. Progress has been slow however, as the extent of state-owned land has declined sharply, and privately owned land is very expensive, making subsidized housing unviable. Over the last three decades the government of

Libya has made housing provision a priority and general housing, investment housing, agricultural housing and low-income housing projects have been implemented. Nearly 400,000 units were constructed in various locations, and efforts are well under way to deliver a further 60,000.

Case Study: Settlement Upgrading in Agadir

The Moroccan National Shelter Upgrading Agency's project in Agadir has been recognised as one of Habitat's Best Practices for Human Settlements. Agadir was devastated by an earthquake in 1960 and housing for lower-income families has been insufficient ever since. In 1992, there were 77 separate shanty areas with 12,500 households (13 percent of Agadir's population). The Upgrading Agency's project has provided approximately 13,000 serviced housing lots, housing units, apartment units, or lots for apartment units, has provided utilities connections for an additional 3,600 families, and created 25,000 jobs per year in construction. Reasons for the project's success include active participation of the clients in project design, implementation and monitoring, recognition of the clients' rights to adequate shelter by local authorities, open dialogue between the parties, and integration of former squatter into cosmopolitan neighbourhoods.

Source: "Africa Environment Outlook: Past, Present and Future Perspectives," as published by the United National Environment Programme posted at www.unep.org/aeo/207.htm

2

from *Aramco World Magazine*

Cairo: Inside the Megacity

By Dick Doughty
March/April 1996

"For the first time, wherever you go, it seems that the city is still there," says Dr. Mohammed Taher al-Sadek, editor of *Al Memar,* the Egyptian Architectural Association's quarterly, and professor of city and regional planning at Cairo University. "Even I can't grasp it any more."

Like any great city, wrote anthropologist Janet Abu-Lughod, Cairo is "a mosaic of subcities," each the product of a different social order, a different technological era and a different economy. Cairo is thus much more than a "new city" added to an "old city."

Much of today's Cairo is of course a product of the industrial age, from neighborhoods of quotidian family-run shops all the way out to huge government or privately-owned factories on the city's outskirts. Almost eclipsed now are the old suqs and capillary alleys where shops are still organized in clusters by trade, just as they were in the Middle Ages. In Cairo's multiple commercial centers, franchise stores offer the abundance of the global consumer economy and the streets are peppered with corporate advertising, sometimes in Arabic, sometimes in English. And only a few kilometers away, there still lie quiet, ox-plowed fields edged by brick-house neighborhoods organized along village lines, the remains of Cairo's once-dominant agricultural economy.

Every one of these sub-cities has a square or a neighborhood or a mosque that is its hub, where the faces of Egypt mingle with those of other parts of the Arab world and Africa and those of the West. When plodding wooden carts, ingeniously repaired Fiats and tinted-window BMW's clog the streets, and when thick crowds of people from poor to rich, illiterate to university-trained, wage laborers and civil servants, merchants and mothers, students and peddlers fill the sidewalks and the squares, every morning and well on into the night, then Cairo pulses as the heady center of uncountable human universes. Intoxicating yet exhausting, it is this seemingly infinite human wealth that lends magnificence to Cairo's soul and confronts it with its greatest paradox.

"No other Arab city is so alive," says a local expert who has studied Cairo's varying neighborhoods. A Cairo-based US advisor to the Government of Egypt observes that international news from Cairo frequently fails to communicate the "tremendous vibrancy here."

Cairo has earned an enviable reputation as one of the friendliest of the world's great cities. Relaxed attitudes, a tradition of hospitality and the renowned Egyptian sense of humor have smoothed daily troubles for centuries. But Cairo is also the heart of Egypt, where 60 million citizens today face the unprecedented challenge of housing, schooling and employing more than 1.3 million newborn compatriots every year. This must be done using little more than four percent of Egypt's territory, the ribbon of Nile-watered land whose total area is only two times that of tiny Kuwait. And of these new citizens, more than 350,000 are born each year in Cairo.

The face of "modern" Cairo began to appear in the late 1860s. Only 300,000 people lived in the city then, half as many as in the 14th century, when it was the most populous capital in all of Europe and the Middle East until the plague of 1326. In the 1860s and early 1870s, the new Suez Canal and the Cairo-Alexandria railroad reinforced Cairo's role as an East-West trade center of the early industrial age. Civil war in the US drove a surge in world demand for cotton, already Egypt's top export.

Flush with this trade, Khedive Ismail, ruler of Egypt, decided in 1867 that Cairo deserved the same wide, radial boulevards that had just reshaped and modernized Paris. Ismail drained the shallow lakes that separated Cairo from the Nile, tore through a jumble of traffic-clogged streets and laid down the network of boulevards and maydans—small, circular plazas that are modern Cairo's geometry.

The expansion of Cairo's population, however, did not begin in earnest until nearly half a century later. Underground sewers, installed in 1915, brought dramatic improvements in public health that nosed the birthrate above the death rate for the first time. By 1930, Cairo's population passed the one-million mark. A wartime manufacturing boom brought a wave of migration that nearly doubled this figure by 1945.

But nothing has been like the last two decades. In the early 1970s, climbing oil prices brought an indirect windfall to the economy. At the same time, farms throughout Egypt were becoming too subdivided to satisfy family needs as they had for centuries. As a result, men sought opportunities in the Arabian Gulf in record numbers. The wages they wired home allowed families once accustomed to subsistence living to step into a nascent consumer economy, buying televisions, home appliances and the like. Many also forsook the village for the city, carrying with them the dreams of education and prosperity that would reshape Cairo.

In 1974 the late President Anwar Sadat's infitah, or "open door" policy, allowed foreign investment in Egypt largely banned since the country's 1952 socialist revolution and ushered in what city planner al-Sadek calls "a new philosophical era." It was a turning point. The boom was on. Over the next 20 years, more than seven million people poured into Cairo. They filled the Nile valley to overflowing and spilled out into the desert, settling in a few years more than twice as much land as Cairo had covered in all its thousand previous years of history.

Now, in the Middle East, Istanbul and Tehran are each only half of Cairo's size. If Cairo were sovereign, it would be the fifth-largest Arab country. Even some Cairo neighborhoods have nation-sized populations: Shubra, north of the city center, houses three million people, comparable to the population of Lebanon.

Cairo is also in every respect the center of Egypt, as it has been almost since its founding in 969. One quarter of all Egyptians live there. The majority of the nation's commerce is generated there, or passes through the city. The great majority of publishing houses and media outlets and nearly all film studios are there, as are half of the nation's hospital beds and university desks. The population density of Cairo is exceeded only by the cities of India. And just how new is Cairo? Look out a window:

Stone-crafted minarets still grace the sky here and there but one building in five is less than 15 years old. And now, fewer than one in eight Cairenes goes to sleep each night in the historic quarters that knew the ways of the city's Fatimid founders.

This astonishing growth until recently surged well ahead of city services. Homes, roads, electricity, telephones and sewer services were all suddenly in short supply. Analysts trying to grasp the magnitude of the change coined terms like "hyper-urbanization." On the ground, planners struggled: Only bits of the Cairo Master Plan of 1970 and the Greater Cairo Master Scheme of 1982 were ever implemented.

"Planning has not coped with the migrations. How could it have?" says Abdel Rahim Shehata, governor of Giza, which encompasses all of Greater Cairo west of the Nile. "We are busy providing the minimum services. It is a race against time."

And what a race it is. In a 1990 report, the United Nations called the city's response "one of the most ambitious physical planning efforts in the developing world." As a result, Cairo is catching up with itself.

A drive around the city can be as impressive as the figures. Seven bridges now span the Nile. Downtown expressways and lesser flyovers bypass coagulated traffic circles. New roads knit outlying suburbs into the urban fabric. A 100-kilometer (62-mile) ring-road expressway is largely complete. Donkey carts are banned from the city center, and traffic police keep everyone moving. Trips that used to be measured in whole hours are now, on good days, a matter of minutes, even as the number of cars in the city threatens to triple, since 1980, to nearly one million.

Like vertebrae along Cairo's riverine spine, 33 stations now articulate the first of three planned Metro rail lines, open since 1988; the second line is slated to open next year. The trains run on time. Along well-swept platforms in downtown stations, locally produced music videos and advertising play on television monitors. For thousands of factory workers who live in the north of the city and work in the south, the Metro has sliced commuting time in half.

The number of telephones has more than quintupled since the early 1980s, years that Cairenes recall with wry comments about the legions of couriers who threaded the gridlock afoot and on bicycles to deliver business messages, because telephones were so scarce. Now, businesses and homes often have lines for voice, fax and modem, and cellular phones are expected to be introduced soon.

The scarcity of land with access to water has historically favored apartment-style housing. For the middle and upper classes, private developers and land speculators who have sought to capitalize on a tenfold rise in land values since 1970 have built a surplus of comfortable housing. For the majority of Cairenes, however, demand for modest, low-cost housing has run well ahead of the massive, government-sponsored construction programs. This has led to the third and ultimately most traditional type of Cairene construction: the owner-built, brick-and-concrete houses of one to five stories that now shelter more than four million people, or one fourth of the city. One Cairo-based US development specialist calls this wave of "informal" housing "the most unique phenomenon of modern Cairo."

Many of these nearly 100 unplatted neighborhoods, now sub-cities in their own right, have their origins in the wages sent home from the Arabian Gulf countries before the 1990 Gulf crisis. Without that flow of money into the economy, experts say, Cairo might have been ringed by shantytowns, a fate that nearly alone among the developing world's megacities it has avoided.

"Informal housing is not a 'problem.' It's a creative response," says anthropologist Linda Oldham, co-author of a 1985 study of these communities and a resident of Cairo since 1976. She points out that even though such a home often lacks utilities for a number of years after it is built, it is nearly always structurally sound. The owners usually supervise the construction, and because the home is often built piecemeal as the family acquires cash, contractors, too, have clear incentives to do good work in order to be hired to put up the next story. "It's a very solid social process," Oldham says.

But Sahal Abou Ezz, undersecretary in the Ministry of Development, New Lands, Housing and Public Utilities, believes that instead of building informally, most people should apply for government-built apartments that unlike many informal buildings and communities are supported with needed utilities and services. Since 1977, he points out, the government has built nearly a million apartment units throughout Greater Cairo. Some of the greatest concentrations lie in the deserts northeast and west of the city, where hundreds of buildings in styled clusters of a dozen or more rise in unbroken, serried ranks, bristling with matchbox balconies.

Many Cairenes, however, hold different views. "People stay in [downtown] Cairo because it has spirit. Thousands of years of dwelling has created a pattern of relationships that is very valuable," says architect Abdel Halim Ibrahim Abdel Halim, whose recent work has been aimed at restoring key points in Cairo's urban core. The families of many of the city's unskilled and semiskilled workers, he explains, often come either from the old parts of Cairo or from villages. "In their presence and their culture the people embody the geometry of the old city," he says. This can make a move to what planners call a "modern" housing complex a culturally difficult experience.

For example, points out sociologist Nawal Hassan, head of the Center for the Study of Egyptian Civilization, laws barring market stands on the streets of public housing areas enacted to keep streets free of congestion often impose an unfamiliar pattern of neighborhood organization on residents who are accustomed to purchasing modest food needs from in front of their homes, or at most within several blocks of home. Accordingly, "people have to adapt to the plan, rather than vice versa," she says, adding that the same situation occurs frequently in public housing projects the world over.

But for "Chef" Gamal, who commutes from an "informal" community west of the Nile to supervise two cooks at a sandwich stand along a canyon-like street lined with apartments near the city center, such mass-produced housing is simply practical. "These towers take in the rising population," he says glancing upward. "How else are we supposed to do that here?"

Since the 1970s, the government has taken the idea of public housing a step further and built half a dozen "satellite cities" or "new towns" in the deserts surrounding Cairo. The one named Fifteenth of May, to the south, began in 1979 as a commuter city serving nearby factories. Of the six new cities, it is now the closest to being successful, and the key, says Hosni Abu Elenin, chairman of the Fifteenth of May Authority, was that the 80,000 residents didn't have to change jobs when they moved.

Other new cities, however, remain sparsely inhabited. A 1989 government report estimated the population in all the satellite cities around Cairo at one-fifth of that called for in the original plans. But this is changing: Tax incentives have prompted a slow but promising migration of private industry, and Cairo's steady expansion may yet weave all into a single cosmopolitan carpet. "As soon as a major international hotel gets built [in Tenth of Ramadan city]," says a German owner of a diaper factory in that new city, "business will really take off. When that happens, people will flock to where the jobs are. I think it is inevitable."

Demographer Hoda Rashad, director of the Social Research Institute at the American University in Cairo, lives in Madinat Nasr, "Victory City," one of the largest and earliest of the desert development areas of the 1970s. Back then, she says, Madinat Nasr was just as bare as the new satellite cities are now, and people spoke of it just as disdainfully. But today, it is a part of Greater Cairo and one with status, to boot. "I was walking home the other night along one of the main streets, with all the shops and lights and cars, and I found myself wondering, 'Am I really still in Egypt?' The change has been so astonishing. [Madinat Nasr] is one of the most popular places now!

"People's feelings about the desert are changing," she adds. "Sometime in the last five years it became a less frightening or lonely place. We Egyptians are people of the river, you know, historically. But now people are more willing to invest [in the new cities]."

Salah Al-Shakhs, who headed the government's Division of General Planning in the late 1960s, maintains that despite the overly optimistic early projections, the new cities are bound to fill up. He points out that it took more than 20 years for Brasilia, Brazil's new capital city, to become popular. "Now it has twice as many people as it was supposed to," he says. "It just takes time."

One of Cairo's most urgent and newly acknowledged tasks is to reverse growth-related environmental degradation. Until recently, pollution, from litter to choking smog, had been largely tolerated as a cost of economic growth. But this attitude too is changing.

"We now view pollution as a cost to Egypt," says Salah Hafez, director of the Egyptian Environmental Affairs Agency. "Our main mission is to institutionalize this new way [of thinking]."

In 1994, Egypt's legislature passed the Law on the Protection of the Environment, which is, according to Hafez, "unprecedented and very promising." He points out that, for the first time, air- and water-quality standards are set under one law. Automobile emissions testing will be enforced gradually over the next three years and, by 2000, Cairo's standards will be based on those commonly used in European cities. Efforts to reduce lead in gasoline have already cut airborne lead emissions by 24 percent. All industries will be held to emissions standards for both air and water. "Enforcement is a matter of education, and of building alliances," Hafez explains.

Cleanup is also going on underground, as the first segments of the world's largest sewage engineering feat opened their sluice-gates in 1994. The $4-billion Greater Cairo Wastewater Project is moling tunnels, some wide enough to drive a bus through, under nearly every section of central Cairo. This spells relief for the pre-World War II sewer system, which, according to Talat Abu Seida, vice-chairman of the project, used to leave up to half the city's daily output untreated.

Few in Cairo speak ill of these physical improvements, or even take them for granted. But opinions and hopes for the future time and again turn on the more personal axis of economics. "The roads, the new highways, the telephones, these are very nice," says one public-sector architect who moonlights as a taxi driver. "But the standard of living has not kept up." Since 1982, inflation has gnawed away 40 percent of the purchasing power of wages.

"Unless the economy is sufficiently activated to provide one-half million new jobs each year for the next 10 years," observes Governor Shehata of Giza candidly, "this country will be in trouble. And we must do this at a time when the role of government is shrinking" through privatization. To this end, many of Egypt's 400 state-owned companies are in the process of being sold to private investors, and the government is working to smooth a variety of legal and regulatory paths for entrepreneurs (see *Aramco World,* January–February 1996). This process of privatization, which now largely dominates economic philosophy in Egypt, both drives and reflects a cultural shift in attitudes toward work in Cairo: Today, a job is often viewed less as a social right as it was in the 1950s and 1960s and more as a reward for individual achievement.

"Government jobs are not the status symbols they once were," says Abdel Salam Hasan, who came to Cairo with his parents 20 years ago from a Delta village and now is a business administrator with a Swedish-owned firm. "Even public sector employees work after hours [in private businesses]. People have more initiative these days."

Until the Gulf crisis, one popular field of employment lay in the oil-producing countries of the Arabian Gulf. When those jobs temporarily dried up, increasing competition in the streets of Cairo led many to look west instead of east for jobs outside Egypt. But in Europe, Canada and the US, languages other than Arabic are spoken, Islam is a minority faith, the skills required tend to be high, and uprooting oneself even temporarily is rarely a simple matter, either practically or emotionally. "The West has a reputation for being cold," says Azli Islam, a recent graduate of the American University in Cairo. "Of my [English-speaking] friends, about half of them want to go and half of them don't." There can be a lot of opportunity, she adds, "but we are very aware of what we would be leaving: our families and the warmth of our culture here."

Among Cairo's most cherished characteristics, many put this ineffable sense of warmth high on today's endangered list. "There is a growing feeling that people are fending for themselves," says demographer Rashad, who says she notices this tendency particularly among young people, who worry about future job prospects.

"It seems that people dealt with each other more like family before," says a 21-year-old medical student at Cairo University, when asked how her city is changing. "Now people deal with more their own problems. Sometimes everybody seems only to care for money."

These concerns lie at the heart of long-running controversy over what "modernization" [*al-tahdith* in Arabic] means on the streets of Cairo. For years, as in many other non-Western cities, it has been understood, too simply, as "old" giving way to "new" terms that are often used to refer, on the one hand, to all that is traditional, based on religion and centered on the extended family, or, on the other hand, to things that are commercial, secular and centered on the individual. The result is a mosaic of frequently competing values. "There used to be a sharp line between wrong and right," says planner al-Sadek. "Now it's mixing up."

"This 'modernization' is a problematic term," says business administrator Abdel Salam Hasan, who is taking a break from work to relax on a newly installed bench alongside a busy street. "'To become modern,' to me, doesn't mean machines or computers but more that, say, someone like me can sit here in a public place and care about it, feel some investment in it. We have to modernize ourselves before we modernize our machines. Real modernization is a human affair."

One who recognizes this daily is Magdy Fahim Aly, MD, who is, outside his medical practice, a community organizer in government-built housing in the northern district of Shubra. He witnessed a breakdown of community when apartments in his housing project were assigned randomly, rather than according to residents' pre-existing family and social affiliations. As in many such new complexes, "people didn't know each other anymore. You didn't know your neighbor in the same building," he says.

For 12 years Aly worked with a local committee to build cooperation in what became one of Cairo's most effective citizen groups. According to Dr. Waffa'a Abdullah, director of the Social and Cultural Center of the National Institute for Planning, Aly and the 1000 families of the Khalafawi public housing complex now affectionately called "Khalafawi Gardens" show that Cairenes can live and even thrive within an often-criticized housing system.

Garbage, Aly says, filled the areas between buildings for years after the complex was built in 1969. But it was when leaky sewer pipes contaminated the drinking water that Aly, then a medical student, joined with six others who decided that something had to change.

In 1979, the group collected donations and laid new sewer lines to their own building. They hauled off the garbage themselves and began planting shrubs and trees. Within two months, 14 other buildings in the complex had followed suit. Five years later, "Khalafawi Gardens" had been transformed by more than a dozen islands of trees, a mosque, a volunteer health clinic, a weekly cinema, a sewing center and a playground that now is packed with several hundred kids every evening.

"At first it was very hard to change people's minds," Aly says. "But conditions forced us to change." "It was not just a physical change, but a social change," adds Abdullah. Now, under Abdullah's guidance, Khalafawi residents are helping train similar groups in eight other government-built housing complexes.

Architect Abdel Halim is similarly committed to nourishing the best in Cairo's urban character by preserving and upgrading its public places. Because land is so scarce, he points out, the city has historically offered little common recreational space.

"Common land is the most vital part of any city. People need a place where they can be aware of their common bonds," he says.

His most recent project is a children's park tucked into a bit more than a hectare (about three acres) of the historic district of Al-Saiyydah Zaynab, just south of the modern city center. Its several plazas, small amphitheater and craft shops offer "a setting to open up activities," he says. His success led the Aga Khan Commission to assign him to develop a 30-hectare (75-acre) park, which will become the city's largest when it is completed in 1999. Set on a small hill overlooking Al-Azhar university and the oldest part of Cairo, it will provide "a point from which the city can look at itself," he says. It will feature a promenade atop Cairo's restored city wall, an amusement park, an Islamic garden, an outdoor amphitheater and spacious lawns. One section is being designed exclusively by residents of the neighborhood it adjoins.

Efforts such as Aly's and Abdel Halim's are increasingly reflected in official programs. Neighborhood self-help associations are more common than ever, and often work in tandem with government and international agencies. There is "real effort," Abdel Halim says, to nurture cultural institutions, including theater, film and literature.

Now that the boom-town growth has slackened and Cairenes have grown increasingly acclimated to an accelerated pace of change, he says, there is more talk, in political and intellectual circles, of the years to come. "The question of sustainability is now being discussed widely," he says. "We have to understand how to get and manage resources as government shrinks."

According to Governor Shehata of Giza, one of the city's most innovative programs, "The Educated Village," has employed 3000 university graduates who otherwise had no work to teach evening classes in adult literacy that will help people find and create jobs. "And we are opening one new school every three days, on the average," he adds. "Not every Egyptian is going to have a yacht on the sea, but it is possible for the majority of people to have basic services, keeping in mind that our resource base is very fragile."

Encouraged by this rise in official support, Abdel Halim remains doggedly hopeful about Cairo's future. "Cairo is on a threshold. It will cost a lot to clean the air and the water, and to provide people with jobs and places to live."

But Cairo's effort to address unprecedented challenges by drawing on its vast cultural wellsprings, he says, is ultimately a creative process. "The old city is an endless reservoir of culture. Something is going to come out of Cairo that is really revealing, that will produce beauty of a different order altogether. Then we can share it with other cities."

But in the remote Tenth District of Madinat Nasr, which presses against open desert far from Cairo's center, Sherif Shehata Muhammad, an honors graduate in transportation engineering from Cairo University, finds that such vision demands a daunting leap of faith. Newly married and 27, he faces hard, practical choices. His eyes widen as he sketches his ambitions for himself and his city. But a moment later, he concedes these may prove beyond reach due to scarcity of resources, and the overabundance of talent for too few jobs. When asked if he might consider applying his own talents outside Egypt, his answer comes as slowly as the evening's desert breeze curling among the ranks of apartment towers.

"My friends and I all work hard. I want to help," he says. "I want this place to be good for my children, too. We are like a fire. But if something throws water on us, then what?"

But if he stays in Egypt, then perhaps Abdel Halim can be right after all.

Source: Dick Doughty, "Cairo: Inside the Megacity," *Aramco World,* March–April 1996.

from the *Cairo Times*

3

Tomb Busting

The City of the Dead will eventually be sacrificed for the sake of progress, but where will the living go?

Mahmoud Kassem
September 2000

Sayyed Ali has lived with his thirteen children inside a tomb the size of a small room for most of his life in the so-called City of the Dead, a massive cemetery on the eastern desert outskirts of Cairo. But the 48-year-old tire repair shop owner might lose his nesting place and livelihood if the punishing forces of progress get their way.

Along with some 30,000 other people who have lived alongside the dead for decades, Ali and his family might eventually be relocated to new cities as part of a twenty-year plan that will turn the cemetery into a lush public garden, a Cairo governate official told the *Cairo Times* this week.

"We are trying to remedy old problems. These tombs are for dead people. We are going to get rid of the people who are living in these tombs," said General Mostafa Mazoun, head of the local council of Manshiyet Nasr, a slum district of Cairo that encompasses parts of the City of the Dead. "I think we need some greenery in the area to let the living breathe," he added. "It's civilized progress."

Cairo holds tens of thousands of tombs in up to two dozen locations around the city; many of them date back to the 14th century, but the majority are from the 19th and 20th centuries. Most of the cemeteries are inhabited by victims of the country's chronic housing shortage. The government promises new apartments for the living and new graves for the dead but local residents affected by the upheaval doubt that they will get either. Islamic law dictates that the cemetery must lay vacant for 15 years before it can be put to any other use.

"We already have 2100 new apartments in the 'new city' in Manshiyet Nasr," said Mazoun, but it is not clear how much the rent will cost. The new apartment blocks, located near the homes built for 1992 earthquake victims, will be ready by October, he said. Workmen are still laboring on the concrete highrises that boast brightly painted shutters.

Meanwhile, in front of Islamic Cairo's Bab Al Nasr, where the city's medieval walls are under restoration, 1800 tombs have already been demolished by the government in the past year to make room for a 25-meter road extension, according to local residents. Some of the inhabitants have been given shelter, though many still remain camped out amid the ruins of broken tombs.

"They gave me a room the size of a kiosk," said Mohammed Abdel Gawwad, who was given notice to vacate the tomb he is living on. "I've got a family of seven, so I had to come back and live with the dead."

"We're staying here because we have no alternative at the moment," said Samir Mohammed, who's still waiting for the bulldozers to move in.

In Manshiyet Nasr, the local council says they are walling in the City of the Dead where the main road meets the tombs to reduce crime and protect tourists, but they have come under fire from local residents who say the project is sacrilegious and harming the local community's business interests.

Mazoun stated that no tombs have been unearthed in the building of the wall, but irate local residents say while digging for the foundation of the wall, workmen unearthed tens of graves without transporting them to new ones.

"I reburied some of the remains myself," said one man, as he stopped some of his colleagues from redigging the tombs to show the bones to the *Cairo Times.* "They showed a complete lack of respect for the dead. They didn't even take any of the bones to be reburied, they just poured in the cement."

The wall has also infuriated local residents like Ali who say it will destroy their roadside business by preventing people who have tires in need of repair from bringing their cars to his shop. The morale of Ali's mechanics—who sit around contemplating early retirement or forays into other trades and curse the idea of a garden—is low. Several workshops have already closed down due to the wall, Ali said.

"This wall is going to push the youth of this area into drugs and terrorism," said one mechanic. Manshiyet Nasr has a notorious reputation for crime and is reputed to be a favorite hideout for criminals, who find the labyrinth-like tombs ideal to hide in when on the run from the law.

"What do we need a garden for? Are we going to grow apples and mangoes?" asked mechanic Khaled Abdel Meguid. "A little beautification is going to destroy our livelihood."

Sayyed says that the two-meter high wall, which so far spans some two kilometers and is slowly encroaching towards his workshop, will also force hundreds of local residents to make lengthy detours in order to cross the main road. "The people are angry, they don't want the wall," said Sayyed. "We're going to be trapped. We won't be able to cross the roads anymore."

Opposition Wafd MP Ayman Nour briefly took up the cause of local residents here, but they're skeptical that any politician will take further interest in the matter until after the November parliamentary elections—by which time the wall will most likely have already been erected.

"We've taken our case to the State Council and parliamentarians but they've all done nothing. There's no compensation, no real interest in our rights," said Sayyed.

The land technically belongs to the government who have over the years awarded licenses to local residents to set up businesses in the City of the Dead, according to Mazoun. "Just as the government gives, it also takes," he said. "We're not obliged to renew licenses. In any case many of these 'workshops' are fronts for drug dealing."

Resigned, most local residents are now pressing for a meter-wide gap in the wall that will allow them to continue their roadside business and facilitate pedestrian traffic. "There is no way that they're going to make gaps in the wall," said Mazoun. "If these people are upset about losing their livelihood because of the wall they can take their case to a court of law."

He maintains that the project has been thoroughly studied and that it will ultimately work for the benefit of the community at large. "We're not doing something against the people, we're doing it for the people."

Source: Mahmoud Kassem, "Tomb Busting," *Cairo Times,* September 2000.

4

from The World Bank

News Release: Loan to Tunisia

World Bank Loan to Help Boost Basic Services, Infrastructure, for Country's Urban Population

WASHINGTON, December 5, 2002—The World Bank today approved a $78 million loan to the Government of Tunisia aimed at improving the quality of life for urban residents by supporting local governments in delivering basic municipal services and infrastructure. The Municipal Development Project (MDP) III is the third in a series of World Bank projects which paves the way for institutional and policy reforms to decentralize expenditure and service responsibilities from Tunisia's central government to local authorities.

With 64 percent of the population living in cities and towns, Tunisia is among the most urbanized countries in the Middle East and North Africa. By the year 2020, the urban population is expected to soar to 70 percent. With unemployment growing along the fringes of urban centers, municipalities are expected to face increased pressure in providing and maintaining basic services.

At the request of the Tunisian government, which recently adopted its 10th Development Plan for the period 2002–2006, the Bank will continue to support municipal, financial and managerial capacity—one of the key priorities of the national strategy.

The MDP adopts a two-pronged strategy to enable Tunisia's municipalities to better manage their resources and assume a greater role in financing investments at the local level. One component focuses on strengthening the Municipal Development Fund's (CPSCL) ability to respond to the increasing needs of municipalities through technical assistance, training and studies supporting its transformation into an autonomous institution.

The second component of the project strategy seeks to strengthen the ability of local governments and related agencies to better deliver and manage public services through training, computerization of tax management and reforming working procedures.

The loan falls in line with the World Bank's country assistance strategy (CAS) for Tunisia which places emphasis on fighting poverty through various means, including municipal and urban development. Since Tunisia's growth will be centered in urban areas and along the coast in the long run, the CAS anticipates a greater need for urban development in low-income areas in the future.

Source: "Tunisia: World Bank Loan to Help Boost Basic Services, Infrastructure, for Country's Urban Population," posted Dec. 5, 2002.

from the Natural History Museum of Los Angeles

5

Who Are the Tuareg?

The Tuareg are a group of mostly nomadic people who live in northern Africa. They divide themselves into five strictly defined social classes. Nobles own land and camels, and rule the confederations or political alliances. Vassals farm and graze their herds on land managed by nobles. And, although slavery was abolished in the Sahara in the 1940s, many descendants of slaves still work with their former owners.

The Tuareg have two other specialized groups. Scholars and religious leaders offer guidance. Finally, an outcast group of artisans, known as blacksmiths because they work with metals, makes everything needed for life in the desert, including camel gear, amulets, jewelry, weapons and most household items.

Married Women Own Their Dwellings

Most Tuareg live in tents made of goat skins or palm fiber mats. When a woman marries, her family makes a tent for her. The tent belongs exclusively to her; she's responsible for its care and upkeep. In fact, men are considered guests in their wives' homes.

Tuareg Tents Make Mobile Homes

How often a nomadic Tuareg family moves depends on the weather, season, and quality of pasture. Though they don't stay put, families remain within designated territories.

Before colonial rule, slaves did the hard work of moving. Now even noble women often move themselves. Usually a Tuareg woman can pack up and move her entire household on two donkeys or one camel. It takes about two hours. Light, non-breakable baskets, enamelware and plastic help make moving easier.

An Oasis Is a Wet Rest Stop on a Dry Desert Highway

As animals and humans travel desert trade routes, they stop at oases to stock up on water. Some oases are just wells in the desert; others support bustling towns. People almost always grow crops, bargain over goods and share news at oases.

As caravaneers plan their trips across the Sahara, they rely on the knowledge of oasis locations passed down through their families, as well as on their own experience. As Tuareg women move their herds of goats, sheep and donkeys from one of the Sahara's dry pastures to another, they make sure that they're never far from water.

Source: "Who Are the Tuareg?" part of the AFRICA: One Continent, Many Worlds Project sponsored by The Field Museum, Chicago, IL.

6

from the *Washington Report on Middle East Affairs*

Tunisia's Berbers under Threat

By Lucy Jones
August/September 2001

Few cars these days climb the road to stony Ghommrassen, located 300 kilometers south of Tunis in the heart of Tunisia's Berber Ksour region. An "Open" sign in English creaks in the slow breeze above a deserted teashop. Tiny rooms carved into the hillside are empty, save for a few discarded possessions—an old pair of shoes, a blanket, broken cooking utensils. The mosque looks freshly painted but, like the other abandoned buildings, is gradually gathering piles of red dust.

In the distance it is possible to see to where the people of this mountain community have moved. Shimmering below on an arid beige plain is "Novi Ghommrassen," or New Ghommrassen, a small town of white buildings with running water, air conditioners, hi-fi stores and—most importantly—health facilities, schools and jobs. Last year, Novi Ghommrassen finally enticed the last of Ghommrassen's Berbers. Instead of living in homes chiseled from the rock, these mountain dwellers now inhabit hastily built apartments, cook on electric stoves instead of open fires and send their children to school on a local bus.

Old Ghommrassen used to be a thriving Berber settlement, boasting several stores, including one for tourists, an olive oil press, a mosque and a small café. The residents of the village lived mainly on the proceeds of olive oil production. But, one by one, the families of Ghommrassen gradually decided to make the move to nearby towns.

The men left first, to look for work, as these Berber strongholds offered little in the way of employment. Their families followed them. The old were the last to leave. "There was no longer any community life," said Mustafa Hazaris, a retired storekeeper who now lives in a stuffy apartment block in New Ghommrassen. "We were away from our friends and relatives. There was nothing left but empty buildings, so we had to move."

The original Ghommrassen is not the only mountain settlement to be abandoned by the Berbers for towns with modern conveniences. Some 10 other villages in southern Tunisia—a few more than two centuries old—have emptied out, and others are on the brink of survival. "As Berbers become assimilated into the towns," said Mohammad Bezara, a Berber historian in Tataouine, the Ksour region's modern administrative center, "we have to question whether the very notion of Berberism in Tunisia is under threat."

The Berbers inhabit swathes of land across North Africa, predominantly in Morocco, Algeria and Tunisia, but also in Libya and Egypt. Algeria's Berbers have been in the news recently because of riots in the country's northern Kabylia Berber region. More than 80 Berbers were killed by government forces trying to quell the violence, which ignited unrest elsewhere in the country. Berbers account for around 20 percent of Algeria's population but face discrimination, and are pressing to have their language and culture recognized and for improved economic conditions.

Although the Berbers across North Africa are related, their origins long have been debated. Some historians, citing archeological evidence dating to the 10th millennium BC, think the Berbers are the indigenous people of North Africa. Other experts say the Berbers came from elsewhere, probably the Mediterranean. According to this theory, ancient immigration resulted in the formation of a group of people whom the ancient Greeks called "barboroi" (barbarians) because their customs varied so greatly from Greek behavior. (The word "Berber" is derived from that derogatory term.)

The Arabs first came to Tunisia in 647, returning to conquer the area 23 years later. Unlike previous invaders, the Arabs were not interested simply in acquiring an empire—they also wanted to introduce Islam and the Arabic language. Since these goals could not be achieved until they established themselves as rulers, the Arabs waged war against the Byzantines and the Berbers, defeating a large group of Berbers led by a Berber princess called Kahina in 702.

Kahina's followers were the first to embrace Islam. With their help, the Arabs went on to conquer all of North Africa and most of Spain, and met with amazing success in establishing their religion everywhere they went. The spread of the Arabic language, however, was much slower. In some isolated areas Arabic was not adopted until recently, and in most Berber areas the Berber language still is widely spoken.

In Tunisia, of the country's population of 9 million an estimated 90,000 people are thought to be "pure" Berber. The Berbers are known for their green eyes, ginger hair and pale skin—although many Berbers with Arabic and sub-Saharan African features also regard themselves as members of this ancient people. Many of the sub-Saharan African Berbers, in fact, originally were slaves of the Berbers. When their masters adopted Islam, however, which bans slavery, the Africans were released. Few returned home, choosing instead to remain within the Berber communities.

Because Berbers in Tunisia always have been a force to reckon with, they have not faced the level of discrimination suffered by Berbers in Morocco and, especially, Algeria. Indeed, French colonizers championed the Berber culture and language as a means of creating a division between Berbers and other Tunisians.

In recent years the government has been quick to recognize the advantages of promoting Berber culture as a tourist attraction. Every day, air-conditioned buses deliver European tourists from Tunisia's coastal areas to Chenini, a Berber settlement of 3,000 inhabitants located in the Ksour region and redolent of olive oil. To encourage the village's traditionally costumed inhabitants to remain, the government has built a sparkling new clinic and a primary school.

At Matmata, cave dwellings have been transformed into luxury hotel accommodation. This has provided locals with a much-needed source of employment, although some residents, weary of the thousands of tourists, have erected barbed wire in front of their caves to keep the visitors out. In 1997 the spectacular *ksar*—a fortification built to store grain—located on a hilltop near Ghommrassen was used by American filmmaker George Lucas to create the set for the last "Star Wars" film.

Even the residents in these thriving Berber settlements face problems, however. A severe lack of employment has led the majority of men from these communities to look for work elsewhere. "As soon as the men reach 20, they leave," said Ahmad Fadel, 34, one of the few young men to stay in Chenini. "They go to France to work in agriculture or to Libya to work in *patisserie* factories. They don't have a choice."

Often the men never return. As a result, the demography of these settlements has changed dramatically in the past two decades. In Chenini, more than 80 percent of the inhabitants are now female. This can be witnessed on Thursday nights at the town's mosque, where a large group made up entirely of elderly women engage in prayer until the early hours of the next day.

Many of the young women are unmarried. "It's hard to find a husband. There are fewer men here these days," said Halima Najjar, 22, an olive collector who lives on the hillside of Chenini. "Sometimes our elders try to match young women with men from the towns, but many parents don't want to lose their daughters so they don't attempt to do this. Also, the men in the towns aren't always Berber. Parents don't want their daughters to enter into mixed marriages."

"But we do want to marry, though," Halima's younger sister, Najia, chimed in.

Even the women who are married, however, spend months alone with their children while their husbands earn a living abroad. "This has a detrimental effect on family life," said one Chenini mother. "Of course it would be better if our men could be here."

Norah Fatia, 75, a rug maker who has lived in the settlement all her life, laments the fate of the man-less settlement. "It never used to be like this," she said. "Many people used to live here. Men as well. We were always celebrating weddings. Maybe it's time for us to change," she sighed.

Such talk is unpopular among Berber activists. In Tataouine, Lazhar Harabi, a campaigner for Berber rights, says the government should do as much as it can to keep the Berbers in their original settlements by providing running water, schools and jobs. "The Berber people are an essential part of the country's heritage," he said. "They cannot be abandoned."

If money from tourism is at stake—as in the case of Chenini—the government seems to deem the assistance of Berber settlements worthwhile. In settlements off the tourist track, however, the state shows little inclination to improve Berber living conditions by building much-needed schools and hospitals. "It's a natural process," said Sayed Berhaza, a local government councilor in Tataouine. "No one is forcing them to leave. There are simply better opportunities elsewhere in the country."

That may be true. But it will result in the scattering of the Berber people and, most probably, their gradual assimilation into the population. The colorful Berber settlements gleefully portrayed in holiday brochures soon may be no more than tourist symbols of a way of life that no longer exists.

Source: Lucy Jones, "Tunisia's Berbers Under Threat," *Washington Report on Middle East Affairs,* Aug/Sept 2001.

7

from David Blink

Libya's Tribal Landscape

As in other Middle Eastern tribal societies, Libya's tribes are arranged in a pyramidal lineage scheme of subtribal, clan, and family elements, the basic unit of tribal life. Before Libya became independent in 1951, the country's tribes operated to a large degree as autonomous political, economic, and military entities. Some 70 to 80 percent of the country's approximately one million inhabitants were then identified as members of a tribe, with true bedouin—those leading a nomadic or seminomadic existence—accounting for about a quarter of this total. The loyalties of the smaller urban population were largely confined to the extended family. Whether settled pastoralists, bedouin, or townsfolk, all Libyans found security among—and were intensely loyal to—their kin, the ultimate guarantors of their survival.

By 1992, 76 percent of Libya's estimated population of 4.5 million was urbanized. If we were to believe regime pronouncements or the comments of some Libyan citizens, we could conclude that the tribal underpinnings of Libyan society dissolved during the transition to an urban environment. On the contrary, rural migrants who flocked to Tripoli during the oil boom of the 1960s and 1970s congregated in loosely woven family and clan groupings. Many retained patterns of social organization specific to particular hinterland tribes, sometimes lending areas of the city's more recently populated suburban districts the flavor of a "Warfalla town" or "little Zintan." Family and tribal affiliation became even more highly defined outside the municipal boundaries of Tripoli and Banghazi, Libya's primary urban centers. In many cases bedouin tent settlements, strictly arranged according to lineage affiliation, sprang up on the edges of Libya's numerous smaller towns during the oil boom, creating a settlement pattern that remained intact as tents were replaced by permanent housing. Some Libyans worked to preserve social cohesiveness in government-planned housing areas by reserving adjoining homes for family expansion, using them as shelter for livestock in the interim. In other cases migrants abandoned rural farmhouses or urban apartment dwellings where contiguous family expansion was not an option or where floor plans and community layouts ignored sensibilities of family and clan association.

On a practical level, the average Libyan's loyalty to traditional patronage networks and authority figures survived the urbanization process and in many cases continued to transcend fealty to the state or nontraditional leadership well after the last bedouin tent settlement disappeared in the mid-1970s. Concerns on the part of Libyans polled at the height of the urbanization experience that traditional values and lifestyles would be lost as the country's youth succumbed to modernizing influences appear unfounded. Younger Libyans tend to balance their need for modern urban amenities with a conscious desire to perpetuate traditional social structures. Many prefer to live among relatives for economic and political reasons or possess minimal means, such as good transportation, to maintain close contact with their kin. Members of the newly formed middle class who fill the technocratic ranks of Qadhafi's regime often view this continuity as a source of national strength and stability, some having attributed the social and political ills they witnessed as students in the West to a breakdown in family patterns of restraint.

Urbanization admittedly has worked to attenuate individual loyalties among better-educated Libyans toward the formerly monolithic tribal unit. This group in particular recognizes that professional growth derives from personal abilities, not from tribal patronage networks. Nonetheless, their devotion to the family and in many cases the clan remains strong, and they continue to evaluate professional interlocutors partly on the basis of tribal histories that are honed from adolescence. Hiring practices based on merit may have supplanted parochial interests where teaching, engineering, and other critical skills are needed, but, as this discussion intends to show, traditional recruitment methods prevail where regime security is concerned. One can conclude that Libya's leadership has not necessarily tried to shape modern national institutions at the expense of traditional ones.

Source: "Libya's Tribal Landscape," by David Blink. Posted on www16.brinkster.com/blinkdt/landscape.asp.

Review Questions

1. The second and third readings are specifically about Cairo. What are some of the problems faced by North Africa's largest city?
2. From what you have read in this module, how does rural life differ from urban life in North Africa?
3. What is the City of the Dead?
4. What are the two components of Tunisia's plan to improve urban conditions?
5. Why is Tunisia's Berber/Imazighen population "under threat?"

Discussion/Essay Questions

1. Part of Waffa's job was to study the feasibility of creating planned satellite cities for overcrowded cities such as Benghazi, Libya, or Cairo, Egypt. Her report stated that this would just add to the problems in the future. What do you think she meant by that? That is, what problems would only worsen as a result of building satellite cities near large cities? Why do you think she proposed new cities entirely separate from existing metropolitan areas as a better alternative?
2. According to the final reading, Libya's government derives some of its strength from the way in which it incorporates pre-existing tribal and village loyalties and authority structures. Pointing to that reading, Waffa's boss at the consulting firm has told her that she must revise her report so that the final proposal will be more in line with the political goals of the Libyan government. What about Waffa's proposal might have made her boss feel that her proposal was not politically viable? How might Waffa revise her proposal to make it more politically acceptable to the Libyan government?
3. It has been proposed that Cairo might actually have a housing surplus—that the media exaggerate the extent of the housing crisis. Given some of the personal stories in the first Cairo reading (reading number 2), what do you think about this? Is it possible to have surplus housing and still have large numbers of homeless?

List of Readings

1. "Africa Environment Outlook: Past, Present and Future Perspectives," posted on the website of the United Nations Environment Programme, *http://www.unep.org/aeo/207.htm.*
2. Dick Doughty, "Cairo: Inside the Megacity," in *Aramco World Magazine,* March–April 1996, posted on the website of Al Mashriq, *http://almashriq.hiof.no/egypt/900/megacity/.*
3. Mahmoud Kassem, "Tomb Busting," in *Cairo Times,* September 2000, posted on the website of the *Cairo Times, http://www.cairotimes.com/news/cemetary.html.*
4. "Tunisia: World Bank Loan to Help Boost Basic Services, Infrastructure, for Country's Urban Population," December 5, 2002, posted on the website of the World Bank, *http://web.worldbank.org/WBSITE/EXTERNAL/NEWS/0,contentMDK:20079706~menuPK:34463~pagePK:34370~piPK:34424~theSitePK:4607,00.html.*
5. "Who Are the Tuareg?" posted on the website of the Natural History Museum of Los Angeles County as part of the *AFRICA: One Continent, Many Worlds Project* website,
   a. *http://www.nhm.org/africa/tour/desert/015.htm*
   b. *http://www.nhm.org/africa/tour/desert/018.htm*
   c. *http://www.nhm.org/africa/tour/desert/001.htm.*
6. Lucy Jones, "Tunisia's Berbers under Threat," in *Washington Report on Middle East Affairs,* August/September 2001, posted on the website of the *WRMEA, http://www.wrmea.com/archives/august-september01/ 0108033.html.*
7. David Blink, "Libya's Tribal Landscape," posted on the *Qadhafi's Tribal Woes* website, *http://www16.brinkster.com/blinkdt/landscape.asp.*

Websites for Additional Research

1. The Megacities Project is "a transnational non-profit network of community, academic, government, business, and media leaders dedicated to sharing innovative solutions to urban problems." Its website has information about the project, as well as case studies on sixteen of the world's megacities. The project's general website is *http://www.megacitiesproject.org*, and the Cairo case study can be found at *http://www.megacitiesproject.org/network/cairo.asp*.
2. The Muslim Heritage website, at *http://www.muslimheritage.com*, is run by the British-based Foundation for Science, Technology and Civilisation in order to "popularise, disseminate and promote an accurate account of Muslim Heritage and its contribution to present day science, technology and civilisation." The website contains links to numerous articles about Muslim history and culture, including "Introduction to the Islamic City" at *http://www.muslim-heritage.com/features/default.cfm?ArticleID=206*.
3. For current news on Africa (sub-Saharan as well as North Africa), the AllAfrica.com website, at *http://allafrica.com*, is an excellent source. AllAfrica.com compiles, indexes, and archives Africa-related news stories from more than a hundred news organizations around the world, posting more than seven hundred new stories each day.

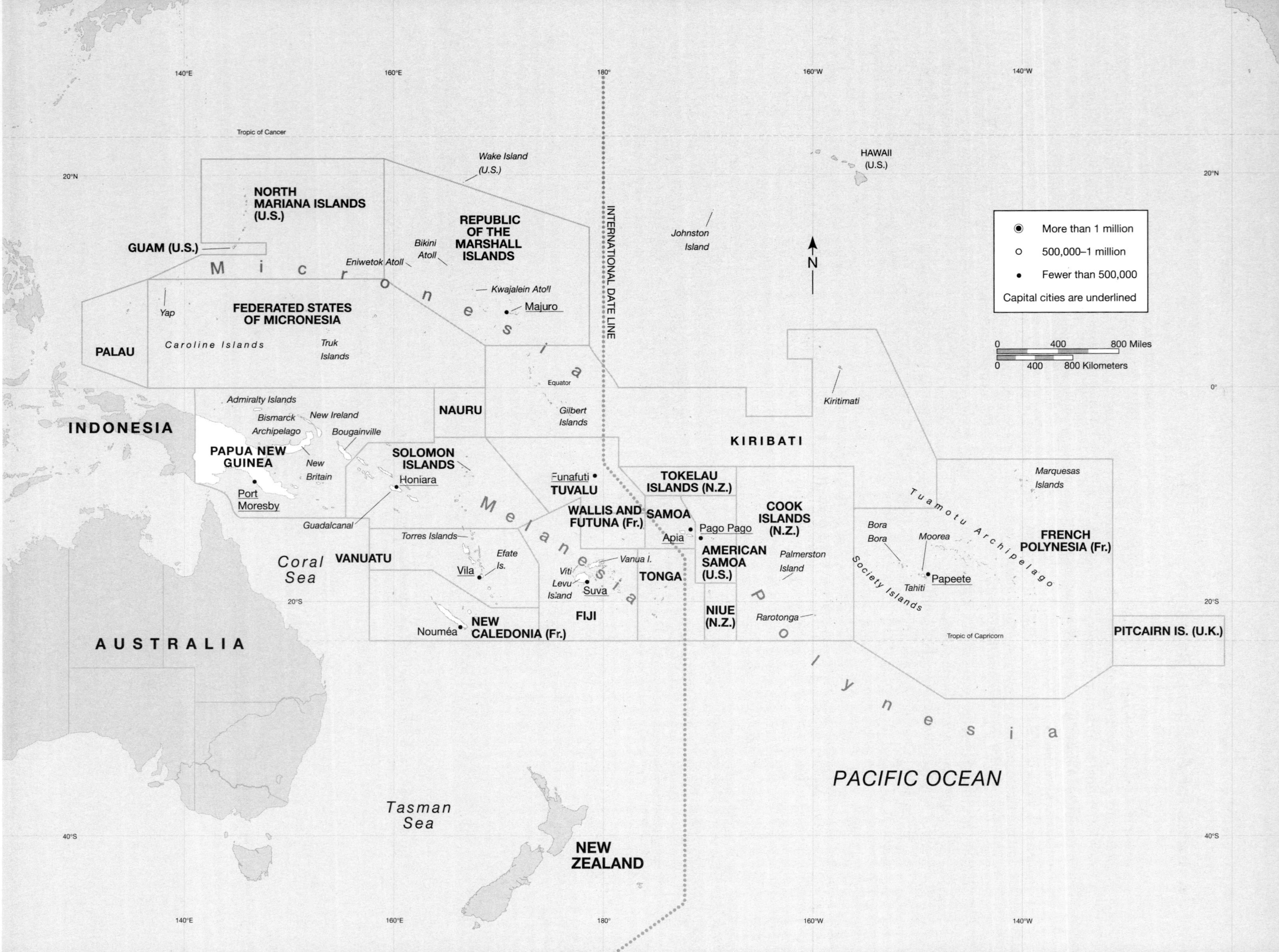

More than 1 million
500,000–1 million
Fewer than 500,000
Capital cities are underlined

0 400 800 Miles
0 400 800 Kilometers

N

140°E
160°E
180°
160°W
140°W
20°N
0°
20°S
40°S

Tropic of Cancer
Equator
Tropic of Capricorn
INTERNATIONAL DATE LINE

HAWAII
(U.S.)
Wake Island
(U.S.)
Johnston
Island
NORTH
MARIANA ISLANDS
(U.S.)
GUAM (U.S.)
REPUBLIC
OF THE
MARSHALL
ISLANDS
Bikini
Atoll
Eniwetok Atoll
Kwajalein Atoll
Majuro
M i c r o n e s i a
FEDERATED STATES
OF MICRONESIA
Yap
Caroline Islands
Truk
Islands
PALAU
Admiralty Islands
Bismarck
Archipelago
New Ireland
Bougainville
INDONESIA
PAPUA NEW
GUINEA
New
Britain
Port
Moresby
NAURU
Gilbert
Islands
KIRIBATI
Kiritimati
SOLOMON
ISLANDS
Honiara
Guadalcanal
Funafuti
TUVALU
M e l a n e s i a
TOKELAU
ISLANDS (N.Z.)
SAMOA
Apia
Pago Pago
AMERICAN
SAMOA
(U.S.)
WALLIS AND
FUTUNA (Fr.)
COOK
ISLANDS
(N.Z.)
Palmerston
Island
Torres Islands
Efate
Is.
Vila
VANUATU
Coral
Sea
Vanua I.
Viti
Levu
Island
Suva
TONGA
FIJI
NIUE
(N.Z.)
Rarotonga
Nouméa
NEW
CALEDONIA (Fr.)
AUSTRALIA
Marquesas
Islands
Tuamotu Archipelago
Bora
Bora
Moorea
Tahiti
Papeete
Society Islands
FRENCH
POLYNESIA (Fr.)
PITCAIRN IS. (U.K.)
P o l y n e s i a
PACIFIC OCEAN
Tasman
Sea
NEW
ZEALAND

Global Warming and Sea-Level Rise in Oceania

Companion to Chapter 14 Resource Issues

The sea is rising around **Ioane Ubaitoi's** island home. As global warming melts the polar ice caps, the rising sea claims sandy beaches, homes and crops in the Republic of Kiribati, and causes havoc with water supplies. Mr. Ubaitoi's Government has sent him to Hamilton [New Zealand] for four months to learn how to save his island from disappearing under the water. Waikato University is hosting a six-month programme sponsored by the South Pacific Regional Environment Programme. It started this week. For the first two months, Mr. Ubaitoi, and 19 other representatives from 10 Pacific countries will learn how to assess the vulnerability of their countries. Then they will go home for two months of field research before returning for a further two months when they will draft proposals to save their homelands.

Ioane Ubaitoi learns techniques for protecting Kiribati's coastline.
Source: Peter Drury/*Waikato Times*

Mr. Ubaitoi is an agriforestry officer in the country of 80,000 people.

He said homeowners had been forced to shift, crops had been destroyed, and beaches had disappeared in the past decade as sea levels started to rise.

Pacific islands are more susceptible to damage from rising sea levels than New Zealand because they are built on coral reefs and are usually less than 3 m [10 ft.] above sea level. The land is also more likely to slump because of its coral base.

Mr. Ubaitoi said sea walls could save the islands—but he hoped by the end of the course he would have an informed opinion on how best to stop the Pacific Ocean washing over Kiribati.

Source: Keri Welham, "Island Man Fights Threat to Sea," *The Waikato Times,* June 19, 1998. ■

Climate Change as a Global Issue

There is mounting evidence that the average temperature on earth is increasing, and it is projected to increase further over the next decades. Much of this climate change is attributed to the greenhouse effect: the increased presence of gases that prevent heat from escaping through the atmosphere, thereby leading to an increase in earth's temperature. The greenhouse effect is not inherently a bad thing. Without it, temperatures on earth would be too cold to support life. The problem is that several of the greenhouse gasses are being produced by humans at record levels, and that is a cause for concern.

The study of global climate change is filled with uncertainties. Even if long-term climate change is occurring (which itself is a matter for debate), scientists are not sure how much of this change is due to an increase in greenhouse gases, because many other natural processes can cause climate change. Scientists debate how much of presently observed change is due to long-term natural processes, how much is due to human-induced factors, and how much of what looks like long-term change is actually "noise" that results from short-term cycles of variation in temperature and precipitation that naturally occur within earth's atmosphere. Further complicating analysis is that the impacts of linked global processes vary from place to place: If current trends continue, some areas will get more rain and some less; some may even become cooler while much of the world becomes warmer.

Nonetheless, most scientists agree that the planet's average temperature is warming at an increasing rate; average temperatures increased by about 1°C during the twentieth century and, at current rates, they will increase by as much as another 5°C during the twenty-first century. Most scientists believe that this climate change is associated with the dramatic increase in carbon dioxide and other greenhouse gases released into the atmosphere during this same period. The increase in these gases, in turn, is associated with increased burning of fossil fuels (the main source of carbon dioxide), production of certain inert gases for industrial use (the main source of fluorocarbons), and destruction of forests that historically have absorbed some of the carbon dioxide that now resides in the atmosphere.

While climate change in most parts of the world will lead to warmer temperatures, the indirect impacts likely will be more complex, and potentially more catastrophic. Global precipitation patterns are likely to tend toward extremes; in general wet areas will become wetter and dry areas will become drier. Severe storms are also likely to increase. These changes will wreak havoc on agricultural regions as species that have adapted to local conditions over the course of thousands of years (or have been nurtured by area inhabitants to meet local conditions) suddenly become incompatible with the local environment. Huge numbers of species of plants and animals will be threatened with extinction as they lose their habitats. The potential costs of these losses are unknown, as it is impossible to place a value on the loss of a species that may be of medical use, or that may provide some other as-yet-unimagined benefit to humanity. Some would argue that the very loss of a species is a terrible loss in itself, aside from the negative impacts that its extinction might have on potential human uses.

Among the various aspects of climate change, the greatest attention has been directed to global warming, and among the various impacts of global warming probably the greatest attention has been directed to the problem of rising sea level. Strong evidence shows that increased temperature already is leading to the melting of polar ice caps. This eventually may lead to a rise in sea level, which would pose an obvious flooding danger to low-lying coastal zones. Increased severity of storms, changes in fishery dynamics due to raised ocean temperatures and changed patterns of currents, and increased saltwater intrusion into coastal aquifers (when these exist as sources of groundwater) are additional ways in which global climate change would have a disproportionate impact on coastal zones. Because cultures have developed in coastal zones that depend on local, coastal ecologies, changes in coastal geophysics and biology would also impact coastal social systems.

Coastal zones are receiving the bulk of attention from climate change researchers because, besides being particularly vulnerable, a large portion of the world's population lives near the coast. Even in large countries with extensive interiors, people disproportionately live in coastal zones. In the forty-eight contiguous states of the United States, for instance, the 673 coastal counties constitute 17 percent of the nation's land area but host 53 percent of the population. In addition, fourteen of the nation's twenty largest cities and seventeen of its twenty fastest-growing counties are located along the coast.

Climate Change Hazards and Pacific Islands

Among the coastal areas vulnerable to climate change, perhaps the most endangered are the islands of the Pacific Ocean. Islands in the Pacific are of two types:

- *Volcanic islands*. These are the tops of volcanoes that have risen from the ocean floor. Examples of volcanic islands include the Hawaiian islands. There, the Pacific plate is moving over a "hot spot" in earth's mantle that allows magma (molten rock) to well up and accumulate over centuries of eruptions and flows. These accumulations form seamounts (underwater mountains), and, if the process continues long enough, the top of the seamount rises above the ocean's surface, forming a volcanic island. Just southeast of Hawaii's

"Big Island" (the southeasternmost of the Hawaiian islands) is a prominent seamount whose top is about 1 kilometer (0.6 miles) underwater. This seamount is expected to surface, forming a new island in the Hawaiian chain, in about 10,000 to 20,000 years. Other islands in the Pacific are volcanoes formed as one plate collides with another plate and sinks under it, also causing magma to rise to the surface.

- *Atolls*. At the heart of the atoll is the coral reef. Coral is one of the oldest types of living systems on earth. It supports some of the planet's most diverse and most fragile ecosystems, sometimes hosting as many as three thousand species of fish. Coral, found in shallow, tropical oceans, requires ocean temperatures between 23 and 25°C to stay alive. Atolls have their origins in the rings of coral reef that build on the outside of islands that are usually volcanic in origin. Once the volcano becomes dormant, the island inside the reef begins to be worn down by erosion or it sinks back into the ocean (a process known as subsidence). The coral grows fast enough to keep up with the subsidence, and thus a shallow ring of dead coral remains, with only a lagoon—no island—in the middle and surrounded by a submerged reef of living coral.

Pacific islands always have been precarious environments for human (and other living) systems. Atolls, in particular, are nearly always small, isolated, and devoid of most of the resources generally considered necessary for survival. The soil is typically thin, freshwater sources are scarce or nonexistent, populations are tiny, tropical cyclones are severe, and distances to major markets or sources of raw materials frequently are enormous. Along with coconut trees and root crops that sometimes grow in the sandy soil, the main source of food is from the fish that populate the adjacent reef. The islands are blessed with idyllic weather (when there is not a storm), but most lack the resources to provide the amenities demanded by all but the most adventurous tourist, and extreme distance from major population centers also hampers their tourism potential. Some islands have turned their isolation into a resource by serving as sites for weapons testing and toxic waste disposal, although the long-term benefits of this resource use are questionable. Bikini Atoll, where the U.S. government tested nuclear weapons in the 1950s, remains partially uninhabitable, as coconuts—a major source of food—still contain residual radioactivity, accumulated from the soil.

Amid the precariousness of the island ecosystem, emigration has been a frequent fact of life. The people of the Pacific, while isolated on their tiny islands, remain connected across the ocean that—almost as much as the land—is their home. Many of the largest Pacific island cities are swelling with populations from outer areas of their island groups. Immigrants from smaller island groups flock to the larger islands, often crossing national borders in the process. Beyond the world of Pacific islands, large Pacific Islander communities have formed in mainland cities around the Pacific Rim, especially in Australia, New Zealand, and the West Coast of the United States. Indeed, funds sent back to small islands from emigrants (whether they are emigrants to capital cities on a country's major island or emigrants to larger cities around the Pacific Rim) often constitute a major portion of local income. Pacific Islanders today debate whether this current trend of migration to distant cities represents a threat to the many distinct, isolated, atoll-based societies of the region or is simply a continuation of the maritime cosmopolitanism that always has characterized the area.

Whether or not the ocean, as opposed to the tiny land masses, is thought of as the true "home" of Pacific Islanders, the people of the Pacific require islands for survival, and these fragile environments face a host of threats from global climate change. The most obvious and direct threat is that of sea-level rise, as the overflow from melting ice caps threatens to swamp low-lying atolls. Tuvalu, for instance, is an independent country made up of nine atolls in the Pacific Ocean, about halfway between Hawaii and Australia, with a population of 11,146 and a total land area of 26 square kilometers (9 square miles). Because the highest point in Tuvalu is just 5 meters (16 feet) above sea level, the country is at risk of being washed off the map. In fact, in 2000 the government of Tuvalu appealed to Australia and New Zealand to take in Tuvaluans if rising sea levels made evacuation necessary.

Threats from global climate change, however, go beyond the fear of being swamped by rising sea levels. Sustenance on these islands typically rests on extraction of resources from fisheries that, in turn, are sustained by the coral reefs that surround almost all tropical islands. These reefs, however, are themselves endangered by a host of forces associated with climate change. Rising water temperatures have led to a process called bleaching, which kills off the coral. Increased storm activity, again at least partly a result of global climate change, also endangers reefs. More intense rain can lead to heightened erosion, which, in areas where agricultural chemicals are used, can cause plumes of chemical-rich soil to be washed out to sea, killing adjacent reefs. On land, as well, changes in the precipitation pattern tending toward extreme drought followed by extreme storms endangers the population, since people on these islands often depend on collected rainwater as their sole freshwater source.

A Global or Local Problem?

How one views nature and interacts with it depends on a number of factors. A subsistence farmer in the Amazon may interact with nature differently than an executive in

Manhattan, and an animist may view nature differently than a Southern Baptist. Nature may be seen as wrapped up in one's everyday life, a set of resources to be exploited as needed, or a sacred pristine world to be revered and preserved. Not only does one's view of nature impact how one perceives environmental problems and designs solutions; it even affects whether one sees the transformation of nature, or hardships caused by nature, as "problems" at all.

Beliefs are also a product of the social context from which they develop. During the expansion of the United States westward, environmental policy was guided by the Judeo-Christian belief that humans are masters over the dominion of earth, and that they should use earth's resources to their benefit. A different set of views about nature, also guided by Judeo-Christian beliefs, formed the basis for the dissenting conservationist movement. According to this view, humans are "stewards" or caretakers of earth, and humans have a duty to care for the planet and its creatures to ensure the future of God's creation. The preservationist movement began in the late nineteenth century, inspired by the writings of Emerson, Thoreau, and John Muir, who founded the Sierra Club. Preservationists believe that earth should be kept in as pristine a condition as possible and that its resources should not be disturbed by humans. Other views on the environment include ecofeminism and deep ecology. Ecofeminists stress the link between the structuring of gender relations in society (wherein women typically are subordinated to men) and the way in which that masculinist society structures the domination of nature. Finally, deep ecologists believe that humans are simply part of earth—no different from plants or animals—and therefore humans have no inherent hierarchical relationship with earth or its creatures. According to deep ecologists, all living things share the same rights of existence.

Turning to climate change, one's position on climate change and what one thinks should be done about it greatly depends not only on one's attitude toward nature, but also on the scale at which one perceives the problem. It is by no means clear whether climate change is a local, national, regional, or global problem (or, as some would assert, whether it is a problem at all). At one level, climate change is a very local issue. Climate-changing pollutants are emitted locally and, especially because the impacts of climate change vary so much from place to place, they are experienced locally as well. Contrasting this perspective, others argue that, while climate change may be both generated and experienced locally, the sources of climate change (large, industrialized mainland regions) typically are distant from those that experience its greatest brunt (small, agricultural islands). In addition, even though the specific impacts of climate change differ from place to place, the overall experience of climate change is global. Therefore, it is often argued, climate change must be considered a global problem, to be dealt with by the world community.

Still others take a middle ground, arguing for a regional or national approach. Our society has been divided into sovereign nation-states, and only nation-state governments have the authority to mandate changes in pollutant emissions. In fact, even when a "global" treaty is signed, the signatories are actually nation-state governments who each agree to change their national laws so as to reflect the new, international mandate. Therefore, some argue, true change in environmental policy can be implemented only at the national level.

Finally, others argue that while the global scale is perhaps too crude for mandating change in environmental behavior, national borders are too artificial for organizing environmental policy, given that most environmental problems (such as airborne pollutants) easily cross national borders within a given region. Perhaps, then, a regional approach would be most effective for coping with issues like climate change.

Each of these approaches has been advocated by one participant or another in the climate change debate, and each is represented in the readings that follow. Probably the dominant voice in debates over global climate change belongs to those who say that global climate change is just that—global—and that a solution must be constructed at the global scale. This is the overall thrust of the 1997 Kyoto Protocol, an agreement sponsored by the United Nations whereby countries agreed to reduce their emissions of greenhouse gases by specified percentages by 2012. The first reading is a statement issued in 2001 by many of the world's national academies of science, in which the signatories argue that climate change is real and global, and that efforts to stop it must be taken on the global level, beginning with the actions mandated by the Kyoto Protocol. This position is opposed in the second reading, a 2001 public letter from President Bush to four U.S. senators who had asked him to clarify his position on the Kyoto Protocol (shortly after releasing this letter, President Bush issued a statement formally announcing that any U.S. efforts to forestall climate change would occur outside the Kyoto framework because the Kyoto framework was "fatally flawed"). While President Bush expresses questions about the science of climate change and Kyoto's exemptions for developing countries, he also takes issue with the way in which the Kyoto Protocol labels climate change as a *global* concern. In his letter, President Bush considers efforts to forestall climate change in the context of existing U.S. laws, economic needs, favored regulatory mechanisms, and energy shortages. For President Bush, the problem may, in part, be global, but the solution must be national.

This position is directly contradicted in the third reading, from the environmental group Greenpeace. Greenpeace asserts that climate change is indeed a global issue demanding global solutions, and that "the world cannot wait" for the United States. The final reading on the Kyoto Protocol, from the British newspaper *The Guardian*, sums up the debate between the United States and Europe on reducing greenhouse gas emissions (and also provides a concise overview of how the Kyoto Protocol would work, albeit with a definite pro-Kyoto slant).

With the fifth reading, the perspective becomes more local, as Janita Pahalad, a climatologist from the Pacific island nation of Fiji, offers her perspective on climate change and sea-level rise. Although Pahalad experiences climate change as a local problem, she asserts that change must occur globally, since Fiji and other Pacific island states generate only a tiny percentage of greenhouse gases. In the next reading, Greenpeace Pacific tries another strategy, arguing that climate change is a *regional* concern and that Australia and New Zealand, as part of the Pacific region, should join with smaller and less powerful Pacific island countries in supporting efforts to limit climate change.

Contrasting both of these perspectives, the seventh reading is from the EnviroTruth.org website, a project of the conservative National Center for Public Policy Research, based in Washington, D.C. Elsewhere on its website, the NCPPR challenges the scientific findings that are cited by most advocates of reducing greenhouse gas emissions. The NCPPR asserts that (a) the global climate is not warming; (b) to the extent that the global climate is warming, it is not due to increased levels of greenhouse gases; and (c) any increased levels of greenhouse gases are not due to human action. In the section of its website reprinted here, the NCPPR argues that even if polar ice caps were to melt due to global warming, it would not lead to a rise in sea level that would threaten islands or coastal regions (and, according to the NCPPR, global warming is not happening anyway). Therefore, the NCPPR concludes, any problems of erosion or subsidence on small islands must be due not to global climate change but to local land-use practices.

The eighth reading, a news release from 2000, announces a talk to be given by Hawaiian researcher Eileen Shea. According to this news release, Shea stresses that, among the various aspects of global climate change, atmospheric warming is among those *least* likely to have a significant impact on Pacific islands. Therefore, she argues, attention should be directed toward other hazards, whose problems can be met through local strategies. Thus, like the NCPPR, Shea focuses on local solutions.

The next reading, an abstract of a paper presented by New Zealand researcher John Hay at a conference held in Samoa in 2000, agrees with Shea that attention should be redirected away from global warming and sea-level rise, and toward other, more immediately threatening impacts of climate change. At the same time, however, Hay agrees with Pahalad that the cause of these local problems is *global* climate change, and that therefore change must be undertaken at the global scale.

In the last reading, the island nation of Tuvalu makes its case that sea-level rise is an extremely pressing *local* problem but that it requires a regional solution. In a broadcast on ABC Radio Australia, several Tuvaluan islanders urge Australia to change its immigration policy so that residents of Tuvalu will have an escape plan should their island nation cease to exist.

Readings

from Sixteen National Academies of Science

1

The Science of Climate Change

17 May 2001

A joint statement issued by the Australian Academy of Sciences, Royal Flemish Academy of Belgium for Sciences and the Arts, Brazilian Academy of Sciences, Royal Society of Canada, Caribbean Academy of Sciences, Chinese Academy of Sciences, French Academy of Sciences, German Academy of Natural Scientists Leopoldina, Indian National Science Academy, Indonesian Academy of Sciences, Royal Irish Academy, Accademia Nazionale dei Lincei (Italy), Academy of Sciences Malaysia, Academy Council of the Royal Society of New Zealand, Royal Swedish Academy of Sciences, and Royal Society (UK).

The work of the Intergovernmental Panel on Climate Change (IPCC) represents the consensus of the international scientific community on climate change science. We recognise IPCC as the world's most reliable source of information on climate change and its causes, and we endorse its method of achieving this consensus. Despite increasing consensus on the science underpinning predictions of global climate change, doubts have been expressed recently about the need to mitigate the risks posed by global climate change. We do not consider such doubts justified.

There will always be some uncertainty surrounding the prediction of changes in such a complex system as the world's climate. Nevertheless, we support the IPCC's conclusion that it is at least 90% certain that temperatures will continue to rise, with average global surface temperature projected to increase by between 1.4 and 5.8 C above 1990 levels by 2100. This increase will be accompanied by rising sea levels, more intense precipitation events in some countries, increased risk of drought in others, and adverse effects on agriculture, health and water resources.

In May 2000, at the InterAcademy Panel (IAP) meeting in Tokyo, 63 academies of science from all parts of the world issued a statement on sustainability in which they noted that "global trends in climate change ... are growing concerns" and pledged themselves to work for sustainability—meeting current human needs while preserving the environment and natural resources needed by future generations. It is now evident that human activities are already contributing adversely to global climate change. Business as usual is no longer a viable option.

We urge everyone—individuals, businesses and governments—to take prompt action to reduce emissions of greenhouse gases. One hundred and eighty-one governments are Parties to the 1992 UN Framework Convention on Climate Change, demonstrating a global commitment to "stabilising atmospheric concentrations of greenhouse gases at safe levels." Eighty-four countries have signed the subsequent 1997 Kyoto Protocol, committing developed countries to reducing their annual aggregate emissions by 5.2% from 1990 levels by 2008–2012.

The ratification of this Protocol represents a small but essential first step towards stabilising atmospheric concentrations of greenhouse gases. It will help create a base on which to build an equitable agreement between all countries in the developed and developing worlds for the more substantial reductions that will be necessary by the middle of the century.

There is much that can be done now to reduce the emissions of greenhouse gases without excessive cost. We believe that there is also a need for a major co-ordinated research effort focusing on the science and technology that underpin mitigation and adaptation strategies related to climate change. This effort should be funded principally by the developed countries and should involve scientists from throughout the world.

The balance of the scientific evidence demands effective steps now to avert damaging changes to the earth's climate.

Source: Greenpeace International, 2001.

2

from George W. Bush

Letter from the President

For Immediate Release
Office of the Press Secretary
March 13, 2001

Thank you for your letter of March 6, 2001, asking for the Administration's views on global climate change, in particular the Kyoto Protocol and efforts to regulate carbon dioxide under the Clean Air Act. My Administration takes the issue of global climate change very seriously.

As you know, I oppose the Kyoto Protocol because it exempts 80 percent of the world, including major population centers such as China and India, from compliance, and would cause serious harm to the U.S. economy. The Senate's vote, 95–0, shows that there is a clear consensus that the Kyoto Protocol is an unfair and ineffective means of addressing global climate change concerns.

As you also know, I support a comprehensive and balanced national energy policy that takes into account the importance of improving air quality. Consistent with this balanced approach, I intend to work with the Congress on a multipollutant strategy to require power plants to reduce emissions of sulfur dioxide, nitrogen oxides, and mercury. Any such strategy would include phasing in reductions over a reasonable period of time, providing regulatory certainty, and offering market-based incentives to help industry meet the targets. I do not believe, however, that the government should impose on power plants mandatory emissions reductions for carbon dioxide, which is not a "pollutant" under the Clean Air Act.

A recently released Department of Energy Report, "Analysis of Strategies for Reducing Multiple Emissions from Power Plants," concluded that including caps on carbon dioxide emissions as part of a multiple emissions strategy would lead to an even more dramatic shift from coal to natural gas for electric power generation and significantly higher electricity prices compared to scenarios in which only sulfur dioxide and nitrogen oxides were reduced.

This is important new information that warrants a reevaluation, especially at a time of rising energy prices and a serious energy shortage. Coal generates more than half of America's electricity supply. At a time when California has already experienced energy shortages, and other Western states are worried about price and availability of energy this summer, we must be very careful not to take actions that could harm consumers. This is especially true given the incomplete state of scientific knowledge of the causes of, and solutions to, global climate change and the lack of commercially available technologies for removing and storing carbon dioxide.

Consistent with these concerns, we will continue to fully examine global climate change issues—including the science, technologies, market-based systems, and innovative options for addressing concentrations of greenhouse gases in the atmosphere. I am very optimistic that, with the proper focus and working with our friends and allies, we will be able to develop technologies, market incentives, and other creative ways to address global climate change.

I look forward to working with you and others to address global climate change issues in the context of a national energy policy that protects our environment, consumers, and economy.

Sincerely,
George W. Bush

from Greenpeace International 3

The Climate Cannot Wait for Bush

But If Bush Doesn't Change, the Climate Will

7 June 2001

A World-Wide Storm of Protest

President George W. Bush's announcement in late March that the United States was abandoning the Kyoto Protocol was met by a storm of protest, both in the US and internationally. Governments, scientists, religious leaders, labour and other public figures, as well as environmental organisations, condemned the move. The US is seen as abandoning its moral, political and legal responsibility to work internationally to address the most pressing international environmental problem of the 21st century: global climate change.

President Bush's upcoming visit to Europe threatens to be marked by outrage at the rejection by the world's worst greenhouse gas polluter of the last 12 years of international climate negotiations.

No Mandate to Wreck the Climate

Greenpeace believes that the Bush administration's isolationist policy will ultimately fail, both domestically and internationally. The recent defection of Senator James Jeffords of Vermont indicates the breadth of opposition to Bush's rejection of Kyoto, his energy policy and the rest of his hard core right wing agenda, even from moderates within his own party. George Bush does not have a mandate from the American people or the Congress to wreck the international climate negotiations. US public opinion and the US Congress are moving inexorably in the right direction. The White House will follow eventually.

Ratify the Climate Treaty With or Without the US

While this right wing drama plays out in Washington, the rest of the world must not be distracted from combating climate change, and the first step is the ratification and entry into force of the Kyoto Protocol.

US Alternative No Alternative

The United States' "alternative," if it ever appears, is very likely to be strong on rhetoric, but very weak on targets and timetables for reducing greenhouse gas emissions, and will try to postpone the hard choices to a time in the future when they will no doubt be much harder and more expensive to take and perhaps to a time when it is too late to reverse the damage that we are doing to the world's climate system.

EU Must Lead Ratification and Implementation of the Climate Treaty

Greenpeace urges the European Union to stand firm in the face of Bush's posturing, and to recognise that the majority of the American people support international action to protect the climate. Europe must show real leadership and fulfil its promise to its own people to ratify the Kyoto Protocol, which must enter into force in time for the Rio + 10 Summit in Johannesburg in September of 2002. Failure to do so will be met with the harshest criticism from the vast majority of Europeans who want to get on with the business of preventing dangerous climate change.

The EU must go on to implement the climate treaty in full, developing the next steps within the convention for further and deeper cuts in greenhouse gases, while waiting for signs that sanity is returning to Washington and a time when the US can be welcomed back into the process.

Waiting for Bush Not an Option

The Kyoto Protocol does not go far enough, it is true, but it was watered down to its present text largely as a result of US demands and corporate intervention. The EU and the rest of the world cannot wait until the political climate in Washington improves, or expect some miraculous "alternative" from Washington. It will not come while the current administration lasts. Waiting for Bush is not an option.

Source: Greenpeace International, 2001.

from *The Guardian*

4

The Heat Is On for a Solution in Bonn

As Crucial Climate Change Talks Open, We Examine the High Cost of Inaction

Paul Brown, Environment Correspondent
Saturday July 14, 2001

What Is the Climate Change Convention?

An agreement made by more than 150 countries at the Earth Summit in Rio in 1992 to limit man-made emissions of greenhouse gases to stop the atmosphere overheating.

What Are Greenhouse Gases?

The main ones are carbon dioxide, methane, and nitrous oxide. They prevent the reflected heat of the sun's rays escaping back into space, like the glass in a greenhouse.

Are Scientists Certain About This?

Scientists agree that global warming is taking place and the vast majority believe it is man made. Burning fossil fuels releases carbon dioxide. Intensive agriculture and rubbish tips release methane. The only uncertainty is the scale of the process and whether we can adapt to it.

How Quickly Is the World Warming?

An average of up to 5°C in 100 years, but more in some areas, notably in the arctic.

How Soon Will We Know?

The weather in the UK is already different, but in the arctic some effects are dramatic: some species, such as the polar bear, face extinction owing to melting ice.

What Is the Kyoto Protocol?

An addition to the Rio convention, first agreed in 1997, to give all developed countries legally binding targets for cuts in emissions from the 1990 levels by 2008–12.

The EU agreed an overall target of 8%, Japan 7%, and 6% for the US.

Why Were the Targets Different?

Some countries found it easier to make cuts than others. The UK had already started the switch to natural gas, Germany had closed many heavy industries, and Japan was already energy efficient. The US found it difficult because of an economic boom in the 90s.

The developed world was to cut its emissions by 5.2%, and it was hoped that developing countries would join in later.

How Can We Keep to the Kyoto Targets without Cutting Domestic Emissions?

There are three ways. Countries can plant forests to absorb and lock in carbon, or change agricultural practices to cut carbon emissions, such as not ploughing, or keeping fewer farm animals which produce methane. They can install clean technology in other countries and claim carbon credits for themselves. They can buy carbon credits from countries such as Russia, where heavy industry has collapsed and national carbon limits are underused.

But Is That Enough to Solve the Problem?

Nowhere near. There is already enough additional greenhouse gas in the atmosphere to alter the climate, but we can stave off the worst if we cut man-made greenhouse gas production by 60% to 80% as soon as possible. The temperature will then stabilise at 5°C higher than now.

Kyoto was meant to be only the beginning, leading to steeper targets by 2020. We need to have cracked the problem by 2050 to avert disaster.

So What Went Wrong?

The rules for how greenhouse gas emissions are measured and how they can be cut were not finalised in Kyoto. It was not agreed to what extent we could rely on planting forests and carbon trading.

Years of wrangling ended in angry exchanges at the Hague in November, and things have gone from bad to worse since then. President George Bush repudiated the protocol, fearing that cutting the use of fossil fuel would damage the US economy.

A former oil man himself, he has been persuaded by the oil industry to dump the Kyoto deal because it will hurt profits and cost jobs.

What Can We Do Now?

The rest of the world could proceed without the US. Most of the EU wants to go it alone and keep to the targets agreed in Kyoto, but the UK and Japan are reluctant without the US. Australia and Canada are against.

Why Is There Such Reluctance?

The US emits a quarter of the world's greenhouse gases and unless this is cut, the efforts of others will not make much difference. It will also be impossible to make developing countries such as China take the problem seriously.

Could the EU Go It Alone, Show the Lead and Then Put Pressure on Mr. Bush?

Yes, but it needs partners. Under the Kyoto rules, 55 countries must ratify the protocol—making it law in their own countries—to make it legally binding across the world.

A second condition is that they must include enough developed countries to make up 55% of total emissions in 1990, from when all targets are calculated.

How Do the Figures Add Up?

The US alone was responsible for 36% of developed world emissions in 1990, so all of the EU, eastern Europe, Japan and Russia are needed to reach 55%.

How Do the Alliances Work?

The EU has persuaded Russia and eastern Europe to ratify but Japan is wavering. There is confusion about the next talks in Bonn on Monday.

So What Will Happen?

The US will continue to destroy the Kyoto deal and suggest new talks. Others will try to make Kyoto easier for the US to accept.

The EU and its allies may forge ahead and ratify it, and challenge Japan to follow suit. Or the talks could collapse.

Who Are Winners and the Losers?

The world's poor countries, and poor people who cannot adapt, will suffer first. There will be flooding, drought and famine. There will be millions of environmental refugees in Africa and Asia. Some northern countries gain marginally from a longer growing season in a warmer climate but the gains will not outweigh the losses.

What Is the Worse-Case Scenario?

Huge tracts of productive land will become submerged, including major cities. Large migrations of people. More natural disasters, triggering a collapse of the insurance market, and a global crash as the world economy collapses.

What Is the Best-Case Scenario?

That man's ingenuity and technology comes to the rescue with hydrogen and solar power replacing fossil fuels to run transport and create electricity.

Does Climate Require Us to Fundamentally Change Our Lifestyles?

Not a lot in the electric hi-tech age. We need to cut fuel consumption, stop flying flowers and vegetables round the world when they can be produced locally, and recycle goods. These changes can be achieved without damaging lifestyles.

If Bonn Collapses, Will It Be a Disaster?

Yes, in the diplomatic sense, and environmentally, too. So far no one has come up with a credible alternative.

Source: Paul Brown, "The Heat Is On for a Solution in Bonn," *The Guardian,* July 14, 2001.

from the South Pacific Regional Environment Programme

5

Climate Change and Sea-Level Rise

by Janita Pahalad
July 1998

Today, climate change and sea-level rise have become household topics, especially in the Pacific region. People are intrigued by these issues, but very few understand what all the fuss is about. Some wish to learn more but few are willing to combat the effects of such phenomena. Climate change and sea-level rise are so interrelated that one cannot talk about sea-level rise without explaining the reason for it. Sea level is a measurable quantity and it can be generally defined as the results of all influences such as daily tides, meteorological, oceanographical and geological effects. For example, climate change and the movement of the earth's crust can change the sea levels significantly.

Many believe that climate change (or global warming, as it is commonly referred to) is mainly due to our desire to progress. Generally, there is a theory in science that climate change and sea-level variation are natural phenomena that occur approximately every ten thousand years or so and something which cannot be avoided. However, our excessive contribution of greenhouse gases into the atmosphere (which did not previously take place) is one of the major factors contributing to global warming. Scientists claim that by the year 2100, expected global temperatures may rise by 1–4° C and the subsequent sea-level rise may be approximately 50 centimetres, although this sea-level rise could be higher in the Pacific region. Research indicates that in the Pacific, temperature has been rising 0.1° C per decade and that sea level has been rising by 2 millimetres per year. Recent data compiled from the NTF's 11 tide gauges in the Pacific show an accelerated sea-level rise of up to 25mm/yr—more than 10 times the global trend this century. This is thought to be related to El Niño-Southern Oscillation (ENSO) variations.

There are many low-lying atolls in the Pacific. Generally, these islands are small and the surrounding waters play a major role in their existence. A 50-centimetre sea-level rise may take away a few kilometres of coastal area from a large island, but it may completely submerge a small island country in the Pacific region. Long before that stage is reached, there may be greater loss of lives and infrastructure due to the enhanced impact of natural disasters such as tropical cyclones, storm surges, floods, tsunamis and so on. The economy of most of the Pacific island countries (PICs) is greatly dependent upon fisheries, agriculture, tourism and overseas aid, and most PICs are struggling to make ends meet.

Leaders of PICs are fully aware of sea-level rise and coastal erosion problems and they are badly in need of applicable advice on how to address the problems. Public awareness of the situation is also important, and this should be conducted in local languages. However, if we are told how to safeguard our coastlines or to reduce the emission of greenhouse gases, the question arises: who should pay the cost? If recent global warming is a man-made problem, PICs are micro-contributors of greenhouse gases yet they are likely to be affected most. Some larger nations have blatantly shown their lack of concern on these issues and they seem to believe that their economy and well-being are much more important than the survival of the people from PICs. This sounds inhumane to us. How can we make our voice heard? Are we over-reacting? One thing is for sure: we are vulnerable.

If climate change and sea-level variation have been natural phenomena in our planet, as suggested by the geological records, do we still need to do anything to protect our future generations? Obviously, this is a difficult question. When and how can we tell with some certainty if we are in danger of losing part or all [of] our homeland? In the mean time, we have to focus upon capacity building for the Pacific community on these issues so we can catch up with current scientific information.

Note: The author, Miss Janita Pahalad, is a Senior Climatologist working at the Fiji Meteorological Services in Nadi, Fiji who visited NTF for three weeks in October 1997 to participate in the Short Term Attachment Workshop, Round III.

Source: Janita Pahalad, "Climate Change and Sea-Level Rise: A Personal View from Fiji," *The South Pacific Sea Level and Climate Change Newsletter*, July 1998.

6

from Greenpeace Pacific

Sea-Level Rise a Big Problem for Tuvalu, Prime Minister Says

SV Rainbow Warrior, Tuesday July 22nd 1997—Sea-level rise and climate chaos caused by global warming are urgent and critical issues for Tuvalu and other low-lying countries, the Tuvalu Prime Minister Rt Hon Bikenibeu Paeniu told Greenpeace yesterday.

Mr. Paeniu made these comments while visiting the SV Rainbow Warrior, the first Greenpeace vessel to ever visit Tuvalu, yesterday (Monday July 21st).

Tuvalu experienced a freak cyclone last month which devastated an outer island. A string of coral atolls no more than two metres above sea level, Tuvalu could be annihilated by the sea-level rise which scientists predict will occur if nothing is done to slow global warming. Greenpeace campaigner Stephanie Mills, on board the Rainbow Warrior, said it was time for the developed countries, who were predominantly responsible for burning the fossil fuels that contribute to global warming, to listen to the voice of Tuvalu.

Greenpeace has been highly critical of Australia and New Zealand's position on climate change. In spite of rhetorical commitments to take action to reduce their greenhouse gas emissions, Australia has recently said it will only agree to an international convention limiting climate change if Australia is allowed to INCREASE its emissions. New Zealand has refused to commit to any target for emissions reductions and is currently increasing its carbon dioxide emissions.

"It is shameful that countries like Australia and New Zealand, which claim to be part of the South Pacific, take no responsibility to reduce their impact on the climate when their own neighbours are at risk," she said. "Australia is lobbying to be declared a special case because of its heavy dependence on the coal industry. But it is Tuvalu, Kiribati and other low-lying coral atolls that should receive the special attention of the rest of the international community, because while contributing little to the climate problem, they are the first to suffer."

Mr. Paeniu also discussed the issue of shipments of high-level nuclear waste and plutonium through the Pacific with Greenpeace representatives. Several shipments of nuclear waste from Japan to France are expected to transit the Pacific every year over the next decade. Greenpeace is campaigning for a regional ban on the shipments and for an end to the international plutonium trade.

The Rainbow Warrior was visiting Tuvalu as part of a Pacific-wide education and information tour against nuclear waste shipments and for environmental protection. It will next call at Honiara in the Solomon Islands (July 29th), then Papua New Guinea, Fiji and the Cook Islands.

Source: "Sea-Level Rise a Big Problem for Tuvalu, Prime Minister Says," Greenpeace Pacific, July 22, 1997.

from the National Center for Public Policy Research

7

Myth #6: Sea Level Is Rising Quickly and It Will Get Worse If the Polar Ice Caps Melt Due to Global Warming

Coastal Settlements and Low-Lying Islands Will Be Submerged

The Envirotruth: Sea level has been rising naturally since the end of the last ice age and this has not accelerated recently. The total rise has been over 120 metres and is still proceeding at a rate of about 18 cm per century. We don't see an increase in this rate during the strong warming that took place between 1900 and 1940 nor did the rate decrease when the climate cooled between 1940 and 1975.

According to Dr. Fred Singer, President of The Science & Environmental Policy Project, Distinguished Research Professor at George Mason University and Professor Emeritus of environmental science at the University of Virginia, ongoing sea level rise is due to the slow melting of Antarctic ice sheets that have been gradually disappearing for about 18,000 years, the date of the last glacial maximum. As far as we can tell from geological data, only temperature variations on a millennial time scale can affect this rate. Climate fluctuations lasting decades or even centuries are too short to affect this rate of melting appreciably. Our best estimate is that these ice sheets will continue to melt for another 5,000 to 7,000 years until they disappear. So unless another ice age commences in the meantime, sea level is bound to keep on rising and there is probably nothing that humans can do about this.

It is also important to understand that, just as the melting of ice cubes in a glass of water does not cause the glass to overflow, the melting of polar sea ice will not result in ocean level changes. Only if massive quantities of inland Antarctic and Greenland glaciers melted would sea levels raise enough to submerge coastal settlements. Dr. Patterson and University of Hawaii Professor of Earth Science Dr. Charles Fletcher maintain that this did not happen 5,500 years ago, when the Earth was three degrees warmer. They also explain that sea level was only two meters higher 120,000 years ago, when temperatures were almost six degrees warmer than now.

Ordinarily, small island-nations like the Maldives and Barbados are not threatened by such a rise. This is because these countries are built entirely on coral and coral fragments. This coral is continually, and quickly, growing upward and, unless something very bad happens to the natural environment in a region, no sea level rise is fast enough to get ahead of coral growth. The Maldivian reefs have been coping with increasing sea level for the past few thousand years and were even able to keep up when the ocean was rising ten times faster than it is now, 10,000 years ago.

Oceanographer Klaus Schwarzer of Christian Albrechts University in Germany explains that today's problems in the Maldives are caused by two factors—local pollution that is killing the reefs (as is the case in Barbados) and inappropriate construction projects. Barriers built out into the ocean to stop the drift of sediment away from the coast are disrupting the circulation of nutrient rich water to the reefs and killing them.

As a result, the Maldives islands are sinking. This has nothing to do with climate change and is the fault of the Maldivian government, which selected a barrier design maladapted for a coral atoll (it was designed for the rock-based Mediterranean Sea coast). Yet, Ismail Shafeeu, the Maldives' Minister of the Environment, still complains, "In the next hundred years or so, what the rest of the world does is going to determine whether we are going to be around or not. We need commitment on the part of people living in countries that are causing this problem. If these countries and the people living in these countries do not change their lifestyles in a way that will allow us to survive, they will have the murder of a nation on their hands." Clearly, Mr. Shafeeu is either misinformed about the science or is engaging in propaganda.

The Barbados has lost nearly all of their reefs due to runoff from their own agriculture. Their wells are becoming more salty simply because they are extracting so much water to irrigate crops that they are actually drawing sea water into their aquifers. As in the Maldives, its problems are caused by flawed domestic practices and have nothing to do with climate change.

If the U.N. and environmental groups are genuinely interested in solving environmental problems in the Maldives and other developing countries, then they should focus on their true causes. To do otherwise virtually guarantees these problems will continue, no matter how sensational an example it provides for climate change alarmists.

Source: "Myth #6: Sea Level Is Rising Quickly and It Will Get Worse if the Polar Ice Caps Melt Due to Global Warming."

Climate Changes Affecting Pacific Islands

Craig DeSilva

HONOLULU Hawai'i (June 27, 2000—PIDP/CPIS)—Small, isolated Pacific Islands have always been vulnerable to global climate changes and weather phenomena.

Although weather forecasters can make broad predictions, it's difficult to pinpoint exactly when or where a typhoon or hurricane will hit.

But Pacific Island countries and territories can prepare themselves in advance to minimize disastrous effects during weather-changing periods.

"By anticipating in advance, you can take some preparatory action," said Eileen Shea, an adjunct fellow of environmental studies at the East-West Center. "I think all of the Pacific Island nations are looking at adapting to climate change as well as reducing greenhouse gases."

Shea will give a talk Wednesday at the Mauna Lani Bay Hotel on the Big Island of Hawai'i on "Consequences of Climate Change: Challenges and Opportunities for Pacific Islands and Hawai'i."

Shea said Pacific Islands that are alerted to major global climate changes can prepare themselves in advance. For example, regarding the 1997–98 El Niño in the Pacific, the Federated States of Micronesia formed a drought task force before it occurred.

The FSM also developed a public awareness campaign to alert the general population to minimize the effects of the drought.

"Water conservation measures were instituted earlier rather than later," Shea said.

In the Republic of Palau, a decision was made to extend the height and width of the main island's principal dam prior to the wet season so it could hold as much rainfall as possible before drought set in.

Shea is also the Climate Project Coordinator at the East-West Center. The EWC's Pacific Islands Regional Assessment program is made up of scientists, researchers, and other officials from the United States and Pacific Islands who are working to deal with climate change issues throughout the Pacific.

The one-year program is being conducted in conjunction with the Office of the U.S. Global Change Research Program, which is looking into climate change. Shea said the Pacific Islands community will have an opportunity through public hearings to present input on a just released draft report entitled "Climate Change Impacts on the United States: The Potential Consequences of Climate Variability and Change." The public hearings end in August.

Shea said although rising sea levels have been a constant issue facing the Pacific, there are other climate changes that have just as important implications.

"For a variety of reasons, sea-level rise has had a hammerlock on discussions of vulnerability to climate change," Shea said. "What we're trying to do is expand our thinking to address other issues beyond sea level."

Some of the climate issues posing a threat to the Pacific include:

- Increase in air temperature, which can possibly increase the intensity of El Niños;
- Changes and increased intensity in tropical cyclone patterns as a result of global warming;
- Changes in ocean circulation patterns and temperatures, which affect fish stock in the Pacific. Some fish in the Pacific, such as tuna, tend to follow warmer water. "If El Niños are more frequent, tuna migratory patterns will change," she said. "That has significant impacts because that means they can move in and out of a nation's jurisdiction. And that's quite significant particularly for (Pacific) jurisdictions that might be thinking tuna fisheries are an economic source in the future."
- Warming of ocean temperatures impacting coral reefs. "The 1997–98 El Niño saw substantial bleaching of coral reefs around the Pacific," Shea said. "They were

much more severe than in the past. That could lead to secondary impacts on surrounding ecosystems and tourism."

"The islands that are most vulnerable are the low-lying atolls," she said. "Any island that doesn't have mountainous relief is problematic. And it varies across the Pacific. Some islands are tectonically growing, such as the Big Island of Hawaii. But others are actually sinking because of tectonics."

Shea adds that a recent study shows that the Pacific is experiencing the same average rate of rising sea levels as the rest of the world.

"They're not seeing an enhancement in sea-level rise any more than (elsewhere) in the world," she said. "The sea-level rise issue is a long-term issue, (one) of whether you're going to have an island, because it (may) be covered over."

Source: Craig DeSilva, "Climate Changes Affecting Pacific Islands," *Pacific Islands Report*, June 27, 2000.

9

from John E. Hay

Climate Change and Small Island States

A Popular Summary of Science-Based Findings and Perspectives, and Their Links with Policy

John E. Hay
International Global Change Institute
University of Waikato
Hamilton
New Zealand

We are certain that human activities result in increased emissions of greenhouse gases into the atmosphere, with a consequent increase in their atmospheric concentrations and thence enhanced radiative forcing on the atmosphere. There is very high certainty that such changes lead to global warming and global sea-level rise.

Characterisations of future climate changes, and their consequences, for the small islands regions of the world have considerably less certainty. This is due, in part, to the inability of global climate models to resolve the spatial patterns consistent with the individual and combined groupings of small islands. However, the available evidence suggests that by the end of this century there will be systematic and significant changes in the mean climate and increases in sea level, resulting in substantial impacts. But these impacts are likely to be small and of less consequence than the aftereffects of the more frequent extreme events and, especially for the Pacific islands region, *relative* to the current ENSO-induced interannual variability in climate and oceanic conditions. Geological processes, leading to uplift and subsidence of the land, also complicate estimates of sea-level rise.

By the end of the century, mean temperatures for the small islands regions may increase by around 3°C, except for the Mediterranean where the increase is likely to be over 4°C. Observed trends in sea level show marked differences between small islands regions and substantial deviations from historic global trends. Despite this, limitations in modelling location specific sea-level rise require the continued use of projections of global sea-level rise in impact and adaptation assessments for the small island countries and regions. The "best estimate" of global sea-level rise is an increase of about 50cm by 2100. The uncertainty in this estimate still implies an increase of 1.5 to 3.5 times over the historic rate of rise.

It is now considered likely that global warming will lead to some increase in maximum tropical cyclone wind speeds and lower central pressures, leading to more damaging storm surges. Sea-level rise and storm surge effects are additive. Thus the combined effects of increases in cyclone intensities and sea-level rise are one of the major threats to the future well-being of small island countries. Model-based studies suggest that by 2080 the number of people flooded by these super storm surges in any typical year will be more than five times higher than present. The islands of the Caribbean and the Indian and Pacific Oceans face the largest relative increase in flood risk, with the number of people at risk being some 200 times higher than in most other parts of the world.

There is no substantive evidence that tropical cyclone numbers will change in a warmer world; nor is a change in regions of formation indicated. But it is possible that changes in the latter may occur in response to long-term changes in ENSO. Spatial patterns of occurrence are unlikely to undergo major changes, except that tropical cyclones may track further polewards. There is low confidence in these projections. Such high uncertainty, along with the large natural interannual variability, makes it extremely difficult to attribute to human interference in the climate system ... to observed and projected changes in atmospheric and oceanic conditions.

A further complication is the conclusion that the interactions, feedbacks and hence indirect effects of global warming are likely to be of greatest consequence for small island countries, given the strong linkages between all natural and human systems in small island countries. One example relates to the impacts of coral bleaching on the social and economic impacts at community and national levels, including reduced supplies of seafood placing greater pressure on terrestrial food sources and the possibility of detrimental changes in land-use and land-cover. Another example relates to the relationship between direct human management of terrestrial carbon stocks and the natural responses of these stocks to climate change. A key reason for this concern is that terrestrial carbon stocks, in the forms of forest and soil carbon, may become progressively degraded as a result of climate or other global changes. Soil carbon losses increase with temperature. Increased plant growth, due to the CO_2 fertilisation effect, will saturate whereas increased respiration losses will not. Moreover, some forests are being established at the limits of their viable range and carbon losses due to forest fires are increasing.

The recently completed national greenhouse gas inventories provide quantified, conclusive evidence that, either on a collective or on a per capita basis, the inhabitants of small island states are minor contributors to elevated atmospheric greenhouse gas concentrations. But this does not mean that the small island states can, or should, sit back and rest on the reputation of being minor emitters of greenhouse gases. Small island states have many good reasons for taking concerted action now that they have more substantive information on which to base their actions. Actions may be taken to increase the efficiency of existing energy supply systems and to consider opportunities for substituting less costly fuels. Information contained in the inventories will help determine the cost effectiveness of the various options and, in turn, guide decision making related to investment and other initiatives. Such rationally based decisions and actions will help countries to achieve sustainable development.

Political factors may also influence the decision to reduce emissions. Any meaningful efforts by minor emitters to reduce emissions would provide a strong message and give impetus for other countries to take domestic action to reduce their overall emissions. The atmosphere is part of the global commons. Thus a country may well decide to act as a good global citizen and reduce its emissions, no matter how small the inventory data show those emissions to be.

As noted previously, for small island countries a variety of factors make it extremely difficult to anticipate the specific national and local impacts of climate change. These include the low resolution of, and confidence in, model-based projections and the sensitivity, complex and hence interactive nature the natural and human systems.

Integrated as opposed to sector-based assessments and responses are essential under such circumstances. Moreover, especially for small island countries, policy development, planning, and implementation should be driven as much by the need to accelerate sustainable development as by the need to adapt to climate change—many adaptation responses will thus be based on "no regrets" policies. Critical to meeting the need for adaptation are both the transfer and assimilation of environmentally sound technologies, and enhancing the use of traditional knowledge and skills. Environmental technology assessment is of growing importance in small island countries.

Policy implications of the foregoing are examined from the perspectives of international negotiations and national development planning. At the international level, the United Nations Framework Convention on Climate Change seeks to prevent "dangerous interference with the climate system." But to date there has been no success in quantifying the specific threshold concentrations of greenhouse gases in the atmosphere that would limit the *integrated and critical* impacts of climate change to a level that avoids "dangerous interference." This is due, in part, to the current inability to anticipate the integrated impacts of climate change at national and community levels in small island countries. Thus, further targetted and integrated vulnerability research and assessments are required, not only to guide national development planning but also to inform international negotiators.

There is also a need to arrest and reverse the current trend whereby the responsibility of Annex 1 countries to reduce their emissions and enhance sinks is, in both cases, being transferred to non-Annex 1 Parties. Reliance on the enhancement of sinks through management of tropical ecosystems is risky. The present uptake of atmospheric CO_2 by the terrestrial biosphere may diminish over time, and increases in terrestrial carbon stocks may bring with it an increased risk of subsequent release of the carbon to the atmosphere. One study of risk reduction through implementation of the Kyoto Protocol has shown that the risk of a 50 cm sea-level rise, or an atmospheric concentration of CO_2 of 560 ppm (associated with a possible reduction of calcification rates in reef communities), would be reduced by less than 10%. Achievement of these thresholds would be delayed by less than a decade. It is clear that the targets in the Kyoto Protocol are incapable of arresting climate change. All Parties to the Convention must take every reasonable step to reduce the concentrations of greenhouse gases in the atmosphere, rather than abdicating responsibility to developing countries.

In national development planning, climate change is only one of many impediments to achievement of environmentally sound and sustainable development—many others are related to high population growth rates and densities, and the migration of people in-country. As with all other sources of pressure on natural and human systems, climate change must be mainstreamed in national development planning.

Five conclusions are derived from the review of science-driven policy and policy-driven science:

- The obvious and relatively well characterised consequences of global warming may not pose the greatest climate-related threat to small island countries—the less well understood extreme events, and the indirect effects of changes in mean conditions, are likely to be of far greater significance;
- International policy positions and negotiating strategies under the Convention are placing a growing emphasis on measures implemented by developing countries (e.g., reduced emissions, enhanced sinks), rather than placing the onus of the main contributors to global warming—this is unjustifiable, on both scientific and moral grounds;
- There is insufficient substantive information on which to base analysis of the sufficiency of response measures—this is due, in part, to the current inability to anticipate the integrated impacts of climate change at national and community levels in small island countries; and
- Integration is key to success in addressing climate change; at the national level, addressing climate change is only one of many policy responses required to achieve

environmentally sound and sustainable development; integrated assessments and the mainstreaming of climate change policies are critical; integration at the international level can, amongst other benefits, result in synergies from compliance with the various environmental legal agreements; and

- There is a need to strengthen still further the capacity of small island developing states to address the preceding challenges, with sustainable outcomes—we already have several success stories, including CCPAC and PICCAP, along with the support of such organisations and initiatives as GEF, its partner organisations, namely the World Bank, UNDP and UNEP, UNDSD, NCSP and UNITAR.

Source: John E. Hay, "Climate Change and Small Island States," paper presented at the Second Alliance of Small Island States and posted at http://sidsnet.org/docshare/climate/hays.doc.

from the Australian Broadcasting Corporation

10

Tuvalu Seeks Mass Migration

July 16, 2003

Kerry O'Brien: The tiny Pacific nation of Tuvalu is pleading for a large-scale exemption from Australia's tough immigration laws.

Tuvalu wants to move its entire population of about 12,000 to Australia to escape an increasingly precarious existence on the coral atolls it inhabits.

The islanders fear the combination of global warming and cyclones will swamp their low-lying homeland and, with strong connections to Australia, they want to come here.

And while the Federal Government has promised help in the event of a disaster, it's not throwing out any lifelines yet.

Natasha Johnson reports.

Natasha Johnson: Golden sunsets and beautiful beaches—Tuvalu looks every bit a piece of paradise.

But its 12,000 residents live in constant fear of the massive expanse of ocean that surrounds them.

Their home measures a mere 26 square kilometres across nine isolated atolls in the middle of the Pacific.

The highest point is just 5m above sea level, meaning they feel decidedly vulnerable to rising sea levels caused by climate change.

Reverend Tofiga Falani, Tuvalu, Congregational Christian Church: Global warming has become a major day-to-day conversation in our life back at home.

So we have been threatened, and our people are so anxious to know, if we are going definitely to be sunk, to sink down, how are we going to survive?

Where else can we live?

Natasha Johnson: Reverend Tofiga Falani is the president of the Tuvaluan Congregational Church, of which 97 percent of Tuvaluans are members.

It's a sister church of the Australian Uniting Church and he's in Melbourne to attend the National Assembly and raise the plight of his nation.

Reverend Tofiga Falani: We are asking our bigger brother Australia, "Please give us an open door so that our people may be evacuated to this land."

Natasha Johnson: The Uniting Church has taken up the cause and outgoing president James Haire met with the Tuvaluan PM earlier this year, who asked him to lobby the Australian Government on their behalf.

Professor James Haire, Outgoing President, Uniting Church: I believe that Australia has three obligations.

One, we are the largest polluter in this part of the world.

Two, we are the largest land mass close to the area.

And, thirdly, there is this long tradition that you rescue those in peril at sea, and you could argue that this is that kind of situation.

Natasha Johnson: They're seeking a graduated migration program, giving special consideration to Tuvaluans.

Professor James Haire: We should increase the number of AusAID scholarships available to the population of Tuvalu.

We should be generous in our application of the rules about expert migrants to this country, migrants with a particular degree of expertise, and allow a greater number of people of Tuvalu in.

Gary Hardgrave, Acting Immigration Minister: There is no immediate crisis.

This is not something that is going to happen this week or next week.

Australia is not in the business of refashioning its migration policy because of something that might be happening 20, 30, 50, 100 or 200 years from now.

Natasha Johnson: While there's broad scientific consensus that sea levels are rising and predictions they'll continue to rise at an accelerated rate over the next century, there's debate about whether Tuvalu will be drowned and, if so, when.

But it does face more immediate risk from cyclones in the region, which, it's feared, are increasing in frequency and intensity because of global warming.

Dr. John Church, Research Scientist, CSIRO: Extreme events associated with cyclones, for example, are causing higher than average surges.

There was a cyclone which passed relatively closely to Tuvalu in 2002.

However, that occurred at a period of neap tides and so, although it caused coastal erosion, didn't have as much impact as it would have had if it had occurred at the time of high spring tides.

Natasha Johnson: Reverend Falani says that storm surges are already causing significant environmental impact, flooding inland pools and contaminating crops with salt water.

Reverend Tofiga Falani: One of our small islets have been washed out, totally washed out, disappeared, because of the continuing cyclones in our area.

So we used to have it in our maps before, but now, if you try to look for the islet, it's totally disappeared, away, no more.

Natasha Johnson: Melbourne and Brisbane have the largest Tuvaluan communities outside Tuvalu.

They've joined the campaign to save family members and friends at home.

An Australian resident, Tito Tapungao, has just returned from four years teaching in Tuvalu and is deeply worried about the fate of his sister and her children still there.

Tito Tapungao: Would the Australian public rather see the Tuvaluans standing on the rocks of the coral reef with the sea to their waist?

I don't think anybody would like to see that time when it comes.

Natasha Johnson: The Australian Government says it's committed $31 million to Pacific nations to combat global warming and will coordinate a rescue and repatriation program with the US and New Zealand should disaster strike.

Professor James Haire: Just cannot wait around, loll around waiting until we pick up the bodies out of the ocean.

We have to take some action now.

Gary Hardgrave: What we don't want to see happen, of course, is anybody hurt by any of those sorts of matters, such as cyclones or inundation.

That's why we've been working with Pacific island states, putting money into programs to deal with the facts as they arise to try and get some kind of accurate forecast.

At this stage, to create a specific visa class for Tuvaluans to come to Australia means that we'd have to introduce a discriminatory migration policy, which would be completely contrary to the way Australia's been working its migration policy for decades.

Reverend Tofiga Falani: I only wish if the Government of Australia start right now to bring our people—don't wait until the last moment when maybe the last boat comes to rescue our people they've all gone down.

So, the sooner the better.

Natasha Johnson: Reverend Falani returns to Tuvalu in a few days and says he'll continue to pray for a change in Australian policy before the rising of the tide.

Review Questions

1. What, according to most scientists, is the connection between increased air pollution in industrialized countries and a rise in sea level in the Pacific?
2. What are some of the reasons the United States gives for not signing the Kyoto Protocol?
3. How does an atoll form, and why might an atoll be sinking for reasons not associated with rising sea level?
4. Why are coral reef ecosystems so important to the survival of Pacific island societies?

Discussion/Essay Questions

1. Consider the case of Ioane Ubaitoi, the man from Kiribati discussed at the beginning of this module, who went to New Zealand in 1998 to take a course in actions that might prevent the submersion of his country from global sea-level rise. Assuming that Mr. Ubaitoi's studies were paid for as development aid by the New Zealand government, was this really the best way for New Zealand to spend its money if it wanted to save Kiribati from eroding? Other possible uses of the money could have included the following:

    - Studying local Kiribati agricultural practices to see how they might be fostering erosion
    - Instructing the people of Kiribati on responding to other aspects of climate change not associated with sea-level rise
    - Lobbying the United States to sign the Kyoto Protocol and reduce emissions of greenhouse gases
    - Developing alternative energy technologies so that industrialized countries could continue to produce energy, but with lowered emissions of greenhouse gases
    - Resettling the people of Kiribati on New Zealand
    - Further studying the relationship between greenhouse gas emissions, climate change, and global warming

    Would any of these projects have been a better use for the money than the way that it was spent? Make your argument relying on one or more of this module's readings.
2. Consider the section of this module that presents a summary of several philosophical perspectives on nature. How would each of these perspectives view

global climate change? For each perspective, would climate change be seen as a "problem" to be "solved"? If so, how would the problem be framed and what kind of actions would be proposed as solutions?

3. Each of the readings differs in the scale at which climate change is seen as a problem. The readings also differ regarding the scale at which solutions should be applied. Building on some of the concepts introduced in the textbook regarding the different scales at which environmental problems occur and also facts learned from the readings about climate change and its impacts, make an argument for the scale of analysis that you think is appropriate for analyzing and impacting climate change.

List of Readings

1. "The Science of Climate Change," joint statement issued by sixteen national academies of science, May 17, 2001, posted on the website of Greenpeace International, *http://archive.greenpeace.org/~climate/climatecountdown/scienceacademies.htm.*
2. George W. Bush, "Text of a Letter from the President to Senators Hagel, Helms, Craig, and Roberts," March 13, 2001, posted on the website of the White House, *http://www.whitehouse.gov/news/releases/2001/03/20010314.html.*
3. "The Climate Cannot Wait for Bush, But If Bush Doesn't Change, the Climate Will," June 7, 2001, posted on the website of Greenpeace International, *http://archive.greenpeace.org/~climate/climatecount down/bushvclimate.htm.*
4. Paul Brown, "The Heat Is On for a Solution in Bonn," July 14, 2001, posted on the website of *The Guardian*, *http://www.guardian.co.uk/theissues/article/0,6512, 522651,00.html.*
5. Janita Pahalad, "Climate Change and Sea-Level Rise: A Personal View from Fiji," in *The South Pacific Sea Level and Climate Change Newsletter*, July 1998, posted on the website of the South Pacific Regional Environment Programme, *http://www.sprep.org.ws/newsletter/ClimateChange/nlcc0303/mf05_.htm.*
6. "Sea-Level Rise a Big Problem for Tuvalu, Prime Minister Says," news release from Greenpeace Pacific, July 22, 1997, posted on the website of the World History Archives, *http://www.hartford-hwp.com/archives/24/023.html.*
7. "Myth #6: 'Sea-Level Is Rising Quickly and It Will Get Worse If the Polar Ice Caps Melt Due to Global Warming: Coastal Settlements and Low-Lying Islands Will Be Submerged'," posted on the website of EnviroTruth.org, a project of the National Center for Public Policy Research, *http://www.envirotruth.org/myth6.cfm.*
8. Craig DeSilva, "Climate Changes Affecting Pacific Islands," news release issued jointly by the Pacific Islands Development Program of the East-West Center and the Center for Pacific Islands Studies of the University of Hawai'i, June 27, 2000, posted on the website of Pacific Islands Reports, *http://166.122.164.43/archive/2000/june/06%2D28%2D07.htm.*
9. John E. Hay, "Climate Change and Small Island States: A Popular Summary of Science-Based Findings and Perspectives, and Their Links with Policy," abstract of a paper presented at the 2nd Alliance of Small Island States (AOSIS) Workshop on Climate Change Negotiations, Management, and Strategy, July 26–August 4, 2000, Apia, Samoa, posted on the website of the Small Island Developing States Network, *http://sidsnet.org/docshare/climate/hays.doc.*
10. ABC Radio Australia, "Tuvalu seeks mass migration," July 16, 2003, posted on the website of ABC Online, *http://www.abc.net.au/7.30/content/2003/s903787.htm.*

Websites for Additional Research

1. Numerous organizations concerned with environmental policy and environmental regulations devote all or parts of their websites to positions on climate change and, more specifically, the Kyoto Protocol. For a particularly clear statement of the pro-Kyoto position, see the website for Greenpeace's climate change campaign at *http://www.greenpeace.org/campaigns/intro?campaign_id=3937*, or find the campaign via the Greenpeace International homepage, *http://www.greenpeace.org*. For an opposing

perspective, visit the website of EnviroTruth.org's climate change campaign at *http://www.envirotruth.org* and click on the link to the climate change campaign.

2. The International Institute for Sustainable Development, a Canadian nongovernmental organization, hosts a page of links to articles representing various positions on the Kyoto Protocol, at *http://www.iisd.ca/linkages/climate/ba/perspectives.html.* The page has not been updated since 1999, but it still provides a nice list of useful articles. For technical material on climate change, visit the website of the United Nations Intergovernmental Panel on Climate Change at *http://www.ipcc.ch.*

3. The United Nations sponsors a network for communication among the world's forty-three small island developing states (SIDS). The network's website, *http://www.sidsnet.org*, serves as a clearinghouse for information of concern to its members, including climate change. The SIDSNET website also hosts the official site of the Association of Small Island States (AOSIS), the organization of small island states' governments, at *http://www.sidsnet.org/aosis.* Another excellent source for news stories and links on island affairs is the Global Islands Network, at *http://www.globalislands.net.*